DISCRETE MATHEMATICS AND ITS APPLICATIONS
Series editor KENNETH H. ROSEN

CRYPTOGRAPHY
THEORY AND PRACTICE
THIRD EDITION

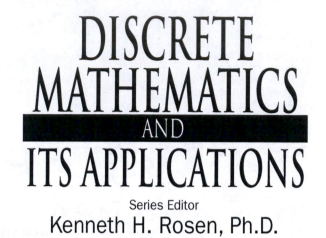

DISCRETE MATHEMATICS AND ITS APPLICATIONS

Series Editor

Kenneth H. Rosen, Ph.D.

Continued Titles

DISCRETE MATHEMATICS AND ITS APPLICATIONS
Series editor KENNETH H. ROSEN

CRYPTOGRAPHY
THEORY AND PRACTICE
THIRD EDITION

DOUGLAS R. STINSON

University of Waterloo
Ontario, Canada

Chapman & Hall/CRC
Taylor & Francis Group
Boca Raton   London   New York

Published in 2006 by
Chapman & Hall/CRC
Taylor & Francis Group
6000 Broken Sound Parkway NW, Suite 300
Boca Raton, FL 33487-2742

Library of Congress Cataloging-in-Publication Data

Catalog record is available from the Library of Congress

Taylor & Francis Group
is the Academic Division of Informa plc.

Visit the Taylor & Francis Web site at
http://www.taylorandfrancis.com

and the CRC Press Web site at
http://www.crcpress.com

Preface

The first edition of this book, which was published in 1995, contained thirteen chapters. My objective was to produce a general textbook that treated all the essential core areas of cryptography, as well as a selection of more advanced topics. In writing the book, I tried to design it to be flexible enough to permit a wide variety of approaches to the subject, so that it could be used for both undergraduate and graduate university courses in cryptography in mathematics, computer science and engineering departments.

The second edition, published in 2002, was focused more tightly on the core areas of cryptography that are most likely to be covered in a course. In contrast to the first edition, the second edition contained only seven chapters. At that time, my intention was to write a companion volume containing updated treatments of other chapters from the first edition, as well as chapters covering new topics.

Eventually, I changed my plans and I decided to proceed directly to a one-volume, expanded third edition. This third edition more closely resembles the first edition in its breadth and scope, but it has been almost completely rewritten. It consists of the seven chapters from the second edition, updated where appropriate, and seven new chapters. Here is a brief synopsis of the fourteen chapters in this third edition of "Cryptography Theory and Practice":

- Chapter 1 is a fairly elementary introduction to simple "classical" cryptosystems. This chapter also presents basic mathematical techniques that are used throughout the book.

- Chapter 2 covers the main elements of Shannon's approach to cryptography, including the concepts of perfect secrecy and entropy and the use of information theory in cryptography.

- Chapter 3 concerns block ciphers. It uses substitution-permutation networks as a mathematical model to introduce many of the concepts of modern block cipher design and analysis, including differential and linear cryptanalysis. There is an emphasis on general principles, and the specific block ciphers that are discussed in this chapter (*DES* and *AES*) serve to illustrate these general principles.

- Chapter 4 contains a unified treatment of keyed and unkeyed hash functions and their application to the construction of message authentication codes. There is an emphasis on mathematical analysis and security proofs. This chapter includes a description of the *Secure Hash Algorithm.*

- Chapter 5 concerns the *RSA Cryptosystem*, together with a considerable amount of background on number-theoretic topics such as primality testing and factoring.

- Chapter 6 discusses public-key cryptosystems, such as the *ElGamal Cryptosystem*, that are based on the **Discrete Logarithm** problem. This chapter also includes material on algorithms for computing discrete logarithms, elliptic curves, and the **Diffie-Hellman** problems.

- Chapter 7 deals with signature schemes. It presents schemes such as the *Digital Signature Algorithm*, and it includes treatment of special types of signature schemes such as undeniable and fail-stop signature schemes.

- Chapter 8 covers pseudorandom bit generation in cryptography. It is based on the corresponding chapter in the first edition.

- Chapter 9 deals with identification (entity authentication). The first part of the chapter discusses schemes that are built from simpler cryptographic "primitives" such as signature schemes or message authentication codes. The second part of the chapter is a treatment of special purpose "zero-knowledge" schemes, based on material from the first edition.

- Chapters 10 and 11 discuss various methods for key establishment. Chapter 10 concerns key distribution and Chapter 11 presents protocols for key agreement. These chapters are significantly expanded from the first edition, which covered this material in abbreviated fashion in one chapter (key establishment was not covered in the second edition). There is a greater emphasis on security models and proofs than before.

- Chapter 12 gives an overview of public-key infrastructures, and it also discusses identity-based cryptography as one possible alternative to PKIs. This is a new chapter.

- The topic of Chapter 13 is secret sharing schemes. It is based on a chapter from the first edition.

- Chapter 14 is a new chapter that discusses some topics in multicast security, including broadcast encryption and copyright protection.

The following features are common to all editions of this book:

- Mathematical background is provided where it was needed, in a "just-in-time" fashion.

- Informal descriptions of the cryptosystems are given along with more precise pseudo-code descriptions.

- Numerical examples are presented to illustrate the workings of most of the algorithms described in the book.

- The mathematical underpinnings of the algorithms and cryptosystems are explained carefully and rigorously.

- Numerous exercises are included, some of them quite challenging.

I believe that these features of the book increase its usefulness as a textbook and a book for independent study.

I have tried to present the material in this book in a logical and natural order. Note that it is possible to omit various earlier chapters in order to concentrate on later material, if so desired. However, there are certain chapters which do depend heavily on some earlier chapters. Some of the more important dependencies are the following:

- Chapter 9 uses material from Chapters 4 (MACs) and 7 (signature schemes).
- Section 13.3.2 uses results on entropy from Chapter 2.
- Chapter 14 uses material from Chapters 10 (key predistribution) and 13 (secret sharing).

There are also many situations where specific mathematical tools introduced in one chapter will be used again in a later chapter, but this should not cause difficulty in using the book in a course.

One of the most difficult things about writing any book in cryptography is deciding how much mathematical background to include. Cryptography is a broad subject, and it requires knowledge of several areas of mathematics, including number theory, groups, rings and fields, linear algebra, probability and information theory. As well, some familiarity with computational complexity, algorithms and NP-completeness theory is useful. In my opinion, it is the breadth of mathematical background required that often creates difficulty for students studying cryptography for the first time.

I have tried not to assume too much mathematical background, and thus I develop mathematical tools as they are needed, for the most part. But it would certainly be helpful for the reader to have some familiarity with basic linear algebra and modular arithmetic.

Many people pointed out typos and errors in the second edition and draft chapters of the third edition, and gave me useful suggestions on new material to include and how various topics should be treated. In particular, I would like to thank Carlisle Adams, Eike Best, Dameng Deng, Shuhong Gao, K. Gopalakrishnan, Pascal Junod, Torleiv Kløve, Jooyoung Lee, Vaclav Matyas, Michael Monagan, James Muir, Phil Rose, Tamir Tassa, and Rebecca Wright. As always, I will appreciate being informed of any errata, which I will post on a web page.

Douglas R. Stinson
Waterloo, Ontario

To my children, Michela and Aiden

Contents

Contents

Contents

1

Classical Cryptography

In this chapter, we provide a gentle introduction to cryptography and cryptanalysis. We present several simple systems, and describe how they can be "broken." Along the way, we discuss various mathematical techniques that will be used throughout the book.

1.1 Introduction: Some Simple Cryptosystems

The fundamental objective of cryptography is to enable two people, usually referred to as Alice and Bob, to communicate over an insecure channel in such a way that an opponent, Oscar, cannot understand what is being said. This channel could be a telephone line or computer network, for example. The information that Alice wants to send to Bob, which we call "plaintext," can be English text, numerical data, or anything at all — its structure is completely arbitrary. Alice encrypts the plaintext, using a predetermined key, and sends the resulting ciphertext over the channel. Oscar, upon seeing the ciphertext in the channel by eavesdropping, cannot determine what the plaintext was; but Bob, who knows the encryption key, can decrypt the ciphertext and reconstruct the plaintext.

These ideas are described formally using the following mathematical notation.

Definition 1.1: A *cryptosystem* is a five-tuple $(\mathcal{P}, \mathcal{C}, \mathcal{K}, \mathcal{E}, \mathcal{D})$, where the following conditions are satisfied:

1. $\mathcal{P}$ is a finite set of possible *plaintexts*;
2. $\mathcal{C}$ is a finite set of possible *ciphertexts*;
3. $\mathcal{K}$, the *keyspace*, is a finite set of possible *keys*;
4. For each $K \in \mathcal{K}$, there is an *encryption rule* $e_K \in \mathcal{E}$ and a corresponding *decryption rule* $d_K \in \mathcal{D}$. Each $e_K : \mathcal{P} \to \mathcal{C}$ and $d_K : \mathcal{C} \to \mathcal{P}$ are functions such that $d_K(e_K(x)) = x$ for every plaintext element $x \in \mathcal{P}$.

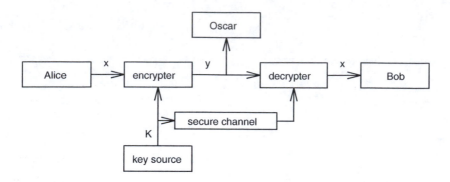

FIGURE 1.1
The communication channel

The main property is property 4. It says that if a plaintext x is encrypted using e_K, and the resulting ciphertext is subsequently decrypted using d_K, then the original plaintext x results.

Alice and Bob will employ the following protocol to use a specific cryptosystem. First, they choose a random key $K \in \mathcal{K}$. This is done when they are in the same place and are not being observed by Oscar, or, alternatively, when they do have access to a secure channel, in which case they can be in different places. At a later time, suppose Alice wants to communicate a message to Bob over an insecure channel. We suppose that this message is a string

$$\mathbf{x} = x_1 x_2 \cdots x_n$$

for some integer $n \geq 1$, where each plaintext symbol $x_i \in \mathcal{P}$, $1 \leq i \leq n$. Each x_i is encrypted using the encryption rule e_K specified by the predetermined key K. Hence, Alice computes $y_i = e_K(x_i)$, $1 \leq i \leq n$, and the resulting ciphertext string

$$\mathbf{y} = y_1 y_2 \cdots y_n$$

is sent over the channel. When Bob receives $y_1 y_2 \cdots y_n$, he decrypts it using the decryption function d_K, obtaining the original plaintext string, $x_1 x_2 \cdots x_n$. See Figure 1.1 for an illustration of the communication channel.

Clearly, it must be the case that each encryption function e_K is an *injective function* (i.e., one-to-one); otherwise, decryption could not be accomplished in an unambiguous manner. For example, if

$$y = e_K(x_1) = e_K(x_2)$$

where $x_1 \neq x_2$, then Bob has no way of knowing whether y should decrypt to x_1 or x_2. Note that if $\mathcal{P} = \mathcal{C}$, it follows that each encryption function is a permutation. That is, if the set of plaintexts and ciphertexts are identical, then each encryption function just rearranges (or permutes) the elements of this set.

1.1.1 The Shift Cipher

In this section, we will describe the *Shift Cipher*, which is based on modular arithmetic. But first we review some basic definitions of modular arithmetic.

Definition 1.2: Suppose a and b are integers, and m is a positive integer. Then we write $a \equiv b \pmod{m}$ if m divides $b-a$. The phrase $a \equiv b \pmod{m}$ is called a *congruence*, and it is read as "a is *congruent* to b modulo m." The integer m is called the *modulus*.

Suppose we divide a and b by m, obtaining integer quotients and remainders, where the remainders are between 0 and $m-1$. That is, $a = q_1 m + r_1$ and $b = q_2 m + r_2$, where $0 \le r_1 \le m - 1$ and $0 \le r_2 \le m - 1$. Then it is not difficult to see that $a \equiv b \pmod{m}$ if and only if $r_1 = r_2$. We will use the notation $a \bmod m$ (without parentheses) to denote the remainder when a is divided by m, i.e., the value r_1 above. Thus $a \equiv b \pmod{m}$ if and only if $a \bmod m = b \bmod m$. If we replace a by $a \bmod m$, we say that a is *reduced* modulo m.

We give a couple of examples. To compute $101 \bmod 7$, we write $101 = 7 \times 14 + 3$. Since $0 \le 3 \le 6$, it follows that $101 \bmod 7 = 3$. As another example, suppose we want to compute $(-101) \bmod 7$. In this case, we write $-101 = 7 \times (-15) + 4$. Since $0 \le 4 \le 6$, it follows that $(-101) \bmod 7 = 4$.

REMARK Many computer programming languages define $a \bmod m$ to be the remainder in the range $-m + 1, \ldots, m - 1$ having the same sign as a. For example, $(-101) \bmod 7$ would be -3, rather than 4 as we defined it above. But for our purposes, it is much more convenient to define $a \bmod m$ always to be non-negative. ∎

We now define arithmetic modulo m: $\mathbb{Z}_m$ is the set $\{0, \ldots, m - 1\}$, equipped with two operations, $+$ and $\times$. Addition and multiplication in $\mathbb{Z}_m$ work exactly like real addition and multiplication, except that the results are reduced modulo m.

For example, suppose we want to compute 11×13 in $\mathbb{Z}_{16}$. As integers, we have $11 \times 13 = 143$. Then we reduce 143 modulo 16 as described above: $143 = 8 \times 16 + 15$, so $143 \bmod 16 = 15$, and hence $11 \times 13 = 15$ in $\mathbb{Z}_{16}$.

These definitions of addition and multiplication in $\mathbb{Z}_m$ satisfy most of the familiar rules of arithmetic. We will list these properties now, without proof:

1. addition is *closed*, i.e., for any $a, b \in \mathbb{Z}_m$, $a + b \in \mathbb{Z}_m$

2. addition is *commutative*, i.e., for any $a, b \in \mathbb{Z}_m$, $a + b = b + a$

3. addition is *associative*, i.e., for any $a, b, c \in \mathbb{Z}_m$, $(a + b) + c = a + (b + c)$

4. 0 is an *additive identity*, i.e., for any $a \in \mathbb{Z}_m$, $a + 0 = 0 + a = a$

Cryptosystem 1.1: *Shift Cipher*

Let $\mathcal{P} = \mathcal{C} = \mathcal{K} = \mathbb{Z}_{26}$. For $0 \leq K \leq 25$, define

$$e_K(x) = (x + K) \bmod 26$$

and

$$d_K(y) = (y - K) \bmod 26$$

$(x, y \in \mathbb{Z}_{26})$.

---

5. the *additive inverse* of any $a \in \mathbb{Z}_m$ is $m-a$, i.e., $a+(m-a) = (m-a)+a = 0$ for any $a \in \mathbb{Z}_m$

6. multiplication is *closed*, i.e., for any $a, b \in \mathbb{Z}_m$, $ab \in \mathbb{Z}_m$

7. multiplication is *commutative*, i.e., for any $a, b \in \mathbb{Z}_m$, $ab = ba$

8. multiplication is *associative*, i.e., for any $a, b, c \in \mathbb{Z}_m$, $(ab)c = a(bc)$

9. 1 is a *multiplicative identity*, i.e., for any $a \in \mathbb{Z}_m$, $a \times 1 = 1 \times a = a$

10. the *distributive property* is satisfied, i.e., for any $a, b, c \in \mathbb{Z}_m$, $(a + b)c = (ac) + (bc)$ and $a(b + c) = (ab) + (ac)$.

Properties 1, 3–5 say that $\mathbb{Z}_m$ forms an algebraic structure called a *group* with respect to the addition operation. Since property 2 also holds, the group is said to be an *abelian group*.

Properties 1–10 establish that $\mathbb{Z}_m$ is, in fact, a *ring*. We will see many other examples of groups and rings in this book. Some familiar examples of rings include the integers, $\mathbb{Z}$; the real numbers, $\mathbb{R}$; and the complex numbers, $\mathbb{C}$. However, these are all infinite rings, and our attention will be confined almost exclusively to finite rings.

Since additive inverses exist in $\mathbb{Z}_m$, we can also subtract elements in $\mathbb{Z}_m$. We define $a - b$ in $\mathbb{Z}_m$ to be $(a - b) \bmod m$. That is, we compute the integer $a - b$ and then reduce it modulo m. For example, to compute $11 - 18$ in $\mathbb{Z}_{31}$, we first subtract 18 from 11, obtaining -7, and then compute $(-7) \bmod 31 = 24$.

We present the *Shift Cipher* as Cryptosystem 1.1. It is defined over $\mathbb{Z}_{26}$ since there are 26 letters in the English alphabet, though it could be defined over $\mathbb{Z}_m$ for any modulus m. It is easy to see that the *Shift Cipher* forms a cryptosystem as defined above, i.e., $d_K(e_K(x)) = x$ for every $x \in \mathbb{Z}_{26}$.

REMARK   For the particular key $K = 3$, the cryptosystem is often called the *Caesar Cipher*, which was purportedly used by Julius Caesar. ∎

We would use the *Shift Cipher* (with a modulus of 26) to encrypt ordinary English text by setting up a correspondence between alphabetic characters and residues modulo 26 as follows: $A \leftrightarrow 0$, $B \leftrightarrow 1$, ..., $Z \leftrightarrow 25$. Since we will be using this correspondence in several examples, let's record it for future use:

A	B	C	D	E	F	G	H	I	J	K	L	M
0	1	2	3	4	5	6	7	8	9	10	11	12

N	O	P	Q	R	S	T	U	V	W	X	Y	Z
13	14	15	16	17	18	19	20	21	22	23	24	25

A small example will illustrate.

Example 1.1 Suppose the key for a *Shift Cipher* is $K = 11$, and the plaintext is

wewillmeetatmidnight.

We first convert the plaintext to a sequence of integers using the specified correspondence, obtaining the following:

$$22 \quad 4 \quad 22 \quad 8 \quad 11 \quad 11 \quad 12 \quad 4 \quad 4 \quad 19$$
$$0 \quad 19 \quad 12 \quad 8 \quad 3 \quad 13 \quad 8 \quad 6 \quad 7 \quad 19$$

Next, we add 11 to each value, reducing each sum modulo 26:

$$7 \quad 15 \quad 7 \quad 19 \quad 22 \quad 22 \quad 23 \quad 15 \quad 15 \quad 4$$
$$11 \quad 4 \quad 23 \quad 19 \quad 14 \quad 24 \quad 19 \quad 17 \quad 18 \quad 4$$

Finally, we convert the sequence of integers to alphabetic characters, obtaining the ciphertext:

HPHTWWXPPELEXTOYTRSE.

To decrypt the ciphertext, Bob will first convert the ciphertext to a sequence of integers, then subtract 11 from each value (reducing modulo 26), and finally convert the sequence of integers to alphabetic characters. □

REMARK In the above example we are using upper case letters for ciphertext and lower case letters for plaintext, in order to improve readability. We will do this elsewhere as well. ∎

If a cryptosystem is to be of practical use, it should satisfy certain properties. We informally enumerate two of these properties now.

1. Each encryption function e_K and each decryption function d_K should be efficiently computable.

2. An opponent, upon seeing a ciphertext string **y**, should be unable to determine the key K that was used, or the plaintext string **x**.

The second property is defining, in a very vague way, the idea of "security." The process of attempting to compute the key K, given a string of ciphertext **y**, is called *cryptanalysis*. (We will make these concepts more precise as we proceed.) Note that, if Oscar can determine K, then he can decrypt **y** just as Bob would, using d_K. Hence, determining K is at least as difficult as determining the plaintext string **x**, given the ciphertext string **y**.

We observe that the *Shift Cipher* (modulo 26) is not secure, since it can be cryptanalyzed by the obvious method of *exhaustive key search*. Since there are only 26 possible keys, it is easy to try every possible decryption rule d_K until a "meaningful" plaintext string is obtained. This is illustrated in the following example.

Example 1.2   Given the ciphertext string

$$\texttt{JBCRCLQRWCRVNBJENBWRWN,}$$

we successively try the decryption keys d_0, d_1, etc. The following is obtained:

$$\texttt{jbcrclqrwcrvnbjenbwrwn}$$
$$\texttt{iabqbkpqvbqumaidmavqvm}$$
$$\texttt{hzapajopuaptlzhclzupul}$$
$$\texttt{gyzozinotzoskygbkytotk}$$
$$\texttt{fxynyhmnsynrjxfajxsnsj}$$
$$\texttt{ewxmxglmrxmqiweziwrmri}$$
$$\texttt{dvwlwfklqwlphvdyhvqlqh}$$
$$\texttt{cuvkvejkpvkogucxgupkpg}$$
$$\texttt{btujudijoujnftbwftojof}$$
$$\texttt{astitchintimesavesnine}$$

At this point, we have determined the plaintext to be the phrase "a stitch in time saves nine," and we can stop. The key is $K = 9$.   □

On average, a plaintext will be computed using this method after trying $26/2 = 13$ decryption rules.

As the above example indicates, a necessary condition for a cryptosystem to be secure is that an exhaustive key search should be infeasible; i.e., the keyspace should be very large. As might be expected, however, a large keyspace is not sufficient to guarantee security.

Cryptosystem 1.2: *Substitution Cipher*

Let $\mathcal{P} = \mathcal{C} = \mathbb{Z}_{26}$. $\mathcal{K}$ consists of all possible permutations of the 26 symbols $0, 1, \ldots, 25$. For each permutation $\pi \in \mathcal{K}$, define

$$e_\pi(x) = \pi(x),$$

and define

$$d_\pi(y) = \pi^{-1}(y),$$

where π^{-1} is the inverse permutation to π.

1.1.2 The Substitution Cipher

Another well-known cryptosystem is the *Substitution Cipher*, which we define now. This cryptosystem has been used for hundreds of years. Puzzle "cryptograms" in newspapers are examples of *Substitution Ciphers*. This cipher is defined as Cryptosystem 1.2.

Actually, in the case of the *Substitution Cipher*, we might as well take $\mathcal{P}$ and $\mathcal{C}$ both to be the 26-letter English alphabet. We used $\mathbb{Z}_{26}$ in the *Shift Cipher* because encryption and decryption were algebraic operations. But in the *Substitution Cipher*, it is more convenient to think of encryption and decryption as permutations of alphabetic characters.

Here is an example of a "random" permutation, π, which could comprise an encryption function. (As before, plaintext characters are written in lower case and ciphertext characters are written in upper case.)

a	b	c	d	e	f	g	h	i	j	k	l	m
X	N	Y	A	H	P	O	G	Z	Q	W	B	T

n	o	p	q	r	s	t	u	v	w	x	y	z
S	F	L	R	C	V	M	U	E	K	J	D	I

Thus, $e_\pi(a) = X$, $e_\pi(b) = N$, etc. The decryption function is the inverse permutation. This is formed by writing the second lines first, and then sorting in alphabetical order. The following is obtained:

A	B	C	D	E	F	G	H	I	J	K	L	M
d	l	r	y	v	o	h	e	z	x	w	p	t

N	O	P	Q	R	S	T	U	V	W	X	Y	Z
b	g	f	j	q	n	m	u	s	k	a	c	i

Hence, $d_\pi(A) = d$, $d_\pi(B) = l$, etc.

As an exercise, the reader might decrypt the following ciphertext using this decryption function:

$$\text{MGZVYZLGHCMHJMYXSSFMNHAHYCDLMHA.}$$

A key for the *Substitution Cipher* just consists of a permutation of the 26 alphabetic characters. The number of possible permutations is 26!, which is more than 4.0×10^{26}, a very large number. Thus, an exhaustive key search is infeasible, even for a computer. However, we shall see later that a *Substitution Cipher* can easily be cryptanalyzed by other methods.

1.1.3 The Affine Cipher

The *Shift Cipher* is a special case of the *Substitution Cipher* which includes only 26 of the 26! possible permutations of 26 elements. Another special case of the *Substitution Cipher* is the *Affine Cipher*, which we describe now. In the *Affine Cipher*, we restrict the encryption functions to functions of the form

$$e(x) = (ax + b) \bmod 26,$$

$a, b \in \mathbb{Z}_{26}$. These functions are called *affine functions*, hence the name *Affine Cipher*. (Observe that when $a = 1$, we have a *Shift Cipher*.)

In order that decryption is possible, it is necessary to ask when an affine function is injective. In other words, for any $y \in \mathbb{Z}_{26}$, we want the congruence

$$ax + b \equiv y \ (\mathrm{mod}\ 26)$$

to have a unique solution for x. This congruence is equivalent to

$$ax \equiv y - b \ (\mathrm{mod}\ 26).$$

Now, as y varies over $\mathbb{Z}_{26}$, so, too, does $y - b$ vary over $\mathbb{Z}_{26}$. Hence, it suffices to study the congruence $ax \equiv y \ (\mathrm{mod}\ 26)$ ($y \in \mathbb{Z}_{26}$).

We claim that this congruence has a unique solution for every y if and only if $\gcd(a, 26) = 1$ (where the gcd function denotes the greatest common divisor of its arguments). First, suppose that $\gcd(a, 26) = d > 1$. Then the congruence $ax \equiv 0 \ (\mathrm{mod}\ 26)$ has (at least) two distinct solutions in $\mathbb{Z}_{26}$, namely $x = 0$ and $x = 26/d$. In this case $e(x) = (ax + b) \bmod 26$ is not an injective function and hence not a valid encryption function.

For example, since $\gcd(4, 26) = 2$, it follows that $4x + 7$ is not a valid encryption function: x and $x + 13$ will encrypt to the same value, for any $x \in \mathbb{Z}_{26}$.

Let's next suppose that $\gcd(a, 26) = 1$. Suppose for some x_1 and x_2 that

$$ax_1 \equiv ax_2 \ (\mathrm{mod}\ 26).$$

Then

$$a(x_1 - x_2) \equiv 0 \ (\mathrm{mod}\ 26),$$

and thus
$$26 \mid a(x_1 - x_2).$$

We now make use of a fundamental property of integer division: if $\gcd(a, b) = 1$ and $a \mid bc$, then $a \mid c$. Since $26 \mid a(x_1 - x_2)$ and $\gcd(a, 26) = 1$, we must therefore have that
$$26 \mid (x_1 - x_2),$$

i.e., $x_1 \equiv x_2 \pmod{26}$.

At this point we have shown that, if $\gcd(a, 26) = 1$, then a congruence of the form $ax \equiv y \pmod{26}$ has, at most, one solution in $\mathbb{Z}_{26}$. Hence, if we let x vary over $\mathbb{Z}_{26}$, then $ax \bmod 26$ takes on 26 distinct values modulo 26. That is, it takes on every value exactly once. It follows that, for any $y \in \mathbb{Z}_{26}$, the congruence $ax \equiv y \pmod{26}$ has a unique solution for x.

There is nothing special about the number 26 in this argument. The following result can be proved in an analogous fashion.

THEOREM 1.1 *The congruence $ax \equiv b \pmod{m}$ has a unique solution $x \in \mathbb{Z}_m$ for every $b \in \mathbb{Z}_m$ if and only if $\gcd(a, m) = 1$.*

Since $26 = 2 \times 13$, the values of $a \in \mathbb{Z}_{26}$ such that $\gcd(a, 26) = 1$ are $a = 1$, 3, 5, 7, 9, 11, 15, 17, 19, 21, 23, and 25. The parameter b can be any element in $\mathbb{Z}_{26}$. Hence the *Affine Cipher* has $12 \times 26 = 312$ possible keys. (Of course, this is much too small to be secure.)

Let's now consider the general setting where the modulus is m. We need another definition from number theory.

Definition 1.3: Suppose $a \geq 1$ and $m \geq 2$ are integers. If $\gcd(a, m) = 1$, then we say that a and m are *relatively prime*. The number of integers in $\mathbb{Z}_m$ that are relatively prime to m is often denoted by $\phi(m)$ (this function is called the *Euler phi-function*).

A well-known result from number theory gives the value of $\phi(m)$ in terms of the prime power factorization of m. (An integer $p > 1$ is *prime* if it has no positive divisors other than 1 and p. Every integer $m > 1$ can be factored as a product of powers of primes in a unique way. For example, $60 = 2^2 \times 3 \times 5$ and $98 = 2 \times 7^2$.)

We record the formula for $\phi(m)$ in the following theorem.

THEOREM 1.2 *Suppose*
$$m = \prod_{i=1}^{n} p_i^{e_i},$$
where the p_i's are distinct primes and $e_i > 0$, $1 \leq i \leq n$. Then
$$\phi(m) = \prod_{i=1}^{n} (p_i^{e_i} - p_i^{e_i - 1}).$$

It follows that the number of keys in the *Affine Cipher* over $\mathbb{Z}_m$ is $m\phi(m)$, where $\phi(m)$ is given by the formula above. (The number of choices for b is m, and the number of choices for a is $\phi(m)$, where the encryption function is $e(x) = ax + b$.) For example, suppose $m = 60$. We have

$$60 = 2^2 \times 3^1 \times 5^1$$

and hence

$$\phi(60) = (4 - 2) \times (3 - 1) \times (5 - 1) = 2 \times 2 \times 4 = 16.$$

The number of keys in the *Affine Cipher* is $60 \times 16 = 960$.

Let's now consider the decryption operation in the *Affine Cipher* with modulus $m = 26$. Suppose that $\gcd(a, 26) = 1$. To decrypt, we need to solve the congruence $y \equiv ax + b \pmod{26}$ for x. The discussion above establishes that the congruence will have a unique solution in $\mathbb{Z}_{26}$, but it does not give us an efficient method of finding the solution. What we require is an efficient algorithm to do this. Fortunately, some further results on modular arithmetic will provide us with the efficient decryption algorithm we seek.

We require the idea of a multiplicative inverse.

Definition 1.4:   Suppose $a \in \mathbb{Z}_m$. The *multiplicative inverse* of a modulo m, denoted $a^{-1} \bmod m$, is an element $a' \in \mathbb{Z}_m$ such that $aa' \equiv a'a \equiv 1 \pmod{m}$. If m is fixed, we sometimes write a^{-1} for $a^{-1} \bmod m$.

By similar arguments to those used above, it can be shown that a has a multiplicative inverse modulo m if and only if $\gcd(a, m) = 1$; and if a multiplicative inverse exists, it is unique modulo m. Also, observe that if $b = a^{-1}$, then $a = b^{-1}$. If p is prime, then every non-zero element of $\mathbb{Z}_p$ has a multiplicative inverse. A ring in which this is true is called a *field*.

In a later section, we will describe an efficient algorithm for computing multiplicative inverses in $\mathbb{Z}_m$ for any m. However, in $\mathbb{Z}_{26}$, trial and error suffices to find the multiplicative inverses of the elements relatively prime to 26:

$$1^{-1} = 1,$$
$$3^{-1} = 9,$$
$$5^{-1} = 21,$$
$$7^{-1} = 15,$$
$$11^{-1} = 19,$$
$$17^{-1} = 23, \text{ and}$$
$$25^{-1} = 25.$$

Cryptosystem 1.3: *Affine Cipher*

Let $\mathcal{P} = \mathcal{C} = \mathbb{Z}_{26}$ and let

$$\mathcal{K} = \{(a, b) \in \mathbb{Z}_{26} \times \mathbb{Z}_{26} : \gcd(a, 26) = 1\}.$$

For $K = (a, b) \in \mathcal{K}$, define

$$e_K(x) = (ax + b) \bmod 26$$

and

$$d_K(y) = a^{-1}(y - b) \bmod 26$$

$(x, y \in \mathbb{Z}_{26})$.

(All of these can be verified easily. For example, $7 \times 15 = 105 \equiv 1 \pmod{26}$, so $7^{-1} = 15$ and $15^{-1} = 7$.)

Consider our congruence $y \equiv ax + b \pmod{26}$. This is equivalent to

$$ax \equiv y - b \pmod{26}.$$

Since $\gcd(a, 26) = 1$, a has a multiplicative inverse modulo 26. Multiplying both sides of the congruence by a^{-1}, we obtain

$$a^{-1}(ax) \equiv a^{-1}(y - b) \pmod{26}.$$

By associativity of multiplication modulo 26, we have that

$$a^{-1}(ax) \equiv (a^{-1}a)x \equiv 1x \equiv x \pmod{26}.$$

Consequently, $x = a^{-1}(y - b) \bmod 26$. This is an explicit formula for x, that is, the decryption function is

$$d(y) = a^{-1}(y - b) \bmod 26.$$

So, finally, the complete description of the *Affine Cipher* is given as Cryptosystem 1.3.

Let's do a small example.

Example 1.3 Suppose that $K = (7, 3)$. As noted above, $7^{-1} \bmod 26 = 15$. The encryption function is

$$e_K(x) = 7x + 3,$$

and the corresponding decryption function is

$$d_K(y) = 15(y - 3) = 15y - 19,$$

where all operations are performed in $\mathbb{Z}_{26}$. It is a good check to verify that $d_K(e_K(x)) = x$ for all $x \in \mathbb{Z}_{26}$. Computing in $\mathbb{Z}_{26}$, we get

$$\begin{aligned} d_K(e_K(x)) &= d_K(7x + 3) \\ &= 15(7x + 3) - 19 \\ &= x + 45 - 19 \\ &= x. \end{aligned}$$

To illustrate, let's encrypt the plaintext *hot*. We first convert the letters h, o, t to residues modulo 26. These are respectively 7, 14, and 19. Now, we encrypt:

$$\begin{aligned} (7 \times 7 + 3) \bmod 26 &= 52 \bmod 26 &= 0 \\ (7 \times 14 + 3) \bmod 26 &= 101 \bmod 26 &= 23 \\ (7 \times 19 + 3) \bmod 26 &= 136 \bmod 26 &= 6. \end{aligned}$$

So the three ciphertext characters are $0, 23$, and 6, which corresponds to the alphabetic string AXG. We leave the decryption as an exercise for the reader. ◻

1.1.4 The Vigenère Cipher

In both the *Shift Cipher* and the *Substitution Cipher*, once a key is chosen, each alphabetic character is mapped to a unique alphabetic character. For this reason, these cryptosystems are called *monoalphabetic cryptosystems*. We now present a cryptosystem which is not monoalphabetic, the well-known *Vigenère Cipher*, as Cryptosystem 1.4. This cipher is named after Blaise de Vigenère, who lived in the sixteenth century.

Using the correspondence $A \leftrightarrow 0, B \leftrightarrow 1, \ldots, Z \leftrightarrow 25$ described earlier, we can associate each key K with an alphabetic string of length m, called a *keyword*. The *Vigenère Cipher* encrypts m alphabetic characters at a time: each plaintext element is equivalent to m alphabetic characters.

Let's do a small example.

Example 1.4   Suppose $m = 6$ and the keyword is $CIPHER$. This corresponds to the numerical equivalent $K = (2, 8, 15, 7, 4, 17)$. Suppose the plaintext is the string

thiscryptosystemisnotsecure.

We convert the plaintext elements to residues modulo 26, write them in groups of six, and then "add" the keyword modulo 26, as follows:

19	7	8	18	2	17	24	15	19	14	18	24
2	8	15	7	4	17	2	8	15	7	4	17
21	15	23	25	6	8	0	23	8	21	22	15

Cryptosystem 1.4: *Vigenère Cipher*

Let m be a positive integer. Define $\mathcal{P} = \mathcal{C} = \mathcal{K} = (\mathbb{Z}_{26})^m$. For a key $K = (k_1, k_2, \ldots, k_m)$, we define

$$e_K(x_1, x_2, \ldots, x_m) = (x_1 + k_1, x_2 + k_2, \ldots, x_m + k_m)$$

and

$$d_K(y_1, y_2, \ldots, y_m) = (y_1 - k_1, y_2 - k_2, \ldots, y_m - k_m),$$

where all operations are performed in $\mathbb{Z}_{26}$.

18	19	4	12	8	18	13	14	19	18	4	2
2	8	15	7	4	17	2	8	15	7	4	17
20	1	19	19	12	9	15	22	8	25	8	19

20	17	4
2	8	15
22	25	19

The alphabetic equivalent of the ciphertext string would thus be:

$$\text{VPXZGIAXIVWPUBTTMJPWIZITWZT.}$$

To decrypt, we can use the same keyword, but we would subtract it modulo 26 from the ciphertext, instead of adding. ▯

Observe that the number of possible keywords of length m in a *Vigenère Cipher* is 26^m, so even for relatively small values of m, an exhaustive key search would require a long time. For example, if we take $m = 5$, then the keyspace has size exceeding 1.1×10^7. This is already large enough to preclude exhaustive key search by hand (but not by computer).

In a *Vigenère Cipher* having keyword length m, an alphabetic character can be mapped to one of m possible alphabetic characters (assuming that the keyword contains m distinct characters). Such a cryptosystem is called a *polyalphabetic cryptosystem*. In general, cryptanalysis is more difficult for polyalphabetic than for monoalphabetic cryptosystems.

1.1.5 The Hill Cipher

In this section, we describe another polyalphabetic cryptosystem called the *Hill Cipher*. This cipher was invented in 1929 by Lester S. Hill. Let m be a positive integer, and define $\mathcal{P} = \mathcal{C} = (\mathbb{Z}_{26})^m$. The idea is to take m linear combinations of the m alphabetic characters in one plaintext element, thus producing the m alphabetic characters in one ciphertext element.

For example, if $m = 2$, we could write a plaintext element as $x = (x_1, x_2)$ and a ciphertext element as $y = (y_1, y_2)$. Here, y_1 would be a linear combination of x_1 and x_2, as would y_2. We might take

$$y_1 = (11x_1 + 3x_2) \bmod 26$$

$$y_2 = (8x_1 + 7x_2) \bmod 26.$$

Of course, this can be written more succinctly in matrix notation as follows:

$$(y_1, y_2) = (x_1, x_2) \begin{pmatrix} 11 & 8 \\ 3 & 7 \end{pmatrix},$$

where all operations are performed in $\mathbb{Z}_{26}$. In general, we will take an $m \times m$ matrix K as our key. If the entry in row i and column j of K is $k_{i,j}$, then we write $K = (k_{i,j})$. For $x = (x_1, \ldots, x_m) \in \mathcal{P}$ and $K \in \mathcal{K}$, we compute $y = e_K(x) = (y_1, \ldots, y_m)$ as follows:

$$(y_1, y_2, \ldots, y_m) = (x_1, x_2, \ldots, x_m) \begin{pmatrix} k_{1,1} & k_{1,2} & \cdots & k_{1,m} \\ k_{2,1} & k_{2,2} & \cdots & k_{2,m} \\ \vdots & \vdots & & \vdots \\ k_{m,1} & k_{m,2} & \cdots & k_{m,m} \end{pmatrix}.$$

In other words, using matrix notation, $y = xK$.

We say that the ciphertext is obtained from the plaintext by means of a *linear transformation*. We have to consider how decryption will work, that is, how x can be computed from y. Readers familiar with linear algebra will realize that we will use the inverse matrix K^{-1} to decrypt. The ciphertext is decrypted using the matrix equation $x = yK^{-1}$.

Here are the definitions of necessary concepts from linear algebra. If $A = (a_{i,j})$ is an $\ell \times m$ matrix and $B = (b_{j,k})$ is an $m \times n$ matrix, then we define the *matrix product* $AB = (c_{i,k})$ by the formula

$$c_{i,k} = \sum_{j=1}^{m} a_{i,j} b_{j,k}$$

for $1 \leq i \leq \ell$ and $1 \leq k \leq n$. That is, the entry in row i and column k of AB is formed by taking the ith row of A and the kth column of B, multiplying corresponding entries together, and summing. Note that AB is an $\ell \times n$ matrix.

Matrix multiplication is associative (that is, $(AB)C = A(BC)$) but not, in general, commutative (it is not always the case that $AB = BA$, even for square matrices A and B).

The $m \times m$ *identity matrix*, denoted by I_m, is the $m \times m$ matrix with 1's on the main diagonal and 0's elsewhere. Thus, the 2×2 identity matrix is

$$I_2 = \begin{pmatrix} 1 & 0 \\ 0 & 1 \end{pmatrix}.$$

I_m is termed an identity matrix since $AI_m = A$ for any $\ell \times m$ matrix A and $I_m B = B$ for any $m \times n$ matrix B. Now, the *inverse matrix* of an $m \times m$ matrix A (if it exists) is the matrix A^{-1} such that $AA^{-1} = A^{-1}A = I_m$. Not all matrices have inverses, but if an inverse exists, it is unique.

With these facts at hand, it is easy to derive the decryption formula given above, assuming that K has an inverse matrix K^{-1}. Since $y = xK$, we can multiply both sides of the formula by K^{-1}, obtaining

$$yK^{-1} = (xK)K^{-1} = x(KK^{-1}) = xI_m = x.$$

(Note the use of the associativity property.)

We can verify that the example encryption matrix defined above has an inverse in $\mathbb{Z}_{26}$:

$$\begin{pmatrix} 11 & 8 \\ 3 & 7 \end{pmatrix}^{-1} = \begin{pmatrix} 7 & 18 \\ 23 & 11 \end{pmatrix}$$

since

$$\begin{pmatrix} 11 & 8 \\ 3 & 7 \end{pmatrix}\begin{pmatrix} 7 & 18 \\ 23 & 11 \end{pmatrix} = \begin{pmatrix} 11 \times 7 + 8 \times 23 & 11 \times 18 + 8 \times 11 \\ 3 \times 7 + 7 \times 23 & 3 \times 18 + 7 \times 11 \end{pmatrix}$$
$$= \begin{pmatrix} 261 & 286 \\ 182 & 131 \end{pmatrix}$$
$$= \begin{pmatrix} 1 & 0 \\ 0 & 1 \end{pmatrix}.$$

(Remember that all arithmetic operations are done modulo 26.)

Let's now do an example to illustrate encryption and decryption in the *Hill Cipher.*

Example 1.5 Suppose the key is

$$K = \begin{pmatrix} 11 & 8 \\ 3 & 7 \end{pmatrix}.$$

From the computations above, we have that

$$K^{-1} = \begin{pmatrix} 7 & 18 \\ 23 & 11 \end{pmatrix}.$$

Suppose we want to encrypt the plaintext *july*. We have two elements of plaintext to encrypt: $(9, 20)$ (corresponding to *ju*) and $(11, 24)$ (corresponding to *ly*). We compute as follows:

$$(9, 20)\begin{pmatrix} 11 & 8 \\ 3 & 7 \end{pmatrix} = (99 + 60, 72 + 140) = (3, 4)$$

and
$$(11, 24) \begin{pmatrix} 11 & 8 \\ 3 & 7 \end{pmatrix} = (121 + 72, 88 + 168) = (11, 22).$$

Hence, the encryption of *july* is *DELW*. To decrypt, Bob would compute:

$$(3, 4) \begin{pmatrix} 7 & 18 \\ 23 & 11 \end{pmatrix} = (9, 20)$$

and
$$(11, 22) \begin{pmatrix} 7 & 18 \\ 23 & 11 \end{pmatrix} = (11, 24).$$

Hence, the correct plaintext is obtained.   ☐

At this point, we have shown that decryption is possible if K has an inverse. In fact, for decryption to be possible, it is necessary that K has an inverse. (This follows fairly easily from elementary linear algebra, but we will not give a proof here.) So we are interested precisely in those matrices K that are invertible.

The invertibility of a (square) matrix depends on the value of its determinant, which we define now.

Definition 1.5:   Suppose that $A = (a_{i,j})$ is an $m \times m$ matrix. For $1 \le i \le m$, $1 \le j \le m$, define A_{ij} to be the matrix obtained from A by deleting the ith row and the jth column. The *determinant* of A, denoted det A, is the value $a_{1,1}$ if $m = 1$. If $m > 1$, then det A is computed recursively from the formula

$$\det A = \sum_{j=1}^{m} (-1)^{i+j} a_{i,j} \det A_{ij},$$

where i is any fixed integer between 1 and m.

It is not at all obvious that the value of det A is independent of the choice of i in the formula given above, but it can be proved that this is indeed the case. It will be useful to write out the formulas for determinants of 2×2 and 3×3 matrices. If $A = (a_{i,j})$ is a 2×2 matrix, then

$$\det A = a_{1,1}a_{2,2} - a_{1,2}a_{2,1}.$$

If $A = (a_{i,j})$ is a 3×3 matrix, then

$$\det A = a_{1,1}a_{2,2}a_{3,3} + a_{1,2}a_{2,3}a_{3,1} + a_{1,3}a_{2,1}a_{3,2}$$
$$- (a_{1,1}a_{2,3}a_{3,2} + a_{1,2}a_{2,1}a_{3,3} + a_{1,3}a_{2,2}a_{3,1}).$$

For large m, the recursive formula given in the definition above is not usually a very efficient method of computing the determinant of an $m \times m$ square matrix.

A preferred method is to compute the determinant using so-called "elementary row operations"; see any text on linear algebra.

Two important properties of determinants that we will use are $\det I_m = 1$; and the multiplication rule $\det(AB) = \det A \times \det B$.

A real matrix K has an inverse if and only if its determinant is non-zero. However, it is important to remember that we are working over $\mathbb{Z}_{26}$. The relevant result for our purposes is that a matrix K has an inverse modulo 26 if and only if $\gcd(\det K, 26) = 1$. To see that this condition is necessary, suppose K has an inverse, denoted K^{-1}. By the multiplication rule for determinants, we have

$$1 = \det I = \det(KK^{-1}) = \det K \det K^{-1}.$$

Hence, $\det K$ is invertible in $\mathbb{Z}_{26}$, which is true if and only if $\gcd(\det K, 26) = 1$.

Sufficiency of this condition can be established in several ways. We will give an explicit formula for the inverse of the matrix K. Define a matrix K^* to have as its (i, j)-entry the value $(-1)^{i+j} \det K_{ji}$. (Recall that K_{ji} is obtained from K by deleting the jth row and the ith column.) K^* is called the *adjoint matrix* of K. We state the following theorem, concerning inverses of matrices over $\mathbb{Z}_n$, without proof.

THEOREM 1.3 *Suppose $K = (k_{i,j})$ is an $m \times m$ matrix over $\mathbb{Z}_n$ such that $\det K$ is invertible in $\mathbb{Z}_n$. Then $K^{-1} = (\det K)^{-1} K^*$, where K^* is the adjoint matrix of K.*

REMARK The above formula for K^{-1} is not very efficient computationally, except for small values of m (e.g., $m = 2, 3$). For larger m, the preferred method of computing inverse matrices would involve performing elementary row operations on the matrix K. ∎

In the 2×2 case, we have the following formula, which is an immediate corollary of Theorem 1.3.

COROLLARY 1.4 *Suppose*

$$K = \begin{pmatrix} k_{1,1} & k_{1,2} \\ k_{2,1} & k_{2,2} \end{pmatrix}$$

is a matrix having entries in $\mathbb{Z}_n$, and $\det K = k_{1,1}k_{2,2} - k_{1,2}k_{2,1}$ is invertible in $\mathbb{Z}_n$. Then

$$K^{-1} = (\det K)^{-1} \begin{pmatrix} k_{2,2} & -k_{1,2} \\ -k_{2,1} & k_{1,1} \end{pmatrix}.$$

Let's look again at the example considered earlier. First, we have

$$\det \begin{pmatrix} 11 & 8 \\ 3 & 7 \end{pmatrix} = (11 \times 7 - 8 \times 3) \bmod 26$$

$$= (77 - 24) \bmod 26$$

$$= 53 \bmod 26$$

$$= 1.$$

Now, $1^{-1} \bmod 26 = 1$, so the inverse matrix is

$$\begin{pmatrix} 11 & 8 \\ 3 & 7 \end{pmatrix}^{-1} = \begin{pmatrix} 7 & 18 \\ 23 & 11 \end{pmatrix},$$

as we verified earlier.

Here is another example, using a 3×3 matrix.

Example 1.6   Suppose that

$$K = \begin{pmatrix} 10 & 5 & 12 \\ 3 & 14 & 21 \\ 8 & 9 & 11 \end{pmatrix},$$

where all entries are in $\mathbb{Z}_{26}$. The reader can verify that $\det K = 7$. In $\mathbb{Z}_{26}$, we have that $7^{-1} \bmod 26 = 15$. The adjoint matrix is

$$K^* = \begin{pmatrix} 17 & 1 & 15 \\ 5 & 14 & 8 \\ 19 & 2 & 21 \end{pmatrix}.$$

Finally, the inverse matrix is

$$K^{-1} = 15 K^* = \begin{pmatrix} 21 & 15 & 17 \\ 23 & 2 & 16 \\ 25 & 4 & 3 \end{pmatrix}.$$

◻

As mentioned above, encryption in the *Hill Cipher* is done by multiplying the plaintext by the matrix K, while decryption multiplies the ciphertext by the inverse matrix K^{-1}. We now give a precise mathematical description of the *Hill Cipher* over $\mathbb{Z}_{26}$; see Cryptosystem 1.5.

Cryptosystem 1.5: *Hill Cipher*

Let $m \geq 2$ be an integer. Let $\mathcal{P} = \mathcal{C} = (\mathbb{Z}_{26})^m$ and let

$$\mathcal{K} = \{m \times m \text{ invertible matrices over } \mathbb{Z}_{26}\}.$$

For a key K, we define

$$e_K(x) = xK$$

and

$$d_K(y) = yK^{-1},$$

where all operations are performed in $\mathbb{Z}_{26}$.

1.1.6 The Permutation Cipher

All of the cryptosystems we have discussed so far involve substitution: plaintext characters are replaced by different ciphertext characters. The idea of a permutation cipher is to keep the plaintext characters unchanged, but to alter their positions by rearranging them using a permutation.

A *permutation* of a finite set X is a bijective function $\pi : X \to X$. In other words, the function π is one-to-one (injective) and onto (*surjective*). It follows that, for every $x \in X$, there is a unique element $x' \in X$ such that $\pi(x') = x$. This allows us to define the *inverse permutation,* $\pi^{-1} : X \to X$ by the rule

$$\pi^{-1}(x) = x' \quad \text{if and only if} \quad \pi(x') = x.$$

Then π^{-1} is also a permutation of X.

The *Permutation Cipher* (also known as the *Transposition Cipher*) is defined formally as Cryptosystem 1.6. This cryptosystem has been in use for hundreds of years. In fact, the distinction between the *Permutation Cipher* and the *Substitution Cipher* was pointed out as early as 1563 by Giovanni Porta.

As with the *Substitution Cipher*, it is more convenient to use alphabetic characters as opposed to residues modulo 26, since there are no algebraic operations being performed in encryption or decryption.

Here is an example to illustrate:

Example 1.7 Suppose $m = 6$ and the key is the following permutation π:

x	1	2	3	4	5	6
$\pi(x)$	3	5	1	6	4	2

Note that the first row of the above diagram lists the values of x, $1 \leq x \leq 6$, and the second row lists the corresponding values of $\pi(x)$. Then the inverse permuta-

---

Cryptosystem 1.6: *Permutation Cipher*

Let m be a positive integer. Let $\mathcal{P} = \mathcal{C} = (\mathbb{Z}_{26})^m$ and let $\mathcal{K}$ consist of all permutations of $\{1, \ldots, m\}$. For a key (i.e., a permutation) π, we define

$$e_\pi(x_1, \ldots, x_m) = (x_{\pi(1)}, \ldots, x_{\pi(m)})$$

and

$$d_\pi(y_1, \ldots, y_m) = (y_{\pi^{-1}(1)}, \ldots, y_{\pi^{-1}(m)}),$$

where π^{-1} is the inverse permutation to π.

---

tion π^{-1} can be constructed by interchanging the two rows, and rearranging the columns so that the first row is in increasing order. Carrying out these operations, we see that the permutation π^{-1} is the following:

x	1	2	3	4	5	6
$\pi^{-1}(x)$	3	6	1	5	2	4

Now, suppose we are given the plaintext

shesellsseashellsbytheseashore.

We first partition the plaintext into groups of six letters:

shesel | lsseas | hellsb | ythese | ashore

Now each group of six letters is rearranged according to the permutation π, yielding the following:

EESLSH | SALSES | LSHBLE | HSYEET | HRAEOS

So, the ciphertext is:

EESLSHSALSESLSHBLEHSYEETHRAEOS.

The ciphertext can be decrypted in a similar fashion, using the inverse permutation π^{-1}. □

We now show that the *Permutation Cipher* is a special case of the *Hill Cipher*. Given a permutation π of the set $\{1, \ldots, m\}$, we can define an associated $m \times m$ permutation matrix $K_\pi = (k_{i,j})$ according to the formula

$$k_{i,j} = \begin{cases} 1 & \text{if } i = \pi(j) \\ 0 & \text{otherwise.} \end{cases}$$

(A *permutation matrix* is a matrix in which every row and column contains exactly one "1," and all other values are "0." A permutation matrix can be obtained from an identity matrix by permuting rows or columns.)

It is not difficult to see that Hill encryption using the matrix K_π is, in fact, equivalent to permutation encryption using the permutation π. Moreover, $K_\pi^{-1} = K_{\pi^{-1}}$, i.e., the inverse matrix to K_π is the permutation matrix defined by the permutation π^{-1}. Thus, Hill decryption is equivalent to permutation decryption.

For the permutation π used in the example above, the associated permutation matrices are

$$K_\pi = \begin{pmatrix} 0 & 0 & 1 & 0 & 0 & 0 \\ 0 & 0 & 0 & 0 & 0 & 1 \\ 1 & 0 & 0 & 0 & 0 & 0 \\ 0 & 0 & 0 & 0 & 1 & 0 \\ 0 & 1 & 0 & 0 & 0 & 0 \\ 0 & 0 & 0 & 1 & 0 & 0 \end{pmatrix}$$

and

$$K_\pi^{-1} = \begin{pmatrix} 0 & 0 & 1 & 0 & 0 & 0 \\ 0 & 0 & 0 & 0 & 1 & 0 \\ 1 & 0 & 0 & 0 & 0 & 0 \\ 0 & 0 & 0 & 0 & 0 & 1 \\ 0 & 0 & 0 & 1 & 0 & 0 \\ 0 & 1 & 0 & 0 & 0 & 0 \end{pmatrix}.$$

The reader can verify that the product of these two matrices is the identity matrix.

1.1.7 Stream Ciphers

In the cryptosystems we have studied so far, successive plaintext elements are encrypted using the same key, K. That is, the ciphertext string $\mathbf{y}$ is obtained as follows:

$$\mathbf{y} = y_1 y_2 \cdots = e_K(x_1) e_K(x_2) \cdots .$$

Cryptosystems of this type are often called *block ciphers*.

An alternative approach is to use what are called stream ciphers. The basic idea is to generate a keystream $\mathbf{z} = z_1 z_2 \cdots$, and use it to encrypt a plaintext string $\mathbf{x} = x_1 x_2 \cdots$ according to the rule

$$\mathbf{y} = y_1 y_2 \cdots = e_{z_1}(x_1) e_{z_2}(x_2) \cdots .$$

The simplest type of stream cipher is one in which the keystream is constructed from the key, independent of the plaintext string, using some specified algorithm. This type of stream cipher is called "synchronous" and can be defined formally as follows:

Definition 1.6:   A *synchronous stream cipher* is a tuple $(\mathcal{P}, \mathcal{C}, \mathcal{K}, \mathcal{L}, \mathcal{E}, \mathcal{D})$, together with a function g, such that the following conditions are satisfied:

1. $\mathcal{P}$ is a finite set of possible *plaintexts*

2. $\mathcal{C}$ is a finite set of possible *ciphertexts*

3. $\mathcal{K}$, the *keyspace*, is a finite set of possible *keys*

4. $\mathcal{L}$ is a finite set called the *keystream alphabet*

5. g is the *keystream generator*. g takes a key K as input, and generates an infinite string $z_1 z_2 \cdots$ called the *keystream*, where $z_i \in \mathcal{L}$ for all $i \geq 1$.

6. For each $z \in \mathcal{L}$, there is an *encryption rule* $e_z \in \mathcal{E}$ and a corresponding *decryption rule* $d_z \in \mathcal{D}$. $e_z : \mathcal{P} \to \mathcal{C}$ and $d_z : \mathcal{C} \to \mathcal{P}$ are functions such that $d_z(e_z(x)) = x$ for every plaintext element $x \in \mathcal{P}$.

To illustrate this definition, we show how the *Vigenère Cipher* can be defined as a synchronous stream cipher. Suppose that m is the keyword length of a *Vigenère Cipher*. Define $\mathcal{K} = (\mathbb{Z}_{26})^m$ and $\mathcal{P} = \mathcal{C} = \mathcal{L} = \mathbb{Z}_{26}$; and define $e_z(x) = (x + z) \bmod 26$ and $d_z(y) = (y - z) \bmod 26$. Finally, define the keystream $z_1 z_2 \cdots$ as follows:

$$z_i = \begin{cases} k_i & \text{if } 1 \leq i \leq m \\ z_{i-m} & \text{if } i \geq m + 1, \end{cases}$$

where $K = (k_1, \ldots, k_m)$. This generates the keystream

$$k_1 k_2 \cdots k_m k_1 k_2 \cdots k_m k_1 k_2 \cdots$$

from the key $K = (k_1, k_2, \ldots, k_m)$.

REMARK   We can think of a block cipher as a special case of a stream cipher where the keystream is constant: $z_i = K$ for all $i \geq 1$.   ∎

A stream cipher is a *periodic stream cipher* with period d if $z_{i+d} = z_i$ for all integers $i \geq 1$. The *Vigenère Cipher* with keyword length m, as described above, can be thought of as a periodic stream cipher with period m.

Stream ciphers are often described in terms of binary alphabets, i.e., $\mathcal{P} = \mathcal{C} = \mathcal{L} = \mathbb{Z}_2$. In this situation, the encryption and decryption operations are just addition modulo 2:

$$e_z(x) = (x + z) \bmod 2$$

and

$$d_z(y) = (y + z) \bmod 2.$$

If we think of "0" as representing the boolean value "false" and "1" as representing "true," then addition modulo 2 corresponds to the *exclusive-or* operation.

Hence, encryption (and decryption) can be implemented very efficiently in hardware.

Let's look at another method of generating a (synchronous) keystream. We will work over binary alphabets. Suppose we start with a binary m-tuple $(k_1, \ldots, k_m)$ and let $z_i = k_i$, $1 \leq i \leq m$ (as before). Now we generate the keystream using a *linear recurrence* of degree m:

$$z_{i+m} = \sum_{j=0}^{m-1} c_j z_{i+j} \bmod 2,$$

for all $i \geq 1$, where $c_0, \ldots, c_{m-1} \in \mathbb{Z}_2$ are specified constants.

REMARK This recurrence is said to have *degree* m since each term depends on the previous m terms. It is *linear* because z_{i+m} is a linear function of previous terms. Note that we can take $c_0 = 1$ without loss of generality, for otherwise the recurrence will be of degree (at most) $m - 1$. ∎

Here, the key K consists of the $2m$ values $k_1, \ldots, k_m, c_0, \ldots, c_{m-1}$. If

$$(k_1, \ldots, k_m) = (0, \ldots, 0),$$

then the keystream consists entirely of 0's. Of course, this should be avoided, as the ciphertext will then be identical to the plaintext. However, if the constants $c_0, \ldots, c_{m-1}$ are chosen in a suitable way, then any other initialization vector $(k_1, \ldots, k_m)$ will give rise to a periodic keystream having period $2^m - 1$. So a "short" key can give rise to a keystream having a very long period. This is certainly a desirable property: we will see in a later section how the *Vigenère Cipher* can be cryptanalyzed by exploiting the fact that the keystream has a short period.

Here is an example to illustrate.

Example 1.8 Suppose $m = 4$ and the keystream is generated using the linear recurrence

$$z_{i+4} = (z_i + z_{i+1}) \bmod 2,$$

$i \geq 1$. If the keystream is initialized with any vector other than $(0, 0, 0, 0)$, then we obtain a keystream of period 15. For example, starting with $(1, 0, 0, 0)$, the keystream is

$$1\,0\,0\,0\,1\,0\,0\,1\,1\,0\,1\,0\,1\,1\,1 \cdots.$$

Any other non-zero initialization vector will give rise to a cyclic permutation of the same keystream. ☐

Another appealing aspect of this method of keystream generation is that the keystream can be produced efficiently in hardware using a *linear feedback shift*

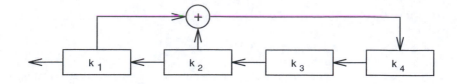

FIGURE 1.2
A linear feedback shift register

register, or LFSR. We would use a shift register with m stages. The vector $(k_1, \ldots, k_m)$ would be used to initialize the shift register. At each time unit, the following operations would be performed concurrently:

1. k_1 would be tapped as the next keystream bit
2. $k_2, \ldots, k_m$ would each be shifted one stage to the left
3. the "new" value of k_m would be computed to be

$$\sum_{j=0}^{m-1} c_j k_{j+1}$$

(this is the "linear feedback").

At any given point in time, the shift register contains m consecutive keystream elements, say $z_i, \ldots, z_{i+m-1}$. After one time unit, the shift register contains $z_{i+1}, \ldots, z_{i+m}$.

Observe that the linear feedback is carried out by tapping certain stages of the register (as specified by the constants c_j having the value "1") and computing a sum modulo 2 (which is an exclusive-or). This is illustrated in Figure 1.2, where we depict the LFSR that will generate the keystream of Example 1.8.

A *non-synchronous stream cipher* is a stream cipher in which each keystream element z_i depends on previous plaintext or ciphertext elements $(x_1, \ldots, x_{i-1}$ and/or $y_1, \ldots, y_{i-1})$ as well as the key K. A simple type of non-synchronous stream cipher, known as the *Autokey Cipher*, is presented as Cryptosystem 1.7. It is apparently due to Vigenère. The reason for the terminology "autokey" is that the plaintext is used to construct the keystream (aside from the initial "priming key" K). Of course, the *Autokey Cipher* is insecure since there are only 26 possible keys.

Here is an example to illustrate:

Example 1.9 Suppose the key is $K = 8$, and the plaintext is

rendezvous.

Cryptosystem 1.7: *Autokey Cipher*

Let $\mathcal{P} = \mathcal{C} = \mathcal{K} = \mathcal{L} = \mathbb{Z}_{26}$. Let $z_1 = K$, and define $z_i = x_{i-1}$ for all $i \geq 2$. For $0 \leq z \leq 25$, define

$$e_z(x) = (x + z) \bmod 26$$

and

$$d_z(y) = (y - z) \bmod 26$$

$(x, y \in \mathbb{Z}_{26})$.

We first convert the plaintext to a sequence of integers:

$$17 \quad 4 \quad 13 \quad 3 \quad 4 \quad 25 \quad 21 \quad 14 \quad 20 \quad 18$$

The keystream is as follows:

$$8 \quad 17 \quad 4 \quad 13 \quad 3 \quad 4 \quad 25 \quad 21 \quad 14 \quad 20$$

Now we add corresponding elements, reducing modulo 26:

$$25 \quad 21 \quad 17 \quad 16 \quad 7 \quad 3 \quad 20 \quad 9 \quad 8 \quad 12$$

In alphabetic form, the ciphertext is:

$$\text{ZVRQHDUJIM.}$$

Now let's look at how the ciphertext would be decrypted. First, we convert the alphabetic string to the numeric string

$$25 \quad 21 \quad 17 \quad 16 \quad 7 \quad 3 \quad 20 \quad 9 \quad 8 \quad 12$$

Then we compute

$$x_1 = d_8(25) = (25 - 8) \bmod 26 = 17.$$

Next,

$$x_2 = d_{17}(21) = (21 - 17) \bmod 26 = 4,$$

and so on. Each time we obtain another plaintext character, we also use it as the next keystream element. $\square$

In the next section, we discuss methods that can be used to cryptanalyze the various cryptosystems we have presented.

1.2 Cryptanalysis

In this section, we discuss some techniques of cryptanalysis. The general assumption that is usually made is that the opponent, Oscar, knows the cryptosystem being used. This is usually referred to as *Kerckhoffs' principle*. Of course, if Oscar does not know the cryptosystem being used, that will make his task more difficult. But we do not want to base the security of a cryptosystem on the (possibly shaky) premise that Oscar does not know what system is being employed. Hence, our goal in designing a cryptosystem will be to obtain security while assuming that Kerckhoffs' principle holds.

First, we want to differentiate between different attack models on cryptosystems. The *attack model* specifies the information available to the adversary when he mounts his attack. The most common types of attack models are enumerated as follows.

ciphertext only attack

> The opponent possesses a string of ciphertext, y.

known plaintext attack

> The opponent possesses a string of plaintext, x, and the corresponding ciphertext, y.

chosen plaintext attack

> The opponent has obtained temporary access to the encryption machinery. Hence he can choose a plaintext string, x, and construct the corresponding ciphertext string, y.

chosen ciphertext attack

> The opponent has obtained temporary access to the decryption machinery. Hence he can choose a ciphertext string, y, and construct the corresponding plaintext string, x.

In each case, the objective of the adversary is to determine the key that was used. This would allow the opponent to decrypt a specific "target" ciphertext string, and further, to decrypt any additional ciphertext strings that are encrypted using the same key.[1]

We first consider the weakest type of attack, namely a ciphertext-only attack. We also assume that the plaintext string is ordinary English text, without punctuation or "spaces." (This makes cryptanalysis more difficult than if punctuation and spaces were encrypted.)

---

[1] At first glance, a chosen ciphertext attack may seem to be a bit artificial. For, if there is only one ciphertext string of interest to the opponent, then the opponent can obviously decrypt that ciphertext string if a chosen ciphertext attack is permitted. However, we are suggesting that the opponent's objective normally includes determining the key that is used by Alice and Bob, so that other ciphertext strings can be decrypted (at a later time, perhaps). A chosen ciphertext attack makes sense in this context.

TABLE 1.1
Probabilities of occurrence of the 26 letters

letter	probability	letter	probability
A	.082	N	.067
B	.015	O	.075
C	.028	P	.019
D	.043	Q	.001
E	.127	R	.060
F	.022	S	.063
G	.020	T	.091
H	.061	U	.028
I	.070	V	.010
J	.002	W	.023
K	.008	X	.001
L	.040	Y	.020
M	.024	Z	.001

Many techniques of cryptanalysis use statistical properties of the English language. Various people have estimated the relative frequencies of the 26 letters by compiling statistics from numerous novels, magazines, and newspapers. The estimates in Table 1.1 were obtained by Beker and Piper. On the basis of these probabilities, Beker and Piper partition the 26 letters into five groups as follows:

1. E, having probability about 0.120
2. T, A, O, I, N, S, H, R, each having probability between 0.06 and 0.09
3. D, L, each having probability around 0.04
4. $C, U, M, W, F, G, Y, P, B$, each having probability between 0.015 and 0.028
5. V, K, J, X, Q, Z, each having probability less than 0.01.

It is also useful to consider sequences of two or three consecutive letters, called *digrams* and *trigrams*, respectively. The 30 most common digrams are (in decreasing order):

$$TH, HE, IN, ER, AN, RE, ED, ON, ES, ST,$$
$$EN, AT, TO, NT, HA, ND, OU, EA, NG, AS,$$
$$OR, TI, IS, ET, IT, AR, TE, SE, HI, OF.$$

The twelve most common trigrams are:

$$THE, ING, AND, HER, ERE, ENT,$$
$$THA, NTH, WAS, ETH, FOR, DTH.$$

1.2.1 Cryptanalysis of the Affine Cipher

As a simple illustration of how cryptanalysis can be performed using statistical data, let's look first at the *Affine Cipher*. Suppose Oscar has intercepted the fol-

TABLE 1.2
Frequency of occurrence of the 26 ciphertext letters

letter	frequency	letter	frequency
A	2	N	1
B	1	O	1
C	0	P	2
D	7	Q	0
E	5	R	8
F	4	S	3
G	0	T	0
H	5	U	2
I	0	V	4
J	0	W	0
K	5	X	2
L	2	Y	1
M	2	Z	0

lowing ciphertext:

Example 1.10   Ciphertext obtained from an *Affine Cipher*

```
FMXVEDKAPHFERBNDKRXRSREFMORUDSDKDVSHVUFEDK
APRKDLYEVLRHHRH
```

The frequency analysis of this ciphertext is given in Table 1.2.

There are only 57 characters of ciphertext, but this is usually sufficient to cryptanalyze an *Affine Cipher*. The most frequent ciphertext characters are: R (8 occurrences), D (7 occurrences), E, H, K (5 occurrences each), and F, S, V (4 occurrences each). As a first guess, we might hypothesize that R is the encryption of e and D is the encryption of t, since e and t are (respectively) the two most common letters. Expressed numerically, we have $e_K(4) = 17$ and $e_K(19) = 3$. Recall that $e_K(x) = ax + b$, where a and b are unknowns. So we get two linear equations in two unknowns:

$$4a + b = 17$$

$$19a + b = 3.$$

This system has the unique solution $a = 6$, $b = 19$ (in $\mathbb{Z}_{26}$). But this is an illegal key, since $\gcd(a, 26) = 2 > 1$. So our hypothesis must be incorrect.

Our next guess might be that R is the encryption of e and E is the encryption of t. Proceeding as above, we obtain $a = 13$, which is again illegal. So we try the next possibility, that R is the encryption of e and H is the encryption of t. This yields $a = 8$, again impossible. Continuing, we suppose that R is the encryption of e and K is the encryption of t. This produces $a = 3$, $b = 5$, which is at least a legal key. It remains to compute the decryption function corresponding to $K = (3, 5)$, and then to decrypt the ciphertext to see if we get a meaningful string of English, or nonsense. This will confirm the validity of $(3, 5)$.

If we perform these operations, we obtain $d_K(y) = 9y - 19$ and the given ciphertext decrypts to yield:

```
algorithmsarequitegeneraldefinitionsofarit
hmeticprocesses
```

We conclude that we have determined the correct key. ⬜

1.2.2 Cryptanalysis of the Substitution Cipher

Here, we look at the more complicated situation, the *Substitution Cipher*. Consider the ciphertext in the following example:

Example 1.11 Ciphertext obtained from a *Substitution Cipher*

```
YIFQFMZRWQFYVECFMDZPCVMRZWNMDZVEJBTXCDDUMJ
NDIFEFMDZCDMQZKCEYFCJMYRNCWJCSZREXCHZUNMXZ
NZUCDRJXYYSMRTMEYIFZWDYVZVYFZUMRZCRWNZDZJJ
XZWGCHSMRNMDHNCMFQCHZJMXJZWIEJYUCFWDJNZDIR
```

The frequency analysis of this ciphertext is given in Table 1.3.

Since Z occurs significantly more often than any other ciphertext character, we might conjecture that $d_K(Z) = e$. The remaining ciphertext characters that occur at least ten times (each) are C, D, F, J, M, R, Y. We might expect that these letters are encryptions of (a subset of) t, a, o, i, n, s, h, r, but the frequencies really do not vary enough to tell us what the correspondence might be.

At this stage we might look at digrams, especially those of the form $-Z$ or $Z-$, since we conjecture that Z decrypts to e. We find that the most common digrams of this type are DZ and ZW (four times each); NZ and ZU (three times each); and RZ, HZ, XZ, FZ, ZR, ZV, ZC, ZD, and ZJ (twice each). Since ZW occurs four times and WZ not at all, and W occurs less often than many other characters, we might guess that $d_K(W) = d$. Since DZ occurs four times and

TABLE 1.3
Frequency of occurrence of the 26 ciphertext letters

letter	frequency	letter	frequency
A	0	N	9
B	1	O	0
C	15	P	1
D	13	Q	4
E	7	R	10
F	11	S	3
G	1	T	2
H	4	U	5
I	5	V	5
J	11	W	8
K	1	X	6
L	0	Y	10
M	16	Z	20

ZD occurs twice, we would think that $d_K(D) \in \{r, s, t\}$, but it is not clear which of the three possibilities is the correct one.

If we proceed on the assumption that $d_K(Z) = e$ and $d_K(W) = d$, we might look back at the ciphertext and notice that we have ZRW occurring near the beginning of the ciphertext, and RW occurs again later on. Since R occurs frequently in the ciphertext and nd is a common digram, we might try $d_K(R) = n$ as the most likely possibility.

At this point, we have the following:

```
------end---------e----ned---e-----------
YIFQFMZRWQFYVECFMDZPCVMRZWNMDZVEJBTXCDDUMJ

--------e----e---------n--d---en----e----e
NDIFEFMDZCDMQZKCEYFCJMYRNCWJCSZREXCHZUNMXZ

-e---n------n------ed---e---e--ne-nd-e-e--
NZUCDRJXYYSMRTMEYIFZWDYVZVYFZUMRZCRWNZDZJJ

-ed-----n-----------e----ed-------d---e--n
XZWGCHSMRNMDHNCMFQCHZJMXJZWIEJYUCFWDJNZDIR
```

Our next step might be to try $d_K(N) = h$, since NZ is a common digram and ZN is not. If this is correct, then the segment of plaintext $ne - ndhe$ suggests that $d_K(C) = a$. Incorporating these guesses, we have:

```
------end-----a---e-a--nedh--e------a-----
YIFQFMZRWQFYVECFMDZPCVMRZWNMDZVEJBTXCDDUMJ

h-------ea---e-a---a---nhad-a-en--a-e-h--e
NDIFEFMDZCDMQZKCEYFCJMYRNCWJCSZREXCHZUNMXZ

he-a-n------n------ed---e---e--neandhe-e--
NZUCDRJXYYSMRTMEYIFZWDYVZVYFZUMRZCRWNZDZJJ

-ed-a---nh---ha---a-e----ed-----a-d--he--n
XZWGCHSMRNMDHNCMFQCHZJMXJZWIEJYUCFWDJNZDIR
```

Now, we might consider M, the second most common ciphertext character. The ciphertext segment RNM, which we believe decrypts to $nh-$, suggests that $h-$ begins a word, so M probably represents a vowel. We have already accounted for a and e, so we expect that $d_K(M) = i$ or o. Since ai is a much more likely digram than ao, the ciphertext digram CM suggests that we try $d_K(M) = i$ first. Then we have:

```
-----iend-----a-i-e-a-inedhi-e------a---i-
YIFQFMZRWQFYVECFMDZPCVMRZWNMDZVEJBTXCDDUMJ

h-----i-ea-i-e-a---a-i-nhad-a-en--a-e-hi-e
NDIFEFMDZCDMQZKCEYFCJMYRNCWJCSZREXCHZUNMXZ

he-a-n-----in-i----ed---e---e-ineandhe-e--
NZUCDRJXYYSMRTMEYIFZWDYVZVYFZUMRZCRWNZDZJJ

-ed-a--inhi--hai--a-e-i--ed-----a-d--he--n
XZWGCHSMRNMDHNCMFQCHZJMXJZWIEJYUCFWDJNZDIR
```

Next, we might try to determine which letter is the encryption of o. Since o is a common plaintext character, we guess that the corresponding ciphertext character is one of D, F, J, Y. Y seems to be the most likely possibility; otherwise, we would get long strings of vowels, namely aoi from CFM or CJM. Hence, let's suppose $d_K(Y) = o$.

The three most frequent remaining ciphertext letters are D, F, J, which we conjecture could decrypt to r, s, t in some order. Two occurrences of the trigram NMD suggest that $d_K(D) = s$, giving the trigram *his* in the plaintext (this is consistent with our earlier hypothesis that $d_K(D) \in \{r, s, t\}$). The segment $HNCMF$ could be an encryption of *chair*, which would give $d_K(F) = r$ (and $d_K(H) = c$) and so we would then have $d_K(J) = t$ by process of elimination. Now, we have:

```
o-r-riend-ro--arise-a-inedhise--t---ass-it
YIFQFMZRWQFYVECFMDZPCVMRZWNMDZVEJBTXCDDUMJ

hs-r-riseasi-e-a-orationhadta-en--ace-hi-e
NDIFEFMDZCDMQZKCEYFCJMYRNCWJCSZREXCHZUNMXZ

he-asnt-oo-in-i-o-redso-e-ore-ineandhesett
NZUCDRJXYYSMRTMEYIFZWDYVZVYFZUMRZCRWNZDZJJ

-ed-ac-inhischair-aceti-ted--to-ardsthes-n
XZWGCHSMRNMDHNCMFQCHZJMXJZWIEJYUCFWDJNZDIR
```

It is now very easy to determine the plaintext and the key for Example 1.11. The complete decryption is the following:

> Our friend from Paris examined his empty glass with surprise, as if evaporation had taken place while he wasn't looking. I poured some more wine and he settled back in his chair, face tilted up towards the sun.[2]

⧠

1.2.3   Cryptanalysis of the Vigenère Cipher

In this section we describe some methods for cryptanalyzing the *Vigenère Cipher*. The first step is to determine the keyword length, which we denote by m. There are a couple of techniques that can be employed. The first of these is the so-called Kasiski test and the second uses the index of coincidence.

The *Kasiski test* was described by Friedrich Kasiski in 1863; however, it was apparently discovered earlier, around 1854, by Charles Babbage. It is based on the observation that two identical segments of plaintext will be encrypted to the same ciphertext whenever their occurrence in the plaintext is δ positions apart, where $\delta \equiv 0 \pmod{m}$. Conversely, if we observe two identical segments of ciphertext, each of length at least three, say, then there is a good chance that they correspond to identical segments of plaintext.

The Kasiski test works as follows. We search the ciphertext for pairs of identical segments of length at least three, and record the distance between the starting positions of the two segments. If we obtain several such distances, say $\delta_1, \delta_2, \ldots$, then we would conjecture that m divides all of the δ_i's, and hence m divides the greatest common divisor of the δ_i's.

Further evidence for the value of m can be obtained by the index of coincidence. This concept was defined by William Friedman in 1920, as follows.

---

[2]P. Mayle, *A Year in Provence*, A. Knopf, Inc., 1989.

Definition 1.7:   Suppose $\mathbf{x} = x_1 x_2 \cdots x_n$ is a string of n alphabetic characters. The *index of coincidence* of $\mathbf{x}$, denoted $I_c(\mathbf{x})$, is defined to be the probability that two random elements of $\mathbf{x}$ are identical.

Suppose we denote the frequencies of $A, B, C, \ldots, Z$ in $\mathbf{x}$ by $f_0, f_1, \ldots, f_{25}$ (respectively). We can choose two elements of $\mathbf{x}$ in $\binom{n}{2}$ ways.[3] For each i, $0 \le i \le 25$, there are $\binom{f_i}{2}$ ways of choosing both elements to be i. Hence, we have the formula

$$I_c(\mathbf{x}) = \frac{\sum_{i=0}^{25} \binom{f_i}{2}}{\binom{n}{2}} = \frac{\sum_{i=0}^{25} f_i(f_i - 1)}{n(n-1)}.$$

Suppose $\mathbf{x}$ is a string of English language text. Denote the expected probabilities of occurrence of the letters $A, B, \ldots, Z$ in Table 1.1 by $p_0, \ldots, p_{25}$, respectively. Then, we would expect that

$$I_c(\mathbf{x}) \approx \sum_{i=0}^{25} p_i^2 = 0.065,$$

since the probability that two random elements both are A is p_0^2, the probability that both are B is p_1^2, etc. The same reasoning applies if $\mathbf{x}$ is a ciphertext string obtained using any monoalphabetic cipher. In this case, the individual probabilities will be permuted, but the quantity $\sum p_i^2$ will be unchanged.

Now, suppose we start with a ciphertext string $\mathbf{y} = y_1 y_2 \cdots y_n$ that has been constructed by using a *Vigenère Cipher*. Define m substrings of $\mathbf{y}$, denoted $\mathbf{y}_1, \mathbf{y}_2, \ldots, \mathbf{y}_m$, by writing out the ciphertext, in columns, in a rectangular array of dimensions $m \times (n/m)$. The rows of this matrix are the substrings $\mathbf{y}_i$, $1 \le i \le m$. In other words, we have that

$$\mathbf{y}_1 = y_1 y_{m+1} y_{2m+1} \cdots ,$$

$$\mathbf{y}_2 = y_2 y_{m+2} y_{2m+2} \cdots ,$$

$$\vdots \quad \vdots \quad \vdots$$

$$\mathbf{y}_m = y_m y_{2m} y_{3m} \cdots .$$

If $\mathbf{y}_1, \mathbf{y}_2, \ldots, \mathbf{y}_m$ are constructed in this way, and m is indeed the keyword length, then each value $I_c(\mathbf{y}_i)$ should be roughly equal to 0.065. On the other hand, if m is not the keyword length, then the substrings $\mathbf{y}_i$ will look much more random, since they will have been obtained by shift encryption with different keys. Observe that a completely random string will have

$$I_c \approx 26 \left(\frac{1}{26} \right)^2 = \frac{1}{26} = 0.038.$$

---

[3] The *binomial coefficient* $\binom{n}{k} = n!/(k!(n-k)!)$ denotes the number of ways of choosing a subset of k objects from a set of n objects.

The two values 0.065 and 0.038 are sufficiently far apart that we will often be able to determine the correct keyword length by this method (or confirm a guess that has already been made using the Kasiski test).

Let us illustrate these two techniques with an example.

Example 1.12 Ciphertext obtained from a *Vigenère Cipher*

```
CHREEVOAHMAERATBIAXXWTNXBEEOPHBSBQMQEQERBW
RVXUOAKXAOSXXWEAHBWGJMMQMNKGRFVGXWTRZXWIAK
LXFPSKAUTEMNDCMGTSXMXBTUIADNGMGPSRELXNJELX
VRVPRTULHDNQWTWDTYGBPHXTFALJHASVBFXNGLLCHR
ZBWELEKMSJIKNBHWRJGNMGJSGLXFEYPHAGNRBIEQJT
AMRVLCRREMNDGLXRRIMGNSNRWCHRQHAEYEVTAQEBBI
PEEWEVKAKOEWADREMXMTBHHCHRTKDNVRZCHRCLQOHP
WQAIIWXNRMGWOIIFKEE
```

First, let's try the Kasiski test. The ciphertext string CHR occurs in five places in the ciphertext, beginning at positions 1, 166, 236, 276 and 286. The distances from the first occurrence to the other four occurrences are (respectively) 165, 235, 275 and 285. The greatest common divisor of these four integers is 5, so that is very likely the keyword length.

Let's see if computation of indices of coincidence gives the same conclusion. With $m = 1$, the index of coincidence is 0.045. With $m = 2$, the two indices are 0.046 and 0.041. With $m = 3$, we get 0.043, 0.050, 0.047. With $m = 4$, we have indices 0.042, 0.039, 0.045, 0.040. Then, trying $m = 5$, we obtain the values 0.063, 0.068, 0.069, 0.061 and 0.072. This also provides strong evidence that the keyword length is five. □

Assuming that we have determined the correct value of m, how do we determine the actual key, $K = (k_1, k_2, \ldots, k_m)$? We describe a simple and effective method now. Let $1 \leq i \leq m$, and let $f_0, \ldots, f_{25}$ denote the frequencies of $A, B, \ldots, Z$, respectively, in the string $\mathbf{y}_i$. Also, let $n' = n/m$ denote the length of the string $\mathbf{y}_i$. Then the probability distribution of the 26 letters in $\mathbf{y}_i$ is

$$\frac{f_0}{n'}, \ldots, \frac{f_{25}}{n'}.$$

Now, recall that the substring $\mathbf{y}_i$ is obtained by shift encryption of a subset of the plaintext elements using a shift k_i. Therefore, we would hope that the shifted probability distribution

$$\frac{f_{k_i}}{n'}, \ldots, \frac{f_{25+k_i}}{n'}$$

would be "close to" the ideal probability distribution $p_0, \ldots, p_{25}$ tabulated in Table 1.1, where subscripts in the above formula are evaluated modulo 26.

TABLE 1.4
Values of M_g

i	value of $M_g(\mathbf{y}_i)$								
1	.035	.031	.036	.037	.035	.039	.028	.028	.048
	.061	.039	.032	.040	.038	.038	.045	.036	.030
	.042	.043	.036	.033	.049	.043	.042	.036	
2	**.069**	.044	.032	.035	.044	.034	.036	.033	.029
	.031	.042	.045	.040	.045	.046	.042	.037	.032
	.034	.037	.032	.034	.043	.032	.026	.047	
3	.048	.029	.042	.043	.044	.034	.038	.035	.032
	.049	.035	.031	.035	**.066**	.035	.038	.036	.045
	.027	.035	.034	.034	.036	.035	.046	.040	
4	.045	.032	.033	.038	**.060**	.034	.034	.034	.050
	.033	.033	.043	.040	.033	.029	.036	.040	.044
	.037	.050	.034	.034	.039	.044	.038	.035	
5	.034	.031	.035	.044	.047	.037	.043	.038	.042
	.037	.033	.032	.036	.037	.036	.045	.032	.029
	.044	**.072**	.037	.027	.031	.048	.036	.037	

Suppose that $0 \leq g \leq 25$, and define the quantity

$$M_g = \sum_{i=0}^{25} \frac{p_i f_{i+g}}{n'}. \tag{1.1}$$

If $g = k_i$, then we would expect that

$$M_g \approx \sum_{i=0}^{25} p_i{}^2 = 0.065,$$

as in the consideration of the index of coincidence. If $g \neq k_i$, then M_g will usually be significantly smaller than 0.065 (see the Exercises for a justification of this statement). Hopefully this technique will allow us to determine the correct value of k_i for each value of i, $1 \leq i \leq m$.

Let us illustrate by returning to Example 1.12.

Example 1.12 (Cont.) We have hypothesized that the keyword length is 5. We now compute the values M_g as described above, for $1 \leq i \leq 5$. These values are tabulated in Table 1.4. For each i, we look for a value of M_g that is close to 0.065. These g's determine the shifts $k_1, \ldots, k_5$.

From the data in Table 1.4, we see that the key is likely to be $K = (9, 0, 13, 4, 19)$, and hence the keyword likely is *JANET*. This is correct, and the complete decryption of the ciphertext is the following:

The almond tree was in tentative blossom. The days were longer, often ending with magnificent evenings of corrugated pink skies. The hunting season was over, with hounds and guns put away for six months. The vineyards were busy again as the well-organized farmers treated their vines and the more lackadaisical neighbors hurried to do the pruning they should have done in November.[4]

⬜

1.2.4 Cryptanalysis of the Hill Cipher

The *Hill Cipher* can be difficult to break with a ciphertext-only attack, but it succumbs easily to a known plaintext attack. Let us first assume that the opponent has determined the value of m being used. Suppose he has at least m distinct plaintext-ciphertext pairs, say

$$x_j = (x_{1,j}, x_{2,j}, \ldots, x_{m,j})$$

and

$$y_j = (y_{1,j}, y_{2,j}, \ldots, y_{m,j}),$$

for $1 \leq j \leq m$, such that $y_j = e_K(x_j)$, $1 \leq j \leq m$. If we define two $m \times m$ matrices $X = (x_{i,j})$ and $Y = (y_{i,j})$, then we have the matrix equation $Y = XK$, where the $m \times m$ matrix K is the unknown key. Provided that the matrix X is invertible, Oscar can compute $K = X^{-1}Y$ and thereby break the system. (If X is not invertible, then it will be necessary to try other sets of m plaintext-ciphertext pairs.)

Let's look at a simple example.

Example 1.13   Suppose the plaintext *friday* is encrypted using a *Hill Cipher* with $m = 2$, to give the ciphertext *PQCFKU*.

We have that $e_K(5, 17) = (15, 16)$, $e_K(8, 3) = (2, 5)$ and $e_K(0, 24) = (10, 20)$. From the first two plaintext-ciphertext pairs, we get the matrix equation

$$\begin{pmatrix} 15 & 16 \\ 2 & 5 \end{pmatrix} = \begin{pmatrix} 5 & 17 \\ 8 & 3 \end{pmatrix} K.$$

Using Corollary 1.4, it is easy to compute

$$\begin{pmatrix} 5 & 17 \\ 8 & 3 \end{pmatrix}^{-1} = \begin{pmatrix} 9 & 1 \\ 2 & 15 \end{pmatrix},$$

so

$$K = \begin{pmatrix} 9 & 1 \\ 2 & 15 \end{pmatrix} \begin{pmatrix} 15 & 16 \\ 2 & 5 \end{pmatrix} = \begin{pmatrix} 7 & 19 \\ 8 & 3 \end{pmatrix}.$$

This can be verified by using the third plaintext-ciphertext pair.   ⬜

---

[4]P. Mayle, *A Year in Provence*, A. Knopf, Inc., 1989.

What would the opponent do if he does not know m? Assuming that m is not too big, he could simply try $m = 2, 3, \ldots$, until the key is found. If a guessed value of m is incorrect, then an $m \times m$ matrix found by using the algorithm described above will not agree with further plaintext-ciphertext pairs. In this way, the value of m can be determined if it is not known ahead of time.

1.2.5 Cryptanalysis of the LFSR Stream Cipher

Recall that the ciphertext is the sum modulo 2 of the plaintext and the keystream, i.e., $y_i = (x_i + z_i) \bmod 2$. The keystream is produced from an initial m-tuple, $(z_1, \ldots, z_m) = (k_1, \ldots, k_m)$, using the linear recurrence

$$z_{m+i} = \sum_{j=0}^{m-1} c_j z_{i+j} \bmod 2,$$

$i \geq 1$, where $c_0, \ldots, c_{m-1} \in \mathbb{Z}_2$.

Since all operations in this cryptosystem are linear, we might suspect that the cryptosystem is vulnerable to a known-plaintext attack, as is the case with the *Hill Cipher*. Suppose Oscar has a plaintext string $x_1 x_2 \cdots x_n$ and the corresponding ciphertext string $y_1 y_2 \cdots y_n$. Then he can compute the keystream bits $z_i = (x_i + y_i) \bmod 2$, $1 \leq i \leq n$. Let us also suppose that Oscar knows the value of m. Then Oscar needs only to compute $c_0, \ldots, c_{m-1}$ in order to be able to reconstruct the entire keystream. In other words, he needs to be able to determine the values of m unknowns.

Now, for any $i \geq 1$, we have

$$z_{m+i} = \sum_{j=0}^{m-1} c_j z_{i+j} \bmod 2,$$

which is a linear equation in the m unknowns. If $n \geq 2m$, then there are m linear equations in m unknowns, which can subsequently be solved.

The system of m linear equations can be written in matrix form as follows:

$$(z_{m+1}, z_{m+2}, \ldots, z_{2m}) = (c_0, c_1, \ldots, c_{m-1}) \begin{pmatrix} z_1 & z_2 & \cdots & z_m \\ z_2 & z_3 & \cdots & z_{m+1} \\ \vdots & \vdots & & \vdots \\ z_m & z_{m+1} & \cdots & z_{2m-1} \end{pmatrix}.$$

If the coefficient matrix has an inverse (modulo 2), we obtain the solution

$$(c_0, c_1, \ldots, c_{m-1}) = (z_{m+1}, z_{m+2}, \ldots, z_{2m}) \begin{pmatrix} z_1 & z_2 & \cdots & z_m \\ z_2 & z_3 & \cdots & z_{m+1} \\ \vdots & \vdots & & \vdots \\ z_m & z_{m+1} & \cdots & z_{2m-1} \end{pmatrix}^{-1}.$$

In fact, the matrix will have an inverse if m is the degree of the recurrence used to generate the keystream (see the Exercises for a proof).

Let's illustrate with an example.

Example 1.14   Suppose Oscar obtains the ciphertext string

$$101101011110010$$

corresponding to the plaintext string

$$011001111111000.$$

Then he can compute the keystream bits:

$$110100100001010.$$

Suppose also that Oscar knows that the keystream was generated using a 5-stage LFSR. Then he would solve the following matrix equation, which is obtained from the first 10 keystream bits:

$$(0, 1, 0, 0, 0) = (c_0, c_1, c_2, c_3, c_4) \begin{pmatrix} 1 & 1 & 0 & 1 & 0 \\ 1 & 0 & 1 & 0 & 0 \\ 0 & 1 & 0 & 0 & 1 \\ 1 & 0 & 0 & 1 & 0 \\ 0 & 0 & 1 & 0 & 0 \end{pmatrix}.$$

It can be verified that

$$\begin{pmatrix} 1 & 1 & 0 & 1 & 0 \\ 1 & 0 & 1 & 0 & 0 \\ 0 & 1 & 0 & 0 & 1 \\ 1 & 0 & 0 & 1 & 0 \\ 0 & 0 & 1 & 0 & 0 \end{pmatrix}^{-1} = \begin{pmatrix} 0 & 1 & 0 & 0 & 1 \\ 1 & 0 & 0 & 1 & 0 \\ 0 & 0 & 0 & 0 & 1 \\ 0 & 1 & 0 & 1 & 1 \\ 1 & 0 & 1 & 1 & 0 \end{pmatrix},$$

by checking that the product of the two matrices, computed modulo 2, is the identity matrix. This yields

$$(c_0, c_1, c_2, c_3, c_4) = (0, 1, 0, 0, 0) \begin{pmatrix} 0 & 1 & 0 & 0 & 1 \\ 1 & 0 & 0 & 1 & 0 \\ 0 & 0 & 0 & 0 & 1 \\ 0 & 1 & 0 & 1 & 1 \\ 1 & 0 & 1 & 1 & 0 \end{pmatrix}$$

$$= (1, 0, 0, 1, 0).$$

Thus the recurrence used to generate the keystream is

$$z_{i+5} = (z_i + z_{i+3}) \bmod 2.$$

$\square$

1.3 Notes

Material on classical cryptography is covered in various textbooks and mono-graphs, such as "Decrypted Secrets, Methods and Maxims of Cryptology" by Bauer [9]; "Cipher Systems, The Protection of Communications" by Beker and Piper [13]; "Cryptology" by Beutelspacher [32]; "Cryptography and Data Security" by Denning [109]; "Code Breaking, A History and Exploration" by Kippenhahn [192]; "Cryptography, A Primer" by Konheim [203]; and "Basic Methods of Cryptography" by van der Lubbe [222].

We have used the statistical data on frequency of English letters that is reported in Beker and Piper [13].

A good reference for elementary number theory is "Elementary Number Theory and its Applications" by Rosen [284]. Background in linear algebra can be found in "Elementary Linear Algebra" by Anton [4].

Two very enjoyable and readable books that provide interesting histories of cryptography are "The Codebreakers" by Kahn [185] and "The Code Book" by Singh [307].

Exercises

1.1 Evaluate the following:
   (a) $7503 \bmod 81$
   (b) $(-7503) \bmod 81$
   (c) $81 \bmod 7503$
   (d) $(-81) \bmod 7503$.

1.2 Suppose that $a, m > 0$, and $a \not\equiv 0 \pmod{m}$. Prove that
$$(-a) \bmod m = m - (a \bmod m).$$

1.3 Prove that $a \bmod m = b \bmod m$ if and only if $a \equiv b \pmod{m}$.

1.4 Prove that $a \bmod m = a - \lfloor \frac{a}{m} \rfloor m$, where $\lfloor x \rfloor = \max\{y \in \mathbb{Z} : y \le x\}$.

1.5 Use exhaustive key search to decrypt the following ciphertext, which was encrypted using a *Shift Cipher*:

   BEEAKFYDJXUQYHYJIQRYHTYJIQFBQDUYJIIKFUHCQD.

1.6 If an encryption function e_K is identical to the decryption function d_K, then the key K is said to be an *involutory key*. Find all the involutory keys in the *Shift Cipher* over $\mathbb{Z}_{26}$.

1.7 Determine the number of keys in an *Affine Cipher* over $\mathbb{Z}_m$ for $m = 30, 100$ and 1225.

1.8 List all the invertible elements in $\mathbb{Z}_m$ for $m = 28, 33$ and 35.

1.9 For $1 \le a \le 28$, determine $a^{-1} \bmod 29$ by trial and error.

1.10 Suppose that $K = (5, 21)$ is a key in an *Affine Cipher* over $\mathbb{Z}_{29}$.
   (a) Express the decryption function $d_K(y)$ in the form $d_K(y) = a'y + b'$, where $a', b' \in \mathbb{Z}_{29}$.

(b) Prove that $d_K(e_K(x)) = x$ for all $x \in \mathbb{Z}_{29}$.

1.11   (a) Suppose that $K = (a, b)$ is a key in an *Affine Cipher* over $\mathbb{Z}_n$. Prove that K is an involutory key if and only if $a^{-1} \bmod n = a$ and $b(a + 1) \equiv 0 \pmod{n}$.

(b) Determine all the involutory keys in the *Affine Cipher* over $\mathbb{Z}_{15}$.

(c) Suppose that $n = pq$, where p and q are distinct odd primes. Prove that the number of involutory keys in the *Affine Cipher* over $\mathbb{Z}_n$ is $n + p + q + 1$.

1.12   (a) Let p be prime. Prove that the number of 2×2 matrices that are invertible over $\mathbb{Z}_p$ is $(p^2 - 1)(p^2 - p)$.

> **HINT**   Since p is prime, $\mathbb{Z}_p$ is a field. Use the fact that a matrix over a field is invertible if and only if its rows are linearly independent vectors (i.e., there does not exist a non-zero linear combination of the rows whose sum is the vector of all 0's).

(b) For p prime and $m \geq 2$ an integer, find a formula for the number of $m \times m$ matrices that are invertible over $\mathbb{Z}_p$.

1.13   For $n = 6, 9$ and 26, how many 2×2 matrices are there that are invertible over $\mathbb{Z}_n$?

1.14   (a) Prove that $\det A \equiv \pm 1 \pmod{26}$ if A is a matrix over $\mathbb{Z}_{26}$ such that $A = A^{-1}$.

(b) Use the formula given in Corollary 1.4 to determine the number of involutory keys in the *Hill Cipher* (over $\mathbb{Z}_{26}$) in the case $m = 2$.

1.15   Determine the inverses of the following matrices over $\mathbb{Z}_{26}$:

(a) $\begin{pmatrix} 2 & 5 \\ 9 & 5 \end{pmatrix}$

(b) $\begin{pmatrix} 1 & 11 & 12 \\ 4 & 23 & 2 \\ 17 & 15 & 9 \end{pmatrix}$

1.16   (a) Suppose that π is the following permutation of $\{1, \dots, 8\}$:

x	1	2	3	4	5	6	7	8
$\pi(x)$	4	1	6	2	7	3	8	5

Compute the permutation π^{-1}.

(b) Decrypt the following ciphertext, for a *Permutation Cipher* with $m = 8$, which was encrypted using the key π:

TGEEMNELNNTDROEOAAHDOETCSHAEIRLM.

1.17   (a) Prove that a permutation π in the *Permutation Cipher* is an involutory key if and only if $\pi(i) = j$ implies $\pi(j) = i$, for all $i, j \in \{1, \dots, m\}$.

(b) Determine the number of involutory keys in the *Permutation Cipher* for $m = 2, 3, 4, 5$ and 6.

1.18   Consider the following linear recurrence over $\mathbb{Z}_2$ of degree four:

$$z_{i+4} = (z_i + z_{i+1} + z_{i+2} + z_{i+3}) \bmod 2,$$

$i \geq 0$. For each of the 16 possible initialization vectors $(z_0, z_1, z_2, z_3) \in (\mathbb{Z}_2)^4$, determine the period of the resulting keystream.

1.19   Redo the preceding question, using the recurrence

$$z_{i+4} = (z_i + z_{i+3}) \bmod 2,$$

$i \geq 0$.

1.20 Suppose we construct a keystream in a synchronous stream cipher using the following method. Let $K \in \mathcal{K}$ be the key, let $\mathcal{L}$ be the keystream alphabet, and let Σ be a finite set of *states*. First, an *initial state* $\sigma_0 \in \Sigma$ is determined from K by some method. For all $i \geq 1$, the state σ_i is computed from the previous state σ_{i-1} according to the following rule:

$$\sigma_i = f(\sigma_{i-1}, K),$$

where $f : \Sigma \times \mathcal{K} \to \Sigma$. Also, for all $i \geq 1$, the keystream element z_i is computed using the following rule:

$$z_i = g(\sigma_i, K),$$

where $g : \Sigma \times \mathcal{K} \to \mathcal{L}$. Prove that any keystream produced by this method has period at most $|\Sigma|$.

1.21 Below are given four examples of ciphertext, one obtained from a *Substitution Cipher*, one from a *Vigenère Cipher*, one from an *Affine Cipher*, and one unspecified. In each case, the task is to determine the plaintext.

Give a clearly written description of the steps you followed to decrypt each ciphertext. This should include all statistical analysis and computations you performed.

The first two plaintexts were taken from "The Diary of Samuel Marchbanks," by Robertson Davies, Clarke Irwin, 1947; the fourth was taken from "Lake Wobegon Days," by Garrison Keillor, Viking Penguin, Inc., 1985.

(a) *Substitution Cipher*:

```
EMGLOSUDCGDNCUSWYSFHNSFCYKDPUMLWGYICOXYSIPJCK
QPKUGKMGOLICGINCGACKSNISACYKZSCKXECJCKSHYSXCG
OIDPKZCNKSHICGIWYGKKGKGOLDSILKGOIUSIGLEDSPWZU
GFZCCNDGYYSFUSZCNXEOJNCGYEOWEUPXEZGACGNFGLKNS
ACIGOIYCKXCJUCIUZCFZCCNDGYYSFEUEKUZCSOCFZCCNC
IACZEJNCSHFZEJZEGMXCYHCJUMGKUCY
```

HINT F decrypts to w.

(b) *Vigenère Cipher*:

```
KCCPKBGUFDPHQTYAVINRRTMVGRKDNBVFDETDGILTXRGUD
DKOTFMBPVGEGLTGCKQRACQCWDNAWCRXIZAKFTLEWRPTYC
QKYVXCHKFTPONCQQRHJVAJUWETMCMSPKQDYHJVDAHCTRL
SVSKCGCZQQDZXGSFRLSWCWSJTBHAFSIASPRJAHKJRJUMV
GKMITZHFPDISPZLVLGWTFPLKKEBDPGCEBSHCTJRWXBAFS
PEZQNRWXCVYCGAONWDDKACKAWBBIKFTIOVKCGGHJVLNHI
FFSQESVYCLACNVRWBBIREPBBVFEXOSCDYGZWPFDTKFQIY
CWHJVLNHIQIBTKHJVNPIST
```

(c) *Affine Cipher*:

```
KQEREJEBCPPCJCRKIEACUZBKRVPKRBCIBQCARBJCVFCUP
KRIOFKPACUZQEPBKRXPEIIEABDKPBCPFCDCCAFIEABDKP
BCPFEQPKAZBKRHAIBKAPCCIBURCCDKDCCJCIDFUIXPAFF
ERBICZDFKABICBBENEFCUPJCVKABPCYDCCDPKBCOCPERK
IVKSCPICBRKIJPKABI
```

(d) unspecified cipher:

BNVSNSIHQCEELSSKKYERIFJKXUMBGYKAMQLJTYAVFBKVT
DVBPVVRJYYLAOKYMPQSCGDLFSRLLPROYGESEBUUALRWXM
MASAZLGLEDFJBZAVVPXWICGJXASCBYEHOSNMULKCEAHTQ
OKMFLEBKFXLRRFDTZXCIWBJSICBGAWDVYDHAVFJXZIBKC
GJIWEAHTTOEWTUHKRQVVRGZBXYIREMMASCSPBNLHJMBLR
FFJELHWEYLWISTFVVYFJCMHYUYRUFSFMGESIGRLWALSWM
NUHSIMYYITCCQPZSICEHBCCMZFEGVJYOCDEMMPGHVAAUM
ELCMOEHVLTIPSUYILVGFLMVWDVYDBTHFRAYISYSGKVSUU
HYHGGCKTMBLRX

1.22   (a) Suppose that $p_1, \ldots, p_n$ and $q_1, \ldots, q_n$ are both probability distributions, and $p_1 \geq \cdots \geq p_n$. Let $q_1', \ldots, q_n'$ be any permutation of $q_1, \ldots, q_n$. Prove that the quantity

$$\sum_{i=1}^{n} p_i q_i'$$

is maximized when $q_1' \geq \cdots \geq q_n'$.

(b) Explain why the expression in Equation (1.1) is likely to be maximized when $g = k_i$.

1.23 Suppose we are told that the plaintext

breathtaking

yields the ciphertext

RUPOTENTOIFV

where the *Hill Cipher* is used (but m is not specified). Determine the encryption matrix.

1.24 An *Affine-Hill Cipher* is the following modification of a *Hill Cipher*: Let m be a positive integer, and define $\mathcal{P} = \mathcal{C} = (\mathbb{Z}_{26})^m$. In this cryptosystem, a key K consists of a pair (L, b), where L is an $m \times m$ invertible matrix over $\mathbb{Z}_{26}$, and $b \in (\mathbb{Z}_{26})^m$. For $x = (x_1, \ldots, x_m) \in \mathcal{P}$ and $K = (L, b) \in \mathcal{K}$, we compute $y = e_K(x) = (y_1, \ldots, y_m)$ by means of the formula $y = xL + b$. Hence, if $L = (\ell_{i,j})$ and $b = (b_1, \ldots, b_m)$, then

$$(y_1, \ldots, y_m) = (x_1, \ldots, x_m) \begin{pmatrix} \ell_{1,1} & \ell_{1,2} & \cdots & \ell_{1,m} \\ \ell_{2,1} & \ell_{2,2} & \cdots & \ell_{2,m} \\ \vdots & \vdots & & \vdots \\ \ell_{m,1} & \ell_{m,2} & \cdots & \ell_{m,m} \end{pmatrix} + (b_1, \ldots, b_m).$$

Suppose Oscar has learned that the plaintext

adisplayedequation

is encrypted to give the ciphertext

DSRMSIOPLXLJBZULLM

and Oscar also knows that $m = 3$. Determine the key, showing all computations.

1.25 Here is how we might cryptanalyze the *Hill Cipher* using a ciphertext-only attack. Suppose that we know that $m = 2$. Break the ciphertext into blocks of length two letters (digrams). Each such digram is the encryption of a plaintext digram using the unknown encryption matrix. Pick out the most frequent ciphertext digram and assume it is the encryption of a common digram in the list following Table 1.1 (for example, TH or ST). For each such guess, proceed as in the known-plaintext attack, until the correct encryption matrix is found.

Here is a sample of ciphertext for you to decrypt using this method:

LMQETXYEAGTXCTUIEWNCTXLZEWUAISPZYVAPEWLMGQWYA
XFTCJMSQCADAGTXLMDXNXSNPJQSYVAPRIQSMHNOCVAXFV

1.26 We describe a special case of a *Permutation Cipher*. Let m, n be positive integers. Write out the plaintext, by rows, in $m \times n$ rectangles. Then form the ciphertext by taking the columns of these rectangles. For example, if $m = 3, n = 4$, then we would encrypt the plaintext "*cryptography*" by forming the following rectangle:

cryp
togr
aphy

The ciphertext would be "*CTAROPYGHPRY*."

(a) Describe how Bob would decrypt a ciphertext string (given values for m and n).

(b) Decrypt the following ciphertext, which was obtained by using this method of encryption:

MYAMRARUYIQTENCTORAHROYWDSOYEOUARRGDERNOGW

1.27 The purpose of this exercise is to prove the statement made in Section 1.2.5 that the $m \times m$ coefficient matrix is invertible. This is equivalent to saying that the rows of this matrix are linearly independent vectors over $\mathbb{Z}_2$.

Suppose that the recurrence has the form

$$z_{m+i} = \sum_{j=0}^{m-1} c_j z_{i+j} \bmod 2,$$

where $(z_1, \ldots, z_m)$ comprises the initialization vector. For $i \geq 1$, define

$$v_i = (z_i, \ldots, z_{i+m-1}).$$

Note that the coefficient matrix has the vectors $v_1, \ldots, v_m$ as its rows, so our objective is to prove that these m vectors are linearly independent.

Prove the following assertions:

(a) For any $i \geq 1$,

$$v_{m+i} = \sum_{j=0}^{m-1} c_j v_{i+j} \bmod 2.$$

(b) Choose h to be the minimum integer such that there exists a non-trivial linear combination of the vectors $v_1, \ldots, v_h$ which sums to the vector $(0, \ldots, 0)$ modulo 2. Then

$$v_h = \sum_{j=0}^{h-2} a_j v_{j+1} \bmod 2,$$

and not all the a_j's are zero. Observe that $h \leq m + 1$, since any $m + 1$ vectors in an m-dimensional vector space are dependent.

(c) Prove that the keystream must satisfy the recurrence

$$z_{h-1+i} = \sum_{j=0}^{h-2} a_j z_{j+i} \bmod 2$$

for any $i \geq 1$.

(d) If $h \leq m$, then the keystream satisfies a linear recurrence of degree less than m. Show that this is impossible, by considering the initialization vector $(0, \ldots, 0, 1)$. Hence, conclude that $h = m + 1$, and therefore the matrix must be invertible.

1.28 Decrypt the following ciphertext, obtained from the *Autokey Cipher*, by using exhaustive key search:

MALVVMAFBHBUQPTSOXALTGVWWRG

1.29 We describe a stream cipher that is a modification of the *Vigenère Cipher*. Given a keyword $(K_1, \ldots, K_m)$ of length m, construct a keystream by the rule $z_i = K_i$ $(1 \leq i \leq m)$, $z_{i+m} = (z_i + 1) \bmod 26$ $(i \geq 1)$. In other words, each time we use the keyword, we replace each letter by its successor modulo 26. For example, if $SUMMER$ is the keyword, we use $SUMMER$ to encrypt the first six letters, we use $TVNNFS$ for the next six letters, and so on.

    (a) Describe how you can use the concept of index of coincidence to first determine the length of the keyword, and then actually find the keyword.

    (b) Test your method by cryptanalyzing the following ciphertext:

IYMYSILONRFNCQXQJEDSHBUIBCJUZBOLFQYSCHATPEQGQ
JEJNGNXZWHHGWFSUKULJQACZKKJOAAHGKEMTAFGMKVRDO
PXNEHEKZNKFSKIFRQVHHOVXINPHMRTJPYWQGJWPUUVKFP
OAWPMRKKQZWLQDYAZDRMLPBJKJOBWIWPSEPVVQMBCRYVC
RUZAAOUMBCHDAGDIEMSZFZHALIGKEMJJFPCIWKRMLMPIN
AYOFIREAOLDTHITDVRMSE

The plaintext was taken from "The Codebreakers," by D. Kahn, Scribner, 1996.

1.30 We describe another stream cipher, which incorporates one of the ideas from the "Enigma" system used by Germany in World War II. Suppose that π is a fixed permutation of $\mathbb{Z}_{26}$. The key is an element $K \in \mathbb{Z}_{26}$. For all integers $i \geq 1$, the keystream element $z_i \in \mathbb{Z}_{26}$ is defined according to the rule $z_i = (K + i - 1) \bmod 26$. Encryption and decryption are performed using the permutations π and π^{-1}, respectively, as follows:

$$e_z(x) = \pi(x) + z \bmod 26$$

and

$$d_z(y) = \pi^{-1}(y - z \bmod 26),$$

where $z \in \mathbb{Z}_{26}$.

    Suppose that π is the following permutation of $\mathbb{Z}_{26}$:

x	0	1	2	3	4	5	6	7	8	9	10	11	12
$\pi(x)$	23	13	24	0	7	15	14	6	25	16	22	1	19

x	13	14	15	16	17	18	19	20	21	22	23	24	25
$\pi(x)$	18	5	11	17	2	21	12	20	4	10	9	3	8

The following ciphertext has been encrypted using this stream cipher; use exhaustive key search to decrypt it:

WRTCNRLDSAFARWKXFTXCZRNHNYPDTZUUKMPLUSOXNEUDO
KLXRMCBKGRCCURR

2

Shannon's Theory

2.1 Introduction

In 1949, Claude Shannon published a paper entitled "Communication Theory of Secrecy Systems" in the *Bell Systems Technical Journal*. This paper had a great influence on the scientific study of cryptography. In this chapter, we discuss several of Shannon's ideas. First, however, we consider some of the various approaches to evaluating the security of a cryptosystem. We define some of the most useful criteria now.

computational security

This measure concerns the computational effort required to break a cryptosystem. We might define a cryptosystem to be *computationally secure* if the best algorithm for breaking it requires at least N operations, where N is some specified, very large number. The problem is that no known practical cryptosystem can be proved to be secure under this definition. In practice, people often study the computational security of a cryptosystem with respect to certain specific types of attacks (e.g., an exhaustive key search). Of course, security against one specific type of attack does not guarantee security against some other type of attack.

provable security

Another approach is to provide evidence of security by means of a reduction. In other words, we show that if the cryptosystem can be "broken" in some specific way, then it would be possible to efficiently solve some well-studied problem that is thought to be difficult. For example, it may be possible to prove a statement of the type "a given cryptosystem is secure if a given integer n cannot be factored." Cryptosystems of this type are sometimes termed *provably secure*, but it must be understood that this approach only provides a proof of security relative to some other problem, not an absolute proof of security. This is a similar situation to proving that a problem is NP-complete: it proves that the given problem is at least as difficult as any other NP-complete problem, but it does not provide an absolute proof of the computational difficulty of the problem.

unconditional security

This measure concerns the security of cryptosystems when there is no bound placed on the amount of computation that Oscar is allowed to do. A cryptosystem is defined to be *unconditionally secure* if it cannot be broken, even with infinite computational resources.

When we discuss the security of a cryptosystem, we should also specify the type of attack that is being considered. For example, in Chapter 1, we saw that neither the *Shift Cipher*, the *Substitution Cipher* nor the *Vigenère Cipher* is computationally secure against a ciphertext-only attack (given a sufficient amount of ciphertext).

What we will do in Section 2.3 is to develop a theory of cryptosystems that are unconditionally secure against a ciphertext-only attack. This theory allows us to prove mathematically that certain cryptosystems are secure if the amount of ciphertext is sufficiently small. For example, it turns out that the *Shift Cipher* and the *Substitution Cipher* are both unconditionally secure if a single element of plaintext is encrypted with a given key. Similarly, the *Vigenère Cipher* with keyword length m is unconditionally secure if the key is used to encrypt only one element of plaintext (which consists of m alphabetic characters).

2.2   Elementary Probability Theory

The unconditional security of a cryptosystem obviously cannot be studied from the point of view of computational complexity because we allow computation time to be infinite. The appropriate framework in which to study unconditional security is probability theory. We need only elementary facts concerning probability; the main definitions are reviewed now. First, we define the idea of a random variable.

Definition 2.1:   A *discrete random variable*, say $\mathbf{X}$, consists of a finite set X and a *probability distribution* defined on X. The probability that the random variable $\mathbf{X}$ takes on the value x is denoted $\mathbf{Pr}[\mathbf{X} = x]$; sometimes we will abbreviate this to $\mathbf{Pr}[x]$ if the random variable $\mathbf{X}$ is fixed. It must be the case that $0 \leq \mathbf{Pr}[x]$ for all $x \in X$, and

$$\sum_{x \in X} \mathbf{Pr}[x] = 1.$$

As an example, we could consider a coin toss to be a random variable defined on the set $\{heads, tails\}$. The associated probability distribution would be

$\mathbf{Pr}[heads] = \mathbf{Pr}[tails] = 1/2$.

Suppose we have random variable $\mathbf{X}$ defined on X, and $E \subseteq X$. The probability that $\mathbf{X}$ takes on a value in the subset E is computed to be

$$\mathbf{Pr}[x \in E] = \sum_{x \in E} \mathbf{Pr}[x]. \tag{2.1}$$

The subset E is often called an *event*.

Example 2.1 Suppose we consider a random throw of a pair of dice. This can be modeled by a random variable $\mathbf{Z}$ defined on the set

$$Z = \{1, 2, 3, 4, 5, 6\} \times \{1, 2, 3, 4, 5, 6\},$$

where $\mathbf{Pr}[(i, j)] = 1/36$ for all $(i, j) \in Z$. Let's consider the sum of the two dice. Each possible sum defines an event, and the probabilities of these events can be computed using equation (2.1). For example, suppose that we want to compute the probability that the sum is 4. This corresponds to the event

$$S_4 = \{(1, 3), (2, 2), (3, 1)\},$$

and therefore $\mathbf{Pr}[S_4] = 3/36 = 1/12$.

The probabilities of all the sums can be computed in a similar fashion. If we denote by S_j the event that the sum is j, then we obtain the following: $\mathbf{Pr}[S_2] = \mathbf{Pr}[S_{12}] = 1/36$, $\mathbf{Pr}[S_3] = \mathbf{Pr}[S_{11}] = 1/18$, $\mathbf{Pr}[S_4] = \mathbf{Pr}[S_{10}] = 1/12$, $\mathbf{Pr}[S_5] = \mathbf{Pr}[S_9] = 1/9$, $\mathbf{Pr}[S_6] = \mathbf{Pr}[S_8] = 5/36$, and $\mathbf{Pr}[S_7] = 1/6$.

Since the events $S_2, \ldots, S_{12}$ partition the set S, it follows that we can consider the value of the sum of a pair of dice to be a random variable in its own right, which has the probability distribution computed above. $\square$

We next consider the concepts of joint and conditional probabilities.

Definition 2.2: Suppose $\mathbf{X}$ and $\mathbf{Y}$ are random variables defined on finite sets X and Y, respectively. The *joint probability* $\mathbf{Pr}[x, y]$ is the probability that $\mathbf{X}$ takes on the value x and $\mathbf{Y}$ takes on the value y. The *conditional probability* $\mathbf{Pr}[x|y]$ denotes the probability that $\mathbf{X}$ takes on the value x given that $\mathbf{Y}$ takes on the value y. The random variables $\mathbf{X}$ and $\mathbf{Y}$ are said to be *independent random variables* if $\mathbf{Pr}[x, y] = \mathbf{Pr}[x]\mathbf{Pr}[y]$ for all $x \in X$ and $y \in Y$.

Joint probability can be related to conditional probability by the formula

$$\mathbf{Pr}[x, y] = \mathbf{Pr}[x|y]\mathbf{Pr}[y].$$

Interchanging x and y, we have that

$$\mathbf{Pr}[x, y] = \mathbf{Pr}[y|x]\mathbf{Pr}[x].$$

From these two expressions, we immediately obtain the following result, which is known as Bayes' theorem.

THEOREM 2.1   (Bayes' theorem)   *If* $\mathbf{Pr}[y] > 0$, *then*

$$\mathbf{Pr}[x|y] = \frac{\mathbf{Pr}[x]\mathbf{Pr}[y|x]}{\mathbf{Pr}[y]}.$$

COROLLARY 2.2   **X** *and* **Y** *are independent random variables if and only if* $\mathbf{Pr}[x|y] = \mathbf{Pr}[x]$ *for all* $x \in X$ *and* $y \in Y$.

Example 2.2   Suppose we consider a random throw of a pair of dice. Let **X** be the random variable defined on the set $X = \{2, \ldots, 12\}$, obtained by considering the sum of two dice, as in Example 2.1. Further, suppose that **Y** is a random variable which takes on the value D if the two dice are the same (i.e., if we throw "doubles"), and the value N, otherwise. Then we have that $\mathbf{Pr}[D] = 1/6$, $\mathbf{Pr}[N] = 5/6$.

It is straightforward to compute joint and conditional probabilities for these random variables. For example, the reader can check that $\mathbf{Pr}[D|4] = 1/3$ and $\mathbf{Pr}[4|D] = 1/6$, so

$$\mathbf{Pr}[D|4]\mathbf{Pr}[4] = \mathbf{Pr}[D]\mathbf{Pr}[4|D],$$

as stated by Bayes' theorem.   ◻

2.3   Perfect Secrecy

Throughout this section, we assume that a cryptosystem $(\mathcal{P}, \mathcal{C}, \mathcal{K}, \mathcal{E}, \mathcal{D})$ is specified, and a particular key $K \in \mathcal{K}$ is used for only one encryption. Let us suppose that there is a probability distribution on the plaintext space, $\mathcal{P}$. Thus the plaintext element defines a random variable, denoted **x**. We denote the *a priori* probability that plaintext x occurs by $\mathbf{Pr}[\mathbf{x} = x]$. We also assume that the key K is chosen (by Alice and Bob) using some fixed probability distribution (often a key is chosen at random, so all keys will be equiprobable, but this need not be the case). So the key also defines a random variable, which we denote by **K**. Denote the probability that key K is chosen by $\mathbf{Pr}[\mathbf{K} = K]$. Recall that the key is chosen before Alice knows what the plaintext will be. Hence, we make the reasonable assumption that the key and the plaintext are independent random variables.

The two probability distributions on $\mathcal{P}$ and $\mathcal{K}$ induce a probability distribution on $\mathcal{C}$. Thus, we can also consider the ciphertext element to be a random variable, say $\mathbf{y}$. It is not hard to compute the probability $\mathbf{Pr}[\mathbf{y} = y]$ that y is the ciphertext that is transmitted. For a key $K \in \mathcal{K}$, define

$$C(K) = \{e_K(x) : x \in \mathcal{P}\}.$$

That is, $C(K)$ represents the set of possible ciphertexts if K is the key. Then, for every $y \in \mathcal{C}$, we have that

$$\mathbf{Pr}[\mathbf{y} = y] = \sum_{\{K : y \in C(K)\}} \mathbf{Pr}[\mathbf{K} = K]\mathbf{Pr}[\mathbf{x} = d_K(y)].$$

We also observe that, for any $y \in \mathcal{C}$ and $x \in \mathcal{P}$, we can compute the conditional probability $\mathbf{Pr}[\mathbf{y} = y | \mathbf{x} = x]$ (i.e., the probability that y is the ciphertext, given that x is the plaintext) to be

$$\mathbf{Pr}[\mathbf{y} = y | \mathbf{x} = x] = \sum_{\{K : x = d_K(y)\}} \mathbf{Pr}[\mathbf{K} = K].$$

It is now possible to compute the conditional probability $\mathbf{Pr}[\mathbf{x} = x | \mathbf{y} = y]$ (i.e., the probability that x is the plaintext, given that y is the ciphertext) using Bayes' theorem. The following formula is obtained:

$$\mathbf{Pr}[\mathbf{x} = x | \mathbf{y} = y] = \frac{\mathbf{Pr}[\mathbf{x} = x] \times \sum_{\{K : x = d_K(y)\}} \mathbf{Pr}[\mathbf{K} = K]}{\sum_{\{K : y \in C(K)\}} \mathbf{Pr}[\mathbf{K} = K]\mathbf{Pr}[\mathbf{x} = d_K(y)]}.$$

Observe that all these calculations can be performed by anyone who knows the probability distributions.

We present a toy example to illustrate the computation of these probability distributions.

Example 2.3 Let $\mathcal{P} = \{a, b\}$ with $\mathbf{Pr}[a] = 1/4, \mathbf{Pr}[b] = 3/4$. Let $\mathcal{K} = \{K_1, K_2, K_3\}$ with $\mathbf{Pr}[K_1] = 1/2, \mathbf{Pr}[K_2] = \mathbf{Pr}[K_3] = 1/4$. Let $\mathcal{C} = \{1, 2, 3, 4\}$, and suppose the encryption functions are defined to be $e_{K_1}(a) = 1, e_{K_1}(b) = 2$; $e_{K_2}(a) = 2, e_{K_2}(b) = 3$; and $e_{K_3}(a) = 3, e_{K_3}(b) = 4$. This cryptosystem can be represented by the following *encryption matrix*:

	a	b
K_1	1	2
K_2	2	3
K_3	3	4

We now compute the probability distribution on $\mathcal{C}$. We obtain the following:

$$\mathbf{Pr}[1] = \frac{1}{8}$$

$$\mathbf{Pr}[2] = \frac{3}{8} + \frac{1}{16} = \frac{7}{16}$$

$$\mathbf{Pr}[3] = \frac{3}{16} + \frac{1}{16} = \frac{1}{4}$$

$$\mathbf{Pr}[4] = \frac{3}{16}.$$

Now we can compute the conditional probability distributions on the plaintext, given that a certain ciphertext has been observed. We have:

$$\mathbf{Pr}[a|1] = 1 \qquad\qquad \mathbf{Pr}[b|1] = 0$$

$$\mathbf{Pr}[a|2] = \frac{1}{7} \qquad\qquad \mathbf{Pr}[b|2] = \frac{6}{7}$$

$$\mathbf{Pr}[a|3] = \frac{1}{4} \qquad\qquad \mathbf{Pr}[b|3] = \frac{3}{4}$$

$$\mathbf{Pr}[a|4] = 0 \qquad\qquad \mathbf{Pr}[b|4] = 1.$$

□

We are now ready to define the concept of perfect secrecy. Informally, perfect secrecy means that Oscar can obtain no information about the plaintext by observing the ciphertext. This idea is made precise by formulating it in terms of the probability distributions we have defined, as follows.

Definition 2.3:   A cryptosystem has *perfect secrecy* if $\mathbf{Pr}[x|y] = \mathbf{Pr}[x]$ for all $x \in \mathcal{P}$, $y \in \mathcal{C}$. That is, the *a posteriori* probability that the plaintext is x, given that the ciphertext y is observed, is identical to the *a priori* probability that the plaintext is x.

In Example 2.3, the perfect secrecy property is satisfied for the ciphertext $y = 3$, but not for the other three ciphertexts.

We now prove that the *Shift Cipher* provides perfect secrecy. This seems quite obvious intuitively. For, if we are given any ciphertext element $y \in \mathbb{Z}_{26}$, then any plaintext element $x \in \mathbb{Z}_{26}$ is a possible decryption of y, depending on the value of the key. The following theorem gives the formal statement and proof using probability distributions.

THEOREM 2.3 *Suppose the 26 keys in the Shift Cipher are used with equal probability 1/26. Then for any plaintext probability distribution, the Shift Cipher has perfect secrecy.*

PROOF Recall that $\mathcal{P} = \mathcal{C} = \mathcal{K} = \mathbb{Z}_{26}$, and for $0 \leq K \leq 25$, the encryption rule e_K is defined as $e_K(x) = (x + K) \bmod 26$ ($x \in \mathbb{Z}_{26}$). First, we compute the probability distribution on $\mathcal{C}$. Let $y \in \mathbb{Z}_{26}$; then

$$\mathbf{Pr}[\mathbf{y} = y] = \sum_{K \in \mathbb{Z}_{26}} \mathbf{Pr}[\mathbf{K} = K]\mathbf{Pr}[\mathbf{x} = d_K(y)]$$

$$= \sum_{K \in \mathbb{Z}_{26}} \frac{1}{26}\mathbf{Pr}[\mathbf{x} = y - K]$$

$$= \frac{1}{26} \sum_{K \in \mathbb{Z}_{26}} \mathbf{Pr}[\mathbf{x} = y - K].$$

Now, for fixed y, the values $(y - K) \bmod 26$ comprise a permutation of $\mathbb{Z}_{26}$. Hence we have that

$$\sum_{K \in \mathbb{Z}_{26}} \mathbf{Pr}[\mathbf{x} = y - K] = \sum_{x \in \mathbb{Z}_{26}} \mathbf{Pr}[\mathbf{x} = x]$$

$$= 1.$$

Consequently,

$$\mathbf{Pr}[y] = \frac{1}{26}$$

for any $y \in \mathbb{Z}_{26}$.

Next, we have that

$$\mathbf{Pr}[y|x] = \mathbf{Pr}[\mathbf{K} = (y - x) \bmod 26]$$

$$= \frac{1}{26}$$

for every x, y. (This is true because, for every x, y, the unique key K such that $e_K(x) = y$ is $K = (y - x) \bmod 26$.) Now, using Bayes' theorem, it is trivial to compute

$$\mathbf{Pr}[x|y] = \frac{\mathbf{Pr}[x]\mathbf{Pr}[y|x]}{\mathbf{Pr}[y]}$$

$$= \frac{\mathbf{Pr}[x]\frac{1}{26}}{\frac{1}{26}}$$

$$= \mathbf{Pr}[x],$$

so we have perfect secrecy. ∎

Hence, the *Shift Cipher* is "unbreakable" provided that a new random key is used to encrypt every plaintext character.

Let us next investigate perfect secrecy in general. If $\mathbf{Pr}[x_0] = 0$ for some $x_0 \in \mathcal{P}$, then it is trivially the case that $\mathbf{Pr}[x_0|y] = \mathbf{Pr}[x_0]$ for all $y \in \mathcal{C}$. So we need only consider those plaintext elements $x \in \mathcal{P}$ such that $\mathbf{Pr}[x] > 0$. For such plaintexts, we observe that, using Bayes' theorem, the condition that $\mathbf{Pr}[x|y] = \mathbf{Pr}[x]$ for all $y \in \mathcal{C}$ is equivalent to $\mathbf{Pr}[y|x] = \mathbf{Pr}[y]$ for all $y \in \mathcal{C}$. Now, let us make the reasonable assumption that $\mathbf{Pr}[y] > 0$ for all $y \in \mathcal{C}$ (if $\mathbf{Pr}[y] = 0$, then ciphertext y is never used and can be omitted from $\mathcal{C}$).

Fix any $x \in \mathcal{P}$. For each $y \in \mathcal{C}$, we have $\mathbf{Pr}[y|x] = \mathbf{Pr}[y] > 0$. Hence, for each $y \in \mathcal{C}$, there must be at least one key K such that $e_K(x) = y$. It follows that $|\mathcal{K}| \geq |\mathcal{C}|$. In any cryptosystem, we must have $|\mathcal{C}| \geq |\mathcal{P}|$ since each encoding rule is injective. In the case of equality, where $|\mathcal{K}| = |\mathcal{C}| = |\mathcal{P}|$, we can give a nice characterization of when perfect secrecy can be obtained. This characterization is originally due to Shannon.

THEOREM 2.4 *Suppose* $(\mathcal{P}, \mathcal{C}, \mathcal{K}, \mathcal{E}, \mathcal{D})$ *is a cryptosystem where* $|\mathcal{K}| = |\mathcal{C}| = |\mathcal{P}|$. *Then the cryptosystem provides perfect secrecy if and only if every key is used with equal probability* $1/|\mathcal{K}|$, *and for every* $x \in \mathcal{P}$ *and every* $y \in \mathcal{C}$, *there is a unique key* K *such that* $e_K(x) = y$.

PROOF Suppose the given cryptosystem provides perfect secrecy. As observed above, for each $x \in \mathcal{P}$ and $y \in \mathcal{C}$, there must be at least one key K such that $e_K(x) = y$. So we have the inequalities:

$$|\mathcal{C}| = |\{e_K(x) : K \in \mathcal{K}\}|$$
$$\leq |\mathcal{K}|.$$

But we are assuming that $|\mathcal{C}| = |\mathcal{K}|$. Hence, it must be the case that

$$|\{e_K(x) : K \in \mathcal{K}\}| = |\mathcal{K}|.$$

That is, there do not exist two distinct keys K_1 and K_2 such that $e_{K_1}(x) = e_{K_2}(x) = y$. Hence, we have shown that for any $x \in \mathcal{P}$ and $y \in \mathcal{C}$, there is exactly one key K such that $e_K(x) = y$.

Denote $n = |\mathcal{K}|$. Let $\mathcal{P} = \{x_i : 1 \leq i \leq n\}$ and fix a ciphertext element $y \in \mathcal{C}$. We can name the keys $K_1, K_2, \ldots, K_n$, in such a way that $e_{K_i}(x_i) = y$, $1 \leq i \leq n$. Using Bayes' theorem, we have

$$\mathbf{Pr}[x_i|y] = \frac{\mathbf{Pr}[y|x_i]\mathbf{Pr}[x_i]}{\mathbf{Pr}[y]}$$
$$= \frac{\mathbf{Pr}[\mathbf{K} = K_i]\mathbf{Pr}[x_i]}{\mathbf{Pr}[y]}.$$

Consider the perfect secrecy condition $\mathbf{Pr}[x_i|y] = \mathbf{Pr}[x_i]$. From this, it follows that $\mathbf{Pr}[K_i] = \mathbf{Pr}[y]$, for $1 \leq i \leq n$. This says that all the keys are used with

---

Cryptosystem 2.1: *One-time Pad*

Let $n \geq 1$ be an integer, and take $\mathcal{P} = \mathcal{C} = \mathcal{K} = (\mathbb{Z}_2)^n$. For $K \in (\mathbb{Z}_2)^n$, define $e_K(x)$ to be the vector sum modulo 2 of K and x (or, equivalently, the exclusive-or of the two associated bitstrings). So, if $x = (x_1, \ldots, x_n)$ and $K = (K_1, \ldots, K_n)$, then

$$e_K(x) = (x_1 + K_1, \ldots, x_n + K_n) \bmod 2.$$

Decryption is identical to encryption. If $y = (y_1, \ldots, y_n)$, then

$$d_K(y) = (y_1 + K_1, \ldots, y_n + K_n) \bmod 2.$$

---

equal probability (namely, $\mathbf{Pr}[y]$). But since the number of keys is $|\mathcal{K}|$, we must have that $\mathbf{Pr}[K] = 1/|\mathcal{K}|$ for every $K \in \mathcal{K}$.

Conversely, suppose the two hypothesized conditions are satisfied. Then the cryptosystem is easily seen to provide perfect secrecy for any plaintext probability distribution, in a manner similar to the proof of Theorem 2.3. We leave the details for the reader. ∎

One well-known realization of perfect secrecy is the *One-time Pad*, which was first described by Gilbert Vernam in 1917 for use in automatic encryption and decryption of telegraph messages. It is interesting that the *One-time Pad* was thought for many years to be an "unbreakable" cryptosystem, but there was no mathematical proof of this until Shannon developed the concept of perfect secrecy over 30 years later. The *One-time Pad* is presented as Cryptosystem 2.1.

Using Theorem 2.4, it is easily seen that the *One-time Pad* provides perfect secrecy. The system is also attractive because of the ease of encryption and decryption. Vernam patented his idea in the hope that it would have widespread commercial use. Unfortunately, there are major disadvantages to unconditionally secure cryptosystems such as the *One-time Pad*. The fact that $|\mathcal{K}| \geq |\mathcal{P}|$ means that the amount of key that must be communicated securely is at least as big as the amount of plaintext. For example, in the case of the *One-time Pad*, we require n bits of key to encrypt n bits of plaintext. This would not be a major problem if the same key could be used to encrypt different messages; however, the security of unconditionally secure cryptosystems depends on the fact that each key is used for only one encryption. (This is the reason for the adjective "one-time" in the *One-time Pad*.)

For example, the *One-time Pad* is vulnerable to a known-plaintext attack, since K can be computed as the exclusive-or of the bitstrings x and $e_K(x)$. Hence, a new key needs to be generated and communicated over a secure channel for every message that is going to be sent. This creates severe key management problems, which has limited the use of the *One-time Pad* in commercial applications. However, the *One-time Pad* has been employed in military and diplomatic contexts, where unconditional security may be of great importance.

The historical development of cryptography has been to try to design cryptosystems where one key can be used to encrypt a relatively long string of plaintext (i.e., one key can be used to encrypt many messages) and still maintain some measure of computational security. Cryptosystems of this type include the *Data Encryption Standard* and the *Advanced Encryption Standard*, which we will discuss in the next chapter.

2.4   Entropy

In the previous section, we discussed the concept of perfect secrecy. We restricted our attention to the special situation where a key is used for only one encryption. We now want to look at what happens as more and more plaintexts are encrypted using the same key, and how likely a cryptanalyst will be able to carry out a successful ciphertext-only attack, given sufficient time.

The basic tool in studying this question is the idea of entropy, a concept from information theory introduced by Shannon in 1948. Entropy can be thought of as a mathematical measure of information or uncertainty, and is computed as a function of a probability distribution.

Suppose we have a discrete random variable $\mathbf{X}$ which takes values from a finite set X according to a specified probability distribution. What is the information gained by the outcome of an experiment which takes place according to this probability distribution? Equivalently, if the experiment has not (yet) taken place, what is the uncertainty about the outcome? This quantity is called the entropy of $\mathbf{X}$ and is denoted by $H(\mathbf{X})$.

These ideas may seem rather abstract, so let's look at a more concrete example. Suppose our random variable $\mathbf{X}$ represents the toss of a coin. As mentioned earlier, the associated probability distribution is $\mathbf{Pr}[heads] = \mathbf{Pr}[tails] = 1/2$. It would seem reasonable to say that the information, or entropy, of a coin toss is one bit, since we could encode *heads* by 1 and *tails* by 0, for example. In a similar fashion, the entropy of n independent coin tosses is n, since the n coin tosses can be encoded by a bitstring of length n.

As a slightly more complicated example, suppose we have a random variable $\mathbf{X}$ that takes on three possible values x_1, x_2, x_3 with probabilities $1/2, 1/4, 1/4$ respectively. Suppose we encode the three possible outcomes as follows: x_1 is encoded as 0, x_2 is encoded as 10, and x_3 is encoded as 11. (This is an example of a Huffman encoding; see Section 2.4.1). Then the (weighted) average number of bits in this encoding of $\mathbf{X}$ is

$$\frac{1}{2} \times 1 + \frac{1}{4} \times 2 + \frac{1}{4} \times 2 = \frac{3}{2}.$$

The above examples suggest that an event which occurs with probability 2^{-n} could perhaps be encoded as a bitstring of length n. More generally, we could plausibly imagine that an outcome occurring with probability p might be encoded by a bitstring of length approximately $- \log_2 p$. Given an arbitrary probability distribution, taking on the values $p_1, p_2, \ldots, p_n$ for a random variable $\mathbf{X}$, we take the weighted average of the quantities $- \log_2 p_i$ to be our measure of information. This motivates the following formal definition.

Definition 2.4: Suppose $\mathbf{X}$ is a discrete random variable which takes on values from a finite set X. Then, the *entropy* of the random variable $\mathbf{X}$ is defined to be the quantity

$$H(\mathbf{X}) = - \sum_{x \in X} \mathbf{Pr}[x] \log_2 \mathbf{Pr}[x].$$

REMARK Observe that $\log_2 y$ is undefined if $y = 0$. Hence, entropy is sometimes defined to be the relevant sum over all the non-zero probabilities. However, since $\lim_{y \to 0} y \log_2 y = 0$, there is no real difficulty with allowing $\mathbf{Pr}[x] = 0$ for some x's.

Also, we note that the choice of two as the base of the logarithms is arbitrary: another base would only change the value of the entropy by a constant factor. ∎

Note that if $|X| = n$ and $\mathbf{Pr}[x] = 1/n$ for all $x \in X$, then $H(\mathbf{X}) = \log_2 n$. Also, it is easy to see that $H(\mathbf{X}) \geq 0$ for any random variable $\mathbf{X}$, and $H(\mathbf{X}) = 0$ if and only if $\mathbf{Pr}[x_0] = 1$ for some $x_0 \in X$ and $\mathbf{Pr}[x] = 0$ for all $x \neq x_0$.

Let us look at the entropy of the various components of a cryptosystem. We can think of the key as being a random variable $\mathbf{K}$ that takes on values in $\mathcal{K}$, and hence we can compute the entropy $H(\mathbf{K})$. Similarly, we can compute entropies $H(\mathbf{P})$ and $H(\mathbf{C})$ of random variables associated with the plaintext and ciphertext, respectively.

To illustrate, we compute the entropies of the cryptosystem of Example 2.3.

Example 2.3 **(Cont.)**   We compute as follows:

$$
\begin{aligned}
H(\mathbf{P}) &= -\frac{1}{4}\log_2 \frac{1}{4} - \frac{3}{4}\log_2 \frac{3}{4} \\
&= -\frac{1}{4}(-2) - \frac{3}{4}(\log_2 3 - 2) \\
&= 2 - \frac{3}{4}(\log_2 3) \\
&\approx 0.81.
\end{aligned}
$$

Similar calculations yield $H(\mathbf{K}) = 1.5$ and $H(\mathbf{C}) \approx 1.85$.   ⊓

2.4.1   Huffman Encodings

In this section, we discuss briefly the connection between entropy and Huffman encodings. As the results in this section are not relevant to the cryptographic applications of entropy, it may be skipped without loss of continuity. However, this discussion may serve to further motivate the concept of entropy.

We introduced entropy in the context of encodings of random events which occur according to a specified probability distribution. We first make these ideas more precise. As before, $\mathbf{X}$ is a random variable which takes on values from a finite set X and p is the associated probability distribution.

An *encoding* of $\mathbf{X}$ is any mapping

$$ f : X \to \{0,1\}^*, $$

where $\{0,1\}^*$ denotes the set of all finite strings of 0's and 1's. Given a finite list (or string) of events $x_1 \cdots x_n$, where each $x_i \in X$, we can extend the encoding f in an obvious way by defining

$$ f(x_1 \cdots x_n) = f(x_1) \, \| \cdots \| \, f(x_n), $$

where $\|$ denotes concatenation. In this way, we can think of f as a mapping

$$ f : X^* \to \{0,1\}^*. $$

Now, suppose a string $x_1 \ldots x_n$ is produced by a *memoryless source*, such that each x_i in the string occurs according to a specified probability distribution on X. This means that the probability of any string $x_1 \cdots x_n$ is computed to be

$$ \mathbf{Pr}[x_1 \cdots x_n] = \mathbf{Pr}[x_1] \times \cdots \times \mathbf{Pr}[x_n]. $$

REMARK   The string $x_1 \cdots x_n$ need not consist of distinct values, since the source is memoryless. As a simple example, consider a sequence of n tosses of a fair coin. If we encode "*heads*" as "1" and "*tails*" as "0," then every binary string of length n corresponds to a sequence of n coin tosses.   ▮

Now, given that we are going to encode strings using the mapping f, it is important that we are able to decode in an unambiguous fashion. Thus it should be the case that the encoding f is injective.

Example 2.4 Suppose $X = \{a, b, c, d\}$, and consider the following three encodings:

$$
\begin{array}{llll}
f(a) = 1 & f(b) = 10 & f(c) = 100 & f(d) = 1000 \\
g(a) = 0 & g(b) = 10 & g(c) = 110 & g(d) = 111 \\
h(a) = 0 & h(b) = 01 & h(c) = 10 & h(d) = 11
\end{array}
$$

It can be seen that f and g are injective encodings, but h is not. Any encoding using f can be decoded by starting at the end and working backwards: every time a 1 is encountered, it signals the beginning of the current element.

An encoding using g can be decoded by starting at the beginning and proceeding sequentially. At any point where we have a substring that is an encoding of $a, b, c,$ or d, we decode it and chop off the substring. For example, given the string 10101110, we decode 10 to b, then 10 to b, then 111 to d, and finally 0 to a. So the decoded string is *bbda*.

To see that h is not injective, it suffices to give an example:

$$h(ac) = h(ba) = 010.$$

☐

From the point of view of ease of decoding, we would prefer the encoding g to f. This is because decoding can be done sequentially from beginning to end if g is used, so no memory is required. The property that allows the simple sequential decoding of g is called the prefix-free property. (An encoding g is a *prefix-free encoding* if there do not exist two elements $x, y \in X$, and a string $z \in \{0, 1\}^*$ such that $g(x) = g(y) \| z$.)

The discussion to this point has not involved entropy. Not surprisingly, entropy is related to the efficiency of an encoding. We will measure the efficiency of an encoding f as we did before: it is the weighted average length (denoted by $\ell(f)$) of an encoding of an element of $\mathbf{X}$. So we have the following definition:

$$\ell(f) = \sum_{x \in X} \mathbf{Pr}[x] |f(x)|,$$

where $|y|$ denotes the length of a string y.

Now, our fundamental problem is to find an injective encoding, f, that minimizes $\ell(f)$. There is a well-known algorithm, known as Huffman's algorithm, that accomplishes this goal. Moreover, the encoding f produced by Huffman's algorithm is prefix-free, and

$$H(\mathbf{X}) \leq \ell(f) < H(\mathbf{X}) + 1.$$

Thus, the value of the entropy of $\mathbf{X}$ provides a close estimate to the average length of the optimal injective encoding.

We will not prove the results stated above, but we will give a short, informal description of Huffman's algorithm. Huffman's algorithm begins with the probability distribution on the set X, and the code of each element is initially empty. In each iteration, the two elements having lowest probability are combined into one element having as its probability the sum of the two smaller probabilities. The element having lowest probability is assigned the value "0" and the element having next lowest probability is assigned the value "1." When only one element remains, the coding for each $x \in X$ can be constructed by following the sequence of elements "backwards" from the final element to the initial element x.

This is easily illustrated with an example.

Example 2.5   Suppose $X = \{a, b, c, d, e\}$ has the following probability distribution: $\mathbf{Pr}[a] = .05$, $\mathbf{Pr}[b] = .10$, $\mathbf{Pr}[c] = .12$, $\mathbf{Pr}[d] = .13$ and $\mathbf{Pr}[e] = .60$. Huffman's algorithm would proceed as indicated in the following table:

a	b	c	d	e
.05	.10	.12	.13	.60
0	1			
.15		.12	.13	.60
		0	1	
.15		.25		.60
0		1		
.40				.60
0				1
1.0				

This leads to the following encodings:

x	$f(x)$
a	000
b	001
c	010
d	011
e	1

Thus, the average length encoding is

$$\ell(f) = .05 \times 3 + .10 \times 3 + .12 \times 3 + .13 \times 3 + .60 \times 1$$
$$= 1.8.$$

Compare this to the entropy:

$$H(\mathbf{X}) = .2161 + .3322 + .3671 + .3842 + .4422$$
$$= 1.7402.$$

It is seen that the average length encoding is very close to the entropy. ▯

2.5   Properties of Entropy

In this section, we prove some fundamental results concerning entropy. First, we state a fundamental result, known as Jensen's inequality, that will be very useful to us. Jensen's inequality involves concave functions, which we now define.

Definition 2.5:   A real-valued function f is a *concave function* on an interval I if

$$f\left(\frac{x+y}{2}\right) \geq \frac{f(x) + f(y)}{2}$$

for all $x, y \in I$. f is a *strictly concave function* on an interval I if

$$f\left(\frac{x+y}{2}\right) > \frac{f(x) + f(y)}{2}$$

for all $x, y \in I$, $x \neq y$.

Here is Jensen's inequality, which we state without proof.

THEOREM 2.5   (Jensen's inequality)   *Suppose f is a continuous strictly concave function on the interval I. Suppose further that*

$$\sum_{i=1}^{n} a_i = 1$$

and $a_i > 0$, $1 \leq i \leq n$. Then

$$\sum_{i=1}^{n} a_i f(x_i) \leq f\left(\sum_{i=1}^{n} a_i x_i\right),$$

where $x_i \in I$, $1 \leq i \leq n$. Further, equality occurs if and only if $x_1 = \cdots = x_n$.

We now proceed to derive several results on entropy. In the next theorem, we make use of the fact that the function $\log_2 x$ is strictly concave on the interval $(0, \infty)$. (In fact, this follows easily from elementary calculus since the second derivative of the logarithm function is negative on the interval $(0, \infty)$.)

THEOREM 2.6 *Suppose* $\mathbf{X}$ *is a random variable having a probability distribution which takes on the values* $p_1, p_2, \ldots, p_n$, *where* $p_i > 0$, $1 \le i \le n$. *Then* $H(\mathbf{X}) \le \log_2 n$, *with equality if and only if* $p_i = 1/n$, $1 \le i \le n$.

PROOF Applying Jensen's inequality, we have the following:

$$H(\mathbf{X}) = -\sum_{i=1}^{n} p_i \log_2 p_i$$

$$= \sum_{i=1}^{n} p_i \log_2 \frac{1}{p_i}$$

$$\le \log_2 \sum_{i=1}^{n} \left(p_i \times \frac{1}{p_i} \right)$$

$$= \log_2 n.$$

Further, equality occurs if and only if $p_i = 1/n$, $1 \le i \le n$. ∎

THEOREM 2.7 $H(\mathbf{X}, \mathbf{Y}) \le H(\mathbf{X}) + H(\mathbf{Y})$, *with equality if and only if* $\mathbf{X}$ *and* $\mathbf{Y}$ *are independent random variables.*

PROOF Suppose $\mathbf{X}$ takes on values x_i, $1 \le i \le m$, and $\mathbf{Y}$ takes on values y_j, $1 \le j \le n$. Denote $p_i = \mathbf{Pr}[\mathbf{X} = x_i]$, $1 \le i \le m$, and $q_j = \mathbf{Pr}[\mathbf{Y} = y_j]$, $1 \le j \le n$. Then define $r_{ij} = \mathbf{Pr}[\mathbf{X} = x_i, \mathbf{Y} = y_j]$, $1 \le i \le m, 1 \le j \le n$ (this is the joint probability distribution).

Observe that

$$p_i = \sum_{j=1}^{n} r_{ij}$$

$(1 \le i \le m)$, and

$$q_j = \sum_{i=1}^{m} r_{ij}$$

$(1 \le j \le n)$. We compute as follows:

$$H(\mathbf{X}) + H(\mathbf{Y}) = -\left(\sum_{i=1}^{m} p_i \log_2 p_i + \sum_{j=1}^{n} q_j \log_2 q_j \right)$$

$$= -\left(\sum_{i=1}^{m} \sum_{j=1}^{n} r_{ij} \log_2 p_i + \sum_{j=1}^{n} \sum_{i=1}^{m} r_{ij} \log_2 q_j \right)$$

$$= -\sum_{i=1}^{m} \sum_{j=1}^{n} r_{ij} \log_2 p_i q_j.$$

On the other hand,

$$H(\mathbf{X}, \mathbf{Y}) = -\sum_{i=1}^{m}\sum_{j=1}^{n} r_{ij} \log_2 r_{ij}.$$

Combining, we obtain the following:

$$H(\mathbf{X}, \mathbf{Y}) - H(\mathbf{X}) - H(\mathbf{Y}) = \sum_{i=1}^{m}\sum_{j=1}^{n} r_{ij} \log_2 \frac{1}{r_{ij}} + \sum_{i=1}^{m}\sum_{j=1}^{n} r_{ij} \log_2 p_i q_j$$

$$= \sum_{i=1}^{m}\sum_{j=1}^{n} r_{ij} \log_2 \frac{p_i q_j}{r_{ij}}$$

$$\leq \log_2 \sum_{i=1}^{m}\sum_{j=1}^{n} p_i q_j$$

$$= \log_2 1$$

$$= 0.$$

(In the above computations, we apply Jensen's inequality, using the fact that the r_{ij}'s are positive real numbers that sum to 1.)

We can also say when equality occurs: it must be the case that there is a constant c such that $p_i q_j / r_{ij} = c$ for all i, j. Using the fact that

$$\sum_{j=1}^{n}\sum_{i=1}^{m} r_{ij} = \sum_{j=1}^{n}\sum_{i=1}^{m} p_i q_j = 1,$$

it follows that $c = 1$. Hence, equality occurs if and only if $r_{ij} = p_i q_j$, i.e., if and only if

$$\mathbf{Pr}[\mathbf{X} = x_i, \mathbf{Y} = y_j] = \mathbf{Pr}[\mathbf{X} = x_i]\mathbf{Pr}[\mathbf{Y} = y_j],$$

$1 \leq i \leq m, 1 \leq j \leq n$. But this says that $\mathbf{X}$ and $\mathbf{Y}$ are independent. ∎

We next define the idea of conditional entropy.

Definition 2.6: Suppose **X** and **Y** are two random variables. Then for any fixed value y of **Y**, we get a (conditional) probability distribution on X; we denote the associated random variable by $\mathbf{X}|y$. Clearly,

$$H(\mathbf{X}|y) = -\sum_{x} \mathbf{Pr}[x|y] \log_2 \mathbf{Pr}[x|y].$$

We define the *conditional entropy*, denoted $H(\mathbf{X}|\mathbf{Y})$, to be the weighted average (with respect to the probabilities $\mathbf{Pr}[y]$) of the entropies $H(\mathbf{X}|y)$ over all possible values y. It is computed to be

$$H(\mathbf{X}|\mathbf{Y}) = -\sum_{y}\sum_{x} \mathbf{Pr}[y]\mathbf{Pr}[x|y] \log_2 \mathbf{Pr}[x|y].$$

The conditional entropy measures the average amount of information about **X** that is not revealed by **Y**.

The next two results are straightforward; we leave the proofs as exercises.

THEOREM 2.8 $H(\mathbf{X}, \mathbf{Y}) = H(\mathbf{Y}) + H(\mathbf{X}|\mathbf{Y})$.

COROLLARY 2.9 $H(\mathbf{X}|\mathbf{Y}) \leq H(\mathbf{X})$, *with equality if and only if* **X** *and* **Y** *are independent.*

2.6   Spurious Keys and Unicity Distance

In this section, we apply the entropy results we have proved to cryptosystems. First, we show a fundamental relationship exists among the entropies of the components of a cryptosystem. The conditional entropy $H(\mathbf{K}|\mathbf{C})$ is called the *key equivocation*; it is a measure of the amount of uncertainty of the key remaining when the ciphertext is known.

THEOREM 2.10   *Let* $(\mathcal{P}, \mathcal{C}, \mathcal{K}, \mathcal{E}, \mathcal{D})$ *be a cryptosystem. Then*

$$H(\mathbf{K}|\mathbf{C}) = H(\mathbf{K}) + H(\mathbf{P}) - H(\mathbf{C}).$$

PROOF   First, observe that $H(\mathbf{K}, \mathbf{P}, \mathbf{C}) = H(\mathbf{C}|\mathbf{K}, \mathbf{P}) + H(\mathbf{K}, \mathbf{P})$. Now, the key and plaintext determine the ciphertext uniquely, since $y = e_K(x)$. This implies that $H(\mathbf{C}|\mathbf{K}, \mathbf{P}) = 0$. Hence, $H(\mathbf{K}, \mathbf{P}, \mathbf{C}) = H(\mathbf{K}, \mathbf{P})$. But **K** and **P** are independent, so $H(\mathbf{K}, \mathbf{P}) = H(\mathbf{K}) + H(\mathbf{P})$. Hence,

$$H(\mathbf{K}, \mathbf{P}, \mathbf{C}) = H(\mathbf{K}, \mathbf{P}) = H(\mathbf{K}) + H(\mathbf{P}).$$

In a similar fashion, since the key and ciphertext determine the plaintext uniquely (i.e., $x = d_K(y)$), we have that $H(\mathbf{P}|\mathbf{K}, \mathbf{C}) = 0$ and hence $H(\mathbf{K}, \mathbf{P}, \mathbf{C}) = H(\mathbf{K}, \mathbf{C})$.

Now, we compute as follows:

$$H(\mathbf{K}|\mathbf{C}) = H(\mathbf{K}, \mathbf{C}) - H(\mathbf{C})$$
$$= H(\mathbf{K}, \mathbf{P}, \mathbf{C}) - H(\mathbf{C})$$
$$= H(\mathbf{K}) + H(\mathbf{P}) - H(\mathbf{C}),$$

giving the desired formula.   ∎

Let us return to Example 2.3 to illustrate this result.

Example 2.3   (Cont.)   We have already computed $H(\mathbf{P}) \approx 0.81, H(\mathbf{K}) = 1.5$ and $H(\mathbf{C}) \approx 1.85$. Theorem 2.10 tells us that $H(\mathbf{K}|\mathbf{C}) \approx 1.5 + 0.81 - 1.85 \approx 0.46$. This can be verified directly by applying the definition of conditional entropy, as follows. First, we need to compute the probabilities $\mathbf{Pr}[\mathbf{K} = K_i | \mathbf{y} = j]$, $1 \le i \le 3, 1 \le j \le 4$. This can be done using Bayes' theorem, and the following values result:

$\mathbf{Pr}[K_1|1] = 1$	$\mathbf{Pr}[K_2|1] = 0$	$\mathbf{Pr}[K_3|1] = 0$
$\mathbf{Pr}[K_1|2] = \dfrac{6}{7}$	$\mathbf{Pr}[K_2|2] = \dfrac{1}{7}$	$\mathbf{Pr}[K_3|2] = 0$
$\mathbf{Pr}[K_1|3] = 0$	$\mathbf{Pr}[K_2|3] = \dfrac{3}{4}$	$\mathbf{Pr}[K_3|3] = \dfrac{1}{4}$
$\mathbf{Pr}[K_1|4] = 0$	$\mathbf{Pr}[K_2|4] = 0$	$\mathbf{Pr}[K_3|4] = 1.$

Now we compute

$$H(\mathbf{K}|\mathbf{C}) = \frac{1}{8} \times 0 + \frac{7}{16} \times 0.59 + \frac{1}{4} \times 0.81 + \frac{3}{16} \times 0 = 0.46,$$

agreeing with the value predicted by Theorem 2.10.   □

Suppose $(\mathcal{P}, \mathcal{C}, \mathcal{K}, \mathcal{E}, \mathcal{D})$ is the cryptosystem being used, and a string of plaintext

$$x_1 x_2 \cdots x_n$$

is encrypted with one key, producing a string of ciphertext

$$y_1 y_2 \cdots y_n.$$

Recall that the basic goal of the cryptanalyst is to determine the key. We are looking at ciphertext-only attacks, and we assume that Oscar has infinite computational resources. We also assume that Oscar knows that the plaintext is a "natural" language, such as English. In general, Oscar will be able to rule out certain keys, but many "possible" keys may remain, only one of which is the correct key. The remaining possible, but incorrect, keys are called *spurious keys*.

For example, suppose Oscar obtains the ciphertext string $WNAJW$, which has been obtained by encryption using a shift cipher. It is easy to see that there are only two "meaningful" plaintext strings, namely *river* and *arena*, corresponding respectively to the possible encryption keys F ($= 5$) and W ($= 22$). Of these two keys, one will be the correct key and the other will be spurious. (Actually, it is moderately difficult to find a ciphertext of length 5 for the *Shift Cipher* that has two meaningful decryptions; the reader might search for other examples.)

Our goal is to prove a bound on the expected number of spurious keys. First, we have to define what we mean by the entropy (per letter) of a natural language L, which we denote H_L. H_L should be a measure of the average information per letter in a "meaningful" string of plaintext. (Note that a random string of alphabetic characters would have entropy (per letter) equal to $\log_2 26 \approx 4.70$.) As a "first-order" approximation to H_L, we could take $H(\mathbf{P})$. In the case where L is the English language, we get $H(\mathbf{P}) \approx 4.19$ by using the probability distribution given in Table 1.1.

Of course, successive letters in a language are not independent, and correlations among successive letters reduce the entropy. For example, in English, the letter "Q" is always followed by the letter "U." For a "second-order" approximation, we would compute the entropy of the probability distribution of all digrams and then divide by 2. In general, define $\mathbf{P}^n$ to be the random variable that has as its probability distribution that of all n-grams of plaintext. We make use of the following definitions.

Definition 2.7: Suppose L is a natural language. The *entropy* of L is defined to be the quantity

$$H_L = \lim_{n \to \infty} \frac{H(\mathbf{P}^n)}{n}$$

and the *redundancy* of L is defined to be

$$R_L = 1 - \frac{H_L}{\log_2 |\mathcal{P}|}.$$

REMARK H_L measures the entropy per letter of the language L. A random language would have entropy $\log_2 |\mathcal{P}|$. So the quantity R_L measures the fraction of "excess characters," which we think of as redundancy. ∎

In the case of the English language, a tabulation of a large number of digrams and their frequencies would produce an estimate for $H(\mathbf{P}^2)$. $H(\mathbf{P}^2)/2 \approx 3.90$ is one estimate obtained in this way. One could continue, tabulating trigrams, etc. and thus obtain an estimate for H_L. In fact, various experiments have yielded the empirical result that $1.0 \leq H_L \leq 1.5$. That is, the average information content in English is something like one to one-and-a-half bits per letter!

Using 1.25 as our estimate of H_L gives a redundancy of about 0.75. This means that the English language is 75% redundant! (This is not to say that one can arbitrarily remove three out of every four letters from English text and hope to still be able to read it. What it does mean is that it is possible to find a Huffman encoding of n-grams, for a large enough value of n, which will compress English text to about one quarter of its original length.)

Given probability distributions on $\mathcal{K}$ and $\mathcal{P}^n$, we can define the induced probability distribution on $\mathcal{C}^n$, the set of n-grams of ciphertext (we already did this in the case $n = 1$). We have defined $\mathbf{P}^n$ to be a random variable representing an n-gram of plaintext. Similarly, define $\mathbf{C}^n$ to be a random variable representing an n-gram of ciphertext.

Given $\mathbf{y} \in \mathbf{C}^n$, define

$$K(\mathbf{y}) = \{K \in \mathcal{K} : \exists \mathbf{x} \in \mathcal{P}^n \text{ such that } \mathbf{Pr}[\mathbf{x}] > 0 \text{ and } e_K(\mathbf{x}) = \mathbf{y}\}.$$

That is, $K(\mathbf{y})$ is the set of keys K for which $\mathbf{y}$ is the encryption of a meaningful string of plaintext of length n, i.e., the set of "possible" keys, given that $\mathbf{y}$ is the ciphertext. If $\mathbf{y}$ is the observed string of ciphertext, then the number of spurious keys is $|K(\mathbf{y})| - 1$, since only one of the "possible" keys is the correct key. The average number of spurious keys (over all possible ciphertext strings of length n) is denoted by $\bar{s}_n$. Its value is computed to be

$$\bar{s}_n = \sum_{\mathbf{y} \in \mathcal{C}^n} \mathbf{Pr}[\mathbf{y}](|K(\mathbf{y})| - 1)$$

$$= \sum_{\mathbf{y} \in \mathcal{C}^n} \mathbf{Pr}[\mathbf{y}]|K(\mathbf{y})| - \sum_{\mathbf{y} \in \mathcal{C}^n} \mathbf{Pr}[\mathbf{y}]$$

$$= \sum_{\mathbf{y} \in \mathcal{C}^n} \mathbf{Pr}[\mathbf{y}]|K(\mathbf{y})| - 1.$$

From Theorem 2.10, we have that

$$H(\mathbf{K}|\mathbf{C}^n) = H(\mathbf{K}) + H(\mathbf{P}^n) - H(\mathbf{C}^n).$$

Also, we can use the estimate

$$H(\mathbf{P}^n) \approx nH_L = n(1 - R_L)\log_2 |\mathcal{P}|,$$

provided n is reasonably large. Certainly,

$$H(\mathbf{C}^n) \leq n \log_2 |\mathcal{C}|.$$

Then, if $|\mathcal{C}| = |\mathcal{P}|$, it follows that

$$H(\mathbf{K}|\mathbf{C}^n) \geq H(\mathbf{K}) - nR_L \log_2 |\mathcal{P}|. \tag{2.2}$$

Next, we relate the quantity $H(\mathbf{K}|\mathbf{C}^n)$ to the number of spurious keys, $\bar{s}_n$. We compute as follows:

$$
\begin{aligned}
H(\mathbf{K}|\mathbf{C}^n) &= \sum_{\mathbf{y} \in \mathcal{C}^n} \mathbf{Pr}[\mathbf{y}] H(\mathbf{K}|\mathbf{y}) \\
&\leq \sum_{\mathbf{y} \in \mathcal{C}^n} \mathbf{Pr}[\mathbf{y}] \log_2 |K(\mathbf{y})| \\
&\leq \log_2 \sum_{\mathbf{y} \in \mathcal{C}^n} \mathbf{Pr}[\mathbf{y}]|K(\mathbf{y})| \\
&= \log_2(\bar{s}_n + 1),
\end{aligned}
$$

where we apply Jensen's inequality (Theorem 2.5) with $f(x) = \log_2 x$. Thus we obtain the inequality

$$H(\mathbf{K}|\mathbf{C}^n) \leq \log_2(\bar{s}_n + 1). \tag{2.3}$$

Combining the two inequalities (2.2) and (2.3), we get that

$$\log_2(\bar{s}_n + 1) \geq H(\mathbf{K}) - nR_L \log_2 |\mathcal{P}|.$$

In the case where keys are chosen equiprobably (which maximizes $H(\mathbf{K})$), we have the following result.

THEOREM 2.11 *Suppose $(\mathcal{P}, \mathcal{C}, \mathcal{K}, \mathcal{E}, \mathcal{D})$ is a cryptosystem where $|\mathcal{C}| = |\mathcal{P}|$ and keys are chosen equiprobably. Let R_L denote the redundancy of the underlying language. Then given a string of ciphertext of length n, where n is sufficiently large, the expected number of spurious keys, $\bar{s}_n$, satisfies*

$$\bar{s}_n \geq \frac{|\mathcal{K}|}{|\mathcal{P}|^{nR_L}} - 1.$$

The quantity $|\mathcal{K}|/|\mathcal{P}|^{nR_L} - 1$ approaches 0 exponentially quickly as n increases. Also, note that the estimate may not be accurate for small values of n, especially since $H(\mathbf{P}^n)/n$ may not be a good estimate for H_L if n is small.

We have one more concept to define.

Definition 2.8:   The *unicity distance* of a cryptosystem is defined to be the value of n, denoted by n_0, at which the expected number of spurious keys becomes zero; i.e., the average amount of ciphertext required for an opponent to be able to uniquely compute the key, given enough computing time.

If we set $\bar{s}_n = 0$ in Theorem 2.11 and solve for n, we get an estimate for the unicity distance, namely

$$n_0 \approx \frac{\log_2 |\mathcal{K}|}{R_L \log_2 |\mathcal{P}|}.$$

As an example, consider the *Substitution Cipher*. In this cryptosystem, $|\mathcal{P}| = 26$ and $|\mathcal{K}| = 26!$. If we take $R_L = 0.75$, then we get an estimate for the unicity distance of

$$n_0 \approx 88.4/(0.75 \times 4.7) \approx 25.$$

This suggests that, given a ciphertext string of length at least 25, (usually) a unique decryption is possible.

2.7 Product Cryptosystems

Another innovation introduced by Shannon in his 1949 paper was the idea of combining cryptosystems by forming their "product." This idea has been of fundamental importance in the design of present-day cryptosystems such as the *Advanced Encryption Standard*, which we study in the next chapter.

For simplicity, we will confine our attention in this section to cryptosystems in which $\mathcal{C} = \mathcal{P}$: a cryptosystem of this type is called an *endomorphic cryptosystem*. Suppose $\mathbf{S}_1 = (\mathcal{P}, \mathcal{P}, \mathcal{K}_1, \mathcal{E}_1, \mathcal{D}_1)$ and $\mathbf{S}_2 = (\mathcal{P}, \mathcal{P}, \mathcal{K}_2, \mathcal{E}_2, \mathcal{D}_2)$ are two endomorphic cryptosystems which have the same plaintext (and ciphertext) spaces. Then the *product cryptosystem* of $\mathbf{S}_1$ and $\mathbf{S}_2$, denoted by $\mathbf{S}_1 \times \mathbf{S}_2$, is defined to be the cryptosystem

$$(\mathcal{P}, \mathcal{P}, \mathcal{K}_1 \times \mathcal{K}_2, \mathcal{E}, \mathcal{D}).$$

A key of the product cryptosystem has the form $K = (K_1, K_2)$, where $K_1 \in \mathcal{K}_1$ and $K_2 \in \mathcal{K}_2$. The encryption and decryption rules of the product cryptosystem are defined as follows: For each $K = (K_1, K_2)$, we have an encryption rule e_K defined by the formula

$$e_{(K_1, K_2)}(x) = e_{K_2}(e_{K_1}(x)),$$

and a decryption rule defined by the formula

$$d_{(K_1, K_2)}(y) = d_{K_1}(d_{K_2}(y)).$$

That is, we first encrypt x with e_{K_1}, and then "re-encrypt" the resulting ciphertext with e_{K_2}. Decrypting is similar, but it must be done in the reverse order:

$$d_{(K_1, K_2)}(e_{(K_1, K_2)}(x)) = d_{(K_1, K_2)}(e_{K_2}(e_{K_1}(x)))$$
$$= d_{K_1}(d_{K_2}(e_{K_2}(e_{K_1}(x))))$$
$$= d_{K_1}(e_{K_1}(x))$$
$$= x.$$

Cryptosystem 2.2: *Multiplicative Cipher*

Let $\mathcal{P} = \mathcal{C} = \mathbb{Z}_{26}$ and let

$$\mathcal{K} = \{a \in \mathbb{Z}_{26} : \gcd(a, 26) = 1\}.$$

For $a \in \mathcal{K}$, define

$$e_a(x) = ax \bmod 26$$

and

$$d_a(y) = a^{-1}y \bmod 26$$

$(x, y \in \mathbb{Z}_{26})$.

Recall also that cryptosystems have probability distributions associated with their keyspaces. Thus we need to define the probability distribution for the keyspace $\mathcal{K}$ of the product cryptosystem. We do this in a very natural way:

$$\mathbf{Pr}[(K_1, K_2)] = \mathbf{Pr}[K_1] \times \mathbf{Pr}[K_2].$$

In other words, choose K_1 and K_2 independently, using the probability distributions defined on $\mathcal{K}_1$ and $\mathcal{K}_2$, respectively.

Here is a simple example to illustrate the definition of a product cryptosystem. We present the *Multiplicative Cipher* as Cryptosystem 2.2.

Suppose **M** is the *Multiplicative Cipher* (with keys chosen equiprobably) and **S** is the *Shift Cipher* (with keys chosen equiprobably). Then it is very easy to see that **M** × **S** is nothing more than the *Affine Cipher* (again, with keys chosen equiprobably). It is slightly more difficult to show that **S** × **M** is also the *Affine Cipher* with equiprobable keys.

Let's prove these assertions. A key in the *Shift Cipher* is an element $K \in \mathbb{Z}_{26}$, and the corresponding encryption rule is $e_K(x) = (x + K) \bmod 26$. A key in the *Multiplicative Cipher* is an element $a \in \mathbb{Z}_{26}$ such that $\gcd(a, 26) = 1$; the corresponding encryption rule is $e_a(x) = ax \bmod 26$. Hence, a key in the product cipher **M** × **S** has the form (a, K), where

$$e_{(a,K)}(x) = (ax + K) \bmod 26.$$

But this is precisely the definition of a key in the *Affine Cipher*. Further, the probability of a key in the *Affine Cipher* is $1/312 = 1/12 \times 1/26$, which is the product of the probabilities of the keys a and K, respectively. Thus **M** × **S** is the *Affine Cipher*.

Now let's consider **S** × **M**. A key in this cipher has the form (K, a), where

$$e_{(K,a)}(x) = a(x + K) \bmod 26 = (ax + aK) \bmod 26.$$

Thus the key (K, a) of the product cipher $\mathbf{S} \times \mathbf{M}$ is identical to the key (a, aK) of the *Affine Cipher*. It remains to show that each key of the *Affine Cipher* arises with the same probability $1/312$ in the product cipher $\mathbf{S} \times \mathbf{M}$. Observe that $aK = K_1$ if and only if $K = a^{-1}K_1$ (recall that $\gcd(a, 26) = 1$, so a has a multiplicative inverse modulo 26). In other words, the key (a, K_1) of the *Affine Cipher* is equivalent to the key $(a^{-1}K_1, a)$ of the product cipher $\mathbf{S} \times \mathbf{M}$. We thus have a bijection between the two key spaces. Since each key is equiprobable, we conclude that $\mathbf{S} \times \mathbf{M}$ is indeed the *Affine Cipher*.

We have shown that $\mathbf{M} \times \mathbf{S} = \mathbf{S} \times \mathbf{M}$. Thus we would say that the two cryptosystems $\mathbf{M}$ and $\mathbf{S}$ *commute*. But not all pairs of cryptosystems commute; it is easy to find counterexamples. On the other hand, the product operation is always *associative*: $(\mathbf{S}_1 \times \mathbf{S}_2) \times \mathbf{S}_3 = \mathbf{S}_1 \times (\mathbf{S}_2 \times \mathbf{S}_3)$.

If we take the product of an (endomorphic) cryptosystem $\mathbf{S}$ with itself, we obtain the cryptosystem $\mathbf{S} \times \mathbf{S}$, which we denote by $\mathbf{S}^2$. If we take the n-fold product, the resulting cryptosystem is denoted by $\mathbf{S}^n$.

A cryptosystem $\mathbf{S}$ is defined to be an *idempotent cryptosystem* if $\mathbf{S}^2 = \mathbf{S}$. Many of the cryptosystems we studied in Chapter 1 are idempotent. For example, the *Shift, Substitution, Affine, Hill, Vigenère* and *Permutation Ciphers* are all idempotent. Of course, if a cryptosystem $\mathbf{S}$ is idempotent, then there is no point in using the product system $\mathbf{S}^2$, as it requires an extra key but provides no more security.

If a cryptosystem is not idempotent, then there is a potential increase in security by iterating it several times. This idea is used in the *Data Encryption Standard*, which consists of 16 iterations. But, of course, this approach requires a non-idempotent cryptosystem to start with. One way in which simple non-idempotent cryptosystems can sometimes be constructed is to take the product of two different (simple) cryptosystems.

REMARK   It is not hard to show that if $\mathbf{S}_1$ and $\mathbf{S}_2$ are both idempotent, and they commute, then $\mathbf{S}_1 \times \mathbf{S}_2$ will also be idempotent. This follows from the following algebraic manipulations:

$$(\mathbf{S}_1 \times \mathbf{S}_2) \times (\mathbf{S}_1 \times \mathbf{S}_2) = \mathbf{S}_1 \times (\mathbf{S}_2 \times \mathbf{S}_1) \times \mathbf{S}_2$$
$$= \mathbf{S}_1 \times (\mathbf{S}_1 \times \mathbf{S}_2) \times \mathbf{S}_2$$
$$= (\mathbf{S}_1 \times \mathbf{S}_1) \times (\mathbf{S}_2 \times \mathbf{S}_2)$$
$$= \mathbf{S}_1 \times \mathbf{S}_2.$$

(Note the use of the associative property in this proof.)

So, if $\mathbf{S}_1$ and $\mathbf{S}_2$ are both idempotent, and we want $\mathbf{S}_1 \times \mathbf{S}_2$ to be non-idempotent, then it is necessary that $\mathbf{S}_1$ and $\mathbf{S}_2$ not commute.   ∎

Fortunately, many simple cryptosystems are suitable building blocks in this type of approach. Taking the product of substitution-type ciphers with permutation-type ciphers is a commonly used technique. We will see several realizations of this in the next chapter.

2.8 Notes

The idea of perfect secrecy and the use of entropy techniques in cryptography was pioneered by Shannon [300]. Product cryptosystems are also discussed in this paper. The concept of entropy was defined by Shannon in [299]. Good introductions to entropy, Huffman coding and related topics can be found in the books by Welsh [342] and Goldie and Pinch [158].

The results of Section 2.6 are due to Beauchemin and Brassard [10], who generalized earlier results of Shannon.

Exercises

2.1 Referring to Example 2.2, determine all the joint and conditional probabilities, $\mathbf{Pr}[x, y]$, $\mathbf{Pr}[x|y]$ and $\mathbf{Pr}[y|x]$, where $x \in \{2, \ldots, 12\}$ and $y \in \{D, N\}$.

2.2 Let n be a positive integer. A *Latin square* of order n is an $n \times n$ array L of the integers $1, \ldots, n$ such that every one of the n integers occurs exactly once in each row and each column of L. An example of a Latin square of order 3 is as follows:

1	2	3
3	1	2
2	3	1

Given any Latin square L of order n, we can define a related cryptosystem. Take $\mathcal{P} = \mathcal{C} = \mathcal{K} = \{1, \ldots, n\}$. For $1 \leq i \leq n$, the encryption rule e_i is defined to be $e_i(j) = L(i, j)$. (Hence each row of L gives rise to one encryption rule.)

Give a complete proof that this *Latin Square Cryptosystem* achieves perfect secrecy provided that every key is used with equal probability.

2.3 (a) Prove that the *Affine Cipher* achieves perfect secrecy if every key is used with equal probability $1/312$.

  (b) More generally, suppose we are given a probability distribution on the set

$$\{a \in \mathbb{Z}_{26} : \gcd(a, 26) = 1\}.$$

  Suppose that every key (a, b) for the *Affine Cipher* is used with probability $\mathbf{Pr}[a]/26$. Prove that the *Affine Cipher* achieves perfect secrecy when this probability distribution is defined on the keyspace.

2.4 Suppose a cryptosystem achieves perfect secrecy for a particular plaintext probability distribution. Prove that perfect secrecy is maintained for any plaintext probability distribution.

2.5 Prove that if a cryptosystem has perfect secrecy and $|\mathcal{K}| = |\mathcal{C}| = |\mathcal{P}|$, then every ciphertext is equally probable.

2.6 Suppose that y and y' are two ciphertext elements (i.e., binary n-tuples) in the *One-time Pad* that were obtained by encrypting plaintext elements x and x', respectively, using the same key, K. Prove that $x + x' \equiv y + y' \pmod{2}$.

2.7   (a) Construct the encryption matrix (as defined in Example 2.3) for the *One-time Pad* with $n = 3$.

      (b) For any positive integer n, give a direct proof that the encryption matrix of a *One-time Pad* defined over $(\mathbb{Z}_2)^n$ is a Latin square of order 2^n, in which the symbols are the elements of $(\mathbb{Z}_2)^n$.

2.8 Suppose X is a set of cardinality n, where $2^k \leq n < 2^{k+1}$, and $\mathbf{Pr}[x] = 1/n$ for all $x \in X$.

    (a) Find a prefix-free encoding of X, say f, such that $\ell(f) = k + 2 - 2^{k+1}/n$.

      HINT   Encode $2^{k+1} - n$ elements of X as strings of length k, and encode the remaining elements as strings of length $k + 1$.

    (b) Illustrate your construction for $n = 6$. Compute $\ell(f)$ and $H(\mathbf{X})$ in this case.

2.9 Suppose $X = \{a, b, c, d, e\}$ has the following probability distribution: $\mathbf{Pr}[a] = .32$, $\mathbf{Pr}[b] = .23$, $\mathbf{Pr}[c] = .20$, $\mathbf{Pr}[d] = .15$ and $\mathbf{Pr}[e] = .10$. Use Huffman's algorithm to find the optimal prefix-free encoding of X. Compare the length of this encoding to $H(\mathbf{X})$.

2.10 Prove that $H(\mathbf{X}, \mathbf{Y}) = H(\mathbf{Y}) + H(\mathbf{X}|\mathbf{Y})$. Then show as a corollary that $H(\mathbf{X}|\mathbf{Y}) \leq H(\mathbf{X})$, with equality if and only if $\mathbf{X}$ and $\mathbf{Y}$ are independent.

2.11 Prove that a cryptosystem has perfect secrecy if and only if $H(\mathbf{P}|\mathbf{C}) = H(\mathbf{P})$.

2.12 Prove that, in any cryptosystem, $H(\mathbf{K}|\mathbf{C}) \geq H(\mathbf{P}|\mathbf{C})$. (Intuitively, this result says that, given a ciphertext, the opponent's uncertainty about the key is at least as great as his uncertainty about the plaintext.)

2.13 Consider a cryptosystem in which $\mathcal{P} = \{a, b, c\}$, $\mathcal{K} = \{K_1, K_2, K_3\}$ and $\mathcal{C} = \{1, 2, 3, 4\}$. Suppose the encryption matrix is as follows:

	a	b	c
K_1	1	2	3
K_2	2	3	4
K_3	3	4	1

Given that keys are chosen equiprobably, and the plaintext probability distribution is $\mathbf{Pr}[a] = 1/2$, $\mathbf{Pr}[b] = 1/3$, $\mathbf{Pr}[c] = 1/6$, compute $H(\mathbf{P})$, $H(\mathbf{C})$, $H(\mathbf{K})$, $H(\mathbf{K}|\mathbf{C})$ and $H(\mathbf{P}|\mathbf{C})$.

2.14 Compute $H(\mathbf{K}|\mathbf{C})$ and $H(\mathbf{K}|\mathbf{P}, \mathbf{C})$ for the *Affine Cipher*, assuming that keys are used equiprobably and the plaintexts are equiprobable.

2.15 Consider a *Vigenère Cipher* with keyword length m. Show that the unicity distance is $1/R_L$, where R_L is the redundancy of the underlying language. (This result is interpreted as follows. If n_0 denotes the number of alphabetic characters being encrypted, then the "length" of the plaintext is n_0/m, since each plaintext element consists of m alphabetic characters. So, a unicity distance of $1/R_L$ corresponds to a plaintext consisting of m/R_L alphabetic characters.)

2.16 Show that the unicity distance of the *Hill Cipher* (with an $m \times m$ encryption matrix) is less than m/R_L. (Note that the number of alphabetic characters in a plaintext of this length is m^2/R_L.)

2.17 A *Substitution Cipher* over a plaintext space of size n has $|\mathcal{K}| = n!$ Stirling's formula gives the following estimate for $n!$:

$$n! \approx \sqrt{2\pi n}\left(\frac{n}{e}\right)^n.$$

   (a) Using Stirling's formula, derive an estimate of the unicity distance of the *Substitution Cipher*.
   (b) Let $m \geq 1$ be an integer. The *m-gram Substitution Cipher* is the *Substitution Cipher* where the plaintext (and ciphertext) spaces consist of all 26^m *m*-grams. Estimate the unicity distance of the *m-gram Substitution Cipher* if $R_L = 0.75$.

2.18 Prove that the *Shift Cipher* is idempotent, assuming that keys are equiprobable.

2.19 Suppose $\mathbf{S}_1$ is the *Shift Cipher* (with equiprobable keys, as usual) and $\mathbf{S}_2$ is the *Shift Cipher* where keys are chosen with respect to some probability distribution $p_{\mathcal{K}}$ (which need not be equiprobable). Prove that $\mathbf{S}_1 \times \mathbf{S}_2 = \mathbf{S}_1$.

2.20 Suppose $\mathbf{S}_1$ and $\mathbf{S}_2$ are *Vigenère Ciphers* with keyword lengths m_1, m_2 respectively, where $m_1 > m_2$. Assume that the keys in each of these ciphers are equiprobable.

   (a) If $m_2 \mid m_1$, then show that $\mathbf{S}_2 \times \mathbf{S}_1 = \mathbf{S}_1$.
   (b) One might try to generalize the previous result by conjecturing that $\mathbf{S}_2 \times \mathbf{S}_1 = \mathbf{S}_3$, where $\mathbf{S}_3$ is the *Vigenère Cipher* with keyword length $\text{lcm}(m_1, m_2)$. Prove that this conjecture is false.

   HINT If $m_1 \not\equiv 0 \pmod{m_2}$, then the number of keys in the product cryptosystem $\mathbf{S}_2 \times \mathbf{S}_1$ is less than the number of keys in $\mathbf{S}_3$.

3

Block Ciphers and the Advanced Encryption Standard

3.1 Introduction

Most modern-day block ciphers are product ciphers (product ciphers were introduced in Section 2.7). Product ciphers frequently incorporate a sequence of permutation and substitution operations. A commonly used design is that of an *iterated cipher*. Here is a description of a typical iterated cipher: The cipher requires the specification of a *round function* and a *key schedule*, and the encryption of a plaintext will proceed through Nr similar *rounds*.

Let K be a random binary key of some specified length. K is used to construct Nr *round keys* (also called *subkeys*), which are denoted $K^1, \ldots, K^{Nr}$. The list of round keys, $(K^1, \ldots, K^{Nr})$, is the key schedule. The key schedule is constructed from K using a fixed, public algorithm.

The round function, say g, takes two inputs: a round key (K^r) and a current *state* (which we denote w^{r-1}). The next state is defined as $w^r = g(w^{r-1}, K^r)$. The initial state, w^0, is defined to be the plaintext, x. The ciphertext, y, is defined to be the state after all Nr rounds have been performed. Therefore, the encryption operation is carried out as follows:

$$w^0 \leftarrow x$$
$$w^1 \leftarrow g(w^0, K^1)$$
$$w^2 \leftarrow g(w^1, K^2)$$
$$\vdots \quad \vdots \quad \vdots$$
$$w^{Nr-1} \leftarrow g(w^{Nr-2}, K^{Nr-1})$$
$$w^{Nr} \leftarrow g(w^{Nr-1}, K^{Nr})$$
$$y \leftarrow w^{Nr}.$$

In order for decryption to be possible, the function g must have the property that it is injective (i.e., one-to-one) if its second argument is fixed. This is equivalent to saying that there exists a function g^{-1} with the property that

$$g^{-1}(g(w,y),y) = w$$

for all w and y. Then decryption can be accomplished as follows:

$$w^{\mathrm{Nr}} \leftarrow y$$
$$w^{\mathrm{Nr}-1} \leftarrow g^{-1}(w^{\mathrm{Nr}}, K^{\mathrm{Nr}})$$
$$\vdots \quad \vdots \quad \vdots$$
$$w^1 \leftarrow g^{-1}(w^2, K^2)$$
$$w^0 \leftarrow g^{-1}(w^1, K^1)$$
$$x \leftarrow w^0.$$

In Section 3.2, we describe a simple type of iterated cipher, the substitution-permutation network, which illustrates many of the main principles used in the design of practical block ciphers. Linear and differential attacks on substitution-permutation networks are described in Sections 3.3 and 3.4, respectively. In Section 3.5, we discuss Feistel-type ciphers and the *Data Encryption Standard*. In Section 3.6, we present the *Advanced Encryption Standard*. Finally, modes of operation of block ciphers are the topic of Section 3.7.

3.2 Substitution-Permutation Networks

We begin by defining a *substitution-permutation network*, or *SPN*. (An SPN is a special type of iterated cipher with a couple of small changes that we will indicate.) Suppose that ℓ and m are positive integers. A plaintext and ciphertext will both be binary vectors of length ℓm (i.e., ℓm is the *block length* of the cipher). An SPN is built from two components, which are denoted π_S and π_P.

$$\pi_S : \{0,1\}^\ell \rightarrow \{0,1\}^\ell$$

is a permutation, and

$$\pi_P : \{1,\ldots,\ell m\} \rightarrow \{1,\ldots,\ell m\}$$

is also a permutation. The permutation π_S is called an *S-box* (the letter "S" denotes "substitution"). It is used to replace ℓ bits with a different set of ℓ bits. π_P, on the other hand, is used to permute ℓm bits.

Cryptosystem 3.1: *Substitution-Permutation Network*

Let ℓ, m and Nr be positive integers, let $\pi_S : \{0,1\}^\ell \to \{0,1\}^\ell$ be a permutation, and let $\pi_P : \{1,\ldots,\ell m\} \to \{1,\ldots,\ell m\}$ be a permutation. Let $\mathcal{P} = \mathcal{C} = \{0,1\}^{\ell m}$, and let $\mathcal{K} \subseteq (\{0,1\}^{\ell m})^{\text{Nr}+1}$ consist of all possible key schedules that could be derived from an initial key K using the key scheduling algorithm. For a key schedule $(K^1,\ldots,K^{\text{Nr}+1})$, we encrypt the plaintext x using Algorithm 3.1.

Given an ℓm-bit binary string, say $x = (x_1,\ldots,x_{\ell m})$, we can regard x as the concatenation of m ℓ-bit substrings, which we denote $x_{<1>},\ldots,x_{<m>}$. Thus

$$x = x_{<1>} \| \cdots \| x_{<m>}$$

and for $1 \leq i \leq m$, we have that

$$x_{<i>} = \left(x_{(i-1)\ell+1},\ldots,x_{i\ell}\right).$$

The SPN will consist of Nr rounds. In each round (except for the last round, which is slightly different), we will perform m substitutions using π_S, followed by a permutation using π_P. Prior to each substitution operation, we will incorporate round key bits via a simple exclusive-or operation. We now present an SPN, based on π_S and π_P, as Cryptosystem 3.1.

In Algorithm 3.1, u^r is the input to the S-boxes in round r, and v^r is the output of the S-boxes in round r. w^r is obtained from v^r by applying the permutation π_P, and then u^{r+1} is constructed from w^r by x-or-ing with the round key K^{r+1} (this is called *round key mixing*). In the last round, the permutation π_P is not applied. As a consequence, the encryption algorithm can also be used for decryption, if appropriate modifications are made to the key schedule and the S-boxes are replaced by their inverses (see the Exercises).

Notice that the very first and last operations performed in this SPN are x-ors with subkeys. This is called *whitening*, and is regarded as a useful way to prevent an attacker from even beginning to carry out an encryption or decryption operation if the key is not known.

We illustrate the above general description with a particular SPN.

Example 3.1 Suppose that $\ell = m = \text{Nr} = 4$. Let π_S be defined as follows, where the input (i.e., z) and the output (i.e., $\pi_S(z)$) are written in hexadecimal notation, $(0 \leftrightarrow (0,0,0,0), 1 \leftrightarrow (0,0,0,1),\ldots,9 \leftrightarrow (1,0,0,1), A \leftrightarrow (1,0,1,0),\ldots, F \leftrightarrow (1,1,1,1))$:

z	0	1	2	3	4	5	6	7	8	9	A	B	C	D	E	F
$\pi_S(z)$	E	4	D	1	2	F	B	8	3	A	6	C	5	9	0	7

Algorithm 3.1: $\text{SPN}(x, \pi_S, \pi_P, (K^1, \ldots, K^{\text{Nr}+1}))$

$w^0 \leftarrow x$
for $r \leftarrow 1$ **to** $\text{Nr} - 1$

$\quad$ **do** $\begin{cases} u^r \leftarrow w^{r-1} \oplus K^r \\ \textbf{for } i \leftarrow 1 \textbf{ to } m \\ \quad \textbf{do } v^r_{<i>} \leftarrow \pi_S(u^r_{<i>}) \\ w^r \leftarrow \left(v^r_{\pi_P(1)}, \ldots, v^r_{\pi_P(\ell m)}\right) \end{cases}$

$u^{\text{Nr}} \leftarrow w^{\text{Nr}-1} \oplus K^{\text{Nr}}$
for $i \leftarrow 1$ **to** m
$\quad$ **do** $v^{\text{Nr}}_{<i>} \leftarrow \pi_S(u^{\text{Nr}}_{<i>})$
$y \leftarrow v^{\text{Nr}} \oplus K^{\text{Nr}+1}$
output (y)

Further, let π_P be defined as follows:

z	1	2	3	4	5	6	7	8	9	10	11	12	13	14	15	16
$\pi_P(z)$	1	5	9	13	2	6	10	14	3	7	11	15	4	8	12	16

See Figure 3.1 for a pictorial representation of this particular SPN. (In this diagram, we have named the S-boxes S^r_i ($1 \le i \le 4, 1 \le r \le 4$) for ease of later reference. All 16 S-boxes incorporate the same substitution function based on π_S.)

In order to complete the description of the SPN, we need to specify a key scheduling algorithm. Here is a simple possibility: suppose that we begin with a 32-bit key $K = (k_1, \ldots, k_{32}) \in \{0, 1\}^{32}$. For $1 \le r \le 5$, define K^r to consist of 16 consecutive bits of K, beginning with k_{4r-3}. (This is not a very secure way to define a key schedule; we have just chosen something easy for purposes of illustration.)

Now let's work out a sample encryption using this SPN. We represent all data in binary notation. Suppose the key is

$$K = 0011\ 1010\ 1001\ 0100\ 1101\ 0110\ 0011\ 1111.$$

Then the round keys are as follows:

$$K^1 = 0011\ 1010\ 1001\ 0100$$

$$K^2 = 1010\ 1001\ 0100\ 1101$$

$$K^3 = 1001\ 0100\ 1101\ 0110$$

$$K^4 = 0100\ 1101\ 0110\ 0011$$

$$K^5 = 1101\ 0110\ 0011\ 1111.$$

Suppose that the plaintext is

$$x = 0010\ 0110\ 1011\ 0111.$$

Then the encryption of x proceeds as follows:

$$w^0 = 0010\ 0110\ 1011\ 0111$$
$$K^1 = 0011\ 1010\ 1001\ 0100$$
$$u^1 = 0001\ 1100\ 0010\ 0011$$
$$v^1 = 0100\ 0101\ 1101\ 0001$$
$$w^1 = 0010\ 1110\ 0000\ 0111$$
$$K^2 = 1010\ 1001\ 0100\ 1101$$
$$u^2 = 1000\ 0111\ 0100\ 1010$$
$$v^2 = 0011\ 1000\ 0010\ 0110$$
$$w^2 = 0100\ 0001\ 1011\ 1000$$
$$K^3 = 1001\ 0100\ 1101\ 0110$$
$$u^3 = 1101\ 0101\ 0110\ 1110$$
$$v^3 = 1001\ 1111\ 1011\ 0000$$
$$w^3 = 1110\ 0100\ 0110\ 1110$$
$$K^4 = 0100\ 1101\ 0110\ 0011$$
$$u^4 = 1010\ 1001\ 0000\ 1101$$
$$v^4 = 0110\ 1010\ 1110\ 1001$$
$$K^5 = 1101\ 0110\ 0011\ 1111, \quad \text{and}$$
$$y = 1011\ 1100\ 1101\ 0110$$

is the ciphertext.   ⬜

SPNs have several attractive features. First, the design is simple and very efficient, in both hardware and software. In software, an S-box is usually implemented in the form of a look-up table. Observe that the memory requirement of the S-box $\pi_S : \{0,1\}^\ell \to \{0,1\}^\ell$ is $\ell 2^\ell$ bits, since we have to store 2^ℓ values, each of which needs ℓ bits of storage. Hardware implementations, in particular, necessitate the use of relatively small S-boxes.

In Example 3.1, we used four identical S-boxes in each round. The memory requirement of the S-box is 2^6 bits. If we instead used one S-box which mapped 16 bits to 16 bits, the memory requirement would be increased to 2^{20} bits, which

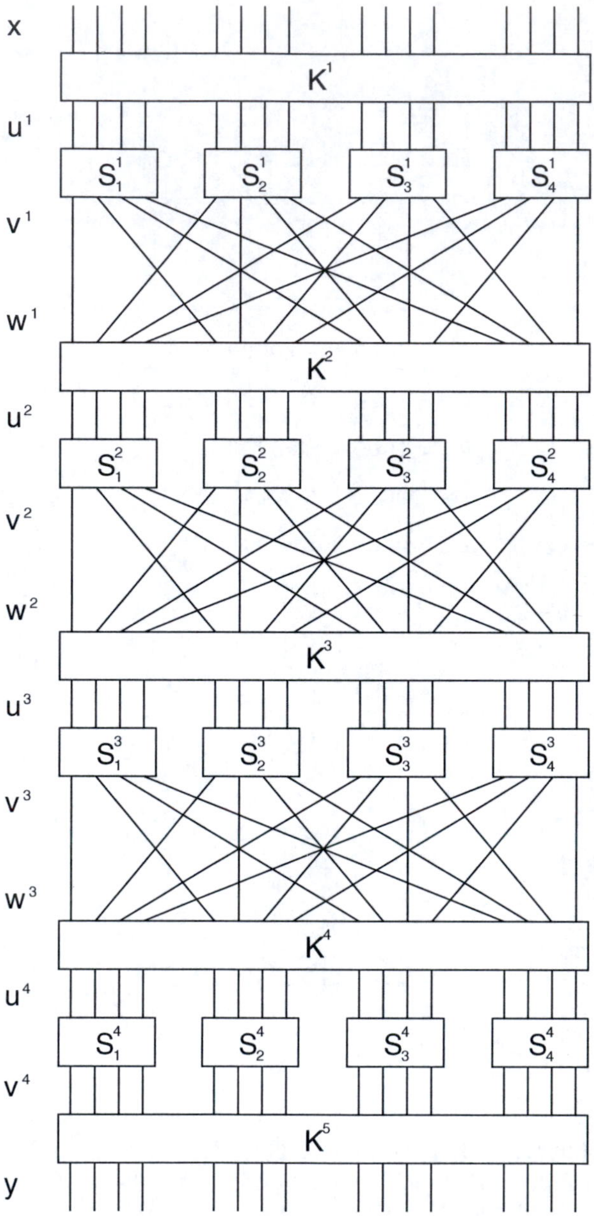

FIGURE 3.1
A substitution-permutation network

would be prohibitively high for some applications. The S-box used in the *Advanced Encryption Standard* (to be discussed in Section 3.6) maps eight bits to eight bits.

The SPN in Example 3.1 is not secure, if for no other reason than the key length (32 bits) is small enough that an exhaustive key search is feasible. However, "larger" SPNs can be designed that are secure against all known attacks. A practical, secure SPN would have a larger key size and block length than Example 3.1, would most likely use larger S-boxes, and would have more rounds. *Rijndael*, which was chosen to be the *Advanced Encryption Standard*, is an example of an SPN that is similar to Example 3.1 in many respects. *Rijndael* has a minimum key size of 128 bits, a block length of 128, a minimum of 10 rounds; and its S-box maps eight bits to eight bits (see Section 3.6 for a complete description).

Many variations of SPNs are possible. One common modification would be to use more than one S-box. In Example, 3.1, we could use four different S-boxes in each round if we so desired, instead of using the same S-box four times. This feature can be found in the *Data Encryption Standard*, which employs eight different S-boxes in each round (see Section 3.5.1). Another popular design strategy is to include an invertible linear transformation in each round, either as a replacement for, or in addition to, the permutation operation. This is done in the *Advanced Encryption Standard* (see Section 3.6.1).

3.3 Linear Cryptanalysis

We begin by informally describing the strategy behind linear cryptanalysis. The idea can be applied, in principle, to any iterated cipher. Suppose that it is possible to find a probabilistic linear relationship between a subset of plaintext bits and a subset of state bits immediately preceding the substitutions performed in the last round. In other words, there exists a subset of bits whose exclusive-or behaves in a non-random fashion (it takes on the value 0, say, with probability bounded away from $1/2$). Now assume that an attacker has a large number of plaintext-ciphertext pairs, all of which are encrypted using the same unknown key K (i.e., we consider a known-plaintext attack). For each of the plaintext-ciphertext pairs, we will begin to decrypt the ciphertext, using all possible candidate keys for the last round of the cipher. For each candidate key, we compute the values of the relevant state bits involved in the linear relationship, and determine if the above-mentioned linear relationship holds. Whenever it does, we increment a counter corresponding to the particular candidate key. At the end of this process, we hope that the candidate key that has a frequency count that is furthest from $1/2$ times the number of pairs contains the correct values for these key bits.

We will illustrate the above description with a detailed example later in this section. First, we need to establish some results from probability theory to provide a (non-rigorous) justification for the techniques involved in the attack.

3.3.1   The Piling-up Lemma

We use terminology and concepts introduced in Section 2.2. Suppose that $\mathbf{X_1}, \mathbf{X_2}, \ldots$ are independent random variables taking on values from the set $\{0, 1\}$. Let $p_1, p_2, \ldots$ be real numbers such that $0 \le p_i \le 1$ for all i, and suppose that

$$\mathbf{Pr}[\mathbf{X_i} = 0] = p_i,$$

$i = 1, 2, \ldots$. Hence,

$$\mathbf{Pr}[\mathbf{X_i} = 1] = 1 - p_i,$$

$i = 1, 2, \ldots$.

Suppose that $i \ne j$. The independence of $\mathbf{X_i}$ and $\mathbf{X_j}$ implies that

$$\mathbf{Pr}[\mathbf{X_i} = 0, \mathbf{X_j} = 0] = p_i p_j$$
$$\mathbf{Pr}[\mathbf{X_i} = 0, \mathbf{X_j} = 1] = p_i(1 - p_j)$$
$$\mathbf{Pr}[\mathbf{X_i} = 1, \mathbf{X_j} = 0] = (1 - p_i)p_j, \quad \text{and}$$
$$\mathbf{Pr}[\mathbf{X_i} = 1, \mathbf{X_j} = 1] = (1 - p_i)(1 - p_j).$$

Now consider the discrete random variable $\mathbf{X_i} \oplus \mathbf{X_j}$ (this is the same thing as $\mathbf{X_i} + \mathbf{X_j} \bmod 2$). It is easy to see that $\mathbf{X_i} \oplus \mathbf{X_j}$ has the following probability distribution:

$$\mathbf{Pr}[\mathbf{X_i} \oplus \mathbf{X_j} = 0] = p_i p_j + (1 - p_i)(1 - p_j)$$
$$\mathbf{Pr}[\mathbf{X_i} \oplus \mathbf{X_j} = 1] = p_i(1 - p_j) + (1 - p_i)p_j.$$

It is often convenient to express a probability distribution of a random variable taking on the values 0 and 1 in terms of a quantity called the bias of the distribution. The *bias* of $\mathbf{X_i}$ is defined to be the quantity

$$\epsilon_i = p_i - \frac{1}{2}.$$

Observe the following facts:

$$-\frac{1}{2} \le \epsilon_i \le \frac{1}{2},$$

$$\mathbf{Pr}[\mathbf{X_i} = 0] = \frac{1}{2} + \epsilon_i, \text{ and}$$

$$\mathbf{Pr}[\mathbf{X_i} = 1] = \frac{1}{2} - \epsilon_i,$$

for $i = 1, 2, \ldots$.

The following result, which gives a formula for the bias of the random variable $\mathbf{X_{i_1}} \oplus \cdots \oplus \mathbf{X_{i_k}}$, is known as the "piling-up lemma."

LEMMA 3.1 **(Piling-up lemma)** *Let $\epsilon_{i_1,i_2,\ldots,i_k}$ denote the bias of the random variable $\mathbf{X_{i_1}} \oplus \cdots \oplus \mathbf{X_{i_k}}$. Then*

$$\epsilon_{i_1,i_2,\ldots,i_k} = 2^{k-1} \prod_{j=1}^{k} \epsilon_{i_j}.$$

PROOF The proof is by induction on k. Clearly the result is true when $k = 1$. We next prove the result for $k = 2$, where we want to determine the bias of $\mathbf{X_{i_1}} \oplus \mathbf{X_{i_2}}$. Using the equations presented above, we have that

$$\mathbf{Pr}[\mathbf{X_{i_1}} \oplus \mathbf{X_{i_2}} = 0] = \left(\frac{1}{2} + \epsilon_{i_1}\right)\left(\frac{1}{2} + \epsilon_{i_2}\right) + \left(\frac{1}{2} - \epsilon_{i_1}\right)\left(\frac{1}{2} - \epsilon_{i_2}\right)$$

$$= \frac{1}{2} + 2\epsilon_{i_1}\epsilon_{i_2}.$$

Hence, the bias of $\mathbf{X_{i_1}} \oplus \mathbf{X_{i_2}}$ is $2\epsilon_{i_1}\epsilon_{i_2}$, as claimed.

Now, as an induction hypothesis, assume that the result is true for $k = \ell$, for some positive integer $\ell \geq 2$. We will prove that the formula is true for $k = \ell + 1$.

We want to determine the bias of $\mathbf{X_{i_1}} \oplus \cdots \oplus \mathbf{X_{i_{\ell+1}}}$. We split this random variable into two parts, as follows:

$$\mathbf{X_{i_1}} \oplus \cdots \oplus \mathbf{X_{i_{\ell+1}}} = (\mathbf{X_{i_1}} \oplus \cdots \oplus \mathbf{X_{i_\ell}}) \oplus \mathbf{X_{i_{\ell+1}}}.$$

The bias of $\mathbf{X_{i_1}} \oplus \cdots \oplus \mathbf{X_{i_\ell}}$ is $2^{\ell-1}\prod_{j=1}^{\ell} \epsilon_{i_j}$ (by induction) and the bias of $\mathbf{X_{i_{\ell+1}}}$ is $\epsilon_{i_{\ell+1}}$. Then, by induction (more specifically, using the formula for $k = 2$), the bias of $\mathbf{X_{i_1}} \oplus \cdots \oplus \mathbf{X_{i_{\ell+1}}}$ is

$$2 \times \left(2^{\ell-1} \prod_{j=1}^{\ell} \epsilon_{i_j}\right) \times \epsilon_{i_{\ell+1}} = 2^{\ell} \prod_{j=1}^{\ell+1} \epsilon_{i_j},$$

as desired. By induction, the proof is complete. ∎

COROLLARY 3.2 *Let $\epsilon_{i_1,i_2,\ldots,i_k}$ denote the bias of the random variable $\mathbf{X_{i_1}} \oplus \cdots \oplus \mathbf{X_{i_k}}$. Suppose that $\epsilon_{i_j} = 0$ for some j. Then $\epsilon_{i_1,i_2,\ldots,i_k} = 0$.*

It is important to realize that Lemma 3.1 holds, in general, only when the relevant random variables are independent. We illustrate this by considering an example. Suppose that $\epsilon_1 = \epsilon_2 = \epsilon_3 = 1/4$. Applying Lemma 3.1, we see that $\epsilon_{1,2} = \epsilon_{2,3} = \epsilon_{1,3} = 1/8$. Now, consider the random variable $\mathbf{X_1} \oplus \mathbf{X_3}$. It is clear that

$$\mathbf{X_1} \oplus \mathbf{X_3} = (\mathbf{X_1} \oplus \mathbf{X_2}) \oplus (\mathbf{X_2} \oplus \mathbf{X_3}).$$

If the two random variables $\mathbf{X_1} \oplus \mathbf{X_2}$ and $\mathbf{X_2} \oplus \mathbf{X_3}$ were independent, then Lemma 3.1 would say that $\epsilon_{1,3} = 2(1/8)^2 = 1/32$. However, we already know that this is not the case: $\epsilon_{1,3} = 1/8$. Lemma 3.1 does not yield the correct value of $\epsilon_{1,3}$ because $\mathbf{X_1} \oplus \mathbf{X_2}$ and $\mathbf{X_2} \oplus \mathbf{X_3}$ are not independent.

3.3.2   Linear Approximations of S-boxes

Consider an S-box $\pi_S : \{0,1\}^m \rightarrow \{0,1\}^n$. (We do not assume that π_S is a permutation, or even that $m = n$.) Let us write an input m-tuple as $X = (x_1, \ldots, x_m)$. This m-tuple is chosen uniformly at random from $\{0,1\}^m$, which means that each co-ordinate x_i defines a random variable $\mathbf{X_i}$ taking on values 0 and 1, having bias $\epsilon_i = 0$. Further, these m random variables are independent.

Now write an output n-tuple as $Y = (y_1, \ldots, y_n)$. Each co-ordinate y_j defines a random variable $\mathbf{Y_j}$ taking on values 0 and 1. These n random variables are, in general, not independent from each other or from the $\mathbf{X_i}$'s. In fact, it is not hard to see that the following formula holds:

$$\mathbf{Pr}[\mathbf{X_1} = x_1, \ldots, \mathbf{X_m} = x_m, \mathbf{Y_1} = y_1, \ldots, \mathbf{Y_n} = y_n] = 0$$

if $(y_1, \ldots, y_n) \neq \pi_S(x_1, \ldots, x_m)$; and

$$\mathbf{Pr}[\mathbf{X_1} = x_1, \ldots, \mathbf{X_m} = x_m, \mathbf{Y_1} = y_1, \ldots, \mathbf{Y_n} = y_n] = 2^{-m}$$

if $(y_1, \ldots, y_n) = \pi_S(x_1, \ldots, x_m)$. (The last formula holds because

$$\mathbf{Pr}[\mathbf{X_1} = x_1, \ldots, \mathbf{X_m} = x_m] = 2^{-m}$$

and

$$\mathbf{Pr}[\mathbf{Y_1} = y_1, \ldots, \mathbf{Y_n} = y_n | \mathbf{X_1} = x_1, \ldots, \mathbf{X_m} = x_m] = 1$$

if $(y_1, \ldots, y_n) = \pi_S(x_1, \ldots, x_m)$.)

It is now relatively straightforward to compute the bias of a random variable of the form

$$\mathbf{X_{i_1}} \oplus \cdots \oplus \mathbf{X_{i_\kappa}} \oplus \mathbf{Y_{j_1}} \oplus \cdots \oplus \mathbf{Y_{j_\ell}}$$

using the formulas stated above. (A linear cryptanalytic attack can potentially be mounted when a random variable of this form has a bias that is bounded away from zero.)

Let's consider a small example.

Example 3.2   We use the S-box from Example 3.1, which is defined by a permutation $\pi_S : \{0,1\}^4 \rightarrow \{0,1\}^4$. We record the possible values taken on by the eight random variables $\mathbf{X_1}, \ldots, \mathbf{X_4}, \mathbf{Y_1}, \ldots, \mathbf{Y_4}$ in the rows of Table 3.1.

Now, consider the random variable $\mathbf{X_1} \oplus \mathbf{X_4} \oplus \mathbf{Y_2}$. The probability that this random variable takes on the value 0 can be determined by counting the number of rows in the above table in which $\mathbf{X_1} \oplus \mathbf{X_4} \oplus \mathbf{Y_2} = 0$, and then dividing by 16 ($16 = 2^4$ is the total number of rows in the table). It is seen that

$$\mathbf{Pr}[\mathbf{X_1} \oplus \mathbf{X_4} \oplus \mathbf{Y_2} = 0] = \frac{1}{2}$$

(and therefore

$$\mathbf{Pr}[\mathbf{X_1} \oplus \mathbf{X_4} \oplus \mathbf{Y_2} = 1] = \frac{1}{2},$$

as well.) Hence, the bias of this random variable is 0.   □

TABLE 3.1
Random variables defined by an S-box

X_1	X_2	X_3	X_4	Y_1	Y_2	Y_3	Y_4
0	0	0	0	1	1	1	0
0	0	0	1	0	1	0	0
0	0	1	0	1	1	0	1
0	0	1	1	0	0	0	1
0	1	0	0	0	0	1	0
0	1	0	1	1	1	1	1
0	1	1	0	1	0	1	1
0	1	1	1	1	0	0	0
1	0	0	0	0	0	1	1
1	0	0	1	1	0	1	0
1	0	1	0	0	1	1	0
1	0	1	1	1	1	0	0
1	1	0	0	0	1	0	1
1	1	0	1	1	0	0	1
1	1	1	0	0	0	0	0
1	1	1	1	0	1	1	1

If we instead analyzed the random variable $X_3 \oplus X_4 \oplus Y_1 \oplus Y_4$, we would find that the bias is $-3/8$. (We suggest that the reader verify this computation.) Indeed, it is not difficult to compute the biases of all $2^8 = 256$ possible random variables of this form.

We record this information using the following notation. We represent each of the relevant random variables in the form

$$\left(\bigoplus_{i=1}^{4} a_i X_i \right) \oplus \left(\bigoplus_{i=1}^{4} b_i Y_i \right),$$

where $a_i \in \{0, 1\}$, $b_i \in \{0, 1\}$, $i = 1, 2, 3, 4$. Then, in order to have a compact notation, we treat each of the binary vectors (a_1, a_2, a_3, a_4) and (b_1, b_2, b_3, b_4) as a hexadecimal digit (these are called the *input sum* and *output sum*, respectively). In this way, each of the 256 random variables is named by a (unique) pair of hexadecimal digits, representing the input and output sum.

As an example, consider the random variable $X_1 \oplus X_4 \oplus Y_2$. The input sum is $(1, 0, 0, 1)$, which is 9 in hexadecimal; the output sum is $(0, 1, 0, 0)$, which is 4 in hexadecimal.

For a random variable having (hexadecimal) input sum a and output sum b (where $a = (a_1, a_2, a_3, a_4)$ and $b = (b_1, b_2, b_3, b_4)$, in binary), let $N_L(a, b)$ denote the number of binary eight-tuples $(x_1, x_2, x_3, x_4, y_1, y_2, y_3, y_4)$ such that

$$(y_1, y_2, y_3, y_4) = \pi_S(x_1, x_2, x_3, x_4)$$

a	0	1	2	3	4	5	6	7	8	9	A	B	C	D	E	F
0	16	8	8	8	8	8	8	8	8	8	8	8	8	8	8	8
1	8	8	6	6	8	8	6	14	10	10	8	8	10	10	8	8
2	8	8	6	6	8	8	6	6	8	8	10	10	8	8	2	10
3	8	8	8	8	8	8	8	8	10	2	6	6	10	10	6	6
4	8	10	8	6	6	4	6	8	8	6	8	10	10	4	10	8
5	8	6	6	8	6	8	12	10	6	8	4	10	8	6	6	8
6	8	10	6	12	10	8	8	10	8	6	10	12	6	8	8	6
7	8	6	8	10	10	4	10	8	6	8	10	8	12	10	8	10
8	8	8	8	8	8	8	8	8	6	10	10	6	10	6	6	2
9	8	8	6	6	8	8	6	6	4	8	6	10	8	12	10	6
A	8	12	6	10	4	8	10	6	10	10	8	8	10	10	8	8
B	8	12	8	4	12	8	12	8	8	8	8	8	8	8	8	8
C	8	6	12	6	6	8	10	8	10	8	10	12	8	10	8	6
D	8	10	10	8	6	12	8	10	4	6	10	8	10	8	8	10
E	8	10	10	8	6	4	8	10	6	8	8	6	4	10	6	8
F	8	6	4	6	6	8	10	8	8	6	12	6	6	8	10	8

FIGURE 3.2
Linear approximation table: values of $N_L(a, b)$

and

$$\left(\bigoplus_{i=1}^{4} a_i x_i \right) \oplus \left(\bigoplus_{i=1}^{4} b_i y_i \right) = 0.$$

The bias of the random variable having input sum a and output sum b is computed as $\epsilon(a, b) = (N_L(a, b) - 8)/16$.

We computed $N_L(9, 4) = 8$, and hence $\epsilon(9, 4) = 0$, in Example 3.2. The table of all values N_L is called the *linear approximation table*; see Figure 3.2.

3.3.3   A Linear Attack on an SPN

Linear cryptanalysis requires finding a set of linear approximations of S-boxes that can be used to derive a linear approximation of the entire SPN (excluding the last round). We will illustrate the procedure using the SPN from Example 3.1. The diagram in Figure 3.3 illustrates the structure of the approximation we will use. This diagram can be interpreted as follows: Lines with arrows correspond to random variables which will be involved in linear approximations. The labeled S-boxes are the ones used in these approximations (they are called the *active S-boxes* in the approximation).

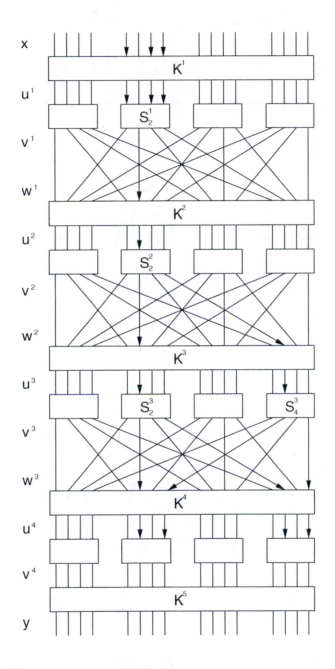

FIGURE 3.3
A linear approximation of a substitution-permutation network

This approximation incorporates four active S-boxes:

- In S_2^1, the random variable $\mathbf{T_1} = \mathbf{U_5^1} \oplus \mathbf{U_7^1} \oplus \mathbf{U_8^1} \oplus \mathbf{V_6^1}$ has bias $1/4$
- In S_2^2, the random variable $\mathbf{T_2} = \mathbf{U_6^2} \oplus \mathbf{V_6^2} \oplus \mathbf{V_8^2}$ has bias $-1/4$
- In S_2^3, the random variable $\mathbf{T_3} = \mathbf{U_6^3} \oplus \mathbf{V_6^3} \oplus \mathbf{V_8^3}$ has bias $-1/4$
- In S_4^3, the random variable $\mathbf{T_4} = \mathbf{U_{14}^3} \oplus \mathbf{V_{14}^3} \oplus \mathbf{V_{16}^3}$ has bias $-1/4$

The four random variables $\mathbf{T_1}, \mathbf{T_2}, \mathbf{T_3}, \mathbf{T_4}$ have biases that are high in absolute value. Further, we will see that their exclusive-or will lead to cancellations of "intermediate" random variables.

If we make the assumption that these four random variables are independent, then we can compute the bias of their x-or using the piling-up lemma (Lemma 3.1). (The random variables are in fact not independent, which means that we cannot provide a mathematical justification of this approximation. Nevertheless, the approximation seems to work in practice, as we shall demonstrate.) We therefore hypothesize that the random variable

$$\mathbf{T_1} \oplus \mathbf{T_2} \oplus \mathbf{T_3} \oplus \mathbf{T_4}$$

has bias equal to $2^3(1/4)(-1/4)^3 = -1/32$.

Now, the random variables $\mathbf{T_1}$, $\mathbf{T_2}$, $\mathbf{T_3}$ and $\mathbf{T_4}$ have the property that their x-or can be expressed in terms of plaintext bits, bits of u^4 (the input to the last round of S-boxes) and key bits. This can be done as follows: First, we have the following relations, which can be easily verified by inspecting Figure 3.3:

$$\mathbf{T_1} = \mathbf{U_5^1} \oplus \mathbf{U_7^1} \oplus \mathbf{U_8^1} \oplus \mathbf{V_6^1} = \mathbf{X_5} \oplus \mathbf{K_5^1} \oplus \mathbf{X_7} \oplus \mathbf{K_7^1} \oplus \mathbf{X_8} \oplus \mathbf{K_8^1} \oplus \mathbf{V_6^1}$$

$$\mathbf{T_2} = \mathbf{U_6^2} \oplus \mathbf{V_6^2} \oplus \mathbf{V_8^2} \qquad = \mathbf{V_6^1} \oplus \mathbf{K_6^2} \oplus \mathbf{V_6^2} \oplus \mathbf{V_8^2}$$

$$\mathbf{T_3} = \mathbf{U_6^3} \oplus \mathbf{V_6^3} \oplus \mathbf{V_8^3} \qquad = \mathbf{V_6^2} \oplus \mathbf{K_6^3} \oplus \mathbf{V_6^3} \oplus \mathbf{V_8^3}$$

$$\mathbf{T_4} = \mathbf{U_{14}^3} \oplus \mathbf{V_{14}^3} \oplus \mathbf{V_{16}^3} \qquad = \mathbf{V_8^2} \oplus \mathbf{K_{14}^3} \oplus \mathbf{V_{14}^3} \oplus \mathbf{V_{16}^3}.$$

If we compute the x-or of the random variables on the right sides of the above equations, we see that the random variable

$$\mathbf{X_5} \oplus \mathbf{X_7} \oplus \mathbf{X_8} \oplus \mathbf{V_6^3} \oplus \mathbf{V_8^3} \oplus \mathbf{V_{14}^3} \oplus \mathbf{V_{16}^3}$$

$$\oplus \mathbf{K_5^1} \oplus \mathbf{K_7^1} \oplus \mathbf{K_8^1} \oplus \mathbf{K_6^2} \oplus \mathbf{K_6^3} \oplus \mathbf{K_{14}^3} \quad (3.1)$$

has bias equal to $-1/32$. The next step is to replace the terms $\mathbf{V_i^3}$ in the above formula by expressions involving $\mathbf{U_i^4}$ and further key bits:

$$\mathbf{V_6^3} = \mathbf{U_6^4} \oplus \mathbf{K_6^4}$$

$$\mathbf{V_8^3} = \mathbf{U_{14}^4} \oplus \mathbf{K_{14}^4}$$

$$\mathbf{V_{14}^3} = \mathbf{U_8^4} \oplus \mathbf{K_8^4}$$

$$\mathbf{V_{16}^3} = \mathbf{U_{16}^4} \oplus \mathbf{K_{16}^4}$$

Now we substitute these four expressions into (3.1), to get the following:

$$\mathbf{X_5} \oplus \mathbf{X_7} \oplus \mathbf{X_8} \oplus \mathbf{U_6^4} \oplus \mathbf{U_8^4} \oplus \mathbf{U_{14}^4} \oplus \mathbf{U_{16}^4}$$

$$\oplus \mathbf{K_5^1} \oplus \mathbf{K_7^1} \oplus \mathbf{K_8^1} \oplus \mathbf{K_6^2} \oplus \mathbf{K_6^3} \oplus \mathbf{K_{14}^3} \oplus \mathbf{K_6^4} \oplus \mathbf{K_8^4} \oplus \mathbf{K_{14}^4} \oplus \mathbf{K_{16}^4} \quad (3.2)$$

This expression only involves plaintext bits, bits of u^4 and key bits. Suppose that the key bits in (3.2) are fixed. Then the random variable

$$\mathbf{K_5^1} \oplus \mathbf{K_7^1} \oplus \mathbf{K_8^1} \oplus \mathbf{K_6^2} \oplus \mathbf{K_6^3} \oplus \mathbf{K_{14}^3} \oplus \mathbf{K_6^4} \oplus \mathbf{K_8^4} \oplus \mathbf{K_{14}^4} \oplus \mathbf{K_{16}^4}$$

has the (fixed) value 0 or 1. It follows that the random variable

$$\mathbf{X_5} \oplus \mathbf{X_7} \oplus \mathbf{X_8} \oplus \mathbf{U_6^4} \oplus \mathbf{U_8^4} \oplus \mathbf{U_{14}^4} \oplus \mathbf{U_{16}^4} \quad (3.3)$$

has bias equal to $\pm 1/32$, where the sign of this bias depends on the values of unknown key bits. Note that the random variable (3.3) involves only plaintext bits and bits of u^4. The fact that (3.3) has bias bounded away from 0 allows us to carry out the linear attack mentioned at the beginning of Section 3.3.

Suppose that we have T plaintext-ciphertext pairs, all of which use the same unknown key, K. (It will turn out that we need $T \approx 8000$ in order for the attack to succeed.) Denote this set of T pairs by $\mathcal{T}$. The attack will allow us to obtain the eight key bits in $K_{<2>}^5$ and $K_{<4>}^5$, namely,

$$K_5^5, K_6^5, K_7^5, K_8^5, K_{13}^5, K_{14}^5, K_{15}^5, \text{ and } K_{16}^5.$$

These are the eight key bits that are exclusive-ored with the output of the S-boxes S_2^4 and S_4^4. Notice that there are $2^8 = 256$ possibilities for this list of eight key bits. We will refer to a binary 8-tuple (comprising values for these eight key bits) as a *candidate subkey*.

For each $(x, y) \in \mathcal{T}$ and for each candidate subkey, it is possible to compute a partial decryption of y and obtain the resulting value for $u_{<2>}^4$ and $u_{<4>}^4$. Then we compute the value

$$x_5 \oplus x_7 \oplus x_8 \oplus u_6^4 \oplus u_8^4 \oplus u_{14}^4 \oplus u_{16}^4 \quad (3.4)$$

taken on by the random variable (3.3). We maintain an array of counters indexed by the 256 possible candidate subkeys, and increment the counter corresponding to a particular subkey whenever (3.4) has the value 0. (This array is initialized to have all values equal to 0.)

At the end of this counting process, we expect that most counters will have a value close to $T/2$, but the counter for the correct candidate subkey will have a value that is close to $T/2 \pm T/32$. This will (hopefully) allow us to identify eight subkey bits.

The algorithm for this particular linear attack is presented as Algorithm 3.2. In this algorithm, the variables L_1 and L_2 take on hexadecimal values. The set $\mathcal{T}$ is

---

Algorithm 3.2: LINEARATTACK($\mathcal{T}, T, \pi_S{}^{-1}$)

for $(L_1, L_2) \leftarrow (0, 0)$ **to** (F, F)
  do $Count[L_1, L_2] \leftarrow 0$
for each $(x, y) \in \mathcal{T}$

do $\begin{cases} \textbf{for } (L_1, L_2) \leftarrow (0, 0) \textbf{ to } (F, F) \\ \quad \textbf{do} \begin{cases} v^4_{<2>} \leftarrow L_1 \oplus y_{<2>} \\ v^4_{<4>} \leftarrow L_2 \oplus y_{<4>} \\ u^4_{<2>} \leftarrow \pi_S{}^{-1}(v^4_{<2>}) \\ u^4_{<4>} \leftarrow \pi_S{}^{-1}(v^4_{<4>}) \\ z \leftarrow x_5 \oplus x_7 \oplus x_8 \oplus u^4_6 \oplus u^4_8 \oplus u^4_{14} \oplus u^4_{16} \\ \textbf{if } z = 0 \\ \quad \textbf{then } Count[L_1, L_2] \leftarrow Count[L_1, L_2] + 1 \end{cases} \end{cases}$

$max \leftarrow -1$
for $(L_1, L_2) \leftarrow (0, 0)$ **to** (F, F)

do $\begin{cases} Count[L_1, L_2] \leftarrow |Count[L_1, L_2] - T/2| \\ \textbf{if } Count[L_1, L_2] > max \\ \quad \textbf{then} \begin{cases} max \leftarrow Count[L_1, L_2] \\ maxkey \leftarrow (L_1, L_2) \end{cases} \end{cases}$

output ($maxkey$)

---

the set of T plaintext-ciphertext pairs used in the attack. $\pi_S{}^{-1}$ is the permutation corresponding to the inverse of the S-box; this is used to partially decrypt the ciphertexts. The output, $maxkey$, contains the "most likely" eight subkey bits identified in the attack.

Algorithm 3.2 is not very complicated. As mentioned previously, we are just computing (3.4) for every plaintext-ciphertext pair $(x, y) \in \mathcal{T}$ and for every possible candidate subkey (L_1, L_2). In order to do this, we refer to Figure 3.3. First, we compute the exclusive-ors $L_1 \oplus y_{<2>}$ and $L_2 \oplus y_{<4>}$. These yield $v^4_{<2>}$ and $v^4_{<4>}$, respectively, when (L_1, L_2) is the correct subkey. $u^4_{<2>}$ and $u^4_{<4>}$ can then be computed from $v^4_{<2>}$ and $v^4_{<4>}$ by using the inverse S-box $\pi_S{}^{-1}$; again, the values obtained are correct if (L_1, L_2) is the correct subkey. Then we compute (3.4) and we increment the counter for the pair (L_1, L_2) if (3.4) has the value 0. After having computed all the relevant counters, we just find the pair (L_1, L_2) corresponding to the maximum counter; this is the output of Algorithm 3.2.

In general, it is suggested that a linear attack based on a linear approximation having bias equal to ϵ will be successful if the number of plaintext-ciphertext pairs, which we denote by T, is approximately $c\,\epsilon^{-2}$, for some "small" constant c. We implemented the attack described in Algorithm 3.2, and found that the attack was usually successful if we took $T = 8000$. Note that $T = 8000$ corresponds to $c \approx 8$, because $\epsilon^{-2} = 1024$.

3.4   Differential Cryptanalysis

Differential cryptanalysis is similar to linear cryptanalysis in many respects. The main difference from linear cryptanalysis is that differential cryptanalysis involves comparing the x-or of two inputs to the x-or of the corresponding two outputs. In general, we will be looking at inputs x and x^* (which are assumed to be binary strings) having a specified (fixed) x-or value denoted by $x' = x \oplus x^*$. Throughout this section, we will use prime markings (') to indicate the x-or of two bitstrings.

Differential cryptanalysis is a chosen-plaintext attack. We assume that an attacker has a large number of tuples (x, x^*, y, y^*), where the x-or value $x' = x \oplus x^*$ is fixed. The plaintext elements (i.e., x and x^*) are encrypted using the same unknown key, K, yielding the ciphertexts y and y^*, respectively. For each of these tuples, we will begin to decrypt the ciphertexts y and y^*, using all possible candidate keys for the last round of the cipher. For each candidate key, we compute the values of certain state bits, and determine if their x-or has a certain value (namely, the most likely value for the given input x-or). Whenever it does, we increment a counter corresponding to the particular candidate key. At the end of this process, we hope that the candidate key that has the highest frequency count contains the correct values for these key bits. (As we did with linear cryptanalysis, we will illustrate the attack with a particular example.)

Definition 3.1:   Let $\pi_S : \{0, 1\}^m \rightarrow \{0, 1\}^n$ be an S-box. Consider an (ordered) pair of bitstrings of length m, say (x, x^*). We say that the *input x-or* of the S-box is $x \oplus x^*$ and the *output x-or* is $\pi_S(x) \oplus \pi_S(x^*)$. Note that the output x-or is a bitstring of length n.

For any $x' \in \{0, 1\}^m$, define the set $\Delta(x')$ to consist of all the ordered pairs (x, x^*) having input x-or equal to x'.

It is easy to see that any set $\Delta(x')$ contains 2^m pairs, and that

$$\Delta(x') = \{(x, x \oplus x') : x \in \{0, 1\}^m\}.$$

For each pair in $\Delta(x')$, we can compute the output x-or of the S-box. Then we can tabulate the resulting distribution of output x-ors. There are 2^m output x-ors, which are distributed among 2^n possible values. A non-uniform output distribution will be the basis for a successful differential attack.

Example 3.3   We again use the S-box from Example 3.1. Suppose we consider input x-or $x' = 1011$. Then

$$\Delta(1011) = \{(0000, 1011), (0001, 1010), \ldots, (1111, 0100)\}.$$

For each ordered pair in the set $\Delta(1011)$, we compute output x-or of π_S. In each row of the following table, we have $x \oplus x^* = 1011$, $y = \pi_S(x)$, $y^* = \pi_S(x^*)$ and $y' = y \oplus y^*$:

x	x^*	y	y^*	y'
0000	1011	1110	1100	0010
0001	1010	0100	0110	0010
0010	1001	1101	1010	0111
0011	1000	0001	0011	0010
0100	1111	0010	0111	0101
0101	1110	1111	0000	1111
0110	1101	1011	1001	0010
0111	1100	1000	0101	1101
1000	0011	0011	0001	0010
1001	0010	1010	1101	0111
1010	0001	0110	0100	0010
1011	0000	1100	1110	0010
1100	0111	0101	1000	1101
1101	0110	1001	1011	0010
1110	0101	0000	1111	1111
1111	0100	0111	0010	0101

Looking at the last column of the above table, we obtain the following distribution of output x-ors:

0000	0001	0010	0011	0100	0101	0110	0111
0	0	8	0	0	2	0	2

1000	1001	1010	1011	1100	1101	1110	1111
0	0	0	0	0	2	0	2

$\square$

In Example 3.3, only five of the 16 possible output x-ors actually occur. This particular example has a very non-uniform distribution.

We can carry out computations, as was done in Example 3.3, for any possible input x-or. It will be convenient to have some notation to describe the distributions of the output x-ors, so we state the following definition. For a bitstring x' of length m and a bitstring y' of length n, define

$$N_D(x', y') = |\{(x, x^*) \in \Delta(x') : \pi_S(x) \oplus \pi_S(x^*) = y'\}|.$$

In other words, $N_D(x', y')$ counts the number of pairs with input x-or equal to x' which also have output x-or equal to y' (for a given S-box). All the values $N_D(a', b')$ for the S-box from Example 3.1 are tabulated in Figure 3.4 (a' and b'

a'	0	1	2	3	4	5	6	7	8	9	A	B	C	D	E	F
0	16	0	0	0	0	0	0	0	0	0	0	0	0	0	0	0
1	0	0	0	2	0	0	0	2	0	2	4	0	4	2	0	0
2	0	0	0	2	0	6	2	2	0	2	0	0	0	0	2	0
3	0	0	2	0	2	0	0	0	0	4	2	0	2	0	0	4
4	0	0	0	2	0	0	6	0	0	2	0	4	2	0	0	0
5	0	4	0	0	0	2	2	0	0	0	4	0	2	0	0	2
6	0	0	0	4	0	4	0	0	0	0	0	0	2	2	2	2
7	0	0	2	2	2	0	2	0	0	2	2	0	0	0	0	4
8	0	0	0	0	0	0	2	2	0	0	0	4	0	4	2	2
9	0	2	0	0	2	0	0	4	2	0	2	2	2	0	0	0
A	0	2	2	0	0	0	0	0	6	0	0	2	0	0	4	0
B	0	0	8	0	0	2	0	2	0	0	0	0	0	2	0	2
C	0	2	0	0	2	2	2	0	0	0	0	2	0	6	0	0
D	0	4	0	0	0	0	0	4	2	0	2	0	2	0	2	0
E	0	0	2	4	2	0	0	0	6	0	0	0	0	0	2	0
F	0	2	0	0	6	0	0	0	0	4	0	2	0	0	2	0

FIGURE 3.4
Difference distribution table: values of $N_D(a', b')$

are the hexadecimal representations of the input and output x-ors, respectively).
Observe that the distribution computed in Example 3.3 corresponds to row "B"
in the table in Figure 3.4.

Recall that the input to the ith S-box in round r of the SPN from Example 3.1
is denoted $u^r_{<i>}$, and

$$u^r_{<i>} = w^{r-1}_{<i>} \oplus K^r_{<i>}.$$

An input x-or is computed as

$$u^r_{<i>} \oplus (u^r_{<i>})^* = (w^{r-1}_{<i>} \oplus K^r_{<i>}) \oplus ((w^{r-1}_{<i>})^* \oplus K^r_{<i>})$$
$$= w^{r-1}_{<i>} \oplus (w^{r-1}_{<i>})^*$$

Therefore, this input x-or does not depend on the subkey bits used in round r; it
is equal to the (permuted) output x-or of round $r - 1$. (However, the output x-or
of round r certainly does depend on the subkey bits in round r.)

Let a' denote an input x-or and let b' denote an output x-or. The pair (a', b')
is called a *differential*. Each entry in the difference distribution table gives rise
to an *x-or propagation ratio* (or more simply, a *propagation ratio*) for the corresponding differential. The propagation ratio $R_p(a', b')$ for the differential (a', b')
is defined as follows:

$$R_p(a', b') = \frac{N_D(a', b')}{2^m}.$$

$R_p(a', b')$ can be interpreted as a conditional probability:

$$\mathbf{Pr}[\text{output x-or} = b' \mid \text{input x-or} = a'] = R_p(a', b').$$

Suppose we find propagation ratios for differentials in consecutive rounds of the SPN, such that the input x-or of a differential in any round is the same as the (permuted) output x-ors of the differentials in the previous round. Then these differentials can be combined to form a *differential trail*. We make the assumption that the various propagation ratios in a differential trail are independent (an assumption which may not be mathematically valid, in fact). This assumption allows us to multiply the propagation ratios of the differentials in order to obtain the propagation ratio of the differential trail.

We illustrate this process by returning to the SPN from Example 3.1. A particular differential trail is shown in Figure 3.5. Arrows are used to highlight the "1" bits in the input and output x-ors of the differentials that are used in the differential trail.

The differential attack arising from Figure 3.5 uses the following propagation ratios of differentials, all of which can be verified from Figure 3.4:

- In S_2^1, $R_p(1011, 0010) = 1/2$
- In S_3^2, $R_p(0100, 0110) = 3/8$
- In S_2^3, $R_p(0010, 0101) = 3/8$
- In S_3^3, $R_p(0010, 0101) = 3/8$

These differentials can be combined to form a differential trail. We therefore obtain a propagation ratio for a differential trail of the first three rounds of the SPN:

$$R_p(0000\ 1011\ 0000\ 0000, 0000\ 0101\ 0101\ 0000) = \frac{1}{2} \times \left(\frac{3}{8}\right)^3 = \frac{27}{1024}.$$

In other words,

$$x' = 0000\ 1011\ 0000\ 0000 \Rightarrow (v^3)' = 0000\ 0101\ 0101\ 0000$$

with probability $27/1024$. However,

$$(v^3)' = 0000\ 0101\ 0101\ 0000 \Leftrightarrow (u^4)' = 0000\ 0110\ 0000\ 0110.$$

Hence, it follows that

$$x' = 0000\ 1011\ 0000\ 0000 \Rightarrow (u^4)' = 0000\ 0110\ 0000\ 0110$$

with probability $27/1024$. Note that $(u^4)'$ is the x-or of two inputs to the last round of S-boxes.

Now we can present an algorithm, for this particular example, based on the informal description at the beginning of this section; see Algorithm 3.3. The input

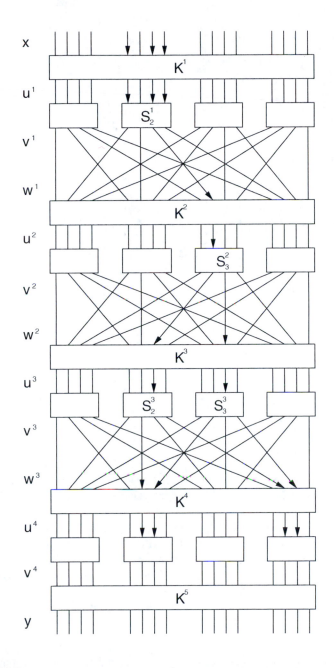

FIGURE 3.5
A differential trail for a substitution-permutation network

Algorithm 3.3: DIFFERENTIAL ATTACK$(\mathcal{T}, T, \pi_S{}^{-1})$

$\textbf{for } (L_1, L_2) \leftarrow (0,0) \textbf{ to } (F, F)$
  $\textbf{do } Count[L_1, L_2] \leftarrow 0$
$\textbf{for each } (x, y, x^*, y^*) \in \mathcal{T}$
$\textbf{do} \begin{cases} \textbf{if } (y_{<1>} = (y_{<1>})^*) \textbf{ and } (y_{<3>} = (y_{<3>})^*) \\ \textbf{then} \begin{cases} \textbf{for } (L_1, L_2) \leftarrow (0,0) \textbf{ to } (F, F) \\ \textbf{do} \begin{cases} v^4_{<2>} \leftarrow L_1 \oplus y_{<2>} \\ v^4_{<4>} \leftarrow L_2 \oplus y_{<4>} \\ u^4_{<2>} \leftarrow \pi_S{}^{-1}(v^4_{<2>}) \\ u^4_{<4>} \leftarrow \pi_S{}^{-1}(v^4_{<4>}) \\ (v^4_{<2>})^* \leftarrow L_1 \oplus (y_{<2>})^* \\ (v^4_{<4>})^* \leftarrow L_2 \oplus (y_{<4>})^* \\ (u^4_{<2>})^* \leftarrow \pi_S{}^{-1}((v^4_{<2>})^*) \\ (u^4_{<4>})^* \leftarrow \pi_S{}^{-1}((v^4_{<4>})^*) \\ (u^4_{<2>})' \leftarrow u^4_{<2>} \oplus (u^4_{<2>})^* \\ (u^4_{<4>})' \leftarrow u^4_{<4>} \oplus (u^4_{<4>})^* \\ \textbf{if } ((u^4_{<2>})' = 0110) \textbf{ and } ((u^4_{<4>})' = 0110) \\ \quad \textbf{then } Count[L_1, L_2] \leftarrow Count[L_1, L_2] + 1 \end{cases} \end{cases} \end{cases}$
$max \leftarrow -1$
$\textbf{for } (L_1, L_2) \leftarrow (0,0) \textbf{ to } (F, F)$
$\textbf{do} \begin{cases} \textbf{if } Count[L_1, L_2] > max \\ \textbf{then} \begin{cases} max \leftarrow Count[L_1, L_2] \\ maxkey \leftarrow (L_1, L_2) \end{cases} \end{cases}$
$\textbf{output } (maxkey)$

and output of this algorithm are similar to linear attack; the main difference is that $\mathcal{T}$ is a set of tuples of the form (x, x^*, y, y^*), where x' is fixed, in the differential attack.

Algorithm 3.3 makes use of a certain *filtering operation*. Tuples (x, x^*, y, y^*) for which the differential holds are often called *right pairs*, and it is the right pairs that allow us to determine the relevant key bits. (Tuples that are not right pairs basically constitute "random noise" that provides no useful information.) A right pair has

$$(u^4_{<1>})' = (u^4_{<3>})' = 0000.$$

Hence, it follows that a right pair must have $y_{<1>} = (y_{<1>})^*$ and $y_{<3>} = (y_{<3>})^*$. If a tuple (x, x^*, y, y^*) does not satisfy these conditions, then we know that it is not a right pair, and we can discard it. This filtering process increases the efficiency of the attack.

The workings of Algorithm 3.3 can be summarized as follows. For each tuple $(x, x^*, y, y^*) \in \mathcal{T}$, we first perform the filtering operation. If (x, x^*, y, y^*) is a right pair, then we test each possible candidate subkey (L_1, L_2) and increment an appropriate counter if a certain x-or is observed. The steps include computing an exclusive-or with candidate subkeys and applying the inverse S-box (as was done in Algorithm 3.2), followed by computation of the relevant xor-value.

A differential attack based on a differential trail having propagation ratio equal to ϵ will often be successful if the number of tuples (x, x^*, y, y^*), which we denote by T, is approximately $c\,\epsilon^{-1}$, for a "small" constant c. We implemented the attack described in Algorithm 3.3, and found that the attack was often successful if we took T between 50 and 100. In this example, $\epsilon^{-1} \approx 38$.

3.5   The Data Encryption Standard

On May 15, 1973, the National Bureau of Standards (now the *National Institute of Standards and Technology*, or *NIST*) published a solicitation for cryptosystems in the Federal Register. This led ultimately to the adoption of the *Data Encryption Standard*, or *DES*, which became the most widely used cryptosystem in the world. *DES* was developed at IBM, as a modification of an earlier system known as *Lucifer*. *DES* was first published in the Federal Register of March 17, 1975. After a considerable amount of public discussion, *DES* was adopted as a standard for "unclassified" applications on January 15, 1977. It was initially expected that DES would only be used as a standard for 10–15 years; however, it proved to be much more durable. *DES* was reviewed approximately every five years after its adoption. Its last renewal was in January 1999; by that time, development of a replacement, the *Advanced Encryption Standard*, had already begun (see Section 3.6).

3.5.1   Description of DES

A complete description of the *Data Encryption Standard* is given in the *Federal Information Processing Standards* (*FIPS*) Publication 46, dated January 15, 1977. *DES* is a special type of iterated cipher called a *Feistel cipher*. We describe the basic form of a Feistel cipher now, using the terminology from Section 3.1. In a Feistel cipher, each state u^i is divided into two halves of equal length, say L^i and R^i. The round function g has the following form: $g(L^{i-1}, R^{i-1}, K^i) = (L^i, R^i)$, where

$$L^i = R^{i-1}$$
$$R^i = L^{i-1} \oplus f(R^{i-1}, K^i).$$

We observe that the function f does not need to satisfy any type of injectivity property. This is because a Feistel-type round function is always invertible, given the round key:

$$L^{i-1} = R^i \oplus f(L^i, K^i)$$
$$R^{i-1} = L^i.$$

DES is a 16-round Feistel cipher having block length 64: it encrypts a plaintext bitstring x (of length 64) using a 56-bit key, K, obtaining a ciphertext bitstring (of length 64). Prior to the 16 rounds of encryption, there is a fixed *initial permutation* **IP** that is applied to the plaintext. We denote

$$\mathbf{IP}(x) = L^0 R^0.$$

After the 16 rounds of encryption, the inverse permutation $\mathbf{IP}^{-1}$ is applied to the bitstring $R^{16} L^{16}$, yielding the ciphertext y. That is,

$$y = \mathbf{IP}^{-1}(R^{16} L^{16})$$

(note that L^{16} and R^{16} are swapped before $\mathbf{IP}^{-1}$ is applied). The application of **IP** and $\mathbf{IP}^{-1}$ has no cryptographic significance, and is often ignored when the security of *DES* is discussed. One round of *DES* encryption is depicted in Figure 3.6.

Each L^i and R^i is 32 bits in length. The function

$$f : \{0, 1\}^{32} \times \{0, 1\}^{48} \to \{0, 1\}^{32}$$

takes as input a 32-bit string (the right half of the current state) and a round key. The key schedule, $(K^1, K^2, \ldots, K^{16})$, consists of 48-bit round keys that are derived from the 56-bit key, K. Each K^i is a certain permuted selection of bits from K.

The f function is shown in Figure 3.7. Basically, it consists of a substitution (using an S-box) followed by a (fixed) permutation, denoted **P**. Suppose we denote the first argument of f by A, and the second argument by J. Then, in order to compute $f(A, J)$, the following steps are executed.

1. A is "expanded" to a bitstring of length 48 according to a fixed *expansion function* **E**. $\mathbf{E}(A)$ consists of the 32 bits from A, permuted in a certain way, with 16 of the bits appearing twice.

2. Compute $\mathbf{E}(A) \oplus J$ and write the result as the concatenation of eight 6-bit strings $B = B_1 B_2 B_3 B_4 B_5 B_6 B_7 B_8$.

3. The next step uses eight S-boxes, denoted $S_1, \ldots, S_8$. Each S-box

$$S_i : \{0, 1\}^6 \to \{0, 1\}^4$$

maps six bits to four bits, and is traditionally depicted as a 4×16 array whose entries come from the integers $0, \ldots, 15$. Given a bitstring of length six, say

$$B_j = b_1 b_2 b_3 b_4 b_5 b_6,$$

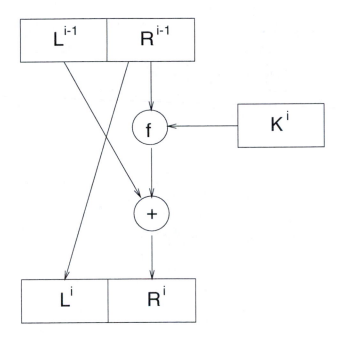

FIGURE 3.6
One round of DES encryption

we compute $S_j(B_j)$ as follows. The two bits $b_1 b_6$ determine the binary representation of a row r of S_j (where $0 \leq r \leq 3$), and the four bits $b_2 b_3 b_4 b_5$ determine the binary representation of a column c of S_j ($0 \leq c \leq 15$). Then $S_j(B_j)$ is defined to be the entry $S_j(r, c)$, written in binary as a bitstring of length four. In this fashion, we compute $C_j = S_j(B_j)$, $1 \leq j \leq 8$.

4. The bitstring

$$C = C_1 C_2 C_3 C_4 C_5 C_6 C_7 C_8$$

of length 32 is permuted according to the permutation $\mathbf{P}$. The resulting bitstring $\mathbf{P}(C)$ is defined to be $f(A, J)$.

For future reference, the eight *DES* S-boxes are now presented:

S_1															
14	4	13	1	2	15	11	8	3	10	6	12	5	9	0	7
0	15	7	4	14	2	13	1	10	6	12	11	9	5	3	8
4	1	14	8	13	6	2	11	15	12	9	7	3	10	5	0
15	12	8	2	4	9	1	7	5	11	3	14	10	0	6	13

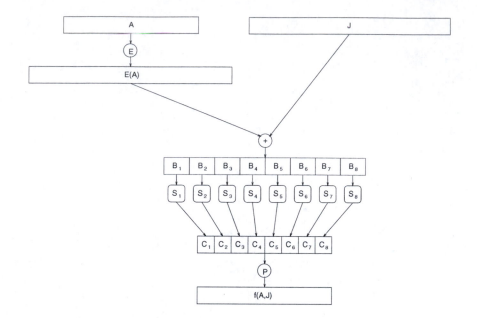

FIGURE 3.7
The DES f function

S_2															
15	1	8	14	6	11	3	4	9	7	2	13	12	0	5	10
3	13	4	7	15	2	8	14	12	0	1	10	6	9	11	5
0	14	7	11	10	4	13	1	5	8	12	6	9	3	2	15
13	8	10	1	3	15	4	2	11	6	7	12	0	5	14	9

S_3															
10	0	9	14	6	3	15	5	1	13	12	7	11	4	2	8
13	7	0	9	3	4	6	10	2	8	5	14	12	11	15	1
13	6	4	9	8	15	3	0	11	1	2	12	5	10	14	7
1	10	13	0	6	9	8	7	4	15	14	3	11	5	2	12

S_4															
7	13	14	3	0	6	9	10	1	2	8	5	11	12	4	15
13	8	11	5	6	15	0	3	4	7	2	12	1	10	14	9
10	6	9	0	12	11	7	13	15	1	3	14	5	2	8	4
3	15	0	6	10	1	13	8	9	4	5	11	12	7	2	14

						S_5									
2	12	4	1	7	10	11	6	8	5	3	15	13	0	14	9
14	11	2	12	4	7	13	1	5	0	15	10	3	9	8	6
4	2	1	11	10	13	7	8	15	9	12	5	6	3	0	14
11	8	12	7	1	14	2	13	6	15	0	9	10	4	5	3

						S_6									
12	1	10	15	9	2	6	8	0	13	3	4	14	7	5	11
10	15	4	2	7	12	9	5	6	1	13	14	0	11	3	8
9	14	15	5	2	8	12	3	7	0	4	10	1	13	11	6
4	3	2	12	9	5	15	10	11	14	1	7	6	0	8	13

						S_7									
4	11	2	14	15	0	8	13	3	12	9	7	5	10	6	1
13	0	11	7	4	9	1	10	14	3	5	12	2	15	8	6
1	4	11	13	12	3	7	14	10	15	6	8	0	5	9	2
6	11	13	8	1	4	10	7	9	5	0	15	14	2	3	12

						S_8									
13	2	8	4	6	15	11	1	10	9	3	14	5	0	12	7
1	15	13	8	10	3	7	4	12	5	6	11	0	14	9	2
7	11	4	1	9	12	14	2	0	6	10	13	15	3	5	8
2	1	14	7	4	10	8	13	15	12	9	0	3	5	6	11

Example 3.4 We show how to compute a sample output of an S-box, using the traditional presentation described above. Consider the S-box S_1, and suppose that the input is the binary 6-tuple 101000. The first and last bits are 10, which is the binary representation of the integer 2. The middle four bits are 0100, which is the binary representation of the integer 4. Row 2 of S_1 is the third row (because the rows are numbered 0, 1, 2, 3); similarly, column 4 is the fifth column. The entry in row 2, column 4 of S_1 is 13, which is 1101 in binary. Therefore 1101 is the output of the S-box S_1, given the input 101000. ⧠

The DES S-boxes are not permutations, of course, because the number of possible inputs (64) exceeds the number of possible outputs (16). However, it can be verified that each row of each of the eight S-boxes is a permutation of the integers 0, ..., 15. This property is one of several design criteria that were required of the S-boxes in order to prevent certain types of cryptanalytic attacks.

The expansion function **E** is specified by the following table:

E bit-selection table					
32	1	2	3	4	5
4	5	6	7	8	9
8	9	10	11	12	13
12	13	14	15	16	17
16	17	18	19	20	21
20	21	22	23	24	25
24	25	26	27	28	29
28	29	30	31	32	1

Given a bitstring of length 32, say $A = (a_1, a_2, \ldots, a_{32})$, $\mathbf{E}(A)$ is the following bitstring of length 48:

$$\mathbf{E}(A) = (a_{32}, a_1, a_2, a_3, a_4, a_5, a_4, \ldots, a_{31}, a_{32}, a_1).$$

The permutation **P** is as follows:

P			
16	7	20	21
29	12	28	17
1	15	23	26
5	18	31	10
2	8	24	14
32	27	3	9
19	13	30	6
22	11	4	25

Denote the bitstring $C = (c_1, c_2, \ldots, c_{32})$. Then the permuted bitstring $\mathbf{P}(C)$ is as follows:

$$\mathbf{P}(C) = (c_{16}, c_7, c_{20}, c_{21}, c_{29}, \ldots, c_{11}, c_4, c_{25}).$$

3.5.2 Analysis of DES

When *DES* was proposed as a standard, there was considerable criticism. One objection to *DES* concerned the S-boxes. All computations in *DES*, with the sole exception of the S-boxes, are linear, i.e., computing the exclusive-or of two outputs is the same as forming the exclusive-or of two inputs and then computing the output. The S-boxes, being the non-linear components of the cryptosystem, are vital to its security. (We saw in Chapter 1 how linear cryptosystems, such as the *Hill Cipher*, could easily be cryptanalyzed by a known plaintext attack.) At the time that *DES* was proposed, several people suggested that its S-boxes might contain hidden "trapdoors" which would allow the National Security Agency to easily decrypt messages while claiming falsely that *DES* is "secure." It is, of course, impossible to disprove such a speculation, but no evidence ever came to light that indicated that trapdoors in *DES* do, in fact, exist.

Actually, it was eventually revealed that the *DES* S-boxes were designed to prevent certain types of attacks. When Biham and Shamir invented the technique of differential cryptanalysis (which we discussed in Section 3.4) in the early 1990s, it was acknowledged that the purpose of certain unpublished design criteria of the S-boxes was to make differential cryptanalysis of *DES* infeasible. Differential cryptanalysis was known to IBM researchers at the time that *DES* was being developed, but it was kept secret for almost 20 years, until Biham and Shamir independently discovered the attack.

The most pertinent criticism of *DES* is that the size of the keyspace, 2^{56}, is too small to be really secure. The IBM *Lucifer* cryptosystem, a predecessor of *DES*, had a 128-bit key. The original proposal for *DES* had a 64-bit key, but this was later reduced to a 56-bit key. IBM claimed that the reason for this reduction was that it was necessary to include eight parity-check bits in the key, meaning that 64 bits of storage could only contain a 56-bit key.

Even in the 1970s, it was argued that a special-purpose machine could be built to carry out a known plaintext attack, which would essentially perform an exhaustive search for the key. That is, given a 64-bit plaintext x and corresponding ciphertext y, every possible key would be tested until a key K is found such that $e_K(x) = y$ (note that there may be more than one such key K). As early as 1977, Diffie and Hellman suggested that one could build a VLSI chip which could test 10^6 keys per second. A machine with 10^6 chips could search the entire key space in about a day. They estimated that such a machine could be built, at that time, for about $20,000,000.

Later, at the CRYPTO '93 Rump Session, Michael Wiener gave a very detailed design of a *DES* key search machine. The machine is based on a key search chip which is pipelined, so that 16 encryptions take place simultaneously. This chip would test 5×10^7 keys per second, and could have been built using 1993 technology for $10.50 per chip. A frame consisting of 5760 chips could be built for $100,000. This would allow a *DES* key to be found in about 1.5 days on average. A machine using ten frames would cost $1,000,000, but would reduce the average search time to about 3.5 hours.

Wiener's machine was never built, but a key search machine costing $250,000 was built in 1998 by the Electronic Frontier Foundation. This computer, called "DES Cracker," contained 1536 chips and could search 88 billion keys per second. It won RSA Laboratory's "DES Challenge II-2" by successfully finding a *DES* key in 56 hours in July 1998. In January 1999, RSA Laboratory's "DES Challenge III" was solved by the DES Cracker working in conjunction with a worldwide network (of 100,000 computers) known as distributed.net. This co-operative effort found a *DES* key in 22 hours, 15 minutes, testing over 245 billion keys per second.

Other than exhaustive key search, the two most important cryptanalytic attacks on *DES* are differential cryptanalysis and linear cryptanalysis. (For SPNs, these attacks were described in Sections 3.4 and 3.3, respectively.) In the case of *DES*, linear cryptanalysis is the more efficient of the two attacks, and an actual implementation of linear cryptanalysis was carried out in 1994 by its inventor, Matsui. This linear cryptanalysis of *DES* is a known-plaintext attack using 2^{43} plaintext-ciphertext pairs, all of which are encrypted using the same (unknown) key. It took 40 days to generate the 2^{43} pairs, and it took 10 days to actually find the key. This cryptanalysis did not have a practical impact on the security of *DES*, however, due to the extremely large number of plaintext-ciphertext pairs that are required to mount the attack: it is unlikely in practice that an adversary will be able to accumulate such a large number of plaintext-ciphertext pairs that are all encrypted using the same key.

3.6 The Advanced Encryption Standard

On January 2, 1997, NIST began the process of choosing a replacement for *DES*. The replacement would be called the *Advanced Encryption Standard*, or *AES*. A formal call for algorithms was made on September 12, 1997. It was required that the *AES* have a block length of 128 bits, and support key lengths of 128, 192 and 256 bits. It was also necessary that the *AES* should be available worldwide on a royalty-free basis.

Submissions were due on June 15, 1998. Of the 21 submitted cryptosystems, 15 met all the necessary criteria and were accepted as *AES* candidates. NIST announced the 15 *AES* candidates at the "First AES Candidate Conference" on August 20, 1998. A "Second AES Candidate Conference" was held in March 1999. Then, in August 1999, five of the candidates were chosen by NIST as finalists: *MARS*, *RC6*, *Rijndael*, *Serpent* and *Twofish*.

The "Third AES Candidate Conference" was held in April 2000. On October 2, 2000, *Rijndael* was selected to be the *Advanced Encryption Standard*. On February 28, 2001, NIST announced that a draft Federal Information Processing Standard for the *AES* was available for public review and comment. *AES* was adopted as a standard on November 26, 2001, and it was published as FIPS 197 in the Federal Register on December 4, 2001.

The selection process for the *AES* was notable for its openness and its international flavor. The three candidate conferences, as well as official solicitations for public comments, provided ample opportunity for feedback and public discussion and analysis of the candidates, and the process was viewed very favorably by everyone involved. The "international" aspect of *AES* is demonstrated by the variety of countries represented by the authors of the 15 candidate ciphers: Australia, Belgium, Canada, Costa Rica, France, Germany, Israel, Japan, Korea, Norway, the United Kingdom and the USA. *Rijndael*, which was ultimately selected as the *AES*, was invented by two Belgian researchers, Daemen and Rijmen.

Another interesting departure from past practice was that the "Second AES Candidate Conference" was held outside the U.S., in Rome, Italy.

AES candidates were evaluated for their suitability according to three main criteria:

- security
- cost
- algorithm and implementation characteristics

Security of the proposed algorithm was absolutely essential, and any algorithm found not to be secure would not be considered further. "Cost" refers to the computational efficiency (speed and memory requirements) of various types of implementations, including software, hardware and smart cards. Algorithm and implementation characteristics include flexibility and algorithm simplicity, among other factors.

In the end, the five finalists were all felt to be secure. *Rijndael* was selected because its combination of security, performance, efficiency, implementability and flexibility was judged to be superior to the other finalists.

3.6.1 Description of AES

As mentioned above, the *AES* has block length 128, and there are three allowable key lengths, namely 128 bits, 192 bits and 256 bits. *AES* is an iterated cipher; the number of rounds, which we denote by Nr, depends on the key length. $Nr = 10$ if the key length is 128 bits; $Nr = 12$ if the key length is 192 bits; and $Nr = 14$ if the key length is 256 bits.

We first give a high-level description of *AES*. The algorithm proceeds as follows:

1. Given a plaintext x, initialize **State** to be x and perform an operation ADD-ROUNDKEY, which x-ors the **RoundKey** with **State**.

2. For each of the first $Nr - 1$ rounds, perform a substitution operation called SUBBYTES on **State** using an S-box; perform a permutation SHIFTROWS on **State**; perform an operation MIXCOLUMNS on **State**; and perform ADD-ROUNDKEY.

3. Perform SUBBYTES; perform SHIFTROWS; and perform ADDROUND-KEY.

4. Define the ciphertext y to be **State**.

From this high-level description, we can see that the structure of the *AES* is very similar in many respects to the SPN discussed in Section 3.2. In every round of both these cryptosystems, we have round key mixing, a substitution step and a permutation step. Both ciphers also include whitening. *AES* is "larger" and it also includes an additional linear transformation (MIXCOLUMNS) in each round.

We now give precise descriptions of all the operations used in the *AES*; describe the structure of **State**; and discuss the construction of the key schedule. All operations in *AES* are byte-oriented operations, and all variables used are considered to be formed from an appropriate number of bytes. The plaintext x consists of 16 bytes, denoted $x_0, \ldots, x_{15}$. **State** is represented as a four by four array of bytes, as follows:

$s_{0,0}$	$s_{0,1}$	$s_{0,2}$	$s_{0,3}$
$s_{1,0}$	$s_{1,1}$	$s_{1,2}$	$s_{1,3}$
$s_{2,0}$	$s_{2,1}$	$s_{2,2}$	$s_{2,3}$
$s_{3,0}$	$s_{3,1}$	$s_{3,2}$	$s_{3,3}$

Initially, **State** is defined to consist of the 16 bytes of the plaintext x, as follows:

$s_{0,0}$	$s_{0,1}$	$s_{0,2}$	$s_{0,3}$		x_0	x_4	x_8	x_{12}
$s_{1,0}$	$s_{1,1}$	$s_{1,2}$	$s_{1,3}$	$\leftarrow$	x_1	x_5	x_9	x_{13}
$s_{2,0}$	$s_{2,1}$	$s_{2,2}$	$s_{2,3}$		x_2	x_6	x_{10}	x_{14}
$s_{3,0}$	$s_{3,1}$	$s_{3,2}$	$s_{3,3}$		x_3	x_7	x_{11}	x_{15}

We will often use hexadecimal notation to represent the contents of a byte. Each byte therefore consists of two hexadecimal digits.

The operation SUBBYTES performs a substitution on each byte of **State** independently, using an S-box, say π_S, which is a permutation of $\{0, 1\}^8$. To present this π_S, we represent bytes in hexadecimal notation. π_S is depicted as a 16 by 16 array, where the rows and columns are indexed by hexadecimal digits. The entry in row X and column Y is $\pi_S(XY)$. The array representation of π_S is presented in Figure 3.8.

In contrast to the S-boxes in *DES*, which are apparently "random" substitutions, the *AES* S-box can be defined algebraically. The algebraic formulation of the *AES* S-box involves operations in a finite field (finite fields are discussed in detail in Section 6.4). We include the following description for the benefit of readers who are already familiar with finite fields (other readers may want to skip this description, or read Section 6.4 first): The permutation π_S incorporates operations in the finite field

$$\mathbb{F}_{2^8} = \mathbb{Z}_2[x]/(x^8 + x^4 + x^3 + x + 1).$$

Let FIELDINV denote the multiplicative inverse of a field element; let BINARY-TOFIELD convert a byte to a field element; and let FIELDTOBINARY perform the inverse conversion. This conversion is done in the obvious way: the field element

$$\sum_{i=0}^{7} a_i x^i$$

corresponds to the byte

$$a_7 a_6 a_5 a_4 a_3 a_2 a_1 a_0,$$

Algorithm 3.4: SUBBYTES($a_7a_6a_5a_4a_3a_2a_1a_0$)

external FIELDINV, BINARYTOFIELD, FIELDTOBINARY
$z \leftarrow$ BINARYTOFIELD($a_7a_6a_5a_4a_3a_2a_1a_0$)
if $z \neq 0$
  then $z \leftarrow$ FIELDINV(z)
($a_7a_6a_5a_4a_3a_2a_1a_0$) $\leftarrow$ FIELDTOBINARY(z)
($c_7c_6c_5c_4c_3c_2c_1c_0$) $\leftarrow$ (01100011)
comment: In the following loop, all subscripts are to be reduced modulo 8

for $i \leftarrow 0$ **to** 7
  do $b_i \leftarrow (a_i + a_{i+4} + a_{i+5} + a_{i+6} + a_{i+7} + c_i)$ mod 2
return ($b_7b_6b_5b_4b_3b_2b_1b_0$)

---

where $a_i \in \mathbb{Z}_2$ for $0 \leq i \leq 7$. Then the permutation π_S is defined according to Algorithm 3.4. In this algorithm, the eight input bits $a_7a_6a_5a_4a_3a_2a_1a_0$ are replaced by the eight output bits $b_7b_6b_5b_4b_3b_2b_1b_0$.

Example 3.5 We do a small example to illustrate Algorithm 3.4, where we also include the conversions to hexadecimal. Suppose we begin with (hexadecimal) 53. In binary, this is

$$01010011,$$

which represents the field element

$$x^6 + x^4 + x + 1.$$

The multiplicative inverse (in the field $\mathbb{F}_{2^8}$) can be shown to be

$$x^7 + x^6 + x^3 + x.$$

Therefore, in binary notation, we have

$$(a_7a_6a_5a_4a_3a_2a_1a_0) = (11001010).$$

Next, we compute

$$b_0 = a_0 + a_4 + a_5 + a_6 + a_7 + c_0 \bmod 2$$
$$= 0 + 0 + 0 + 1 + 1 + 1 \bmod 2$$
$$= 1$$
$$b_1 = a_1 + a_5 + a_6 + a_7 + a_0 + c_1 \bmod 2$$
$$= 1 + 0 + 1 + 1 + 0 + 1 \bmod 2$$
$$= 0,$$

X \ Y	0	1	2	3	4	5	6	7	8	9	A	B	C	D	E	F
0	63	7C	77	7B	F2	6B	6F	C5	30	01	67	2B	FE	D7	AB	76
1	CA	82	C9	7D	FA	59	47	F0	AD	D4	A2	AF	9C	A4	72	C0
2	B7	FD	93	26	36	3F	F7	CC	34	A5	E5	F1	71	D8	31	15
3	04	C7	23	C3	18	96	05	9A	07	12	80	E2	EB	27	B2	75
4	09	83	2C	1A	1B	6E	5A	A0	52	3B	D6	B3	29	E3	2F	84
5	53	D1	00	ED	20	FC	B1	5B	6A	CB	BE	39	4A	4C	58	CF
6	D0	EF	AA	FB	43	4D	33	85	45	F9	02	7F	50	3C	9F	A8
7	51	A3	40	8F	92	9D	38	F5	BC	B6	DA	21	10	FF	F3	D2
8	CD	0C	13	EC	5F	97	44	17	C4	A7	7E	3D	64	5D	19	73
9	60	81	4F	DC	22	2A	90	88	46	EE	B8	14	DE	5E	0B	DB
A	E0	32	3A	0A	49	06	24	5C	C2	D3	AC	62	91	95	E4	79
B	E7	C8	37	6D	8D	D5	4E	A9	6C	56	F4	EA	65	7A	AE	08
C	BA	78	25	2E	1C	A6	B4	C6	E8	DD	74	1F	4B	BD	8B	8A
D	70	3E	B5	66	48	03	F6	0E	61	35	57	B9	86	C1	1D	9E
E	E1	F8	98	11	69	D9	8E	94	9B	1E	87	E9	CE	55	28	DF
F	8C	A1	89	0D	BF	E6	42	68	41	99	2D	0F	B0	54	BB	16

FIGURE 3.8
The AES S-box

etc. The result is that

$$(b_7 b_6 b_5 b_4 b_3 b_2 b_1 b_0) = (11101101).$$

In hexadecimal notation, 11101101 is ED.

This computation can be checked by verifying that the entry in row 5 and column 3 of Figure 3.8 is ED. ∎

The operation SHIFTROWS acts on **State** as shown in the following diagram:

$s_{0,0}$	$s_{0,1}$	$s_{0,2}$	$s_{0,3}$
$s_{1,0}$	$s_{1,1}$	$s_{1,2}$	$s_{1,3}$
$s_{2,0}$	$s_{2,1}$	$s_{2,2}$	$s_{2,3}$
$s_{3,0}$	$s_{3,1}$	$s_{3,2}$	$s_{3,3}$

$\leftarrow$

$s_{0,0}$	$s_{0,1}$	$s_{0,2}$	$s_{0,3}$
$s_{1,1}$	$s_{1,2}$	$s_{1,3}$	$s_{1,0}$
$s_{2,2}$	$s_{2,3}$	$s_{2,0}$	$s_{2,1}$
$s_{3,3}$	$s_{3,0}$	$s_{3,1}$	$s_{3,2}$

The operation MIXCOLUMNS is carried out on each of the four columns of **State**; it is presented as Algorithm 3.5. Each column of **State** is replaced by a new column which is formed by multiplying that column by a certain matrix of elements of the field $\mathbb{F}_{2^8}$. Here, "multiplication" means multiplication in the field $\mathbb{F}_{2^8}$. We assume that the external procedure FIELDMULT takes as input two field elements, and computes their product in the field. In Algorithm 3.5, we are

---

Algorithm 3.5: MixColumn(c)

external FieldMult, BinaryToField, FieldToBinary
for $i \leftarrow 0$ **to** 3
  do $t_i \leftarrow$ BinaryToField($s_{i,c}$)
$u_0 \leftarrow$ FieldMult(x, t_0) $\oplus$ FieldMult($x + 1, t_1$) $\oplus t_2 \oplus t_3$
$u_1 \leftarrow$ FieldMult(x, t_1) $\oplus$ FieldMult($x + 1, t_2$) $\oplus t_3 \oplus t_0$
$u_2 \leftarrow$ FieldMult(x, t_2) $\oplus$ FieldMult($x + 1, t_3$) $\oplus t_0 \oplus t_1$
$u_3 \leftarrow$ FieldMult(x, t_3) $\oplus$ FieldMult($x + 1, t_0$) $\oplus t_1 \oplus t_2$
for $i \leftarrow 0$ **to** 3
  do $s_{i,c} \leftarrow$ FieldToBinary(u_i)

---

multiplying by the field elements x and $x + 1$; these correspond to the bitstrings 00000010 and 00000011, respectively.

Field addition is just componentwise addition modulo 2 (i.e., the x-or of the corresponding bitstrings). This operation is denoted by "$\oplus$" in Algorithm 3.5.

It remains to discuss the key schedule for the *AES*. We describe how to construct the key schedule for the 10-round version of *AES*, which uses a 128-bit key (key schedules for 12- and 14-round versions are similar to 10-round *AES*, but there are some minor differences in the key scheduling algorithm). We need 11 round keys, each of which consists of 16 bytes. The key scheduling algorithm is word-oriented (a *word* consists of 4 bytes, or, equivalently, 32 bits). Therefore each round key is comprised of four words. The concatenation of the round keys is called the *expanded key*, which consists of 44 words. It is denoted $w[0], \ldots, w[43]$, where each $w[i]$ is a word. The expanded key is constructed using the operation KeyExpansion, which is presented as Algorithm 3.6.

The input to this algorithm is the 128-bit key, *key*, which is treated as an array of bytes, $key[0], \ldots, key[15]$; and the output is the array of words, w, that was introduced above.

KeyExpansion incorporates two other operations, which are named RotWord and SubWord. RotWord(B_0, B_1, B_2, B_3) performs a cyclic shift of the four bytes B_0, B_1, B_2, B_3, i.e.,

$$\text{RotWord}(B_0, B_1, B_2, B_3) = (B_1, B_2, B_3, B_0).$$

SubWord(B_0, B_1, B_2, B_3) applies the *AES* S-box to each of the four bytes B_0, B_1, B_2, B_3, i.e.,

$$\text{SubWord}(B_0, B_1, B_2, B_3) = (B_0', B_1', B_2', B_3')$$

where $B_i' = $ SubBytes(B_i), $i = 0, 1, 2, 3$. *RCon* is an array of 10 words, denoted $RCon[1], \ldots, RCon[10]$. These are constants that are defined in hexadecimal notation at the beginning of Algorithm 3.6.

Algorithm 3.6: KEYEXPANSION(key)

external ROTWORD, SUBWORD
$RCon[1] \leftarrow$ 01000000
$RCon[2] \leftarrow$ 02000000
$RCon[3] \leftarrow$ 04000000
$RCon[4] \leftarrow$ 08000000
$RCon[5] \leftarrow$ 10000000
$RCon[6] \leftarrow$ 20000000
$RCon[7] \leftarrow$ 40000000
$RCon[8] \leftarrow$ 80000000
$RCon[9] \leftarrow$ 1B000000
$RCon[10] \leftarrow$ 36000000
for $i \leftarrow 0$ **to** 3
  do $w[i] \leftarrow (key[4i], key[4i+1], key[4i+2], key[4i+3])$
for $i \leftarrow 4$ **to** 43
  do $\begin{cases} temp \leftarrow w[i-1] \\ \textbf{if } i \equiv 0 \pmod 4 \\ \quad \textbf{then } temp \leftarrow \text{SUBWORD}(\text{ROTWORD}(temp)) \oplus RCon[i/4] \\ w[i] \leftarrow w[i-4] \oplus temp \end{cases}$
return $(w[0], \ldots, w[43])$

We have now described all the operations required to perform an encryption operation in the *AES*. In order to decrypt, it is necessary to perform all operations in the reverse order, and use the key schedule in reverse order. Further the operations SHIFTROWS, SUBBYTES and MIXCOLUMNS must be replaced by their inverse operations (the operation ADDROUNDKEY is its own inverse). It is also possible to construct an "equivalent inverse cipher" which performs *AES* decryption by doing a sequence of (inverse) operations in the same order as is done for *AES* encryption. It is suggested that this can lead to implementation efficiencies.

3.6.2 Analysis of AES

Obviously, the *AES* is secure against all known attacks. Various aspects of its design incorporate specific features that help provide security against specific attacks. For example, the use of the finite field inversion operation in the construction of the S-box yields linear approximation and difference distribution tables in which the entries are close to uniform. This provides security against differential and linear attacks. As well, the linear transformation, MIXCOLUMNS, makes it impossible to find differential and linear attacks that involve "few" active S-boxes (the designers refer to this feature as the *wide trail strategy*). There are apparently no known attacks on *AES* that are faster than exhaustive search. The "best" attacks on *AES* apply to variants of the cipher in which the number of rounds is reduced, and are not effective for 10-round *AES*.

3.7 Modes of Operation

Four *modes of operation* were developed for *DES*. They were standardized in FIPS Publication 81 in December 1980. These modes of operation can be used (with minor changes) for any block cipher. More recently, some additional modes of operation have been proposed for *AES*. The following six modes of operation are either approved, or under consideration as standards for *AES*. (The first four of these modes of operation are the ones that were originally adopted for *DES*.)

- *electronic codebook mode* (ECB mode),
- *cipher feedback mode* (CFB mode),
- *cipher block chaining mode* (CBC mode),
- *output feedback mode* (OFB mode),
- *counter mode*, and
- *counter with cipher-block chaining mode* (CCM mode).

Here are short descriptions of these modes of operation

ECB mode

This mode corresponds to the naive use of a block cipher: given a sequence $x_1 x_2 \cdots$ of plaintext blocks, each x_i is encrypted with the same key K, producing a string of ciphertext blocks, $y_1 y_2 \cdots$.

One obvious weakness of ECB mode is that the encryption of identical plaintext blocks yields identical ciphertext blocks. This is a serious weakness if the underlying message blocks are chosen from a "low entropy" plaintext space. To take an extreme example, if a plaintext block always consists entirely of 0's or entirely of 1's, then ECB mode is essentially useless.

CBC mode

In CBC mode, each ciphertext block y_i is x-ored with the next plaintext block, x_{i+1}, before being encrypted with the key K. More formally, we start with an *initialization vector*, denoted by IV, and define $y_0 = \text{IV}$. (Note that IV has the same length as a plaintext block.) Then we construct $y_1, y_2, \ldots$, using the rule

$$y_i = e_K(y_{i-1} \oplus x_i),$$

$i \geq 1$.

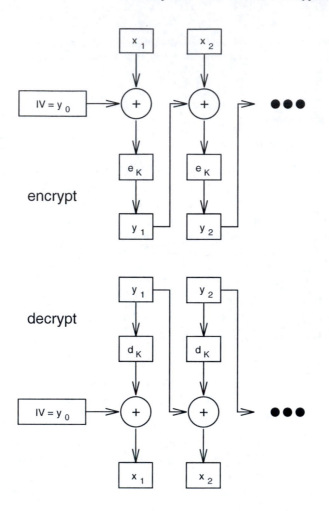

FIGURE 3.9
CBC mode

Encryption and decryption using CBC mode is depicted in Figure 3.9. Observe that, if a plaintext block x_i is changed in CBC mode, then y_i and all subsequent ciphertext blocks will be affected. This property means that CBC mode is useful for purposes of authentication. More specifically, this mode can be used to produce a *message authentication code*, or MAC. The MAC is appended to a sequence of plaintext blocks, and is used to convince Bob that the given sequence of plaintext originated with Alice and was not tampered with by Oscar. Thus the MAC guarantees the integrity (or authenticity) of a message (but it does not provide secrecy, of course). We will say much more about MACs in Chapter 4. The use of CBC modes to construct

MACs is studied further in Section 4.4.2.

OFB mode

In OFB mode, a keystream is generated, which is then x-ored with the plaintext (i.e., it operates as a stream cipher, cf. Section 1.1.7). OFB mode is actually a synchronous stream cipher: the keystream is produced by repeatedly encrypting an initialization vector, IV. We define $z_0 = $ IV, and then compute the keystream $z_1 z_2 \cdots$ using the rule

$$z_i = e_K(z_{i-1}),$$

for all $i \geq 1$. The plaintext sequence $x_1 x_2 \cdots$ is then encrypted by computing

$$y_i = x_i \oplus z_i,$$

for all $i \geq 1$.

Decryption is straightforward. First, recompute the keystream $z_1 z_2 \cdots$, and then compute

$$x_i = y_i \oplus z_i,$$

for all $i \geq 1$. Note that the encryption function e_K is used for both encryption and decryption in OFB mode.

CFB mode

CFB mode also generates a keystream for use in a synchronous stream cipher. We start with $y_0 = $ IV (an initialization vector) and we produce the keystream element z_i by encrypting the previous ciphertext block. That is,

$$z_i = e_K(y_{i-1}),$$

for all $i \geq 1$. As in OFB mode, we encrypt using the formula

$$y_i = x_i \oplus z_i,$$

for all $i \geq 1$. Again, the encryption function e_K is used for both encryption and decryption in CFB mode.

The use of CFB mode is depicted in Figure 3.10.

counter mode

Counter mode is similar to OFB mode; the only difference is in how the keystream is consructed. Suppose that the length of a plaintext block is denoted by m. In counter mode, we choose a *counter*, denoted ctr, which is a bitstring of length m. Then we construct a sequence of bitstrings of length m, denoted $T_1, T_2, \ldots$, defined as follows:

$$T_i = ctr + i - 1 \bmod 2^m$$

for all $i \geq 1$. Then we encrypt the plaintext blocks $x_1, x_2, \ldots$ by computing

$$y_i = x_i \oplus e_K(T_i),$$

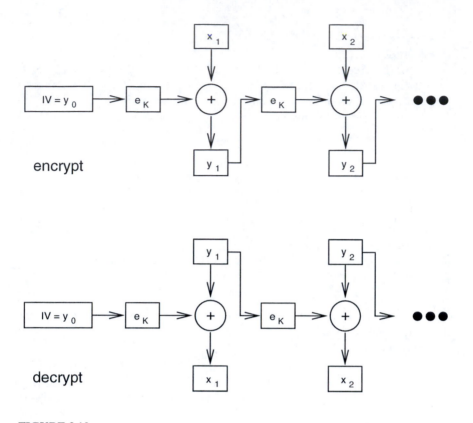

FIGURE 3.10
CFB mode

for all $i \geq 1$. Observe that the keystream in counter mode is obtained by encrypting the sequence of counters using the key K.

As in the case of OFB mode, the keystream in counter mode can be constructed independently of the plaintext. However, in counter mode, there is no need to iteratively compute a sequence of encryptions; each keystream element $e_K(T_i)$ can be computed independently of any other keystream element. (In contrast, OFB mode requires one to compute z_{i-1} prior to computing z_i.) This feature of counter mode permits very efficient implementations in software or hardware by exploiting opportunities for parallelism.

CCM mode

Basically, CCM mode combines the use of counter mode (for encryption) with CBC-mode (for authentication). See Section 4.4.2 for more information.

3.8 Notes and References

A nice article on the history of *DES* was written by Smid and Branstad [310]. A description of *Lucifer* can be found in [130]. Coppersmith [95] discusses aspects of the design of DES that are pertinent to its security against certain types of attacks. Wiener's *DES* key search machine was described at CRYPTO '93 [344]. Landau has written useful articles on *DES* [212] and *AES* [213, 214]. Knudsen [193] is a good recent survey on block ciphers.

Federal Information Processing Standards (FIPS) publications concerning *DES* include the following: description of *DES* [136]; implementing and using *DES* [138]; modes of operation of *DES* [137]; and authentication using *DES* [139].

A time-memory trade-off for *DES* was discovered by Hellman [173]. We present a related method in the Exercises.

A description of the *AES* can be found in FIPS publication 197 [146]. The counter and CCM modes of operation are presented in the NIST special publications 800-38A [121] and 800-38C [123], respectively. Nechvatal *et al.* [250] is a detailed report on the development of the *AES*. *Rijndael* is described in [100]; its predecessor, known as *Square*, is presented in [99]. Daemen and Rijmen have also written a monograph [101] explaining *Rijndael* and the design strategies they incorporated into its design. Information on attacks on reduced-round variants of *Rijndael* can be found in Ferguson *et al.* [131].

There have been some criticisms that the algebraic structure of the S-boxes in *Rijndael* might lead to possible attacks. For example, Ferguson, Schroeppel and Whiting [133] suggested that a certain relatively simple algebraic representation of *Rijndael* could possibly be exploited by a cryptanalyst. See also the papers by Murphy and Robshaw [247] and Courtois and Pieprzyk [98], which discuss approaches for potential algebraic attacks.

The technique of differential cryptanalysis was developed by Biham and Shamir [34] (see also [36] and their book on the differential cryptanalysis of *DES* [35]). Linear cryptanalysis was developed by Matsui [226, 227]. Some works developing the theoretical basis for these attacks are Lai, Massey and Murphy [209] and Nyberg [255]. The results of recent experiments on the effectiveness of linear cryptanalysis of *DES* can be found in Junod [184].

Our treatment of differential and linear cryptanalysis is based on the excellent tutorial by Heys [175]; we have also used the differential and linear attacks on SPNs that are described in [175]. General design principles for SPNs that are resistant to linear and differential cryptanalysis are presented by Heys and Tavares [176]. Recent results on the security of *Rijndael* against linear cryptanalysis have been given by Keliher, Meijer and Tavares [188, 189].

Nyberg [254] suggested the use of field inverses to define S-boxes (the technique which was later used in *Rijndael*). Chabaud and Vaudenay [86] also study the design of S-boxes that are resistant to differential and linear cryptanalysis.

_____

Exercises

3.1 Let y be the output of Algorithm 3.1 on input x, where π_S and π_P are defined as in Example 3.1. In other words,

$$y = \text{SPN}\left(x, \pi_S, \pi_P, (K^1, \ldots, K^{\text{Nr}+1})\right),$$

where $(K^1, \ldots, K^{\text{Nr}+1})$ is the key schedule. Find a substitution π_{S^*} and a permutation π_{P^*} such that

$$x = \text{SPN}\left(y, \pi_{S^*}, \pi_{P^*}, (L^{\text{Nr}+1}, \ldots, L^1)\right),$$

where each L^i is a permutation of K^i.

3.2 Prove that decryption in a Feistel cipher can be done by applying the encryption algorithm to the ciphertext, with the key schedule reversed.

3.3 Let $DES(x, K)$ represent the encryption of plaintext x with key K using the *DES* cryptosystem. Suppose $y = DES(x, K)$ and $y' = DES(c(x), c(K))$, where $c(\cdot)$ denotes the bitwise complement of its argument. Prove that $y' = c(y)$ (i.e., if we complement the plaintext and the key, then the ciphertext is also complemented). Note that this can be proved using only the "high-level" description of *DES* — the actual structure of S-boxes and other components of the system are irrelevant.

3.4 Before the *AES* was developed, it was suggested to increase the security of *DES* by using the product cipher *DES* $\times$ *DES*, as discussed in Section 2.7. This product cipher uses two 56-bit keys.

This exercise considers known-plaintext attacks on product ciphers. In general, suppose that we take the product of any endomorphic cipher $\mathbf{S} = (\mathcal{P}, \mathcal{P}, \mathcal{K}, \mathcal{E}, \mathcal{D})$ with itself. Further, suppose that $\mathcal{K} = \{0, 1\}^n$ and $\mathcal{P} = \{0, 1\}^m$.

Now, assume we have several plaintext-ciphertext pairs for the product cipher $\mathbf{S}^2$, say $(x_1, y_1), \ldots, (x_\ell, y_\ell)$, all of which are obtained using the same unknown key, (K_1, K_2).

(a) Prove that $e_{K_1}(x_i) = d_{K_2}(y_i)$ for all i, $1 \le i \le \ell$. Give a heuristic argument that the expected number of keys (K_1, K_2) such that $e_{K_1}(x_i) = d_{K_2}(y_i)$ for all i, $1 \le i \le \ell$, is roughly $2^{2n-\ell m}$.

(b) Assume that $\ell \ge 2n/m$. A time-memory trade-off can be used to compute the unknown key (K_1, K_2). We compute two lists, each containing 2^n items, where each item contains an ℓ-tuple of elements of $\mathcal{P}$ as well as an element of $\mathcal{K}$. If the two lists are sorted, then a common ℓ-tuple can be identified by means of a linear search through each of the two lists. Show that this algorithm requires $2^{n+1}(m\ell + n)$ bits of memory and $\ell 2^{n+1}$ encryptions and/or decryptions.

(c) Show that the memory requirement of the attack can be reduced by a factor of 2^t if the total number of encryptions is increased by a factor of 2^t.

HINT   Break the problem up into 2^{2t} subcases, each of which is specified by simultaneously fixing t bits of K_1 and t bits of K_2.

3.5 Suppose that we have the following 128-bit *AES* key, given in hexadecimal notation:

$$\text{2B7E151628AED2A6ABF7158809CF4F3C}$$

Construct the complete key schedule arising from this key.

3.6 Compute the encryption of the following plaintext (given in hexadecimal notation) using the 10-round *AES*:

$$\text{3243F6A8885A308D313198A2E0370734}$$

Use the 128-bit key from the previous exercise.

3.7 Suppose a sequence of plaintext blocks, $x_1 \cdots x_n$, yields the ciphertext sequence $y_1 \cdots y_n$. Suppose that one ciphertext block, say y_i, is transmitted incorrectly (i.e., some 1's are changed to 0's and vice versa). Show that the number of plaintext blocks that will be decrypted incorrectly is equal to one if ECB or OFB modes are used for encryption; and equal to two if CBC or CFB modes are used.

3.8 The purpose of this question is to investigate a time-memory trade-off for a chosen plaintext attack on a certain type of cipher. Suppose we have a cryptosystem in which $\mathcal{P} = \mathcal{C} = \mathcal{K}$, which attains perfect secrecy. Then it must be the case that $e_K(x) = e_{K_1}(x)$ implies $K = K_1$. Denote $\mathcal{P} = Y = \{y_1, \ldots, y_N\}$. Let x be a fixed plaintext. Define the function $g : Y \to Y$ by the rule $g(y) = e_y(x)$. Define a directed graph G having vertex set Y, in which the edge set consists of all the directed edges of the form $(y_i, g(y_i))$, $1 \le i \le N$.

Algorithm 3.7: TIME-MEMORY TRADE-OFF(y)

$y_0 \leftarrow y$
$backup \leftarrow$ **false**
while $g(y) \ne y_0$
$\quad$ **do** $\begin{cases} \textbf{if } y = z_j \text{ for some } j \textbf{ and not } backup \\ \quad \textbf{then } \begin{cases} y \leftarrow g^{-T}(z_j) \\ backup \leftarrow \textbf{true} \end{cases} \\ \quad \textbf{else } \begin{cases} y \leftarrow g(y) \\ K \leftarrow y \end{cases} \end{cases}$

(a) Prove that G consists of the union of disjoint directed cycles.

(b) Let T be a desired time parameter. Suppose we have a set of elements $Z = \{z_1, \ldots, z_m\} \subseteq Y$ such that, for every element $y_i \in Y$, either y_i is contained in a cycle of length at most T, or there exists an element $z_j \ne y_i$ such that the distance from y_i to z_j (in G) is at most T. Prove that there exists such a set Z such that

$$|Z| \le \frac{2N}{T},$$

so $|Z|$ is $O(N/T)$.

(c) For each $z_j \in Z$, define $g^{-T}(z_j)$ to be the element y_i such that $g^T(y_i) = z_j$, where g^T is the function that consists of T iterations of g. Construct a table X consisting of the ordered pairs $(z_j, g^{-T}(z_j))$, sorted with respect to their first coordinates.

A pseudo-code description of an algorithm to find K, given $y = e_K(x)$, is presented. Prove that this algorithm finds K in at most T steps. (Hence the time-memory trade-off is $O(N)$.)

(d) Describe a pseudo-code algorithm to construct the desired set Z in time $O(NT)$ without using an array of size N.

3.9 Suppose that X_1, X_2 and X_3 are independent discrete random variables defined on the set $\{0, 1\}$. Let ϵ_i denote the bias of X_i, for $i = 1, 2, 3$. Prove that $X_1 \oplus X_2$ and $X_2 \oplus X_3$ are independent if and only if $\epsilon_1 = 0$, $\epsilon_3 = 0$ or $\epsilon_2 = \pm 1/2$.

3.10 For the each of eight *DES* S-boxes, compute the bias of the random variable

$$X_2 \oplus Y_1 \oplus Y_2 \oplus Y_3 \oplus Y_4.$$

(Note that these biases are all relatively large in absolute value.)

3.11 The *DES* S-box S_4 has some unusual properties:

(a) Prove that the second row of S_4 can be obtained from the first row by means of the following mapping:

$$(y_1, y_2, y_3, y_4) \mapsto (y_2, y_1, y_4, y_3) \oplus (0, 1, 1, 0),$$

where the entries are represented as binary strings.

(b) Show that any row of S_4 can be transformed into any other row by a similar type of operation.

3.12 Suppose that $\pi_S : \{0, 1\}^m \to \{0, 1\}^n$ is an S-box. Prove the following facts about the function N_L.

(a) $N_L(0, 0) = 2^m$.

(b) $N_L(a, 0) = 2^{m-1}$ for all integers a such that $0 < a \le 2^m - 1$.

(c) For all integers b such that $0 \le b \le 2^n - 1$, it holds that

$$\sum_{a=0}^{2^m-1} N_L(a, b) = 2^{2m-1} \pm 2^{m-1}.$$

(d) It holds that

$$\sum_{a=0}^{2^m-1} \sum_{b=0}^{2^n-1} N_L(a, b) \in \{2^{n+2m-1}, 2^{n+2m-1} + 2^{n+m-1}\}.$$

3.13 An S-box $\pi_S : \{0, 1\}^m \to \{0, 1\}^n$ is said to be *balanced* if

$$|\pi_S^{-1}(y)| = 2^{n-m}$$

for all $y \in \{0, 1\}^n$. Prove the following facts about the function N_L for a balanced S-box.

(a) $N_L(0, b) = 2^{m-1}$ for all integers b such that $0 < b \le 2^n - 1$.

(b) For all integers a such that $0 \le a \le 2^m - 1$, it holds that

$$\sum_{b=0}^{2^n-1} N_L(a, b) = 2^{m+n-1} - 2^{m-1} + i2^n,$$

where i is an integer such that $0 \le i \le 2^{m-n}$.

3.14 Suppose that the S-box of Example 3.1 is replaced by the S-box defined by the following substitution $\pi_{S'}$:

z	0	1	2	3	4	5	6	7	8	9	A	B	C	D	E	F
$\pi_{S'}(z)$	8	4	2	1	C	6	3	D	A	5	E	7	F	B	9	0

(a) Compute the table of values N_L for this S-box.

(b) Find a linear approximation using three active S-boxes, and use the piling-up lemma to estimate the bias of the random variable

$$X_{16} \oplus U_1^4 \oplus U_9^4.$$

(c) Describe a linear attack, analogous to Algorithm 3.3, that will find eight subkey bits in the last round.

(d) Implement your attack and test it to see how many plaintexts are required in order for the algorithm to find the correct subkey bits (approximately 1000–1500 plaintexts should suffice; this attack is more efficient than Algorithm 3.3 because the bias is larger by a factor of 2, which means that the number of plaintexts can be reduced by a factor of about 4).

3.15 Suppose that the S-box of Example 3.1 is replaced by the S-box defined by the following substitution $\pi_{S''}$:

z	0	1	2	3	4	5	6	7	8	9	A	B	C	D	E	F
$\pi_{S''}(z)$	E	2	1	3	D	9	0	6	F	4	5	A	8	C	7	B

(a) Compute the table of values N_D for this S-box.

(b) Find a differential trail using four active S-boxes, namely, S_1^1, S_4^1, S_4^2 and S_4^3, that has propagation ratio $27/2048$.

(c) Describe a differential attack, analogous to Algorithm 3.3, that will find eight subkey bits in the last round.

(d) Implement your attack and test it to see how many plaintexts are required in order for the algorithm to find the correct subkey bits (approximately 100–200 plaintexts should suffice; this attack is not as efficient as Algorithm 3.3 because the propagation ratio is smaller by a factor of 2).

3.16 Suppose that we use the SPN presented in Example 3.1, but the S-box is replaced by a function π_T that is not a permutation. This means, in particular, that π_T is not surjective. Use this fact to derive a ciphertext-only attack that can be used to determine the key bits in the last round, given a sufficient number of ciphertexts which all have been encrypted using the same key.

4
Cryptographic Hash Functions

4.1 Hash Functions and Data Integrity

A cryptographic hash function can provide assurance of data integrity. A hash function is used to construct a short "fingerprint" of some data; if the data is altered, then the fingerprint will no longer be valid. Even if the data is stored in an insecure place, its integrity can be checked from time to time by recomputing the fingerprint and verifying that the fingerprint has not changed.

Let h be a hash function and let x be some data. As an illustrative example, x could be a binary string of arbitrary length. The corresponding fingerprint is defined to be $y = h(x)$. This fingerprint is often referred to as a *message digest*. A message digest would typically be a fairly short binary string; 160 bits is a common choice.

We suppose that y is stored in a secure place, but x is not. If x is changed, to x', say, then we hope that the "old" message digest, y, is not also a message digest for x'. If this is indeed the case, then the fact that x has been altered can be detected simply by computing the message digest $y' = h(x')$ and verifying that $y' \neq y$.

A particularly important application of hash functions occurs in the context of digital signature schemes, which will be studied in Chapter 7.

The motivating example discussed above assumes the existence of a single, fixed hash function. It is also useful to study families of keyed hash functions. A keyed hash function is often used as a *message authentication code*, or *MAC*. Suppose that Alice and Bob share a secret key, K, which determines a hash function, say h_K. For a message, say x, the corresponding authentication tag, $y = h_K(x)$, can be computed by Alice or Bob. The pair (x, y) can be transmitted over an insecure channel from Alice to Bob. When Bob receives the pair (x, y), he can verify if $y = h_K(x)$. If this condition holds, then he is confident that neither x nor y was altered by an adversary, provided that the hash family is "secure." In partcular, Bob is assured that the message x originates from Alice.

Notice the distinction between the assurance of data integrity provided by an unkeyed, as opposed to a keyed, hash function. In the case of an unkeyed hash function, the message digest must be securely stored so it cannot be altered. On the other hand, if Alice and Bob use a secret key K to specify the hash function they are using, then they can transmit both the data and the authentication tag over an insecure channel.

In the remainder of this chapter, we will study hash functions, as well as keyed hash families. We begin by giving definitions for a keyed hash family.

Definition 4.1:   A *hash family* is a four-tuple $(\mathcal{X}, \mathcal{Y}, \mathcal{K}, \mathcal{H})$, where the following conditions are satisfied:

1. $\mathcal{X}$ is a set of possible *messages*
2. $\mathcal{Y}$ is a finite set of possible *message digests* or *authentication tags*
3. $\mathcal{K}$, the *keyspace*, is a finite set of possible *keys*
4. For each $K \in \mathcal{K}$, there is a *hash function* $h_K \in \mathcal{H}$. Each $h_K : \mathcal{X} \to \mathcal{Y}$.

In the above definition, $\mathcal{X}$ could be a finite or infinite set; $\mathcal{Y}$ is always a finite set. If $\mathcal{X}$ is a finite set, a hash function is sometimes called a *compression function*. In this situation, we will always assume that $|\mathcal{X}| \geq |\mathcal{Y}|$, and we will often assume the stronger condition that $|\mathcal{X}| \geq 2|\mathcal{Y}|$.

A pair $(x, y) \in \mathcal{X} \times \mathcal{Y}$ is said to be a *valid pair* under the key K if $h_K(x) = y$. Much of what we discuss in this chapter concerns methods to prevent the construction of certain types of valid pairs by an adversary.

Let $\mathcal{F}^{\mathcal{X}, \mathcal{Y}}$ denote the set of all functions from $\mathcal{X}$ to $\mathcal{Y}$. Suppose that $|\mathcal{X}| = N$ and $|\mathcal{Y}| = M$. Then it is clear that $|\mathcal{F}^{\mathcal{X}, \mathcal{Y}}| = M^N$. Any hash family $\mathcal{F} \subseteq \mathcal{F}^{\mathcal{X}, \mathcal{Y}}$ is termed an (N, M)-*hash family*.

An *unkeyed hash function* is a function $h : \mathcal{X} \to \mathcal{Y}$, where $\mathcal{X}$ and $\mathcal{Y}$ are the same as in Definition 4.1. We could think of an unkeyed hash function simply as a hash family in which there is only one possible key, i.e., one in which $|\mathcal{K}| = 1$.

The remaining sections of this chapter are organized as follows. In Section 4.2, we introduce concepts of security for hash functions, in particular, the idea of collision resistance. We also study the exact security of "ideal" hash functions in the random oracle model in this section; and we discuss the birthday paradox, which provides an estimate of the difficulty of finding collisions for an arbitrary hash function. In Section 4.3, we introduce the important design technique of iterated hash functions. We discuss how this method is used in the design of practical hash functions, as well as in the construction of a provably secure hash function from a secure compression function. Section 4.4 provides a treatment of message authentication codes, where we again provide some general constructions and security proofs. Unconditionally secure MACs, and their construction using strongly universal hash families, are considered in Section 4.5.

4.2 Security of Hash Functions

Suppose that $h : X \to Y$ is an unkeyed hash function. Let $x \in X$, and define $y = h(x)$. In many cryptographic applications of hash functions, it is desirable that the only way to produce a valid pair (x, y) is to first choose x, and then compute $y = h(x)$ by applying the function h to x. Other security requirements of hash functions are motivated by their applications in particular protocols, such as signature schemes (see Chapter 7). We now define three problems; if a hash function is to be considered secure, it should be the case that these three problems are difficult to solve.

Problem 4.1: **Preimage**

Instance: A hash function $h : X \to Y$ and an element $y \in Y$.

Find: $x \in X$ such that $h(x) = y$.

Given a (possible) message digest y, the problem **Preimage** asks if x can be found such that $h(x) = y$. If **Preimage** can be solved for a given $y \in Y$, then the pair (x, y) is a valid pair. A hash function for which **Preimage** cannot be efficiently solved is often said to be *one-way* or *preimage resistant*.

Problem 4.2: **Second Preimage**

Instance: A hash function $h : X \to Y$ and an element $x \in X$.

Find: $x' \in X$ such that $x' \neq x$ and $h(x') = h(x)$.

Given a message x, the problem **Second Preimage** asks if $x' \neq x$ can be found such that $h(x') = h(x)$. Note that, if this can be done, then $(x', h(x))$ is a valid pair. A hash function for which **Second Preimage** cannot be efficiently solved is often said to be *second preimage resistant*.

Problem 4.3: **Collision**

Instance: A hash function $h : X \to Y$.

Find: $x, x' \in X$ such that $x' \neq x$ and $h(x') = h(x)$.

The problem **Collision** asks if $x' \neq x$ can be found such that $h(x') = h(x)$. A solution to this problem yields two valid pairs, (x, y) and (x', y), where $y = h(x) = h(x')$. There are various scenarios where we want to avoid such a situation from arising. A hash function for which **Collision** cannot be efficiently solved is often said to be *collision resistant*.

Some of the questions we address in the next sections concern the difficulty of each of these three problems, as well as the relative difficulty of the three problems.

4.2.1   The Random Oracle Model

In this section, we describe a certain idealized model for a hash function, which attempts to capture the concept of an "ideal" hash function. If a hash function h is well designed, it should be the case that the only efficient way to determine the value $h(x)$ for a given x is to actually evaluate the function h at the value x. This should remain true even if many other values $h(x_1), h(x_2), \ldots$ have already been computed.

To illustrate an example where the above property does not hold, suppose that the hash function $h : \mathbb{Z}_n \times \mathbb{Z}_n \to \mathbb{Z}_n$ is a linear function, say

$$h(x, y) = ax + by \bmod n,$$

$a, b \in \mathbb{Z}_n$ and $n \geq 2$ is a positive integer. Suppose that we are given the values

$$h(x_1, y_1) = z_1$$

and

$$h(x_2, y_2) = z_2.$$

Let $r, s \in \mathbb{Z}_n$; then we have that

$$h(rx_1 + sx_2 \bmod n, ry_1 + sy_2 \bmod n) = a(rx_1 + sx_2) + b(ry_1 + sy_2) \bmod n$$
$$= r(ax_1 + by_1) + s(ax_2 + by_2) \bmod n$$
$$= rh(x_1, y_1) + sh(x_2, y_2) \bmod n.$$

Therefore, given the value of function h at two points (x_1, y_1) and (x_2, y_2), we know its value at various other points, without actually having to evaluate h at those points (and note also that we do not even need to know the values of the constants a and b in order to apply the above-described technique).

The *random oracle model*, which was introduced by Bellare and Rogaway, provides a mathematical model of an "ideal" hash function. In this model, a hash function $h : \mathcal{X} \to \mathcal{Y}$ is chosen randomly from $\mathcal{F}^{\mathcal{X}, \mathcal{Y}}$, and we are only permitted *oracle* access to the function h. This means that we are not given a formula or an algorithm to compute values of the function h. Therefore, the only way to compute a value $h(x)$ is to query the oracle. This can be thought of as looking up the value $h(x)$ in a giant book of random numbers such that, for each possible x, there is a completely random value $h(x)$.[1]

Although a true random oracle does not exist in real life, we hope that a well-designed hash function will "behave" like a random oracle. So it is useful to study the random oracle model and its security with respect to the three problems introduced above. This is done in the next section.

As a consequence of the assumptions made in the random oracle model, it is obvious that the following *independence* property holds:

---

[1] In fact, the book *A Million Random Digits with 100,000 Normal Deviates* was published by the RAND Corporation in 1955. This book could be viewed as an approximation to a random oracle, perhaps.

THEOREM 4.1 *Suppose that $h \in \mathcal{F}^{\mathcal{X},\mathcal{Y}}$ is chosen randomly, and let $\mathcal{X}_0 \subseteq \mathcal{X}$. Suppose that the values $h(x)$ have been determined (by querying an oracle for h) if and only if $x \in \mathcal{X}_0$. Then $\mathbf{Pr}[h(x) = y] = 1/M$ for all $x \in \mathcal{X} \backslash \mathcal{X}_0$ and all $y \in \mathcal{Y}$.*

In the above theorem, the probability $\mathbf{Pr}[h(x) = y]$ is in fact a conditional probability that is computed over all functions h that take on the specified values for all $x \in \mathcal{X}_0$. Theorem 4.1 is the key property used in proofs involving complexity of problems in the random oracle model.

4.2.2 Algorithms in the Random Oracle Model

In this section, we consider the complexity of the three problems defined in Section 4.2 in the random oracle model. An algorithm in the random oracle model can be applied to any hash function, since the algorithm needs to know nothing whatsoever about the hash function (except that a method must be specified to evaluate the hash function for arbitrary values of x).

The algorithms we present and analyze are *randomized algorithms*; they can make random choices during their execution. A *Las Vegas algorithm* is a randomized algorithm which may fail to give an answer (i.e., it can terminate with the message "failure"), but if the algorithm does return an answer, then the answer must be correct.

Suppose $0 \leq \epsilon < 1$ is a real number. A randomized algorithm has *worst-case success probability* ϵ if, for every problem instance, the algorithm returns a correct answer with probability at least ϵ. A randomized algorithm has *average-case success probability* ϵ if the probability that the algorithm returns a correct answer, averaged over all problem instances of a specified size, is at least ϵ. Note that, in this latter situation, the probability that the algorithm returns a correct answer for a given problem instance can be greater than or less than ϵ.

In this section, we use the terminology (ϵ, Q)-*algorithm* to denote a Las Vegas algorithm with average-case success probability ϵ, in which the number of oracle queries (i.e., evaluations of h) made by the algorithm is at most Q. The success probability ϵ is the average over all possible random choices of $h \in \mathcal{F}^{\mathcal{X},\mathcal{Y}}$, and all possible random choices of $x \in \mathcal{X}$ or $y \in \mathcal{Y}$, if x and/or y is specified as part of the problem instance.

We analyze the trivial algorithms, which evaluate $h(x)$ for Q values of $x \in \mathcal{X}$, in the random oracle model. In fact, it turns out that the complexity of such an algorithm is independent of the choice of the Q values of x because we are averaging over all functions $h \in \mathcal{F}^{\mathcal{X},\mathcal{Y}}$.

We first consider Algorithm 4.1, which attempts to solve **Preimage** by evaluating h at Q points.

THEOREM 4.2 *For any $\mathcal{X}_0 \subseteq \mathcal{X}$ with $|\mathcal{X}_0| = Q$, the average-case success probability of Algorithm 4.1 is $\epsilon = 1 - (1 - 1/M)^Q$.*

Algorithm 4.1: FIND-PREIMAGE(h, y, Q)

choose any $\mathcal{X}_0 \subseteq \mathcal{X}, |\mathcal{X}_0| = Q$
for each $x \in \mathcal{X}_0$
$\quad$ **do** $\begin{cases} \textbf{if } h(x) = y \\ \quad \textbf{then return } (x) \end{cases}$
return (failure)

Algorithm 4.2: FIND-SECOND-PREIMAGE(h, x, Q)

$y \leftarrow h(x)$
choose $\mathcal{X}_0 \subseteq \mathcal{X} \setminus \{x\}, |\mathcal{X}_0| = Q - 1$
for each $x_0 \in \mathcal{X}_0$
$\quad$ **do** $\begin{cases} \textbf{if } h(x_0) = y \\ \quad \textbf{then return } (x_0) \end{cases}$
return (failure)

PROOF Let $y \in \mathcal{Y}$ be fixed. Let $\mathcal{X}_0 = \{x_1, \ldots, x_Q\}$. For $1 \leq i \leq Q$, let E_i denote the event "$h(x_i) = y$." It follows from Theorem 4.1 that the E_i's are independent events, and $\mathbf{Pr}[E_i] = 1/M$ for all $1 \leq i \leq Q$. Therefore, it holds that

$$\mathbf{Pr}[E_1 \lor E_2 \lor \cdots \lor E_Q] = 1 - \left(1 - \frac{1}{M}\right)^Q.$$

The success probability of Algorithm 4.1, for any fixed y, is constant. Therefore, the success probability averaged over all $y \in \mathcal{Y}$ is identical, too. ∎

Note that the above success probability is approximately Q/M provided that Q is small compared to M.

We now present and analyze a very similar algorithm, Algorithm 4.2, that attempts to solve **Second Preimage**.

The analysis of Algorithm 4.2 is similar to the previous algorithm. The only difference is that we require an "extra" application of h to compute $y = h(x)$ for the input value x.

THEOREM 4.3 *For any $\mathcal{X}_0 \subseteq \mathcal{X} \setminus \{x\}$ with $|\mathcal{X}_0| = Q - 1$, the success probability of Algorithm 4.2 is $\epsilon = 1 - (1 - 1/M)^{Q-1}$.*

Next, we look at an elementary algorithm for **Collision**.

Algorithm 4.3: FIND-COLLISION(h, Q)

choose $\mathcal{X}_0 \subseteq \mathcal{X}, |\mathcal{X}_0| = Q$
for each $x \in \mathcal{X}_0$
  do $y_x \leftarrow h(x)$
if $y_x = y_{x'}$ for some $x' \neq x$
  then return (x, x')
  else return (failure)

In Algorithm 4.3, the test to see if $y_x = y_{x'}$ for some $x' \neq x$ could be done efficiently by sorting the y_x's, for example. This algorithm is analyzed using a probability argument analogous to the standard "birthday paradox." The *birthday paradox* says that in a group of 23 randomly chosen people, at least two will share a birthday with probability at least $1/2$. (Of course this is not a paradox, but it is probably counter-intuitive). This may not appear to be relevant to hash functions, but if we reformulate the problem, the connection will be clear. Suppose that the function h has as its domain the set of all living human beings, and for all x, $h(x)$ denotes the birthday of person x. Then the range of h consists of the 365 days in a year (366 days if we include February 29). Finding two people with the same birthday is the same thing as finding a collision for this particular hash function. In this setting, the birthday paradox is saying that Algorithm 4.3 has success probability at least $1/2$ when $Q = 23$ and $M = 365$.

We now analyze Algorithm 4.3 in general, in the random oracle model. This algorithm is analogous to throwing Q balls randomly into M bins and then checking to see if some bin contains at least two balls. (The Q balls correspond to the Q random x_i's, and the M bins correspond to the M possible elements of $\mathcal{Y}$.)

THEOREM 4.4 *For any $\mathcal{X}_0 \subseteq \mathcal{X}$ with $|\mathcal{X}_0| = Q$, the success probability of Algorithm 4.3 is*

$$\epsilon = 1 - \left(\frac{M-1}{M}\right)\left(\frac{M-2}{M}\right)\cdots\left(\frac{M-Q+1}{M}\right).$$

PROOF Let $\mathcal{X}_0 = \{x_1, \ldots, x_Q\}$. For $1 \leq i \leq Q$, let E_i denote the event

$$\text{"}h(x_i) \notin \{h(x_1), \ldots, h(x_{i-1})\}\text{."}$$

Using induction, it follows from Proposition 4.1 that $\mathbf{Pr}[E_1] = 1$ and

$$\mathbf{Pr}[E_i | E_1 \wedge E_2 \wedge \cdots \wedge E_{i-1}] = \frac{M-i+1}{M},$$

for $2 \leq i \leq Q$. Therefore, we have that

$$\mathbf{Pr}[E_1 \wedge E_2 \wedge \cdots \wedge E_Q] = \left(\frac{M-1}{M}\right)\left(\frac{M-2}{M}\right)\cdots\left(\frac{M-Q+1}{M}\right).$$

The probability that there is at least one collision is $1 - \mathbf{Pr}[E_1 \wedge E_2 \wedge \cdots \wedge E_Q]$, so the desired result follows. ∎

The above theorem shows that the probability of finding no collisions is

$$\left(1 - \frac{1}{M}\right)\left(1 - \frac{2}{M}\right) \cdots \left(1 - \frac{Q-1}{M}\right) = \prod_{i=1}^{Q-1}\left(1 - \frac{i}{M}\right).$$

If x is a small real number, then $1 - x \approx e^{-x}$. This estimate is derived by taking the first two terms of the series expansion

$$e^{-x} = 1 - x + \frac{x^2}{2!} - \frac{x^3}{3!} \cdots.$$

Using this estimate, the probability of finding no collisions is approximately

$$\prod_{i=1}^{Q-1}\left(1 - \frac{i}{M}\right) \approx \prod_{i=1}^{Q-1} e^{\frac{-i}{M}}$$

$$= e^{-\sum_{i=1}^{Q-1}\frac{i}{M}}$$

$$= e^{\frac{-Q(Q-1)}{2M}}.$$

Consequently, we can estimate the probability of finding at least one collision to be

$$1 - e^{\frac{-Q(Q-1)}{2M}}.$$

If we denote this probability by ϵ, then we can solve for Q as a function of M and ϵ:

$$e^{\frac{-Q(Q-1)}{2M}} \approx 1 - \epsilon$$

$$\frac{-Q(Q-1)}{2M} \approx \ln(1 - \epsilon)$$

$$Q^2 - Q \approx 2M \ln\frac{1}{1-\epsilon}.$$

If we ignore the term $-Q$, then we estimate that

$$Q \approx \sqrt{2M \ln\frac{1}{1-\epsilon}}.$$

If we take $\epsilon = .5$, then our estimate is

$$Q \approx 1.17\sqrt{M}.$$

So this says that hashing just over $\sqrt{M}$ random elements of X yields a collision with a probability of 50%. Note that a different choice of ϵ leads to a different constant factor, but Q will still be proportional to $\sqrt{M}$. The algorithm is a $(1/2, O(\sqrt{M}))$-algorithm.

Algorithm 4.4: COLLISION-TO-SECOND-PREIMAGE(h)

external ORACLE-2ND-PREIMAGE
choose $x \in \mathcal{X}$ uniformly at random
if ORACLE-2ND-PREIMAGE$(h, x) = x'$
  then return (x, x')
  else return (failure)

We return to the example we mentioned earlier. Taking $M = 365$ in our estimate, we get $Q \approx 22.3$. Hence, as mentioned earlier, the probability is at least $1/2$ that there will be at least one duplicated birthday among 23 randomly chosen people.

The birthday attack imposes a lower bound on the sizes of secure message digests. A 40-bit message digest would be very insecure, since a collision could be found with probability $1/2$ with just over 2^{20} (about a million) random hashes. It is usually suggested that the minimum acceptable size of a message digest is 128 bits (the birthday attack will require over 2^{64} hashes in this case). In fact, a 160-bit message digest (or larger) is usually recommended.

4.2.3 Comparison of Security Criteria

In the random oracle model, we have seen that solving **Collision** is easier than solving **Preimage** or **Second Preimage**. A related question is whether there exist reductions among the three problems which could be applied to arbitrary hash functions. It is fairly easy to see that we can reduce **Collision** to **Second Preimage** using Algorithm 4.4.

Suppose that ORACLE-2ND-PREIMAGE is an (ϵ, Q)-algorithm that solves **Second Preimage** for a particular, fixed hash function h. If ORACLE-2ND-PREIMAGE returns a value x' when it is given input (h, x), then it must be the case that $x' \neq x$, because ORACLE-2ND-PREIMAGE is assumed to be a Las Vegas algorithm. As a consequence, it is clear that COLLISION-TO-SECOND-PREIMAGE is an (ϵ, Q)-algorithm that solves **Collision** for the same hash function h. This reduction does not make any assumptions about the hash function h. Because of this reduction, we could say that the property of collision resistance implies the property of second preimage resistance.

We are now going to investigate the more interesting question of whether **Collision** can be reduced to **Preimage**. In other words, does collision resistance imply preimage resistance? We will prove that this is indeed the case, at least in some special situations. More specifically, we will prove that an arbitrary algorithm that solves **Preimage** with probability equal to 1 can be used to solve **Collision**. This reduction can be accomplished with a fairly weak assumption on the rel-

---

Algorithm 4.5: COLLISION-TO-PREIMAGE(h)

external ORACLE-PREIMAGE
choose $x \in \mathcal{X}$ uniformly at random
$y \leftarrow h(x)$
if (ORACLE-PREIMAGE(h, y) $= x'$) **and** ($x' \neq x$)
   then return (x, x')
   else return (failure)

---

ative sizes of the domain and range of the hash function. We will assume that the hash function $h : \mathcal{X} \to \mathcal{Y}$, where $\mathcal{X}$ and $\mathcal{Y}$ are finite sets and $|\mathcal{X}| \geq 2|\mathcal{Y}|$. We suppose that ORACLE-PREIMAGE is a $(1, Q)$ algorithm for **Preimage**. ORACLE-PREIMAGE accepts as input a message digest $y \in \mathcal{Y}$, and always finds an element ORACLE-PREIMAGE(y) $\in \mathcal{X}$ such that $h($ORACLE-PREIMAGE(y)$) = y$ (in particular, this implies that h is surjective). We will analyze the algorithm COLLISION-TO-PREIMAGE, which is presented as Algorithm 4.5.

We prove the following theorem.

THEOREM 4.5 *Suppose $h : \mathcal{X} \to \mathcal{Y}$ is a hash function where $|\mathcal{X}|$ and $|\mathcal{Y}|$ are finite and $|\mathcal{X}| \geq 2|\mathcal{Y}|$. Suppose* ORACLE-PREIMAGE *is a $(1, Q)$-algorithm for* **Preimage**, *for the fixed hash function h. Then* COLLISION-TO-PREIMAGE *is a $(1/2, Q + 1)$-algorithm for* **Collision**, *for the fixed hash function h.*

PROOF Clearly COLLISION-TO-PREIMAGE is a probabilistic algorithm of the Las Vegas type, since it either finds a collision or returns "failure." Thus our main task is to compute the average-case probability of success.

For any $x \in \mathcal{X}$, define $x \sim x_1$ if $h(x) = h(x_1)$. It is easy to see that $\sim$ is an equivalence relation. Define $[x] = \{x_1 \in \mathcal{X} : x \sim x_1\}$. Each equivalence class $[x]$ consists of the inverse image of an element of $\mathcal{Y}$, i.e., for every equivalence class $[x]$, there exists a (unique) value $y \in \mathcal{Y}$ such that $[x] = h^{-1}(y)$. We assumed that ORACLE-PREIMAGE always finds a preimage of any element y, which means that $h^{-1}(y) \neq \emptyset$ for all $y \in \mathcal{Y}$. Therefore, the number of equivalence classes $[x]$ is equal to $|\mathcal{Y}|$. Denote the set of these $|\mathcal{Y}|$ equivalence classes by $\mathcal{C}$.

Now, suppose x is the random element of $\mathcal{X}$ chosen by the algorithm COLLISION-TO-PREIMAGE. For this x, there are $|[x]|$ possible x_1's that could be returned as the output of ORACLE-PREIMAGE. $|[x]| - 1$ of these x_1's are different from x and thus yield a collision. (Note that the algorithm ORACLE-PREIMAGE does not know the representative of the equivalence class $[x]$ that was initially chosen by algorithm COLLISION-TO-PREIMAGE.) So, given the element $x \in \mathcal{X}$, the probability of success is $(|[x]| - 1)/|[x]|$.

The probability of success of the algorithm COLLISION-TO-PREIMAGE is computed by averaging over all possible choices for x:

$$\mathbf{Pr}[\text{success}] = \frac{1}{|\mathcal{X}|} \sum_{x \in \mathcal{X}} \frac{|[x]| - 1}{|[x]|}$$

$$= \frac{1}{|\mathcal{X}|} \sum_{C \in \mathcal{C}} \sum_{x \in C} \frac{|C| - 1}{|C|}$$

$$= \frac{1}{|\mathcal{X}|} \sum_{C \in \mathcal{C}} (|C| - 1)$$

$$= \frac{1}{|\mathcal{X}|} \left(\sum_{C \in \mathcal{C}} |C| - \sum_{C \in \mathcal{C}} 1 \right)$$

$$= \frac{|\mathcal{X}| - |\mathcal{Y}|}{|\mathcal{X}|}$$

$$\geq \frac{|\mathcal{X}| - |\mathcal{X}|/2}{|\mathcal{X}|}$$

$$= \frac{1}{2}.$$

Note that we use the fact that $|\mathcal{X}| \geq 2|\mathcal{Y}|$ in the next-to-last line of the computation performed above.

In summary, we have constructed a Las Vegas algorithm with average-case success probability at least $1/2$. ∎

4.3 Iterated Hash Functions

So far, we have considered hash functions with a finite domain (i.e., compression functions). We now study a particular technique by which a compression function, say **compress**, can be extended to a hash function with an infinite domain, h. A hash function h constructed by this method is called an *iterated hash function*.

In this section, we restrict our attention to hash functions whose inputs and outputs are bitstrings (i.e., strings formed of zeroes and ones). We denote the length of a bitstring x by $|x|$, and the concatenation of bitstrings x and y is denoted $x \parallel y$.

Suppose that **compress** : $\{0,1\}^{m+t} \to \{0,1\}^m$ is a compression function (where $t \geq 1$). We will construct an iterated hash function

$$h : \bigcup_{i=m+t+1}^{\infty} \{0,1\}^i \to \{0,1\}^{\ell},$$

based on the compression function **compress**. The evaluation of h consists of the following three main steps:

preprocessing step

Given an input string x, where $|x| \geq m + t + 1$, construct a string y, using a public algorithm, such that $|y| \equiv 0 \pmod{t}$. Denote

$$y = y_1 \parallel y_2 \parallel \cdots \parallel y_r,$$

where $|y_i| = t$ for $1 \leq i \leq r$.

processing step

Let IV be a public initial value which is a bitstring of length m. Then compute the following:

$$z_0 \leftarrow \text{IV}$$

$$z_1 \leftarrow \textbf{compress}(z_0 \parallel y_1)$$

$$z_2 \leftarrow \textbf{compress}(z_1 \parallel y_2)$$

$$\vdots \quad \vdots \quad \vdots$$

$$z_r \leftarrow \textbf{compress}(z_{r-1} \parallel y_r).$$

This processing step is illustrated in Figure 4.1.

output transformation

Let $g : \{0, 1\}^m \to \{0, 1\}^\ell$ be a public function. Define $h(x) = g(z_r)$.

REMARK   This step is optional. If an output transformation is not desired, then define $h(x) = z_r$.   ∎

A commonly used preprocessing step is to construct the string y in the following way:

$$y = x \parallel \textbf{pad}(x),$$

where $\textbf{pad}(x)$ is constructed from x using a *padding function*. A padding function typically incorporates the value of $|x|$, and pads the result with additional bits (zeros, for example) so that the resulting string y has a length that is a multiple of t.

The preprocessing step must ensure that the mapping $x \mapsto y$ is an injection. (If the mapping $x \mapsto y$ is not one-to-one, then it may be possible to find $x \neq x'$ so that $y = y'$. Then $h(x) = h(x')$, and h would not be collision-resistant.) Note also that $|y| = rt \geq |x|$ because of the required injective property.

Most hash functions used in practice are in fact iterated hash functions and can be viewed as special cases of the generic construction described above. In a later section, we will describe one such hash function, the *Secure Hash Algorithm*

FIGURE 4.1
The processing step in an iterated hash function

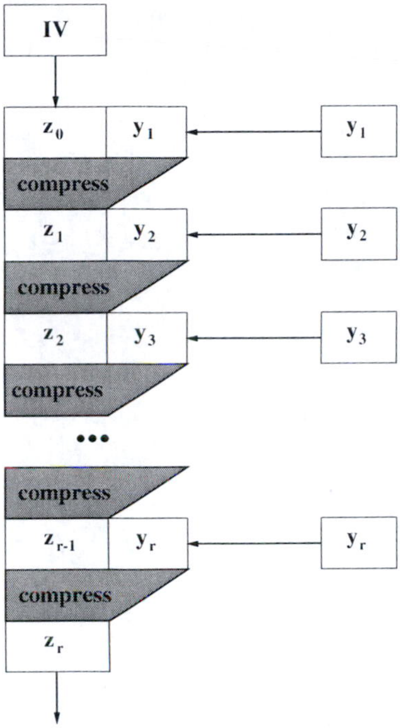

(known as *SHA-1*) in detail. The Merkle-Damgård construction, which we discuss in the next section, is also a construction of an iterated hash function which permits a formal security proof to be given.

4.3.1 The Merkle-Damgård Construction

In this section, we present a particular method of constructing a hash function from a compression function. This construction has the property that the resulting hash function satisfies desired security properties such as collision resistance provided that the compression function does. This technique is often called the *Merkle-Damgård construction*.

Suppose **compress** : $\{0, 1\}^{m+t} \rightarrow \{0, 1\}^m$ is a collision resistant compression function, where $t \geq 1$. We will use **compress** to construct a collision resistant hash function $h : \mathcal{X} \rightarrow \{0, 1\}^m$, where

$$\mathcal{X} = \bigcup_{i=m+t+1}^{\infty} \{0, 1\}^i.$$

Algorithm 4.6: MERKLE-DAMGÅRD(x)

external compress
comment: compress : $\{0,1\}^{m+t} \to \{0,1\}^m$, where $t \geq 2$
$n \leftarrow |x|$
$k \leftarrow \lceil n/(t-1) \rceil$
$d \leftarrow k(t-1) - n$
for $i \leftarrow 1$ **to** $k-1$
   do $y_i \leftarrow x_i$
$y_k \leftarrow x_k \,\|\, 0^d$
$y_{k+1} \leftarrow$ the binary representation of d
$z_1 \leftarrow 0^{m+1} \,\|\, y_1$
$g_1 \leftarrow$ **compress**(z_1)
for $i \leftarrow 1$ **to** k
   do $\begin{cases} z_{i+1} \leftarrow g_i \,\|\, 1 \,\|\, y_{i+1} \\ g_{i+1} \leftarrow \textbf{compress}(z_{i+1}) \end{cases}$
$h(x) \leftarrow g_{k+1}$
return $(h(x))$

We first consider the situation where $t \geq 2$.

We will treat elements of $x \in X$ as bit-strings. Suppose $|x| = n \geq m + t + 1$. We can express x as the concatenation

$$x = x_1 \,\|\, x_2 \,\|\, \cdots \,\|\, x_k,$$

where

$$|x_1| = |x_2| = \cdots = |x_{k-1}| = t - 1$$

and

$$|x_k| = t - 1 - d,$$

where $0 \leq d \leq t - 2$. Hence, we have that

$$k = \left\lceil \frac{n}{t-1} \right\rceil.$$

We define $h(x)$ to be the output of Algorithm 4.6.

Denote

$$y(x) = y_1 \,\|\, y_2 \,\|\, \cdots \,\|\, y_{k+1}.$$

Observe that y_k is formed from x_k by padding on the right with d zeroes, so that all the blocks y_i ($1 \leq i \leq k$) are of length $t - 1$. Also, y_{k+1} should be padded on the left with zeroes so that $|y_{k+1}| = t - 1$.

As was done in the generic construction described in Section 4.3, we hash x by first constructing $y(x)$, and then processing the blocks $y_1, y_2, \ldots, y_{k+1}$ in a particular fashion. y_{k+1} is defined in such a way that the mapping $x \mapsto y(x)$ is an injection, which we observed is necessary for the iterated hash function to be collision-resistant.

The following theorem proves that, if a collision can be found for h, then a collision can be found for **compress**. In other words, h is collision resistant provided that **compress** is collision resistant.

THEOREM 4.6   *Suppose* **compress** $: \{0,1\}^{m+t} \to \{0,1\}^m$ *is a collision resistant compression function, where* $t \geq 2$. *Then the function*

$$h : \bigcup_{i=m+t+1}^{\infty} \{0,1\}^i \to \{0,1\}^m,$$

as constructed in Algorithm 4.6, is a collision resistant hash function.

PROOF   Suppose that we can find $x \neq x'$ such that $h(x) = h(x')$. We will show how we can find a collision for **compress** in polynomial time.

Denote

$$y(x) = y_1 \parallel y_2 \parallel \cdots \parallel y_{k+1}$$

and

$$y(x') = y_1' \parallel y_2' \parallel \cdots \parallel y_{\ell+1}',$$

where x and x' are padded with d and d' 0's, respectively. Denote the g-values computed in the algorithm by $g_1, \ldots, g_{k+1}$ and $g_1', \ldots, g_{\ell+1}'$, respectively.

We identify two cases, depending on whether $|x| \equiv |x'| \pmod{t-1}$ (or not).

case 1: $|x| \not\equiv |x'| \pmod{t-1}$.

Here $d \neq d'$ and $y_{k+1} \neq y_{\ell+1}'$. We have

$$\begin{aligned}
\mathbf{compress}(g_k \parallel 1 \parallel y_{k+1}) &= g_{k+1} \\
&= h(x) \\
&= h(x') \\
&= g_{\ell+1}' \\
&= \mathbf{compress}(g_\ell' \parallel 1 \parallel y_{\ell+1}'),
\end{aligned}$$

which is a collision for **compress** because $y_{k+1} \neq y_{\ell+1}'$.

case 2: $|x| \equiv |x'| \pmod{t-1}$.

It is convenient to split this case into two subcases:

case 2a: $|x| = |x'|$.

Here we have $k = \ell$ and $y_{k+1} = y'_{k+1}$. We begin as in case 1:

$$\mathbf{compress}(g_k \parallel 1 \parallel y_{k+1}) = g_{k+1}$$
$$= h(x)$$
$$= h(x')$$
$$= g'_{k+1}$$
$$= \mathbf{compress}(g'_k \parallel 1 \parallel y'_{k+1}).$$

If $g_k \neq g'_k$, then we find a collision for **compress**, so assume $g_k = g'_k$. Then we have

$$\mathbf{compress}(g_{k-1} \parallel 1 \parallel y_k) = g_k$$
$$= g'_k$$
$$= \mathbf{compress}(g'_{k-1} \parallel 1 \parallel y'_k).$$

Either we find a collision for **compress**, or $g_{k-1} = g'_{k-1}$ and $y_k = y'_k$. Assuming we do not find a collision, we continue working backwards, until finally we obtain

$$\mathbf{compress}(0^{m+1} \parallel y_1) = g_1$$
$$= g'_1$$
$$= \mathbf{compress}(0^{m+1} \parallel y'_1).$$

If $y_1 \neq y'_1$, then we find a collision for **compress**, so we assume $y_1 = y'_1$. But then $y_i = y'_i$ for $1 \leq i \leq k + 1$, so $y(x) = y(x')$. But this implies $x = x'$, because the mapping $x \mapsto y(x)$ is an injection. We assumed $x \neq x'$, so we have a contradiction.

case 2b: $|x| \neq |x'|$.

Without loss of generality, assume $|x'| > |x|$, so $\ell > k$. This case proceeds in a similar fashion as case 2a. Assuming we find no collisions for **compress**, we eventually reach the situation where

$$\mathbf{compress}(0^{m+1} \parallel y_1) = g_1$$
$$= g'_{\ell-k+1}$$
$$= \mathbf{compress}(g'_{\ell-k} \parallel 1 \parallel y'_{\ell-k+1}).$$

But the $(m + 1)$st bit of

$$0^{m+1} \parallel y_1$$

is a 0 and the $(m + 1)$st bit of

$$g'_{\ell-k} \parallel 1 \parallel y'_{\ell-k+1}$$

is a 1. So we find a collision for **compress**.

Algorithm 4.7: MERKLE-DAMGÅRD2(x)

external compress
comment: compress $: \{0,1\}^{m+1} \rightarrow \{0,1\}^m$

$n \leftarrow |x|$
$y \leftarrow 11 \parallel f(x_1) \parallel f(x_2) \parallel \cdots \parallel f(x_n)$
denote $y = y_1 \parallel y_2 \parallel \cdots \parallel y_k$, where $y_i \in \{0,1\}, 1 \leq i \leq k$
$g_1 \leftarrow$ **compress**$(0^m \parallel y_1)$
for $i \leftarrow 1$ **to** $k - 1$
$\quad$ **do** $g_{i+1} \leftarrow$ **compress**$(g_i \parallel y_{i+1})$
return (g_k)

Since we have considered all possible cases, we have proven the desired conclusion. ∎

The construction presented in Algorithm 4.6 can be used only when $t \geq 2$. Let's now look at the situation where $t = 1$. We need to use a different construction for h. Suppose $|x| = n \geq m + 2$. We first encode x in a special way. This will be done using the function f defined as follows:

$$f(0) = 0$$
$$f(1) = 01.$$

The construction of $h(x)$ is presented as Algorithm 4.7.

The encoding $x \mapsto y = y(x)$, as defined in Algorithm 4.7, satisfies two important properties:

1. If $x \neq x'$, then $y(x) \neq y(x')$ (i.e., $x \mapsto y(x)$ is an injection).

2. There do not exist two strings $x \neq x'$ and a string z such that $y(x) = z \parallel y(x')$. (In other words, no encoding is a *postfix* of another encoding. This is easily seen because each string $y(x)$ begins with 11, and there do not exist two consecutive 1's in the remainder of the string.)

THEOREM 4.7 *Suppose* **compress** $: \{0,1\}^{m+1} \rightarrow \{0,1\}^m$ *is a collision resistant compression function. Then the function*

$$h : \bigcup_{i=m+2}^{\infty} \{0,1\}^i \rightarrow \{0,1\}^m,$$

as constructed in Algorithm 4.7, is a collision resistant hash function.

PROOF   Suppose that we can find $x \neq x'$ such that $h(x) = h(x')$. Denote

$$y(x) = y_1 y_2 \cdots y_k$$

and

$$y(x') = y_1' y_2' \cdots y_\ell'.$$

We consider two cases.

case 1: $k = \ell$.

As in Theorem 4.6, either we find a collision for **compress**, or we obtain $y = y'$. But this implies $x = x'$, a contradiction.

case 2: $k \neq \ell$.

Without loss of generality, assume $\ell > k$. This case proceeds in a similar fashion. Assuming we find no collisions for **compress**, we have the following sequence of equalities:

$$y_k = y_\ell'$$

$$y_{k-1} = y_{\ell-1}'$$

$$\vdots \quad \vdots$$

$$y_1 = y_{\ell-k+1}'.$$

But this contradicts the "postfix-free" property stated above.

We conclude that h is collision resistant. ∎

We summarize the two constructions of hash functions in this section, and the number of applications of **compress** needed to compute h, in the following theorem.

THEOREM 4.8   *Suppose* **compress** $: \{0,1\}^{m+t} \to \{0,1\}^m$ *is a collision resistant compression function, where* $t \geq 1$. *Then there exists a collision resistant hash function*

$$h : \bigcup_{i=m+t+1}^{\infty} \{0,1\}^i \to \{0,1\}^m.$$

The number of times **compress** *is computed in the evaluation of h is at most*

$$1 + \left\lceil \frac{n}{t-1} \right\rceil \quad \text{if } t \geq 2$$
$$2n + 2 \quad \text{if } t = 1,$$

where $|x| = n$.

Algorithm 4.8: SHA-1-PAD(x)

comment: $|x| \leq 2^{64} - 1$

$d \leftarrow (447 - |x|) \bmod 512$
$\ell \leftarrow$ the binary representation of $|x|$, where $|\ell| = 64$
$y \leftarrow x \parallel 1 \parallel 0^d \parallel \ell$

4.3.2 The Secure Hash Algorithm

In this section, we describe *SHA-1* (the *Secure Hash Algorithm*), which is an iterated hash function with a 160-bit message digest. *SHA-1* is built from word-oriented operations on bitstrings, where a word consists of 32 bits (or eight hexadecimal digits). The operations used in *SHA-1* are as follows:

$X \wedge Y$	bitwise "and" of X and Y
$X \vee Y$	bitwise "or" of X and Y
$X \oplus Y$	bitwise "xor" of X and Y
$\neg X$	bitwise complement of X
$X + Y$	integer addition modulo 2^{32}
$\mathbf{ROTL}^s(X)$	circular left shift of X by s positions $(0 \leq s \leq 31)$

We first describe the padding scheme used in *SHA-1*; it is presented as Algorithm 4.8. Note that *SHA-1* requires that $|x| \leq 2^{64} - 1$. Therefore the binary representation of $|x|$, which is denoted by ℓ in Algorithm 4.8, has length at most 64 bits. If $|\ell| < 64$, then it is padded on the left with zeroes so that its length is exactly 64 bits.

In the construction of y, we append a single 1 to x, then we concatenate enough 0's so that the length becomes congruent to 448 modulo 512, and finally we concatenate 64 bits that contain the binary representation of the (original) length of x. The resulting string y has length divisible by 512. Then we write y as a concatenation of n blocks, each having 512 bits:

$$y = M_1 \parallel M_2 \parallel \cdots \parallel M_n.$$

Define the functions $\mathbf{f}_0, \ldots, \mathbf{f}_{79}$ as follows:

$$\mathbf{f}_t(B, C, D) = \begin{cases} (B \wedge C) \vee ((\neg B) \wedge D) & \text{if } 0 \leq t \leq 19 \\ B \oplus C \oplus D & \text{if } 20 \leq t \leq 39 \\ (B \wedge C) \vee (B \wedge D) \vee (C \wedge D) & \text{if } 40 \leq t \leq 59 \\ B \oplus C \oplus D & \text{if } 60 \leq t \leq 79. \end{cases}$$

Each function $\mathbf{f}_t$ takes three words B, C and D as input, and produces one word as output.

Cryptosystem 4.1: *SHA-1*(x)

external SHA-1-PAD
global $K_0, \ldots, K_{79}$
$y \leftarrow$ SHA-1-PAD(x)
denote $y = M_1 \parallel M_2 \parallel \cdots \parallel M_n$, where each M_i is a 512-bit block
$H_0 \leftarrow$ 67452301
$H_1 \leftarrow$ EFCDAB89
$H_2 \leftarrow$ 98BADCFE
$H_3 \leftarrow$ 10325476
$H_4 \leftarrow$ C3D2E1F0
for $i \leftarrow 1$ **to** n

$\textbf{do} \begin{cases} \text{denote } M_i = W_0 \parallel W_1 \parallel \cdots \parallel W_{15}, \text{ where each } W_i \text{ is a word} \\ \textbf{for } t \leftarrow 16 \textbf{ to } 79 \\ \quad \textbf{do } W_t \leftarrow \textbf{ROTL}^1(W_{t-3} \oplus W_{t-8} \oplus W_{t-14} \oplus W_{t-16}) \\ A \leftarrow H_0 \\ B \leftarrow H_1 \\ C \leftarrow H_2 \\ D \leftarrow H_3 \\ E \leftarrow H_4 \\ \textbf{for } t \leftarrow 0 \textbf{ to } 79 \\ \quad \textbf{do} \begin{cases} temp \leftarrow \textbf{ROTL}^5(A) + \mathbf{f}_t(B, C, D) + E + W_t + K_t \\ E \leftarrow D \\ D \leftarrow C \\ C \leftarrow \textbf{ROTL}^{30}(B) \\ B \leftarrow A \\ A \leftarrow temp \end{cases} \\ H_0 \leftarrow H_0 + A \\ H_1 \leftarrow H_1 + B \\ H_2 \leftarrow H_2 + C \\ H_3 \leftarrow H_3 + D \\ H_4 \leftarrow H_4 + E \end{cases}$

return $(H_0 \parallel H_1 \parallel H_2 \parallel H_3 \parallel H_4)$

Define the word constants $K_0, \ldots, K_{79}$, which are used in the computation of $SHA\text{-}1(x)$, as follows:

$$
K_t = \begin{cases}
\texttt{5A827999} & \text{if } 0 \le t \le 19 \\
\texttt{6ED9EBA1} & \text{if } 20 \le t \le 39 \\
\texttt{8F1BBCDC} & \text{if } 40 \le t \le 59 \\
\texttt{CA62C1D6} & \text{if } 60 \le t \le 79.
\end{cases}
$$

SHA-1, which is presented as Cryptosystem 4.1, closely follows the general model of an iterated hash function. The padding scheme extends the input x by at most one extra 512-bit block. The compression function maps $160 + 512$ bits to 160 bits, where the 512 bits comprise a block of the message.

SHA-1 is one in a series of related iterated hash functions. The first of these hash functions, *MD4*, was proposed by Rivest in 1990. Rivest then modified *MD4* to produce *MD5* in 1992. *SHA* was proposed as a standard by NIST in 1993, and it was adopted as FIPS 180. *SHA-1* is a minor variation of *SHA*; it was published in 1995 as FIPS 180-1 (and *SHA* is now generally referred to by the name *SHA-0*). The only difference between *SHA-1* and *SHA-0* is that *SHA-0* omits the 1-bit rotations in the construction of $W_{16}, \ldots, W_{79}$.

This progression of hash functions incorporated various modifications to improve the security of the later versions of the hash functions against attacks that were found against earlier versions. For example, collisions in the compression functions of *MD4* and *MD5* were discovered in the mid-1990s. It was shown in 1998 that *SHA-0* had a weakness that would allow collisions to be found in approximately 2^{61} steps (this attack is much more efficient than a birthday attack, which would require about 2^{80} steps).

More recently, in 2004, a collision for *SHA-0* was actually found by Joux and reported at CRYPTO 2004. Collisions for *MD5* and several other popular hash functions were also presented at CRYPTO 2004, by Wang, Feng, Lai and Yu. Finally, it appears that collisions in *SHA-1* might be found in the not-too-distant future. In fact, as of March, 2005, collisions in a 58-round "reduced" *SHA-1* have been discovered, and it is estimated that a collision in the full *SHA-1* could be found using less 2^{69} hash operations. These last results are due to Wang, Yin and Yu.

FIPS 180-2, which was adopted in August, 2002, includes *SHA-1* as well as three new hash functions, which are known as *SHA-256*, *SHA-384* and *SHA-512*. The suffixes "256," "384" and "512" refer to the sizes of the message digests. The three new hash functions are also iterated hash functions, but they have a more complex description than *SHA-1*. Finally, *SHA-224* is yet another proposed addition to the *SHA* family of hash functions.

4.4   Message Authentication Codes

We now turn our attention to message authentication codes, which are keyed hash functions satisfying certain security properties. As we will see, the security properties required by a MAC are rather different than those required by an (unkeyed) hash function.

One common way of constructing a MAC is to incorporate a secret key into an unkeyed hash function, by including it as part of the message to be hashed. This must be done carefully, however, in order to prevent certain attacks from being carried out. We illustrate the possible pitfalls with a couple of simple examples.

As a first attempt, suppose we construct a keyed hash function h_K from an unkeyed iterated hash function, say h, by defining $IV = K$ and keeping its value secret. For simplicity, suppose also that h does not have a preprocessing step or an output transformation. Such a hash function requires that every input message x have length that is a multiple of t, where $\textbf{compress} : \{0,1\}^{m+t} \to \{0,1\}^m$ is the compression function used to build h. Further, the key K is an m-bit key.

We show how an opponent can construct a valid MAC for a certain message, without knowing the secret key K, given any message x and its corresponding MAC, $h_K(x)$. Let x' be any bitstring of length t, and consider the message $x \parallel x'$. The MAC for this message, $h_K(x \parallel x')$, is computed to be

$$h_K(x \parallel x') = \textbf{compress}(h_K(x) \parallel x').$$

Since $h_K(x)$ and x' are both known, it is a simple matter for an opponent to compute $h_K(x \parallel x')$, even though K is secret. This is clearly a security problem.

Even if messages are padded, a modification of the above attack can be carried out. For example, suppose that $y = x \parallel \textbf{pad}(x)$ in the preprocessing step. Note that $|y| = rt$ for some integer r. Let w be any bitstring of length t, and define

$$x' = x \parallel \textbf{pad}(x) \parallel w.$$

In the preprocessing step, we would compute

$$y' = x' \parallel \textbf{pad}(x') = x \parallel \textbf{pad}(x) \parallel w \parallel \textbf{pad}(x'),$$

where $|y'| = r't$ for some integer $r' > r$.

Consider the computation of $h_K(x')$ (this is the same as computing $h(x')$ when $IV = K$). In the processing step, it is clear that $z_r = h_K(x)$. It is therefore possible for an adversary to compute the following:

$$z_{r+1} \leftarrow \textbf{compress}(h_K(x) \parallel y_{r+1})$$

$$z_{r+2} \leftarrow \textbf{compress}(z_{r+1} \parallel y_{r+2})$$

$$\vdots \quad \vdots \quad \vdots$$

$$z_{r'} \leftarrow \textbf{compress}(z_{r'-1} \parallel y_{r'}),$$

and then
$$h_K(x') = z_{r'}.$$

Therefore the adversary can compute $h_K(x')$ even though he doesn't know the secret key K (and notice that the attack makes no assumptions about the length of the pad).

Keeping the above examples in mind, we formulate definitions of what it should mean for a MAC algorithm to be secure. As we saw, the objective of an opponent is to try to produce a pair (x, y) that is valid under an unknown but fixed key, K. The opponent is allowed to request (up to) Q valid MACs on messages $x_1, x_2, \ldots$ of his own choosing. It is convenient to assume that there is a black box (or oracle) that answers the adversary's queries, and we will often use this terminology when it is convenient to do so.

Thus, the adversary obtains a list of valid pairs (under the unknown key K):

$$(x_1, y_1), (x_2, y_2), \ldots, (x_Q, y_Q)$$

by querying the oracle with messages $x_1, \ldots, x_Q$. Later, when the adversary outputs the pair (x, y), it is required that $x \notin \{x_1, \ldots, x_Q\}$. If, in addition, (x, y) is a valid pair, then the pair is said to be a *forgery*. If the probability that the adversary outputs a forgery is at least ϵ, then the adversary is said to be an (ϵ, Q)-*forger* for the given MAC. The probability ϵ could be taken to be either an average-case probability over all the possible keys, or worst-case. To be concrete, in the following sections we will generally take ϵ to be a worst-case probability. This means that the adversary can produce a forgery with probability at least ϵ, regardless of the secret key being used.

Using this terminology, the attacks described above are $(1, 1)$-forgers.

4.4.1 Nested MACs and HMAC

A *nested MAC* builds a MAC algorithm from the composition of two (keyed) hash families. Suppose that $(\mathcal{X}, \mathcal{Y}, \mathcal{K}, \mathcal{G})$ and $(\mathcal{Y}, \mathcal{Z}, \mathcal{L}, \mathcal{H})$ are hash families. The *composition* of these hash families is the hash family $(\mathcal{X}, \mathcal{Z}, \mathcal{M}, \mathcal{G} \circ \mathcal{H})$ in which $\mathcal{M} = \mathcal{K} \times \mathcal{L}$ and

$$\mathcal{G} \circ \mathcal{H} = \{g \circ h : g \in \mathcal{G}, h \in \mathcal{H}\},$$

where $(g \circ h)_{(K,L)}(x) = h_L(g_K(x))$ for all $x \in \mathcal{X}$. In this construction, $\mathcal{Y}$ and $\mathcal{Z}$ are finite sets such that $|\mathcal{Y}| \geq |\mathcal{Z}|$; $\mathcal{X}$ could be finite or infinite. If $\mathcal{X}$ is finite, then $|\mathcal{X}| > |\mathcal{Y}|$.

We are interested in finding situations under which we can guarantee that a nested MAC is secure, assuming that the two hash families from which it is constructed satisfy appropriate security requirements. Roughly speaking, it can be shown that the nested MAC is secure provided that the following two conditions are satisfied:

1. $(\mathcal{Y}, \mathcal{Z}, \mathcal{L}, \mathcal{H})$ is secure as a MAC, given a fixed (unknown) key, and

2. $(\mathcal{X}, \mathcal{Y}, \mathcal{K}, \mathcal{G})$ is collision-resistant, given a fixed (unknown) key.

Intuitively, we are building a secure "big MAC" (namely, the nested MAC) from the composition of a secure "little MAC" (namely, $(\mathcal{Y}, \mathcal{Z}, \mathcal{L}, \mathcal{H})$) and a collision-resistant keyed hash family (namely, $(\mathcal{X}, \mathcal{Y}, \mathcal{K}, \mathcal{G})$). Let's try to make the above conditions more precise, and then present a proof of a specific security result.

The security result will in fact be comparing the relative difficulties of certain types of attacks against the three hash families. We will therefore be considering all three of the following adversaries:

- a forger for the nested MAC (a "big MAC attack"),
- a forger for the little MAC (a "little MAC attack"), and
- a collision-finder for the hash family, when the key is secret (an "unknown-key collision attack").

Here is a more careful description of each of the three adversaries: In a *big MAC attack*, a pair of keys, (K, L), is chosen and kept secret. The adversary is allowed to choose values for x and query a big MAC oracle for values of $h_L(g_K(x))$. Then the adversary attempts to output a pair (x', z) such that $z = h_L(g_K(x'))$, where x' was not one of its previous queries.

In a *little MAC attack*, a key L is chosen and kept secret. The adversary is allowed to choose values for y and query a little MAC oracle for values of $h_L(y)$. Then the adversary attempts to output a pair (y', z) such that $z = h_L(y')$, where y' was not one of its previous queries.

In an *unknown-key collision attack*, a key K is chosen and kept secret. The adversary is allowed to choose values for x and query a hash oracle for values of $g_K(x)$. Then the adversary attempts to output a pair x', x'' such that $x' \neq x''$ and $g_K(x') = g_K(x'')$.

We will assume that there does not exist an $(\epsilon_1, q + 1)$-unknown-key collision attack for a randomly chosen function $g_K \in \mathcal{G}$. (If the key K were not secret, then this would correspond to our usual notion of collision resistance. Since we assume that K is secret, the problem facing the adversary is more difficult, and therefore we are making a weaker security assumption than collision-resistance.) We also assume that there does not exist an (ϵ_2, q)-little MAC attack for a randomly chosen function $h_L \in \mathcal{H}$, where L is secret. Finally, suppose that there exists an (ϵ, q)-big MAC attack for a randomly chosen function $(g \circ h)_{(K, L)} \in \mathcal{G} \circ \mathcal{H}$, where (K, L) is secret.

With probability at least ϵ, the big MAC attack outputs a valid pair (x, z) after making at most Q queries to a big MAC oracle. Let $x_1, \ldots, x_q$ denote the queries made by the adversary, and let $z_1, \ldots, z_q$ be the corresponding responses made by the oracle. After the adversary has finished executing, we have the list of valid pairs $(x_1, z_1), \ldots, (x_q, z_q)$, as well as the possibly valid pair (x, z).

Suppose we now take the values $x_1, \ldots, x_q$, and x, and make $q + 1$ queries to a hash oracle. We obtain the list of values $y_1 = g_K(x_1), \ldots, y_q = g_K(x_q)$, and $y = g_K(x)$. Suppose it happens that $y \in \{y_1, \ldots, y_q\}$, say $y = y_i$. Then we can output the pair x, x_i as a solution to **Collision**. This would be a successful unknown-key collision attack. On the other hand, if $y \notin \{y_1, \ldots, y_q\}$, then we output the pair (y, z), which (possibly) is a valid pair for the little MAC. This would be a forgery constructed after (indirectly) obtaining q answers to q little MAC queries, namely $(y_1, z_1), \ldots, (y_q, z_q)$.

By the assumption we made, any unknown-key collision attack has probability at most ϵ_1 of succeeding. As well, we assumed that the big MAC attack has success probability at least ϵ. Therefore, the probability that (x, z) is a valid pair and $y \notin \{y_1, \ldots, y_q\}$ is at least $\epsilon - \epsilon_1$. The success probability of any little MAC attack is at most ϵ_2, and the success probability of the little MAC attack described above is at least $\epsilon - \epsilon_1$. Hence, it follows that $\epsilon \leq \epsilon_1 + \epsilon_2$.

We have proven the following result.

THEOREM 4.9 *Suppose* $(\mathcal{X}, \mathcal{Z}, \mathcal{M}, \mathcal{G} \circ \mathcal{H})$ *is a nested MAC. Suppose there does not exist an* $(\epsilon_1, q + 1)$-*collision attack for a randomly chosen function* $g_K \in \mathcal{G}$, *when the key* K *is secret. Further, suppose that there does not exist an* (ϵ_2, q)-*forger for a randomly chosen function* $h_L \in \mathcal{H}$, *where* L *is secret. Finally, suppose there exists an* (ϵ, q)-*forger for the nested MAC, for a randomly chosen function* $(g \circ h)_{(K,L)} \in \mathcal{G} \circ \mathcal{H}$. *Then* $\epsilon \leq \epsilon_1 + \epsilon_2$.

HMAC is a nested MAC algorithm that was adopted as a FIPS standard in March, 2002. It constructs a MAC from an (unkeyed) hash function; we describe *HMAC* based on *SHA-1*. This version of *HMAC* uses a 512-bit key, denoted K. x is the message to be authenticated, and *ipad* and *opad* are 512-bit constants, defined in hexadecimal notation as follows:

$$ipad = 3636 \cdots 36$$

$$opad = 5C5C \cdots 5C$$

Then the 160-bit MAC is defined as follows:

$$\text{HMAC}_K(x) = \text{SHA-1}((K \oplus opad) \parallel \text{SHA-1}((K \oplus ipad) \parallel x)).$$

Note that *HMAC* uses *SHA-1* with the value $K \oplus ipad$, which is prepended to x, used as the key. This application of *SHA-1* is assumed to be secure against an unknown-key collision attack. Now the key value $K \oplus opad$ is prepended to the previously constructed message digest, and *SHA-1* is applied again. This second computation of *SHA-1* requires only one application of the compression function, and we are assuming that *SHA-1* when used in this way is secure as a MAC. If these two assumptions are valid, then Theorem 4.9 says that *HMAC* is secure as a MAC.

Cryptosystem 4.2: $CBC\text{-}MAC(x, K)$

denote $x = x_1 \,\|\, \cdots \,\|\, x_n$
IV $\leftarrow 00 \cdots 0$
$y_0 \leftarrow$ IV
for $i \leftarrow 1$ **to** n
   do $y_i \leftarrow e_K(y_{i-1} \oplus x_i)$
return (y_n)

4.4.2 CBC-MAC and Authenticated Encryption

One of the most popular ways to construct a MAC is to use a block cipher in CBC mode with a fixed (public) initialization vector. In CBC mode, recall that each ciphertext block y_i is x-ored with the next plaintext block, x_{i+1}, before being encrypted with the secret key K. More formally, we start with an initialization vector, denoted by IV, and define $y_0 = $ IV. Then we construct $y_1, y_2, \ldots$ using the rule

$$y_i = e_K(y_{i-1} \oplus x_i),$$

for all $i \geq 1$.

Suppose that $(\mathcal{P}, \mathcal{C}, \mathcal{K}, \mathcal{E}, \mathcal{D})$ is an endomorphic cryptosystem, where $\mathcal{P} = \mathcal{C} = \{0, 1\}^t$. Let IV be the bitstring consisting of t zeroes, and let $K \in \mathcal{K}$ be a secret key. Finally, let $x = x_1 \,\|\, \cdots \,\|\, x_n$ be a bitstring of length tn (for some positive integer n), where each x_i is a bitstring of length t. We compute CBC-MAC(x, K) as shown in Algorithm 4.2.

The best known general attack on $CBC\text{-}MAC$ is a birthday (collision) attack. We describe this attack now. Basically, we assume that the adversary can request MACs on a large number of messages. If a duplicated MAC is found, then the adversary can construct an additional message and request its MAC. Finally, the adversary can produce a new message and its corresponding MAC (i.e., a forgery), even though he does not know the secret key. The attack works for messages of any prespecified fixed size.

In preparation for the attack, let $n \geq 3$ be an integer and let $x_3, \ldots, x_n$ be fixed bitstrings of length t. Let $q \approx 1.17 \times 2^{t/2}$ be an integer, and choose any q distinct bitstrings of length t, which we denote $x_1^1, \ldots, x_1^q$. Next, let $x_2^1, \ldots, x_2^q$ be randomly chosen bitstrings of length t. Finally, for $1 \leq i \leq q$ and for $3 \leq k \leq n$, define $x_k^i = x_k$, and then define

$$x^i = x_1^i \,\|\, \cdots \,\|\, x_n^i$$

for $1 \leq i \leq q$. Note that $x^i \neq x^j$ if $i \neq j$, because $x_1^i \neq x_1^j$.

The attack can now be carried out. First the adversary requests the MACs of $x^1, x^2, \ldots, x^q$. In the computation of the MAC of each x^i using Cryptosystem

4.2, values $y_0^i, \ldots, y_n^i$ are computed, and y_n^i is the resulting MAC. Now suppose that x^i and x^j have identical MACs, i.e., $y_n^i = y_n^j$. This happens if and only if $y_2^i = y_2^j$, which in turn happens if and only if

$$y_1^i \oplus x_2^i = y_1^j \oplus x_2^j.$$

Let x_δ be any nonzero bitstring of length t. Define

$$v = x_1^i \, \| \, (x_2^i \oplus x_\delta) \, \| \cdots \| \, x_n^i$$

and

$$w = x_1^j \, \| \, (x_2^j \oplus x_\delta) \, \| \cdots \| \, x_n^j.$$

Then the adversary requests the MAC of v. It is not difficult to see that v and w have identical MACs, so the adversary is able to construct the MAC of w even though he does not know the key K. This attack produces a $(1/2, O(2^{t/2}))$-forger.

It is known that *CBC-MAC* is secure if the underlying encryption satisfies appropriate security properties. That is, if certain plausible but unproved assumptions about the randomness of an encryption scheme are true, then *CBC-MAC* will be secure.

The CCM mode of operation provides *authenticated encryption* by producing a MAC as part of the encryption process. CCM mode uses counter mode encryption in addition to a CBC-MAC. Let K be the encryption key and let $x = x_1 \, \| \cdots \| \, x_n$ be the plaintext. As in counter mode, we choose a counter, ctr. Then we construct the sequence of counters $T_0, T_1, T_2, \ldots, T_n$, defined as follows:

$$T_i = ctr + i \bmod 2^m$$

for $0 \leq i \leq n$, where m is the block length of the cipher. We encrypt the plaintext blocks $x_1, x_2, \ldots, x_n$ by computing

$$y_i = x_i \oplus e_K(T_i),$$

for all $i \geq 1$. Then we compute $temp = \text{CBC-MAC}(x, K)$ and $y' = T_0 \oplus temp$. The ciphertext consists of the string $y = y_1 \, \| \cdots \| \, y_n \, \| \, y'$.

To decrypt and verify y, one would first decrypt $y_1 \, \| \cdots \| \, y_n$ using counter mode decryption with the counter sequence $T_1, T_2, \ldots, T_n$, obtaining the plaintext string x. The second step is to compute $\text{CBC-MAC}(x, K)$ and see if it is equal to $y' \oplus T_0$. The ciphertext is rejected if this condition does not hold.

4.5 Unconditionally Secure MACs

In this section, we define universal hash families and discuss their application to the construction of unconditionally secure MACs. In our study of unconditionally secure MACs, we assume that a key is used to produce only one authentication tag. Therefore, an adversary is limited to making at most one query before he outputs a (possible) forgery. Stated another way, we will construct MACs for which we can prove that there do not exist $(\epsilon, 0)$- and $(\epsilon, 1)$-forgers, for appropriate values of ϵ, even when the adversary possesses infinite computing power.

For any integer $q \geq 0$, we define the *deception probability* Pd_q to be the maximum value of ϵ such that an (ϵ, q)-forger exists. This maximum is computed over all possible values of the secret key, K, assuming that K is chosen uniformly at random from $\mathcal{K}$. In general, we want to construct MACs for which Pd_0 and Pd_1 are small. This means that the adversary has a low probability of successfully carrying out an attack, no matter which key is used. Sometimes we will refer to the attack carried out by an $(\epsilon, 0)$-forger as *impersonation*, and the attack carried out by an $(\epsilon, 1)$-forger will be termed *substitution*.

We illustrate the above concepts by considering a small example of an unconditionally secure MAC.

Example 4.1   Suppose

$$\mathcal{X} = \mathcal{Y} = \mathbb{Z}_3$$

and

$$\mathcal{K} = \mathbb{Z}_3 \times \mathbb{Z}_3.$$

For each $K = (a, b) \in \mathcal{K}$ and each $x \in \mathcal{X}$, define

$$h_{(a,b)}(x) = ax + b \bmod 3,$$

and then define

$$\mathcal{H} = \{h_{(a,b)} : (a, b) \in \mathbb{Z}_3 \times \mathbb{Z}_3\}.$$

It will be useful to study the *authentication matrix* of the hash family $(\mathcal{X}, \mathcal{Y}, \mathcal{K}, \mathcal{H})$, which tabulates all the values $h_{(a,b)}(x)$ as follows. For each key $(a, b) \in \mathcal{K}$ and for each $x \in \mathcal{X}$, place the authentication tag $h_{(a,b)}(x)$ in row (a, b) and column x of a $|\mathcal{K}| \times |\mathcal{X}|$ matrix, say M. The array M is presented in Figure 4.2.

Let's first consider an impersonation attack. Oscar will pick a message x, and attempt to guess the "correct" authentication tag. Denote by K_0 the actual key being used (which is unknown to Oscar). Oscar will succeed in creating a forgery if he guesses the tag $y_0 = h_{K_0}(x)$. However, for any $x \in \mathcal{X}$ and $y \in \mathcal{Y}$, it is easy to verify that there are exactly three (out of nine) keys $K \in \mathcal{K}$ such that $h_K(x) = y$. (In other words, each symbol occurs three times in each column of the authentication matrix.) Thus, any pair (x, y) will be a valid pair with probability $1/3$. Hence, it follows that $Pd_0 = 1/3$.

Substitution is a bit more complicated to analyze. As a specific case, suppose Oscar queries the tag for $x = 0$, and is given the answer $y = 0$. Therefore $(x, y) = (0, 0)$ is a valid pair. This gives Oscar some information about the key: he knows that

$$K_0 \in \{(0, 0), (1, 0), (2, 0)\}.$$

FIGURE 4.2
An authentication matrix

key	0	1	2
$(0,0)$	0	0	0
$(0,1)$	1	1	1
$(0,2)$	2	2	2
$(1,0)$	0	1	2
$(1,1)$	1	2	0
$(1,2)$	2	0	1
$(2,0)$	0	2	1
$(2,1)$	1	0	2
$(2,2)$	2	1	0

Now suppose Oscar outputs the pair $(1, 1)$ as a (possible) forgery. The pair $(1, 1)$ is a forgery if and only if $K_0 = (1, 0)$. The (conditional) probability that K_0 is the key, given that $(0, 0)$ is a valid pair, is $1/3$, since the key is known to be in the set $\{(0, 0), (1, 0), (2, 0)\}$.

A similar analysis can be done for any value of x for which Oscar queries the tag y, and for any pair (x', y') (where $x' \neq x$) that Oscar outputs as his (possible) forgery. In general, knowledge of any valid pair (x, y) restricts the key to one of three possibilities. Then, for each choice of (x', y') (where $x' \neq x$), it can be verified that there is one key (out of the three possible keys) under which y' is the correct authentication tag for x'. Hence, it follows that $Pd_1 = 1/3$. □

We now discuss how to compute the deception probabilities for an arbitrary message authentication code by examining its authentication matrix. First, we consider Pd_0. As above, let K_0 denote the key chosen by Alice and Bob. For $x \in X$ and $y \in Y$, define **payoff**(x, y) to be the probability that the pair (x, y) is valid. It is not difficult to see that

$$\mathbf{payoff}(x, y) = \mathbf{Pr}[y = h_{K_0}(x)]$$
$$= \frac{|\{K \in \mathcal{K} : h_K(x) = y\}|}{|\mathcal{K}|}.$$

That is, **payoff**(x, y) is computed by counting the number of rows of the authentication matrix that have entry y in column x, and dividing the result by the number of possible keys.

In order to maximize his chance of success, Oscar will choose (x, y) such that

payoff (x, y) is a maximum. Hence, we have the following formula:

$$Pd_0 = \max\{\mathbf{payoff}(x, y) : x \in \mathcal{X}, y \in \mathcal{Y}\}. \tag{4.1}$$

Now, we turn our attention to substitution. Suppose we fix $x \in \mathcal{X}$ and $y \in \mathcal{Y}$ such that (x, y) is a valid pair. (x is a possible query by Oscar, and y would be the oracle's reply.) Now let $x' \in \mathcal{X}$, where $x' \neq x$. Define **payoff** $(x', y'; x, y)$ to be the probability that (x', y') is a valid pair, given that (x, y) is a valid pair. As before, let K_0 denote the key chosen by Alice and Bob. Then we can compute the following:

$$\begin{aligned}
\mathbf{payoff}(x', y'; x, y) &= \mathbf{Pr}[y' = h_{K_0}(x') | y = h_{K_0}(x)] \\
&= \frac{\mathbf{Pr}[y' = h_{K_0}(x') \wedge y = h_{K_0}(x)]}{\mathbf{Pr}[y = h_{K_0}(x)]} \\
&= \frac{|\{K \in \mathcal{K} : h_K(x') = y', h_K(x) = y\}|}{|\{K \in \mathcal{K} : h_K(x) = y\}|}.
\end{aligned}$$

The numerator of this fraction is the number of rows of the authentication matrix that have the value y in column x, and also have the value y' in column x'; the denominator is the number of rows that have the value y in column x. Note that the denominator is non-zero because we are assuming that (x, y) is a valid pair under at least one key.

Suppose we define

$$\mathcal{V} = \{(x, y) : |\{K \in \mathcal{K} : h_K(x) = y\}| \geq 1\}.$$

Observe that $\mathcal{V}$ is just the set of all pairs (x, y) that are valid pairs under at least one key. Then the following formula can be used to compute Pd_1:

$$Pd_1 = \max_x \left\{ \min_y \left\{ \max_{(x', y')} \{\mathbf{payoff}(x', y'; x, y)\} \right\} \right\}, \tag{4.2}$$

where $x' \neq x$ and $(x, y) \in \mathcal{V}$.

Some explanation would be helpful, as this is a complicated formula. Recall that Oscar first chooses the value of x in any manner that he wishes. Then Oscar is given the authentication tag y (where $y = h_K(x)$ and K is the unknown key), Finally, Oscar selects (x', y') such that **payoff** $(x', y'; x, y)$ is a maximum. It should be clear that we maximize over the possible choices of x and (x', y') in computing Pd_1 because that is how Oscar will choose them. On the other hand, the value of y is not chosen by Oscar; it depends on the unknown key K. Therefore, in order to be able to say that Oscar's success probability is at least Pd_1, we need to minimize over the possible choices of y.

4.5.1 Strongly Universal Hash Families

Strongly universal hash families are used in several areas of cryptography. We begin with a definition of these important objects.

Definition 4.2:   Suppose that $(X, Y, K, \mathcal{H})$ is an (N, M) hash family. This hash family is *strongly universal* provided that the following condition is satisfied for every $x, x' \in X$ such that $x \neq x'$, and for every $y, y' \in Y$:

$$|\{K \in K : h_K(x) = y, h_K(x') = y'\}| = \frac{|K|}{M^2}.$$

As an example, the reader can verify that the hash family in Example 4.1 is a strongly universal $(3, 3)$-hash family.

Strongly universal hash families immediately yield authentication codes in which Pd_0 and Pd_1 can easily be computed. We prove a theorem on the values of these deception probabilities after stating and proving a simple lemma about strongly universal hash families.

LEMMA 4.10 *Suppose that $(X, Y, K, \mathcal{H})$ is a strongly universal (N, M)-hash family. Then*

$$|\{K \in K : h_K(x) = y\}| = \frac{|K|}{M},$$

for every $x \in X$ and for every $y \in Y$.

PROOF Let $x, x' \in X$ and $y \in Y$ be fixed, where $x \neq x'$. Then we have the following:

$$|\{K \in K : h_K(x) = y\}| = \sum_{y' \in Y} |\{K \in K : h_K(x) = y, h_K(x') = y'\}|$$

$$= \sum_{y' \in Y} \frac{|K|}{M^2}$$

$$= \frac{|K|}{M}.$$

∎

THEOREM 4.11 *Suppose that $(X, Y, K, \mathcal{H})$ is a strongly universal (N, M)-hash family. Then $(X, Y, K, \mathcal{H})$ is an authentication code with $Pd_0 = Pd_1 = 1/M$.*

PROOF We proved in Lemma 4.10 that

$$|\{K \in K : h_K(x) = y\}| = \frac{|K|}{M},$$

for every $x \in X$ and for every $y \in Y$. Therefore **payoff**$(x, y) = 1/M$ for every $x \in X, y \in Y$, and hence $Pd_0 = 1/M$.

Now let $x, x' \in X$ such that $x \neq x'$, and let $y, y' \in Y$, where $(x, y) \in V$. We have that

$$\mathbf{payoff}(x', y'; x, y) = \frac{|\{K \in \mathcal{K} : h_K(x') = y', h_K(x) = y\}|}{|\{K \in \mathcal{K} : h_K(x) = y\}|}$$

$$= \frac{|\mathcal{K}|/M^2}{|\mathcal{K}|/M}$$

$$= \frac{1}{M}.$$

Therefore $Pd_1 = 1/M$. ∎

We now give some constructions of strongly universal hash families. Our first construction generalizes Example 4.1.

THEOREM 4.12 *Let p be prime. For $a, b \in \mathbb{Z}_p$, define $f_{(a,b)} : \mathbb{Z}_p \to \mathbb{Z}_p$ by the rule*

$$f_{(a,b)}(x) = ax + b \bmod p.$$

Then $(\mathbb{Z}_p, \mathbb{Z}_p, \mathbb{Z}_p \times \mathbb{Z}_p, \{f_{(a,b)} : a, b \in \mathbb{Z}_p\})$ is a strongly universal (p, p)-hash family.

PROOF   Suppose that $x, x', y, y' \in \mathbb{Z}_p$, where $x \neq x'$. We will show that there is a unique key $(a, b) \in \mathbb{Z}_p \times \mathbb{Z}_p$ such that $ax + b \equiv y \pmod{p}$ and $ax' + b \equiv y' \pmod{p}$. This is not difficult, as (a, b) is the solution of a system of two linear equations in two unknowns over $\mathbb{Z}_p$. Specifically,

$$a = (y' - y)(x' - x)^{-1} \bmod p, \quad \text{and}$$
$$b = y - x(y' - y)(x' - x)^{-1} \bmod p.$$

(Note that $(x' - x)^{-1} \bmod p$ exists because $x \not\equiv x' \pmod{p}$ and p is prime.)   ∎

Here is a construction for classes of strongly universal hash families in which the domain can have much larger cardinality than the range.

THEOREM 4.13 *Let ℓ be a positive integer and let p be prime. Define*

$$X = \{0, 1\}^{\ell} \backslash \{(0, \ldots, 0)\}.$$

For every $\vec{r} \in (\mathbb{Z}_p)^{\ell}$, define $f_{\vec{r}} : X \to \mathbb{Z}_p$ by the rule

$$f_{\vec{r}}(\vec{x}) = \vec{r} \cdot \vec{x} \bmod p,$$

where $\vec{x} \in X$ and

$$\vec{r} \cdot \vec{x} = \sum_{i=1}^{\ell} r_i x_i$$

is the usual inner product of vectors. Then $(X, \mathbb{Z}_p, (\mathbb{Z}_p)^\ell, \{f_{\vec{r}} : \vec{r} \in (\mathbb{Z}_p)^\ell\})$ *is a strongly universal* $(2^\ell - 1, p)$-*hash family.*

PROOF   Let $\vec{x}, \vec{x}' \in X$, $\vec{x} \neq \vec{x}'$, and let $y, y' \in \mathbb{Z}_p$. We want to show that the number of vectors $\vec{r} \in (\mathbb{Z}_p)^\ell$ such that $\vec{r} \cdot \vec{x} \equiv y \pmod{p}$ and $\vec{r} \cdot \vec{x}' \equiv y' \pmod{p}$ is a constant. The desired vectors $\vec{r}$ are the solution of two linear equations in ℓ unknowns over $\mathbb{Z}_p$. The two equations are linearly independent, and so the number of solutions to the linear system is $p^{\ell-2}$, which is a constant. ∎

4.5.2   Optimality of Deception Probabilities

In this section, we prove some lower bounds on deception probabilities of unconditionally secure MACs, which show that the authentication codes derived from strongly universal hash families have minimum possible deception probabilities.

Suppose (X, Y, K, H) is an (N, M)-hash family. Suppose we fix a message $x \in X$. Then we can compute as follows:

$$\sum_{y \in Y} \textbf{payoff}(x, y) = \sum_{y \in Y} \frac{|\{K \in K : h_K(x) = y\}|}{|K|}$$

$$= \frac{|K|}{|K|}$$

$$= 1.$$

Hence, for every $x \in X$, there exists an authentication tag y (depending on x), such that

$$\textbf{payoff}(x, y) \geq \frac{1}{M}.$$

The following theorem is an easy consequence of the above computations.

THEOREM 4.14   *Suppose* (X, Y, K, H) *is an* (N, M)-*hash family. Then* $Pd_0 \geq 1/M$. *Further,* $Pd_0 = 1/M$ *if and only if*

$$|\{K \in K : h_K(x) = y\}| = \frac{|K|}{M} \tag{4.3}$$

for every $x \in X$, $y \in Y$.

Now, we turn our attention to substitution. Suppose that we fix $x, x' \in X$ and $y, y' \in Y$, where $x \neq x'$ and $(x, y) \in V$. We have the following:

$$\sum_{y' \in Y} \textbf{payoff}(x', y'; x, y) = \sum_{y' \in Y} \frac{|\{K \in K : h_K(x') = y', h_K(x) = y\}|}{|\{K \in K : h_K(x) = y\}|}$$

$$= \frac{|\{K \in K : h_K(x) = y\}|}{|\{K \in K : h_K(x) = y\}|}$$

$$= 1.$$

Hence, for each $(x, y) \in V$ and for each x' such that $x' \neq x$, there exists an authentication tag y' such that

$$\mathbf{payoff}(x', y'; x, y) \geq \frac{1}{M}.$$

We have proven the following theorem.

THEOREM 4.15 *Suppose $(\mathcal{X}, \mathcal{Y}, \mathcal{K}, \mathcal{H})$ is an (N, M)-hash family. Then $Pd_1 \geq 1/M$.*

With a bit more work, we can determine necessary and sufficient conditions such that $Pd_1 = 1/M$.

THEOREM 4.16 *Suppose $(\mathcal{X}, \mathcal{Y}, \mathcal{K}, \mathcal{H})$ is an (N, M)-hash family. Then $Pd_1 = 1/M$ if and only if the hash family is strongly universal.*

PROOF We proved already in Theorem 4.11 that $Pd_1 = 1/M$ if the hash family is strongly universal. We need to prove the converse now; so, we assume that $Pd_1 = 1/M$.

We will show first that $V = \mathcal{X} \times \mathcal{Y}$. Let $(x', y') \in \mathcal{X} \times \mathcal{Y}$; we will show that $(x', y') \in V$. Let $x \in \mathcal{X}$, $x \neq x'$. Choose any $y \in \mathcal{Y}$ such that $(x, y) \in V$. From the discussion preceding Theorem 4.15, it is clear that

$$\frac{|\{K \in \mathcal{K} : h_K(x') = y', h_K(x) = y\}|}{|\{K \in \mathcal{K} : h_K(x) = y\}|} = \frac{1}{M} \tag{4.4}$$

for every $x, x' \in \mathcal{X}$, $x' \neq x$, $y, y' \in \mathcal{Y}$ such that $(x, y) \in V$. Therefore

$$|\{K \in \mathcal{K} : h_K(x') = y', h_K(x) = y\}| > 0,$$

and hence

$$|\{K \in \mathcal{K} : h_K(x') = y'\}| > 0.$$

This proves that $(x', y') \in V$, and hence $V = \mathcal{X} \times \mathcal{Y}$.

Now, let's look again at (4.4). Let $x, x' \in \mathcal{X}$, $x \neq x'$, and let $y, y' \in \mathcal{Y}$. We now know that $(x, y) \in V$ and $(x', y') \in V$, so we can interchange the roles of (x, y) and (x', y') in (4.4). This yields

$$|\{K \in \mathcal{K} : h_K(x) = y\}| = |\{K \in \mathcal{K} : h_K(x') = y'\}|$$

for all such x, x', y, y'. Hence, the quantity

$$|\{K \in \mathcal{K} : h_K(x) = y\}|$$

is a constant. (In other words, the number of occurrences of any symbol y in any column x of the authentication matrix x is a constant.) Now, we can return one last time to (4.4), and it follows that the quantity

$$|\{K \in \mathcal{K} : h_K(x') = y', h_K(x) = y\}|$$

is also a constant. Therefore the hash family is strongly universal. ∎

The following corollary establishes that $Pd_0 = 1/M$ whenever $Pd_1 = 1/M$.

COROLLARY 4.17 *Suppose* $(\mathcal{X}, \mathcal{Y}, \mathcal{K}, \mathcal{H})$ *is an* (N, M)-*hash family such that* $Pd_1 = 1/M$. *Then* $Pd_0 = 1/M$.

PROOF Under the stated hypotheses, Theorem 4.16 says that $(\mathcal{X}, \mathcal{Y}, \mathcal{K}, \mathcal{H})$ is strongly universal. Then $Pd_0 = 1/M$ from Theorem 4.11. ∎

4.6 Notes and References

For a good recent survey on hash functions, see Preneel [274]. Concepts such as preimage resistance and collision resistance have been discussed for some time; see [274] for further details.

The random oracle model was introduced by Bellare and Rogaway in [22]; the analyses in Section 4.2.2 are based on Stinson [324].

The material from Section 4.3 is based on Damgård [102]. Similar methods were discovered by Merkle [239].

Rivest's *MD4* and *MD5* hashing algorithms are described in [281] and [282], respectively. *SHA* (aka *SHA-0*) is the subject of FIPS publication 180 [140]; it was superceded by *SHA-1*, which is described in FIPS publication 180-1 [141]. FIPS publication 180-2 [142] includes the additional hash algorithms *SHA-256*, *SHA-384* and *SHA-512*.

In 1998, Dobbertin [120] found collisions for *MD4*. Collisions for the compression function of *MD5* were previously found, in 1996, by Dobbertin [119]. A collision for the full *MD5* is presented in Wang, Feng, Lai and Yu [338]. The current status of collision search attacks on *SHA-1* (as of March, 2005) is discussed in Wang, Yin and Yu [339].

The *SHA-0* collision found by Joux and announced at CRYPTO 2004 has apparently not been published. It is, however, available from various sources on the web. The two messages are both 2048 bits in length, and they are presented in Table 4.1 using hexadecimal notation. The hash value for both messages is

$$\texttt{c9f160777d4086fe8095fba58b7e20c228a4006b}$$

Two papers describing some of the relevant theory used to discover the above-mentioned collisions are Chabaud and Joux [85] and Biham and Chen [37].

Preneel and van Oorschot [276] is a recent study of iterated message authentication codes. Security proofs for several types of MACs have been published:

TABLE 4.1
A collision for *SHA-0*

First message:

```
a766a602  b65cffe7  73bcf258  26b322b3
d01b1a97  2684ef53  3e3b4b7f  53fe3762
24c08e47  e959b2bc  3b519880  b9286568
247d110f  70f5c5e2  b4590ca3  f55f52fe
effd4c8f  e68de835  329e603c  c51e7f02
545410d1  671d108d  f5a4000d  cf20a439
4949d72c  d14fbb03  45cf3a29  5dcda89f
998f8755  2c9a58b1  bdc38483  5e477185
f96e68be  bb0025d2  d2b69edf  21724198
f688b41d  eb9b4913  fbe696b5  457ab399
21e1d759  1f89de84  57e8613c  6c9e3b24
2879d4d8  783b2d9c  a9935ea5  26a729c0
6edfc501  37e69330  be976012  cc5dfe1c
14c4c68b  d1db3ecb  24438a59  a09b5db4
35563e0d  8bdf572f  77b53065  cef31f32
dc9dbaa0  4146261e  9994bd5c  d0758e3d
```

Second message:

```
a766a602  b65cffe7  73bcf258  26b322b1
d01b1ad7  2684ef51  be3b4b7f  d3fe3762
a4c08e45  e959b2fc  3b519880  39286528
a47d110d  70f5c5e0  34590ce3  755f52fc
6ffd4c8d  668de875  329e603e  451e7f02
d45410d1  e71d108d  f5a4000d  cf20a439
4949d72c  d14fbb01  45cf3a69  5dcda89d
198f8755  ac9a58b1  3dc38481  5e4771c5
796e68fe  bb0025d0  52b69edd  a17241d8
7688b41f  6b9b4911  7be696f5  c57ab399
a1e1d719  9f89de86  57e8613c  ec9e3b26
a879d498  783b2d9e  29935ea7  a6a72980
6edfc503  37e69330  3e976010  4c5dfe5c
14c4c689  51db3ecb  a4438a59  209b5db4
35563e0d  8bdf572f  77b53065  cef31f30
dc9dbae0  4146261c  1994bd5c  50758e3d
```

Bellare, Canetti and Krawczyk [16] proved the security of *HMAC*; Bellare, Kilian and Rogaway [19] showed that *CBC-MAC* is secure; and a MAC known as *XOR-MAC*, due to Bellare, Guerin and Rogaway, is shown to be secure in [18].

The use of cipher block chaining mode for message authentication was first specified in FIPS publication 113 [139] in 1985. A modification of *CBC-MAC* known as *CMAC* is presented in the NIST (draft) special publication 800-38B [122]. *CMAC* is based on *OMAC*, which is due to Iwata and Kurosawa [181]. *HMAC* was adopted as a standard; see FIPS publication 198 [147].

Unconditionally secure authentication codes were invented in 1974 by Gilbert,

MacWilliams and Sloane [156]. Much of the theory of unconditionally secure authentication codes was developed by Simmons, who proved many fundamental results in the area; Simmons [304] is a good survey. Our treatment differs from the model introduced by Simmons in that we are considering active attacks, in which the adversary queries an oracle for an authentication tag before producing a (possible) forgery. In the model considered by Simmons, the attacks are passive attacks: the adversary observes a message with a corresponding authentication tag, but this message is not chosen by the adversary.

Universal hash families were introduced by Carter and Wegman [84, 341]. Their paper [341] was the first to apply strongly universal hash families to authentication. *Almost strongly universal hash families*, which allow the key length of unconditionally secure MACs to be greatly reduced, were formally defined by Stinson [321]. For more on this and related topics, see the expository paper by Stinson [322]. Finally, we note that universal hash families are also used in the construction of efficient computationally secure MACs; one such MAC is *UMAC*, which is described in Black *et al.* [40].

Exercises

4.1 Suppose $h : X \to Y$ is an (N, M)-hash function. For any $y \in Y$, let
$$h^{-1}(y) = \{x : h(x) = y\}$$
and denote $s_y = |h^{-1}(y)|$. Define
$$S = |\{\{x_1, x_2\} : h(x_1) = h(x_2)\}|.$$
Note that S counts the number of unordered pairs in X that collide under h.

(a) Prove that
$$\sum_{y \in Y} s_y = N,$$
so the mean of the s_y's is
$$\bar{s} = \frac{N}{M}.$$

(b) Prove that
$$S = \sum_{y \in Y} \binom{s_y}{2} = \frac{1}{2} \sum_{y \in Y} s_y^2 - \frac{N}{2}.$$

(c) Prove that
$$\sum_{y \in Y} (s_y - \bar{s})^2 = 2S + N - \frac{N^2}{M}.$$

(d) Using the result proved in part (c), prove that
$$S \geq \frac{1}{2} \left(\frac{N^2}{M} - N \right).$$
Further, show that equality is attained if and only if
$$s_y = \frac{N}{M}$$

for every $y \in \mathcal{Y}$.

4.2 As in Exercise 4.1, suppose $h : \mathcal{X} \to \mathcal{Y}$ is an (N, M)-hash function, and let

$$h^{-1}(y) = \{x : h(x) = y\}$$

for any $y \in \mathcal{Y}$. Let ϵ denote the probability that $h(x_1) = h(x_2)$, where x_1 and x_2 are random (not necessarily distinct) elements of $\mathcal{X}$. Prove that

$$\epsilon \geq \frac{1}{M},$$

with equality if and only if

$$|h^{-1}(y)| = \frac{N}{M}$$

for every $y \in \mathcal{Y}$.

4.3 Suppose that $h : \mathcal{X} \to \mathcal{Y}$ is an (N, M)-hash function, let

$$h^{-1}(y) = \{x : h(x) = y\}$$

and let $s_y = |h^{-1}(y)|$ for any $y \in \mathcal{Y}$. Suppose that we try to solve **Preimage** for the function h, using Algorithm 4.1, assuming that we have only oracle access for h. For a given $y \in \mathcal{Y}$, suppose that $\mathcal{X}_0$ is chosen to be a random subset of $\mathcal{X}$ having cardinality q.

(a) Prove that the success probability of Algorithm 4.1, given y, is

$$1 - \frac{\binom{N-s_y}{q}}{\binom{N}{q}}.$$

(b) Prove that the average success probabilty of Algorithm 4.1 (over all $y \in \mathcal{Y}$) is

$$1 - \frac{1}{M} \sum_{y \in \mathcal{Y}} \frac{\binom{N-s_y}{q}}{\binom{N}{q}}.$$

(c) In the case $q = 1$, show that the success probability in part (b) is $1/M$.

4.4 Suppose that $h : \mathcal{X} \to \mathcal{Y}$ is an (N, M)-hash function, let

$$h^{-1}(y) = \{x : h(x) = y\}$$

and let $s_y = |h^{-1}(y)|$ for any $y \in \mathcal{Y}$. Suppose that we try to solve **Second Preimage** for the function h, using Algorithm 4.2, assuming that we have only oracle access for h. For a given $x \in \mathcal{Y}$, suppose that $\mathcal{X}_0$ is chosen to be a random subset of $\mathcal{X}\backslash\{x\}$ having cardinality $q - 1$.

(a) Prove that the success probability of Algorithm 4.2, given x, is

$$1 - \frac{\binom{N-s_y}{q-1}}{\binom{N-1}{q-1}}.$$

(b) Prove that the average success probabilty of Algorithm 4.2 (over all $x \in \mathcal{X}$) is

$$1 - \frac{1}{N} \sum_{y \in \mathcal{Y}} \frac{s_y \binom{N-s_y}{q-1}}{\binom{N-1}{q-1}}.$$

(c) In the case $q = 2$, show that the success probability in part (b) is

$$\frac{\sum_{y \in \mathcal{Y}} s_y^2}{N(N-1)} - \frac{1}{N-1}.$$

4.5 If we define a hash function (or compression function) h that will hash an n-bit binary string to an m-bit binary string, we can view h as a function from $\mathbb{Z}_{2^n}$ to $\mathbb{Z}_{2^m}$. It is tempting to define h using integer operations modulo 2^m. We show in this exercise that some simple constructions of this type are insecure and should therefore be avoided.

   (a) Suppose that $n = m > 1$ and $h : \mathbb{Z}_{2^m} \to \mathbb{Z}_{2^m}$ is defined as

   $$h(x) = x^2 + ax + b \bmod 2^m.$$

   Prove that it is easy to solve **Second Preimage** for any $x \in \mathbb{Z}_{2^m}$ without having to solve a quadratic equation.

   (b) Suppose that $n > m$ and $h : \mathbb{Z}_{2^n} \to \mathbb{Z}_{2^m}$ is defined to be a polynomial of degree d:

   $$h(x) = \sum_{i=0}^{d} a_i x^i \bmod 2^m,$$

   where $a_i \in \mathbb{Z}$ for $0 \le i \le d$. Prove that it is easy to solve **Second Preimage** for any $x \in \mathbb{Z}_{2^n}$ without having to solve a polynomial equation.

4.6 Suppose that $f : \{0, 1\}^m \to \{0, 1\}^m$ is a preimage resistant bijection. Define $h : \{0, 1\}^{2m} \to \{0, 1\}^m$ as follows. Given $x \in \{0, 1\}^{2m}$, write

   $$x = x' \parallel x''$$

   where $x', x'' \in \{0, 1\}^m$. Then define

   $$h(x) = f(x' \oplus x'').$$

   Prove that h is not second preimage resistant.

4.7 For $M = 365$ and $15 \le q \le 30$, compare the exact value of ϵ given by the formula in the statement of Theorem 4.4 with the estimate for ϵ derived after the proof of that theorem.

4.8 Suppose $h : \mathcal{X} \to \mathcal{Y}$ is a hash function where $|\mathcal{X}|$ and $|\mathcal{Y}|$ are finite and $|\mathcal{X}| \ge 2|\mathcal{Y}|$. Suppose that h is balanced (i.e.,

   $$|h^{-1}(y)| = \frac{|\mathcal{X}|}{|\mathcal{Y}|}$$

   for all $y \in \mathcal{Y}$). Finally, suppose ORACLE-PREIMAGE is an (ϵ, Q)-algorithm for **Preimage**, for the fixed hash function h. Prove that COLLISION-TO-PREIMAGE is an $(\epsilon/2, Q + 1)$-algorithm for **Collision**, for the fixed hash function h.

4.9 Suppose $h_1 : \{0, 1\}^{2m} \to \{0, 1\}^m$ is a collision resistant hash function.

   (a) Define $h_2 : \{0, 1\}^{4m} \to \{0, 1\}^m$ as follows:

   1. Write $x \in \{0, 1\}^{4m}$ as $x = x_1 \parallel x_2$, where $x_1, x_2 \in \{0, 1\}^{2m}$.
   2. Define $h_2(x) = h_1(h_1(x_1) \parallel h_1(x_2))$.

   Prove that h_2 is collision resistant.

   (b) For an integer $i \ge 2$, define a hash function $h_i : \{0, 1\}^{2^i m} \to \{0, 1\}^m$ recursively from h_{i-1}, as follows:

   1. Write $x \in \{0, 1\}^{2^i m}$ as $x = x_1 \parallel x_2$, where $x_1, x_2 \in \{0, 1\}^{2^{i-1} m}$.
   2. Define $h_i(x) = h_1(h_{i-1}(x_1) \parallel h_{i-1}(x_2))$.

   Prove that h_i is collision resistant.

4.10 In this exercise, we consider a simplified version of the Merkle-Damgård construction. Suppose

   $$\mathbf{compress} : \{0, 1\}^{m+t} \to \{0, 1\}^m,$$

where $t \geq 1$, and suppose that

$$x = x_1 \parallel x_2 \parallel \cdots \parallel x_k,$$

where

$$|x_1| = |x_2| = \cdots = |x_k| = t.$$

We study the following iterated hash function:

Algorithm 4.9: SIMPLIFIED MERKLE-DAMGÅRD(x, k, t)

external compress
$z_1 \leftarrow 0^m \parallel x_1$
$g_1 \leftarrow \textbf{compress}(z_1)$
for $i \leftarrow 1$ **to** $k - 1$
$\quad \textbf{do} \begin{cases} z_{i+1} \leftarrow g_i \parallel x_{i+1} \\ g_{i+1} \leftarrow \textbf{compress}(z_{i+1}) \end{cases}$
$h(x) \leftarrow g_k$
return $(h(x))$

Suppose that **compress** is collision resistant, and suppose further that **compress** is zero preimage resistant, which means that it is hard to find $z \in \{0,1\}^{m+t}$ such that **compress**$(z) = 0^m$. Under these assumptions, prove that h is collision resistant.

4.11 A message authentication code can be produced by using a block cipher in CFB mode instead of CBC mode. Given a sequence of plaintext blocks, $x_1 \cdots x_n$, suppose we define the initialization vector **IV** to be x_1. Then encrypt the sequence $x_2 \cdots x_n$ using key K in CFB mode, obtaining the ciphertext sequence $y_1 \cdots y_{n-1}$ (note that there are only $n - 1$ ciphertext blocks). Finally, define the MAC to be $e_K(y_{n-1})$. Prove that this MAC is identical to CBC-MAC, which was presented in Section 4.4.2.

4.12 Suppose that $(\mathcal{P}, \mathcal{C}, \mathcal{K}, \mathcal{E}, \mathcal{D})$ is an endomorphic cryptosystem with $\mathcal{P} = \mathcal{C} = \{0,1\}^m$. Let $n \geq 2$ be an integer, and define a hash family $(\mathcal{X}, \mathcal{Y}, \mathcal{K}, \mathcal{H})$, where $\mathcal{X} = (\{0,1\}^m)^n$ and $\mathcal{Y} = \{0,1\}^m$, as follows:

$$h_K(x_1, \ldots x_n) = e_K(x_1) \oplus \cdots \oplus e_K(x_n).$$

Prove that $(\mathcal{X}, \mathcal{Y}, \mathcal{K}, \mathcal{H})$ is not a secure message authentication code as follows.
  (a) Prove the existence of a $(1, 1)$-forger for this hash family.
  (b) Prove the existence of a $(1, 2)$-forger for this hash family which can forge the MAC for an arbitrary message $(x_1, \ldots, x_n)$ (this is called a *forgery!selective*; the forgeries previously considered are examples of *existential forgeries*). Note that the difficult case is when $x_1 = \cdots = x_n$.

4.13 Suppose that $(\mathcal{P}, \mathcal{C}, \mathcal{K}, \mathcal{E}, \mathcal{D})$ is an endomorphic cryptosystem with $\mathcal{P} = \mathcal{C} = \{0,1\}^m$. Let $n \geq 2$ and $m \geq 3$ be integers, and define a hash family $(\mathcal{X}, \mathcal{Y}, \mathcal{K}, \mathcal{H})$, where $\mathcal{X} = (\{0,1\}^m)^n$ and $\mathcal{Y} = \{0,1\}^m$, as follows:

$$h_K(x_1, \ldots, x_n) = e_K(x_1) + 3e_K(x_2) + \cdots + (2n - 1)e_K(x_n) \bmod 2^m.$$

  (a) When n is odd, prove the existence of a $(1, 2)$-forger for this hash family.
  (b) When $n = 2$, prove the existence of a $(1/2, 2)$-forger for this hash family, as follows:

1. Request the MACs of (x_1, x_2) and (x_2, x_1), where $x_1 \neq x_2$. Suppose that $a = h_K(x_1, x_2)$ and $b = h_K(x_2, x_1)$.

2. Show that there are exactly eight ordered pairs (y_1, y_2) such that $y_1 = e_K(x_1)$, $y_2 = e_K(x_2)$ is consistent with the given MAC values a and b.

3. Choose one of the eight possible values for y_1 at random, define $y = 4y_1 \bmod 2^m$, and output the (possible) forgery $(x_1, x_1), y$. Prove that this is a valid forgery with probability $1/2$.

(c) Prove the existence of a $(1,3)$-forger for this hash family which can forge the MAC for an arbitrary message $(y_1, \ldots, y_n)$.

4.14 Suppose that $(\mathcal{X}, \mathcal{Y}, \mathcal{K}, \mathcal{H})$ is a strongly universal (N, M)-hash family.

(a) If $|\mathcal{K}| = M^2$, show that there exists a $(1,2)$-forger for this hash family (i.e., $Pd_2 = 1$).

(b) (This generalizes the result proven in part (a).) Denote $\lambda = |\mathcal{K}|/M^2$. Prove there exists a $(1/\lambda, 2)$-forger for this hash family (i.e., $Pd_2 \geq 1/\lambda$).

4.15 Compute Pd_0 and Pd_1 for the following authentication code, represented in matrix form:

key	1	2	3	4
1	1	1	2	3
2	1	2	3	1
3	2	1	3	1
4	2	3	1	2
5	3	2	1	3
6	3	3	2	1

4.16 Let p be an odd prime. For $a, b \in \mathbb{Z}_p$, define $f_{(a,b)} : \mathbb{Z}_p \to \mathbb{Z}_p$ by the rule

$$f_{(a,b)}(x) = (x + a)^2 + b \bmod p.$$

Prove that $(\mathbb{Z}_p, \mathbb{Z}_p, \mathbb{Z}_p \times \mathbb{Z}_p, \{f_{(a,b)} : a, b \in \mathbb{Z}_p\})$ is a strongly universal (p, p)-hash family.

4.17 Let $k \geq 1$ be an integer. An (N, M) hash family, $(\mathcal{X}, \mathcal{Y}, \mathcal{K}, \mathcal{H})$, is *strongly k-universal* provided that the following condition is satisfied for all choices of k distinct elements $x_1, x_2, \ldots, x_k \in \mathcal{X}$ and for all choices of k (not necessarily distinct) elements $y_1, \ldots, y_k \in \mathcal{Y}$:

$$|\{K \in \mathcal{K} : h_K(x_i) = y_i \text{ for } 1 \leq i \leq k\}| = \frac{|\mathcal{K}|}{M^k}.$$

(a) Prove that a strongly k-universal hash family is strongly ℓ-universal for all ℓ such that $1 \leq \ell \leq k$.

(b) Let p be prime and let $k \geq 1$ be an integer. For all k-tuples $(a_0, \ldots, a_{k-1}) \in (\mathbb{Z}_p)^k$, define $f_{(a_0, \ldots, a_{k-1})} : \mathbb{Z}_p \to \mathbb{Z}_p$ by the rule

$$f_{(a_0, \ldots, a_{k-1})}(x) = \sum_{i=0}^{k-1} a_i x^i \bmod p.$$

Prove that $(\mathbb{Z}_p, \mathbb{Z}_p, (\mathbb{Z}_p)^k, \{f_{(a_0, \ldots, a_{k-1})} : (a_0, \ldots, a_{k-1}) \in (\mathbb{Z}_p)^k\})$ is a strongly k-universal (p, p) hash family.

HINT Use the fact that any degree d polynomial over a field has at most d roots.

5

The RSA Cryptosystem and Factoring Integers

5.1 Introduction to Public-key Cryptography

In the classical model of cryptography that we have been studying up until now, Alice and Bob secretly choose the key K. K then gives rise to an encryption rule e_K and a decryption rule d_K. In the cryptosystems we have seen so far, d_K is either the same as e_K, or easily derived from it (for example, DES decryption is identical to encryption, but the key schedule is reversed). A cryptosystem of this type is known as a *symmetric-key cryptosystem*, since exposure of either of e_K or d_K renders the system insecure.

One drawback of a symmetric-key system is that it requires the prior communication of the key K between Alice and Bob, using a secure channel, before any ciphertext is transmitted. In practice, this may be very difficult to achieve. For example, suppose Alice and Bob live far away from each other and they decide that they want to communicate electronically, using email. In a situation such as this, Alice and Bob may not have access to a reasonable secure channel.

The idea behind a *public-key cryptosystem* is that it might be possible to find a cryptosystem where it is computationally infeasible to determine d_K given e_K. If so, then the encryption rule e_K is a *public key* which could be published in a directory, for example (hence the term public-key system). The advantage of a public-key system is that Alice (or anyone else) can send an encrypted message to Bob (without the prior communication of a shared secret key) by using the public encryption rule e_K. Bob will be the only person that can decrypt the ciphertext, using the decryption rule d_K, which is called the *private key*.

Consider the following analogy: Alice places an object in a metal box, and then locks it with a combination lock left there by Bob. Bob is the only person who can open the box since only he knows the combination.

The idea of a public-key cryptosystem was put forward by Diffie and Hellman in 1976. Then, in 1977, Rivest, Shamir, and Adleman invented the well-known *RSA Cryptosystem* which we study in this chapter. Several public-key systems have since been proposed, whose security rests on different computational problems. Of these, the most important are the *RSA Cryptosystem* (and variations of it), in which the security is based on the difficulty of factoring large integers; and the *ElGamal Cryptosystem* (and variations such as *Elliptic Curve Cryptosystems*) in which the security is based on the discrete logarithm problem. We discuss the *RSA Cryptosystem* and its variants in this chapter, while *ElGamal Cryptosystems* are studied in Chapter 6.

Prior to Diffie and Hellman, the idea of public-key cryptography had already been proposed by James Ellis in January 1970, in a paper entitled "The possibility of non-secret encryption." (The phrase "non-secret encryption" can be read as "public-key cryptography.") James Ellis was a member of the Communication-Electronics Security Group (CESG), which is a special section of the British Government Communications Headquarters (GCHQ). This paper was not published in the open literature, and was one of five papers released by the GCHQ officially in December 1997. Also included in these five papers was a 1973 paper written by Clifford Cocks, entitled "A note on non-secret encryption," in which a public-key cryptosystem is described that is essentially the same as the *RSA Cryptosystem*.

One very important observation is that a public-key cryptosystem can never provide unconditional security. This is because an opponent, on observing a ciphertext y, can encrypt each possible plaintext in turn using the public encryption rule e_K until he finds the unique x such that $y = e_K(x)$. This x is the decryption of y. Consequently, we study the computational security of public-key systems.

It is helpful conceptually to think of a public-key system in terms of an abstraction called a trapdoor one-way function. We informally define this notion now.

Bob's public encryption function, e_K, should be easy to compute. We have just noted that computing the inverse function (i.e., decrypting) should be hard (for anyone other than Bob). Recall from Section 4.2 that a function that is easy to compute but hard to invert is often called a one-way function. In the context of encryption, we desire that e_K be an injective one-way function so that decryption can be performed. Unfortunately, although there are many injective functions that are believed to be one-way, there currently do not exist such functions that can be proved to be one-way.

Here is an example of a function which is believed to be one-way. Suppose n is the product of two large primes p and q, and let b be a positive integer. Then define $f : \mathbb{Z}_n \to \mathbb{Z}_n$ to be

$$f(x) = x^b \bmod n.$$

(If $\gcd(b, \phi(n)) = 1$, then this is in fact an RSA encryption function; we will have much more to say about it later.)

If we are to construct a public-key cryptosystem, then it is not sufficient to find an injective one-way function. We do not want e_K to be one-way from Bob's point of view, because he needs to be able to decrypt messages that he receives in an efficient way. Thus, it is necessary that Bob possesses a *trapdoor*, which consists of secret information that permits easy inversion of e_K. That is, Bob can decrypt efficiently because he has some extra secret knowledge, namely, K, which provides him with the decryption function d_K. So, we say that a function is a *trapdoor one-way function* if it is a one-way function, but it becomes easy to invert with the knowledge of a certain trapdoor.

Let's consider the function $f(x) = x^b \bmod n$ considered above. We will see in Section 5.3 that the inverse function f^{-1} has a similar form: $f(x) = x^a \bmod n$ for an appropriate value of a. The trapdoor is an efficient method for computing the correct exponent a (given b), which makes use of the factorization of n.

It is often convenient to specify a family of trapdoor one-way functions, say $\mathcal{F}$. Then a function $f \in \mathcal{F}$ is chosen at random and used as the public encryption function; the inverse function, f^{-1}, is the private decryption function. This is analogous to choosing a random key from a specified keyspace, as we did with symmetric-key cryptosystems.

The rest of this chapter is organized as follows. Section 5.2 introduces several important number-theoretic results. In Section 5.3, we begin our study of the *RSA Cryptosystem*. Section 5.4 presents some important methods of testing. Section 5.5 is a short section on the existence of square roots modulo n. Then we present several algorithms for factoring in Section 5.6. Section 5.7 considers other attacks against the *RSA Cryptosystem*, and the *Rabin Cryptosystem* is described in Section 5.8. Finally, semantic security of RSA-like cryptosystems is the topic of Section 5.9.

5.2 More Number Theory

Before describing how the *RSA Cryptosystem* works, we need to discuss some more facts concerning modular arithmetic and number theory. Two fundamental tools that we require are the EUCLIDEAN ALGORITHM and the Chinese remainder theorem.

5.2.1 The Euclidean Algorithm

We already observed in Chapter 1 that $\mathbb{Z}_n$ is a ring for any positive integer n. We also proved there that $b \in \mathbb{Z}_n$ has a multiplicative inverse if and only if $\gcd(b, n) = 1$, and that the number of positive integers less than n and relatively prime to n is $\phi(n)$.

The set of residues modulo n that are relatively prime to n is denoted $\mathbb{Z}_n{}^*$. It is not hard to see that $\mathbb{Z}_n{}^*$ forms an abelian group under multiplication. We already have stated that multiplication modulo n is associative and commutative, and that 1 is the multiplicative identity. Any element in $\mathbb{Z}_n{}^*$ will have a multiplicative inverse (which is also in $\mathbb{Z}_n{}^*$). Finally, $\mathbb{Z}_n{}^*$ is closed under multiplication since xy is relatively prime to n whenever x and y are relatively prime to n (prove this!).

Algorithm 5.1: EUCLIDEAN ALGORITHM(a, b)

$r_0 \leftarrow a$
$r_1 \leftarrow b$
$m \leftarrow 1$
while $r_m \neq 0$

$\quad$ **do** $\begin{cases} q_m \leftarrow \lfloor \frac{r_{m-1}}{r_m} \rfloor \\ r_{m+1} \leftarrow r_{m-1} - q_m r_m \\ m \leftarrow m + 1 \end{cases}$

$m \leftarrow m - 1$
return $(q_1, \ldots, q_m; r_m)$
comment: $r_m = \gcd(a, b)$

At this point, we know that any $b \in \mathbb{Z}_n{}^*$ has a multiplicative inverse, b^{-1}, but we do not yet have an efficient algorithm to compute b^{-1}. Such an algorithm exists; it is called the EXTENDED EUCLIDEAN ALGORITHM. However, we first describe the EUCLIDEAN ALGORITHM, in its basic form, which can be used to compute the greatest common divisor of two positive integers, say a and b. The EUCLIDEAN ALGORITHM sets r_0 to be a and r_1 to be b, and performs the following sequence of divisions:

$$
\begin{array}{rcll}
r_0 &=& q_1 r_1 + r_2, & 0 < r_2 < r_1 \\
r_1 &=& q_2 r_2 + r_3, & 0 < r_3 < r_2 \\
\vdots & \vdots & \vdots & \vdots \\
r_{m-2} &=& q_{m-1} r_{m-1} + r_m, & 0 < r_m < r_{m-1} \\
r_{m-1} &=& q_m r_m.
\end{array}
$$

A pseudocode description of the EUCLIDEAN ALGORITHM is presented as Algorithm 5.1.

REMARK　We will make use of the list $(q_1, \ldots, q_m)$ that is computed during the execution of Algorithm 5.1 in a later section of this chapter. ∎

In Algorithm 5.1, it is not hard to show that

$$\gcd(r_0, r_1) = \gcd(r_1, r_2) = \cdots = \gcd(r_{m-1}, r_m) = r_m.$$

Hence, it follows that $\gcd(r_0, r_1) = r_m$.

Since the EUCLIDEAN ALGORITHM computes greatest common divisors, it can be used to determine if a positive integer $b < n$ has a multiplicative inverse modulo n, by calling EUCLIDEAN ALGORITHM(n, b) and checking to see if $r_m = 1$. However, it does not compute the value of b^{-1} mod n (if it exists).

Now, suppose we define two sequences of numbers,

$$t_0, t_1, \ldots, t_m \quad \text{and} \quad s_0, s_1, \ldots, s_m,$$

according to the following recurrences (where the q_j's are defined as in Algorithm 5.1):

$$t_j = \begin{cases} 0 & \text{if } j = 0 \\ 1 & \text{if } j = 1 \\ t_{j-2} - q_{j-1}t_{j-1} & \text{if } j \geq 2 \end{cases}$$

and

$$s_j = \begin{cases} 1 & \text{if } j = 0 \\ 0 & \text{if } j = 1 \\ s_{j-2} - q_{j-1}s_{j-1} & \text{if } j \geq 2. \end{cases}$$

Then we have the following useful result.

THEOREM 5.1 *For $0 \leq j \leq m$, we have that $r_j = s_j r_0 + t_j r_1$, where the r_j's are defined as in Algorithm 5.1, and the s_j's and t_j's are defined in the above recurrence.*

PROOF The proof is by induction on j. The assertion is trivially true for $j = 0$ and $j = 1$. Assume the assertion is true for $j = i - 1$ and $i - 2$, where $i \geq 2$; we will prove the assertion is true for $j = i$. By induction, we have that

$$r_{i-2} = s_{i-2}r_0 + t_{i-2}r_1$$

and

$$r_{i-1} = s_{i-1}r_0 + t_{i-1}r_1.$$

Now, we compute:

$$\begin{aligned} r_i &= r_{i-2} - q_{i-1}r_{i-1} \\ &= s_{i-2}r_0 + t_{i-2}r_1 - q_{i-1}(s_{i-1}r_0 + t_{i-1}r_1) \\ &= (s_{i-2} - q_{i-1}s_{i-1})r_0 + (t_{i-2} - q_{i-1}t_{i-1})r_1 \\ &= s_i r_0 + t_i r_1. \end{aligned}$$

Hence, the result is true, for all integers $j \geq 0$, by induction. ∎

In Algorithm 5.2, we present the EXTENDED EUCLIDEAN ALGORITHM, which takes two integers a and b as input and computes integers r, s and t such that $r = \gcd(a, b)$ and $sa + tb = r$. In this version of the algorithm, we do not keep track of all the q_j's, r_j's, s_j's and t_j's; it suffices to record only the "last" two terms in each of these sequences at any point in the algorithm.

The next corollary is an immediate consequence of Theorem 5.1.

Algorithm 5.2: EXTENDED EUCLIDEAN ALGORITHM(a, b)

$a_0 \leftarrow a$
$b_0 \leftarrow b$
$t_0 \leftarrow 0$
$t \leftarrow 1$
$s_0 \leftarrow 1$
$s \leftarrow 0$
$q \leftarrow \lfloor \frac{a_0}{b_0} \rfloor$
$r \leftarrow a_0 - qb_0$
while $r > 0$

$\text{\textbf{do}} \begin{cases} temp \leftarrow t_0 - qt \\ t_0 \leftarrow t \\ t \leftarrow temp \\ temp \leftarrow s_0 - qs \\ s_0 \leftarrow s \\ s \leftarrow temp \\ a_0 \leftarrow b_0 \\ b_0 \leftarrow r \\ q \leftarrow \lfloor \frac{a_0}{b_0} \rfloor \\ r \leftarrow a_0 - qb_0 \end{cases}$

$r \leftarrow b_0$
return (r, s, t)
comment: $r = \gcd(a, b)$ and $sa + tb = r$

COROLLARY 5.2   *Suppose* $\gcd(r_0, r_1) = 1$. *Then* $r_1^{-1} \bmod r_0 = t_m \bmod r_0$.

PROOF   From Theorem 5.1, we have that

$$1 = \gcd(r_0, r_1) = s_m r_0 + t_m r_1.$$

Reducing this equation modulo r_0, we obtain

$$t_m r_1 \equiv 1 \pmod{r_0}.$$

The result follows.   ∎

We present a small example to illustrate, in which we show the values of all the s_j's, t_j's, q_j's and r_j's.

Example 5.1 Suppose we wish to calculate 28^{-1} mod 75. Then we compute the following:

i	r_i	q_i	s_i	t_i
0	75		1	0
1	28	2	0	1
2	19	1	1	-2
3	9	2	-1	3
4	1	9	3	-8

Therefore, we have found that

$$3 \times 75 - 8 \times 28 = 1.$$

Applying Corollary 5.2, we see that

$$28^{-1} \bmod 75 = -8 \bmod 75 = 67.$$

$\square$

The EXTENDED EUCLIDEAN ALGORITHM immediately yields the value b^{-1} modulo a (if it exists). In fact, the multiplicative inverse $b^{-1} \bmod a = t \bmod a$; this follows immediately from Corollary 5.2. However, a more efficient way to compute multiplicative inverses is to remove the s's from Algorithm 5.2, and to reduce the t's modulo a during each iteration of the main loop. We obtain Algorithm 5.3.

5.2.2 The Chinese Remainder Theorem

The Chinese remainder theorem is really a method of solving certain systems of congruences. Suppose $m_1, \ldots, m_r$ are pairwise relatively prime positive integers (that is, $\gcd(m_i, m_j) = 1$ if $i \neq j$). Suppose $a_1, \ldots, a_r$ are integers, and consider the following system of congruences:

$$x \equiv a_1 \pmod{m_1}$$

$$x \equiv a_2 \pmod{m_2}$$

$$\vdots$$

$$x \equiv a_r \pmod{m_r}.$$

The Chinese remainder theorem asserts that this system has a unique solution modulo $M = m_1 \times m_2 \times \cdots \times m_r$. We will prove this result in this section, and also describe an efficient algorithm for solving systems of congruences of this type.

Algorithm 5.3: MULTIPLICATIVE INVERSE(a, b)

$a_0 \leftarrow a$
$b_0 \leftarrow b$
$t_0 \leftarrow 0$
$t \leftarrow 1$
$q \leftarrow \lfloor \frac{a_0}{b_0} \rfloor$
$r \leftarrow a_0 - q b_0$
while $r > 0$

do $\begin{cases} temp \leftarrow (t_0 - qt) \bmod a \\ t_0 \leftarrow t \\ t \leftarrow temp \\ a_0 \leftarrow b_0 \\ b_0 \leftarrow r \\ q \leftarrow \lfloor \frac{a_0}{b_0} \rfloor \\ r \leftarrow a_0 - q b_0 \end{cases}$

if $b_0 \neq 1$
   then b has no inverse modulo a
   else return (t)

It is convenient to study the function $\chi : \mathbb{Z}_M \rightarrow \mathbb{Z}_{m_1} \times \cdots \times \mathbb{Z}_{m_r}$, which we define as follows:

$$\chi(x) = (x \bmod m_1, \ldots, x \bmod m_r).$$

Example 5.2 Suppose $r = 2$, $m_1 = 5$ and $m_2 = 3$, so $M = 15$. Then the function χ has the following values:

$$
\begin{array}{lllllll}
\chi(0) & = & (0,0) & \chi(1) & = & (1,1) & \chi(2) & = & (2,2) \\
\chi(3) & = & (3,0) & \chi(4) & = & (4,1) & \chi(5) & = & (0,2) \\
\chi(6) & = & (1,0) & \chi(7) & = & (2,1) & \chi(8) & = & (3,2) \\
\chi(9) & = & (4,0) & \chi(10) & = & (0,1) & \chi(11) & = & (1,2) \\
\chi(12) & = & (2,0) & \chi(13) & = & (3,1) & \chi(14) & = & (4,2).
\end{array}
$$

$\square$

Proving the Chinese remainder theorem amounts to proving that the function χ is a bijection. In Example 5.2 this is easily seen to be the case. In fact, we will be able to give an explicit general formula for the inverse function χ^{-1}.

For $1 \leq i \leq r$, define

$$M_i = \frac{M}{m_i}.$$

Then it is not difficult to see that

$$\gcd(M_i, m_i) = 1$$

for $1 \le i \le r$. Next, for $1 \le i \le r$, define

$$y_i = M_i^{-1} \bmod m_i.$$

(This inverse exists because $\gcd(M_i, m_i) = 1$, and it can be found using Algorithm 5.3.) Note that

$$M_i y_i \equiv 1 \pmod{m_i}$$

for $1 \le i \le r$.

Now, define a function $\rho : \mathbb{Z}_{m_1} \times \cdots \times \mathbb{Z}_{m_r} \to \mathbb{Z}_M$ as follows:

$$\rho(a_1, \ldots, a_r) = \sum_{i=1}^{r} a_i M_i y_i \bmod M.$$

We will show that the function $\rho = \chi^{-1}$, i.e., it provides an explicit formula for solving the original system of congruences.

Denote $X = \rho(a_1, \ldots, a_r)$, and let $1 \le j \le r$. Consider a term $a_i M_i y_i$ in the above summation, reduced modulo m_j: If $i = j$, then

$$a_i M_i y_i \equiv a_i \pmod{m_i}$$

because

$$M_i y_i \equiv 1 \pmod{m_i}.$$

On the other hand, if $i \neq j$, then

$$a_i M_i y_i \equiv 0 \pmod{m_j}$$

because $m_j \mid M_i$ in this case. Thus, we have that

$$X \equiv \sum_{i=1}^{r} a_i M_i y_i \pmod{m_j}$$

$$\equiv a_j \pmod{m_j}.$$

Since this is true for all j, $1 \le j \le r$, X is a solution to the system of congruences.

At this point, we need to show that the solution X is unique modulo M. But this can be done by simple counting. The function χ is a function from a domain of cardinality M to a range of cardinality M. We have just proved that χ is a surjective (i.e., onto) function. Hence, χ must also be injective (i.e., one-to-one), since the domain and range have the same cardinality. It follows that χ is a bijection and $\chi^{-1} = \rho$. Note also that χ^{-1} is a linear function of its arguments $a_1, \ldots, a_r$.

Here is a bigger example to illustrate.

Example 5.3   Suppose $r = 3$, $m_1 = 7$, $m_2 = 11$ and $m_3 = 13$. Then $M = 1001$. We compute $M_1 = 143$, $M_2 = 91$ and $M_3 = 77$, and then $y_1 = 5$, $y_2 = 4$ and $y_3 = 12$. Then the function $\chi^{-1} : \mathbb{Z}_7 \times \mathbb{Z}_{11} \times \mathbb{Z}_{13} \to \mathbb{Z}_{1001}$ is the following:

$$\chi^{-1}(a_1, a_2, a_3) = (715a_1 + 364a_2 + 924a_3) \bmod 1001.$$

For example, if $x \equiv 5 \pmod 7$, $x \equiv 3 \pmod{11}$ and $x \equiv 10 \pmod{13}$, then this formula tells us that

$$x = (715 \times 5 + 364 \times 3 + 924 \times 10) \bmod 1001$$

$$= 13907 \bmod 1001$$

$$= 894.$$

This can be verified by reducing 894 modulo 7, 11 and 13.   ▯

For future reference, we record the results of this section as a theorem.

THEOREM 5.3   (Chinese remainder theorem)   *Suppose $m_1, \dots, m_r$ are pairwise relatively prime positive integers, and suppose $a_1, \dots, a_r$ are integers. Then the system of r congruences $x \equiv a_i \pmod{m_i}$ $(1 \le i \le r)$ has a unique solution modulo $M = m_1 \times \cdots \times m_r$, which is given by*

$$x = \sum_{i=1}^{r} a_i M_i y_i \bmod M,$$

where $M_i = M/m_i$ and $y_i = M_i^{-1} \bmod m_i$, for $1 \le i \le r$.

5.2.3   Other Useful Facts

We next mention another result from elementary group theory, called Lagrange's theorem, that will be relevant in our treatment of the *RSA Cryptosystem*. For a (finite) multiplicative group G, define the *order* of an element $g \in G$ to be the smallest positive integer m such that $g^m = 1$. The following result is fairly simple, but we will not prove it here.

THEOREM 5.4   (Lagrange)   *Suppose G is a multiplicative group of order n, and $g \in G$. Then the order of g divides n.*

For our purposes, the following corollaries are essential.

COROLLARY 5.5   *If $b \in \mathbb{Z}_n{}^*$, then $b^{\phi(n)} \equiv 1 \pmod n$.*

PROOF $\mathbb{Z}_n{}^*$ is a multiplicative group of order $\phi(n)$.   ∎

COROLLARY 5.6   (Fermat)   *Suppose p is prime and $b \in \mathbb{Z}_p$. Then $b^p \equiv b$* (mod p).

PROOF   If p is prime, then $\phi(p) = p - 1$. So, for $b \not\equiv 0 \pmod{p}$, the result follows from Corollary 5.5. For $b \equiv 0 \pmod{p}$, the result is also true since $0^p \equiv 0 \pmod{p}$.   ∎

At this point, we know that if p is prime, then $\mathbb{Z}_p^*$ is a group of order $p - 1$, and any element in $\mathbb{Z}_p^*$ has order dividing $p - 1$. In fact, if p is prime, then the group $\mathbb{Z}_p^*$ is a *cyclic group*: there exists an element $\alpha \in \mathbb{Z}_p^*$ having order equal to $p - 1$. We will not prove this very important fact, but we do record it for future reference:

THEOREM 5.7   *If p is prime, then $\mathbb{Z}_p^*$ is a cyclic group.*

An element α having order $p-1$ modulo p is called a *primitive element* modulo p. Observe that α is a primitive element modulo p if and only if

$$\{\alpha^i : 0 \leq i \leq p - 2\} = \mathbb{Z}_p^*.$$

Now, suppose p is prime and α is a primitive element modulo p. Any element $\beta \in \mathbb{Z}_p^*$ can be written as $\beta = \alpha^i$, where $0 \leq i \leq p - 2$, in a unique way. It is not difficult to prove that the order of $\beta = \alpha^i$ is

$$\frac{p - 1}{\gcd(p - 1, i)}.$$

Thus β is itself a primitive element if and only if $\gcd(p - 1, i) = 1$. It follows that the number of primitive elements modulo p is $\phi(p - 1)$.

We do a small example to illustrate.

Example 5.4   Suppose $p = 13$. The results proven above establish that there are exactly four primitive elements modulo 13. First, by computing successive powers of 2, we can verify that 2 is a primitive element modulo 13:

$$2^0 \bmod 13 = 1$$

$$2^1 \bmod 13 = 2$$

$$2^2 \bmod 13 = 4$$

$$2^3 \bmod 13 = 8$$

$$2^4 \bmod 13 = 3$$

$$2^5 \bmod 13 = 6$$

$$2^6 \bmod 13 = 12$$

$$2^7 \bmod 13 = 11$$

$$2^8 \bmod 13 = 9$$

$$2^9 \bmod 13 = 5$$

$$2^{10} \bmod 13 = 10$$

$$2^{11} \bmod 13 = 7.$$

The element 2^i is primitive if and only if $\gcd(i, 12) = 1$, i.e., if and only if $i = 1, 5, 7$ or 11. Hence, the primitive elements modulo 13 are $2, 6, 7$ and 11. $\quad\square$

In the above example, we computed all the powers of 2 in order to verify that it was a primitive element modulo 13. If p is a large prime, however, it would take a long time to compute $p - 1$ powers of an element $\alpha \in \mathbb{Z}_p^*$. Fortunately, if the factorization of $p - 1$ is known, then we can verify whether $\alpha \in \mathbb{Z}_p^*$ is a primitive element much more quickly, by making use of the following result.

THEOREM 5.8 *Suppose that $p > 2$ is prime and $\alpha \in \mathbb{Z}_p^*$. Then α is a primitive element modulo p if and only if $\alpha^{(p-1)/q} \not\equiv 1 \pmod{p}$ for all primes q such that $q \mid (p - 1)$.*

PROOF If α is a primitive element modulo p, then $\alpha^i \not\equiv 1 \pmod{p}$ for all i such that $1 \le i \le p - 2$, so the result follows.

Conversely, suppose that $\alpha \in \mathbb{Z}_p^*$ is not a primitive element modulo p. Let d be the order of α. Then $d \mid (p-1)$ by Lagrange's theorem, and $d < p - 1$ because α is not primitive. Then $(p - 1)/d$ is an integer exceeding 1. Let q be a prime divisor of $(p - 1)/d$. Then d is a divisor of the integer $(p - 1)/q$. Since $\alpha^d \equiv 1 \pmod{p}$ and $d \mid (p - 1)/q$, it follows that $\alpha^{(p-1)/q} \equiv 1 \pmod{p}$. $\quad\blacksquare$

The factorization of 12 is $12 = 2^2 \times 3$. Therefore, in the previous example, we could verify that 2 is a primitive element modulo 13 by verifying that $2^6 \not\equiv 1 \pmod{13}$ and $2^4 \not\equiv 1 \pmod{13}$.

Cryptosystem 5.1: *RSA Cryptosystem*

Let $n = pq$, where p and q are primes. Let $\mathcal{P} = \mathcal{C} = \mathbb{Z}_n$, and define

$$\mathcal{K} = \{(n, p, q, a, b) : ab \equiv 1 \pmod{\phi(n)}\}.$$

For $K = (n, p, q, a, b)$, define

$$e_K(x) = x^b \bmod n$$

and

$$d_K(y) = y^a \bmod n$$

$(x, y \in \mathbb{Z}_n)$. The values n and b comprise the public key, and the values p, q and a form the private key.

5.3   The RSA Cryptosystem

We can now describe the *RSA Cryptosystem*. This cryptosystem uses computations in $\mathbb{Z}_n$, where n is the product of two distinct odd primes p and q. For such an integer n, note that $\phi(n) = (p-1)(q-1)$. The formal description is given as Cryptosystem 5.1.

Let's verify that encryption and decryption are inverse operations. Since

$$ab \equiv 1 \pmod{\phi(n)},$$

we have that

$$ab = t\phi(n) + 1$$

for some integer $t \geq 1$. Suppose that $x \in \mathbb{Z}_n^*$; then we have

$$(x^b)^a \equiv x^{t\phi(n)+1} \pmod{n}$$
$$\equiv (x^{\phi(n)})^t x \pmod{n}$$
$$\equiv 1^t x \pmod{n}$$
$$\equiv x \pmod{n},$$

as desired. We leave it as an Exercise to show that $(x^b)^a \equiv x \pmod{n}$ if $x \in \mathbb{Z}_n \backslash \mathbb{Z}_n^*$.

Here is a small (insecure) example of the *RSA Cryptosystem*.

Example 5.5   Suppose Bob chooses $p = 101$ and $q = 113$. Then $n = 11413$ and $\phi(n) = 100 \times 112 = 11200$. Since $11200 = 2^6 5^2 7$, an integer b can be used as an encryption exponent if and only if b is not divisible by 2, 5 or 7. (In practice, however, Bob will not factor $\phi(n)$. He will verify that $\gcd(\phi(n), b) = 1$ using Algorithm 5.3. If this is the case, then he will compute b^{-1} at the same time.) Suppose Bob chooses $b = 3533$. Then

$$b^{-1} \bmod 11200 = 6597.$$

Hence, Bob's secret decryption exponent is $a = 6597$.

Bob publishes $n = 11413$ and $b = 3533$ in a directory. Now, suppose Alice wants to encrypt the plaintext 9726 to send to Bob. She will compute

$$9726^{3533} \bmod 11413 = 5761$$

and send the ciphertext 5761 over the channel. When Bob receives the ciphertext 5761, he uses his secret decryption exponent to compute

$$5761^{6597} \bmod 11413 = 9726.$$

(At this point, the encryption and decryption operations might appear to be very complicated, but we will discuss efficient algorithms for these operations in the next section.) $\square$

The security of the *RSA Cryptosystem* is based on the belief that the encryption function $e_K(x) = x^b \bmod n$ is a one-way function, so it will be computationally infeasible for an opponent to decrypt a ciphertext. The trapdoor that allows Bob to decrypt a ciphertext is the knowledge of the factorization $n = pq$. Since Bob knows this factorization, he can compute $\phi(n) = (p-1)(q-1)$, and then compute the decryption exponent a using the EXTENDED EUCLIDEAN ALGORITHM. We will say more about the security of the *RSA Cryptosystem* later on.

5.3.1   Implementing RSA

There are many aspects of the *RSA Cryptosystem* to discuss, including the details of setting up the cryptosystem, the efficiency of encrypting and decrypting, and security issues. In order to set up the system, Bob uses the RSA PARAMETER GENERATION algorithm, presented informally as Algorithm 5.4. How Bob carries out the steps of this algorithm will be discussed later in this chapter.

Algorithm 5.4: RSA PARAMETER GENERATION

1. Generate two large primes, p and q, such that $p \neq q$
2. $n \leftarrow pq$ and $\phi(n) \leftarrow (p-1)(q-1)$
3. Choose a random b $(1 < b < \phi(n))$ such that $\gcd(b, \phi(n)) = 1$
4. $a \leftarrow b^{-1} \bmod \phi(n)$
5. The public key is (n, b) and the private key is (p, q, a).

One obvious attack on the *RSA Cryptosystem* is for a cryptanalyst to attempt to factor n. If this can be done, it is a simple manner to compute $\phi(n) = (p-1)(q-1)$ and then compute the decryption exponent a from b exactly as Bob did. (It has been conjectured that breaking the *RSA Cryptosystem* is polynomially equivalent[1] to factoring n, but this remains unproved.)

If the *RSA Cryptosystem* is to be secure, it is certainly necessary that $n = pq$ must be large enough that factoring it will be computationally infeasible. Current factoring algorithms are able to factor numbers having up to 512 bits in their binary representation (for more information on factoring, see Section 5.6). It is generally recommended that, to be on the safe side, one should choose each of p and q to be 512-bit primes; then n will be a 1024-bit modulus. Factoring a number of this size is well beyond the capability of the best current factoring algorithms.

Leaving aside for the moment the question of how to find 512-bit primes, let us look now at the arithmetic operations of encryption and decryption. An encryption (or decryption) involves performing one exponentiation modulo n. Since n is very large, we must use multiprecision arithmetic to perform computations in $\mathbb{Z}_n$, and the time required will depend on the number of bits in the binary representation of n.

Suppose that x and y are positive integers having k and ℓ bits respectively in their binary representations; i.e., $k = \lfloor \log_2 x \rfloor + 1$ and $\ell = \lfloor \log_2 y \rfloor + 1$. Assume that $k \geq \ell$. Using standard "grade-school" arithmetic techniques, it is not difficult to obtain big-oh upper bounds on the amount of time to perform various operations on x and y. We summarize these results now (and we do not claim that these are the best possible bounds).

- $x + y$ can be computed in time $O(k)$
- $x - y$ can be computed in time $O(k)$
- xy can be computed in time $O(k\ell)$
- $\lfloor x/y \rfloor$ can be computed in time $O(\ell(k - \ell))$. $O(k\ell)$ is a weaker bound.
- $\gcd(x, y)$ can be computed in time $O(k^3)$.

In reference to the last item, GCD's can be computed using Algorithm 5.1. It can be shown that the number of iterations required in the EUCLIDEAN AL-

---

[1] Two problems are said to be *polynomially equivalent* if the existence of a polynomial-time algorithm for either problem implies the existence of a polynomial-time algorithm for the other problem.

GORITHM is $O(k)$ (see the Exercises). Each iteration performs a long division requiring time $O(k^2)$; so, the complexity of a GCD computation is seen to be $O(k^3)$. (Actually, a more careful analysis can be used to show that the complexity is, in fact, $O(k^2)$.)

Now we turn to modular arithmetic, i.e., operations in $\mathbb{Z}_n$. Suppose that n is a k-bit integer, and $0 \leq m_1, m_2 \leq n - 1$. Also, let c be a positive integer. We have the following:

- Computing $(m_1 + m_2) \bmod n$ can be done in time $O(k)$.
- Computing $(m_1 - m_2) \bmod n$ can be done in time $O(k)$.
- Computing $(m_1 m_2) \bmod n$ can be done in time $O(k^2)$.
- Computing $(m_1)^{-1} \bmod n$ can be done in time $O(k^3)$ (provided that this inverse exists).
- Computing $(m_1)^c \bmod n$ can be done in time $O((\log c) \times k^2)$.

Most of the above results are not hard to prove. The first three operations (modular addition, subtraction and multiplication) can be accomplished by doing the corresponding integer operation and then performing a single reduction modulo n. Modular inversion (i.e., computing multiplicative inverses) is done using Algorithm 5.3. The complexity is analyzed in a similar fashion as a GCD computation.

We now consider *modular exponentiation*, i.e., computation of a function of the form $x^c \bmod n$. Both the encryption and the decryption operations in the *RSA Cryptosystem* are modular exponentiations. Computation of $x^c \bmod n$ can be done using $c - 1$ modular multiplications; however, this is very inefficient if c is large. Note that c might be as big as $\phi(n) - 1$, which is almost as big as n and exponentially large compared to k.

The well-known SQUARE-AND-MULTIPLY ALGORITHM reduces the number of modular multiplications required to compute $x^c \bmod n$ to at most 2ℓ, where ℓ is the number of bits in the binary representation of c. It follows that $x^c \bmod n$ can be computed in time $O(\ell k^2)$. If we assume that $c < n$ (as it is in the definition of the *RSA Cryptosystem*), then we see that RSA encryption and decryption can both be done in time $O((\log n)^3)$, which is a polynomial function of the number of bits in one plaintext (or ciphertext) character.

The SQUARE-AND-MULTIPLY ALGORITHM assumes that the exponent c is represented in binary notation, say

$$c = \sum_{i=0}^{\ell-1} c_i 2^i,$$

where $c_i = 0$ or 1, $0 \leq i \leq \ell - 1$. The algorithm to compute $z = x^c \bmod n$ is presented as Algorithm 5.5.

The proof of correctness of this algorithm is left as an Exercise. It is easy to count the number of modular multiplications in the algorithm. There are always ℓ squarings performed. The number of modular multiplications of the type $z \leftarrow$

Algorithm 5.5: SQUARE-AND-MULTIPLY(x, c, n)

$z \leftarrow 1$
for $i \leftarrow \ell - 1$ **downto** 0

$\quad$ **do** $\begin{cases} z \leftarrow z^2 \bmod n \\ \textbf{if } c_i = 1 \\ \quad \textbf{then } z \leftarrow (z \times x) \bmod n \end{cases}$

$\quad$ **return** (z)

$(z \times x) \bmod n$ is equal to the number of 1's in the binary representation of c. This is an integer between 0 and ℓ. Thus, the total number of modular multiplications is at least ℓ and at most 2ℓ, as stated above.

We will illustrate the use of the SQUARE-AND-MULTIPLY ALGORITHM by returning to Example 5.5.

Example 5.5 **(Cont.)** Recall that $n = 11413$, and the public encryption exponent is $b = 3533$. Alice encrypts the plaintext 9726 by computing $9726^{3533} \bmod 11413$, using the SQUARE-AND-MULTIPLY ALGORITHM, as follows:

i	b_i	z
11	1	$1^2 \times 9726 = 9726$
10	1	$9726^2 \times 9726 = 2659$
9	0	$2659^2 = 5634$
8	1	$5634^2 \times 9726 = 9167$
7	1	$9167^2 \times 9726 = 4958$
6	1	$4958^2 \times 9726 = 7783$
5	0	$7783^2 = 6298$
4	0	$6298^2 = 4629$
3	1	$4629^2 \times 9726 = 10185$
2	1	$10185^2 \times 9726 = 105$
1	0	$105^2 = 11025$
0	1	$11025^2 \times 9726 = 5761$

Hence, as stated earlier, the ciphertext is 5761. $\qquad\qquad$ □

To this point we have discussed the RSA encryption and decryption operations. Regarding RSA PARAMETER GENERATION, methods to construct the primes p and q (Step 1) will be discussed in the next section. Step 2 is straightforward and can be done in time $O((\log n)^2)$. Steps 3 and 4 utilize Algorithm 5.3, which has complexity $O((\log n)^2)$.

5.4   Primality Testing

In setting up the *RSA Cryptosystem*, it is necessary to generate large "random primes." The way this is done is to generate large random numbers, and then test them for primality. In 2002, it was proven by Agrawal, Kayal and Saxena that there is a polynomial-time deterministic algorithm for primality testing. This was a major breakthrough that solved a longstanding open problem. However, in practice, primality testing is still done mainly by using a randomized polynomial-time Monte Carlo algorithm such as the SOLOVAY-STRASSEN ALGORITHM or the MILLER-RABIN ALGORITHM, both of which we will present in this section. These algorithms are fast (i.e., an integer n can be tested in time that is polynomial in $\log_2 n$, the number of bits in the binary representation of n), but there is a possibility that the algorithm may claim that n is prime when it is not. However, by running the algorithm enough times, the error probability can be reduced below any desired threshold. (We will discuss this in more detail a bit later.)

The other pertinent question is how many random integers (of a specified size) will need to be tested until we find one that is prime. Suppose we define $\pi(N)$ to be the number of primes that are less than or equal to N. A famous result in number theory, called the *Prime number theorem*, states that $\pi(N)$ is approximately $N/\ln N$. Hence, if an integer p is chosen at random between 1 and N, then the probability that it is prime is about $1/\ln N$. For a 1024 bit modulus $n = pq$, p and q will be chosen to be 512 bit primes. A random 512 bit integer will be prime with probability approximately $1/\ln 2^{512} \approx 1/355$. That is, on average, given 355 random 512 bit integers p, one of them will be prime (of course, if we restrict our attention to odd integers, the probability doubles, to about $2/355$). So we can in fact generate sufficiently large random numbers that are "probably prime," and hence parameter generation for the *RSA Cryptosystem* is indeed practical. We proceed to describe how this is done.

A *decision problem* is a problem in which a question is to be answered "yes" or "no." Recall that a randomized algorithm is any algorithm that uses random numbers (in contrast, an algorithm that does not use random numbers is called a *deterministic algorithm*). The following definitions pertain to randomized algorithms for decision problems.

Definition 5.1:   A *yes-biased Monte Carlo algorithm* is a randomized algorithm for a decision problem in which a "yes" answer is (always) correct, but a "no" answer may be incorrect. A *no-biased* Monte Carlo algorithm is defined in the obvious way. We say that a yes-biased Monte Carlo algorithm has *error probability* equal to ϵ if, for any instance in which the answer is "yes," the algorithm will give the (incorrect) answer "no" with probability at most ϵ. (This probability is computed over all possible random choices made by the algorithm when it is run with a given input.)

REMARK A Las Vegas algorithm may not give an answer, but any answer it gives is correct. In contrast, a Monte Carlo algorithm always gives an answer, but the answer may be incorrect. ∎

The decision problem called **Composites** is presented as Problem 5.1.

Problem 5.1: **Composites**

Instance: A positive integer $n \geq 2$.

Question: Is n composite?

Note that an algorithm for a decision problem only has to answer "yes" or "no." In particular, in the case of the problem **Composites**, we do not require the algorithm to find a factorization in the case that n is composite.

We will first describe the SOLOVAY-STRASSEN ALGORITHM, which is a yes-biased Monte Carlo algorithm for **Composites** with error probability $1/2$. Hence, if the algorithm answers "yes," then n is composite; conversely, if n is composite, then the algorithm answers "yes" with probability at least $1/2$.

Although the MILLER-RABIN ALGORITHM (which we will discuss later) is faster than the SOLOVAY-STRASSEN ALGORITHM, we first look at the SOLOVAY-STRASSEN ALGORITHM because it is easier to understand conceptually and because it involves some number-theoretic concepts that will be useful in later chapters of the book. We begin by developing some further background from number theory before describing the algorithm.

5.4.1 Legendre and Jacobi Symbols

Definition 5.2: Suppose p is an odd prime and a is an integer. a is defined to be a *quadratic residue* modulo p if $a \not\equiv 0 \pmod{p}$ and the congruence $y^2 \equiv a \pmod{p}$ has a solution $y \in \mathbb{Z}_p$. a is defined to be a *quadratic non-residue* modulo p if $a \not\equiv 0 \pmod{p}$ and a is not a quadratic residue modulo p.

Example 5.6 In $\mathbb{Z}_{11}$, we have that $1^2 = 1$, $2^2 = 4$, $3^2 = 9$, $4^2 = 5$, $5^2 = 3$, $6^2 = 3$, $7^2 = 5$, $8^2 = 9$, $9^2 = 4$, and $(10)^2 = 1$. Therefore the quadratic residues modulo 11 are $1, 3, 4, 5$ and 9, and the quadratic non-residues modulo 11 are $2, 6, 7, 8$ and 10. ☐

Suppose that p is an odd prime and a is a quadratic residue modulo p. Then there exists $y \in \mathbb{Z}_p{}^*$ such that $y^2 \equiv a \pmod{p}$. Clearly, $(-y)^2 \equiv a \pmod{p}$, and $y \not\equiv -y \pmod{p}$ because p is odd. Now consider the quadratic congruence

$x^2 - a \equiv 0 \pmod{p}$. This congruence can be factored as $(x - y)(x + y) \equiv 0$ $\pmod{p}$, which is the same thing as saying that $p \mid (x-y)(x+y)$. Now, because p is prime, it follows that $p \mid (x - y)$ or $p \mid (x + y)$. In other words, $x \equiv \pm y$ $\pmod{p}$, and we conclude that there are exactly two solutions (modulo p) to the congruence $x^2 - a \equiv 0 \pmod{p}$. Moreover, these two solutions are negatives of each other modulo p.

We now study the problem of determining whether an integer a is quadratic residue modulo p. The decision problem **Quadratic Residues** (Problem 5.2) is defined in the obvious way. Notice that this problem just asks for a "yes" or "no" answer: it does not require us to compute square roots in the case when a is a quadratic residue modulo p.

Problem 5.2: **Quadratic Residues**

Instance: An odd prime p, and an integer a.

Question: Is a a quadratic residue modulo p?

We prove a result, known as Euler's criterion, that will give rise to a polynomial-time deterministic algorithm for **Quadratic Residues**.

THEOREM 5.9   **(Euler's Criterion)**   *Let p be an odd prime. Then a is a quadratic residue modulo p if and only if*

$$a^{(p-1)/2} \equiv 1 \pmod{p}.$$

PROOF   First, suppose $a \equiv y^2 \pmod{p}$. Recall from Corollary 5.6 that if p is prime, then $a^{p-1} \equiv 1 \pmod{p}$ for any $a \not\equiv 0 \pmod{p}$. Thus we have

$$a^{(p-1)/2} \equiv (y^2)^{(p-1)/2} \pmod{p}$$
$$\equiv y^{p-1} \pmod{p}$$
$$\equiv 1 \pmod{p}.$$

Conversely, suppose $a^{(p-1)/2} \equiv 1 \pmod{p}$. Let b be a primitive element modulo p. Then $a \equiv b^i \pmod{p}$ for some positive integer i. Then we have

$$a^{(p-1)/2} \equiv (b^i)^{(p-1)/2} \pmod{p}$$
$$\equiv b^{i(p-1)/2} \pmod{p}.$$

Since b has order $p - 1$, it must be the case that $p - 1$ divides $i(p - 1)/2$. Hence, i is even, and then the square roots of a are $\pm b^{i/2} \bmod p$. ∎

Theorem 5.9 yields a polynomial-time algorithm for **Quadratic Residues**, by using the SQUARE-AND-MULTIPLY ALGORITHM for exponentiation modulo p. The complexity of the algorithm will be $O((\log p)^3)$.

We now need to give some further definitions from number theory.

Definition 5.3: Suppose p is an odd prime. For any integer a, define the *Legendre symbol* $\left(\frac{a}{p}\right)$ as follows:

$$\left(\frac{a}{p}\right) = \begin{cases} 0 & \text{if } a \equiv 0 \pmod{p} \\ 1 & \text{if } a \text{ is a quadratic residue modulo } p \\ -1 & \text{if } a \text{ is a quadratic non-residue modulo } p. \end{cases}$$

We have already seen that $a^{(p-1)/2} \equiv 1 \pmod{p}$ if and only if a is a quadratic residue modulo p. If a is a multiple of p, then it is clear that $a^{(p-1)/2} \equiv 0 \pmod{p}$. Finally, if a is a quadratic non-residue modulo p, then $a^{(p-1)/2} \equiv -1 \pmod{p}$ because

$$(a^{(p-1)/2})^2 \equiv a^{p-1} \equiv 1 \pmod{p}$$

and $a^{(p-1)/2} \not\equiv 1 \pmod{p}$. Hence, we have the following result, which provides an efficient algorithm to evaluate Legendre symbols:

THEOREM 5.10 *Suppose p is an odd prime. Then*

$$\left(\frac{a}{p}\right) \equiv a^{(p-1)/2} \pmod{p}.$$

Next, we define a generalization of the Legendre symbol.

Definition 5.4: Suppose n is an odd positive integer, and the prime power factorization of n is

$$n = \prod_{i=1}^{k} p_i^{e_i}.$$

Let a be an integer. The *Jacobi symbol* $\left(\frac{a}{n}\right)$ is defined to be

$$\left(\frac{a}{n}\right) = \prod_{i=1}^{k} \left(\frac{a}{p_i}\right)^{e_i}.$$

Example 5.7 Consider the Jacobi symbol $\left(\frac{6278}{9975}\right)$. The prime power factorization

---

Algorithm 5.6: SOLOVAY-STRASSEN(n)

choose a random integer a such that $1 \le a \le n - 1$
$x \leftarrow \left(\frac{a}{n}\right)$
if $x = 0$
    then return ("n is composite")
$y \leftarrow a^{(n-1)/2} \pmod{n}$
if $x \equiv y \pmod{n}$
    then return ("n is prime")
    else return ("n is composite")

---

of 9975 is $9975 = 3 \times 5^2 \times 7 \times 19$. Thus we have

$$\left(\frac{6278}{9975}\right) = \left(\frac{6278}{3}\right)\left(\frac{6278}{5}\right)^2\left(\frac{6278}{7}\right)\left(\frac{6278}{19}\right)$$

$$= \left(\frac{2}{3}\right)\left(\frac{3}{5}\right)^2\left(\frac{6}{7}\right)\left(\frac{8}{19}\right)$$

$$= (-1)(-1)^2(-1)(-1)$$

$$= -1.$$

$\square$

Suppose $n > 1$ is odd. If n is prime, then $\left(\frac{a}{n}\right) \equiv a^{(n-1)/2} \pmod{n}$ for any a. On the other hand, if n is composite, it may or may not be the case that $\left(\frac{a}{n}\right) \equiv a^{(n-1)/2} \pmod{n}$. If this congruence holds, then n is called an *Euler pseudo-prime* to the base a. For example, 91 is an Euler pseudo-prime to the base 10, because

$$\left(\frac{10}{91}\right) = -1 \equiv 10^{45} \pmod{91}.$$

It can be shown that, for any odd composite n, n is an Euler pseudoprime to the base a for at most half of the integers $a \in \mathbb{Z}_n^*$ (see the Exercises). It is also easy to see that $\left(\frac{a}{n}\right) = 0$ if and only if $\gcd(a, n) > 1$ (therefore, if $1 \le a \le n - 1$ and $\left(\frac{a}{n}\right) = 0$, it must be the case that n is composite).

5.4.2   The Solovay-Strassen Algorithm

We present the SOLOVAY-STRASSEN ALGORITHM, as Algorithm 5.6. The facts proven in the previous section show that this is is a yes-biased Monte Carlo algorithm with error probability at most $1/2$.

At this point it is not clear that Algorithm 5.6 is a polynomial-time algorithm. We already know how to evaluate $a^{(n-1)/2} \bmod n$ in time $O((\log n)^3)$, but how do we compute Jacobi symbols efficiently? It might appear to be necessary to first factor n, since the Jacobi symbol $\left(\frac{a}{n}\right)$ is defined in terms of the factorization of n. But, if we could factor n, we would already know if it is prime; so this approach ends up in a vicious circle.

Fortunately, we can evaluate a Jacobi symbol without factoring n by using some results from number theory, the most important of which is a generalization of the law of quadratic reciprocity (property 4 below). We now enumerate these properties without proof:

1. If n is a positive odd integer and $m_1 \equiv m_2 \pmod{n}$, then

$$\left(\frac{m_1}{n}\right) = \left(\frac{m_2}{n}\right).$$

2. If n is a positive odd integer, then

$$\left(\frac{2}{n}\right) = \begin{cases} 1 & \text{if } n \equiv \pm 1 \pmod 8 \\ -1 & \text{if } n \equiv \pm 3 \pmod 8. \end{cases}$$

3. If n is a positive odd integer, then

$$\left(\frac{m_1 m_2}{n}\right) = \left(\frac{m_1}{n}\right)\left(\frac{m_2}{n}\right).$$

In particular, if $m = 2^k t$ and t is odd, then

$$\left(\frac{m}{n}\right) = \left(\frac{2}{n}\right)^k \left(\frac{t}{n}\right).$$

4. Suppose m and n are positive odd integers. Then

$$\left(\frac{m}{n}\right) = \begin{cases} -\left(\frac{n}{m}\right) & \text{if } m \equiv n \equiv 3 \pmod 4 \\ \left(\frac{n}{m}\right) & \text{otherwise.} \end{cases}$$

Example 5.8 As an illustration of the application of these properties, we evaluate the Jacobi symbol $\left(\frac{7411}{9283}\right)$ in Figure 5.1. Notice that we successively apply properties 4, 1, 3, and 2 in this computation. □

In general, by applying these four properties in the same manner as was done in the example above, it is possible to compute a Jacobi symbol $\left(\frac{a}{n}\right)$ in polynomial time. The only arithmetic operations that are required are modular reductions and factoring out powers of two. Note that if an integer is represented in binary notation, then factoring out powers of two amounts to determining the number of

FIGURE 5.1
Evaluation of a Jacobi symbol

$$\left(\frac{7411}{9283}\right) = -\left(\frac{9283}{7411}\right) \qquad \text{by property 4}$$

$$= -\left(\frac{1872}{7411}\right) \qquad \text{by property 1}$$

$$= -\left(\frac{2}{7411}\right)^4 \left(\frac{117}{7411}\right) \qquad \text{by property 3}$$

$$= -\left(\frac{117}{7411}\right) \qquad \text{by property 2}$$

$$= -\left(\frac{7411}{117}\right) \qquad \text{by property 4}$$

$$= -\left(\frac{40}{117}\right) \qquad \text{by property 1}$$

$$= -\left(\frac{2}{117}\right)^3 \left(\frac{5}{117}\right) \qquad \text{by property 3}$$

$$= \left(\frac{5}{117}\right) \qquad \text{by property 2}$$

$$= \left(\frac{117}{5}\right) \qquad \text{by property 4}$$

$$= \left(\frac{2}{5}\right) \qquad \text{by property 1}$$

$$= -1 \qquad \text{by property 2.}$$

trailing zeroes. So, the complexity of the algorithm is determined by the number of modular reductions that must be done. It is not difficult to show that at most $O(\log n)$ modular reductions are performed, each of which can be done in time $O((\log n)^2)$. This shows that the complexity is $O((\log n)^3)$, which is polynomial in $\log n$. (In fact, the complexity can be shown to be $O((\log n)^2)$ by more precise analysis.)

Suppose that we have generated a random number n and tested it for primality using the SOLOVAY-STRASSEN ALGORITHM. If we have run the algorithm m times, what is our confidence that n is prime? It is tempting to conclude that the probability that such an integer n is prime is $1 - 2^{-m}$. This conclusion is often stated in both textbooks and technical articles, but it cannot be inferred from the given data.

We need to be careful about our use of probabilities. Suppose we define the following random variables: **a** denotes the event

"a random odd integer n of a specified size is composite,"

and **b** denotes the event

"the algorithm answers 'n is prime' m times in succession."

It is certainly the case that the probability $\mathbf{Pr}[\mathbf{b}|\mathbf{a}] \leq 2^{-m}$. However, the probability that we are really interested is $\mathbf{Pr}[\mathbf{a}|\mathbf{b}]$, which is usually not the same as $\mathbf{Pr}[\mathbf{b}|\mathbf{a}]$.

We can compute $\mathbf{Pr}[\mathbf{a}|\mathbf{b}]$ using Bayes' theorem (Theorem 2.1). In order to do this, we need to know $\mathbf{Pr}[\mathbf{a}]$. Suppose $N \leq n \leq 2N$. Applying the Prime number theorem, the number of (odd) primes between N and $2N$ is approximately

$$\frac{2N}{\ln 2N} - \frac{N}{\ln N} \approx \frac{N}{\ln N}$$

$$\approx \frac{n}{\ln n}.$$

Since there are $N/2 \approx n/2$ odd integers between N and $2N$, we will use the estimate

$$\mathbf{Pr}[\mathbf{a}] \approx 1 - \frac{2}{\ln n}.$$

Then we can compute as follows:

$$\mathbf{Pr}[\mathbf{a}|\mathbf{b}] = \frac{\mathbf{Pr}[\mathbf{b}|\mathbf{a}]\mathbf{Pr}[\mathbf{a}]}{\mathbf{Pr}[\mathbf{b}]}$$

$$= \frac{\mathbf{Pr}[\mathbf{b}|\mathbf{a}]\mathbf{Pr}[\mathbf{a}]}{\mathbf{Pr}[\mathbf{b}|\mathbf{a}]\mathbf{Pr}[\mathbf{a}] + \mathbf{Pr}[\mathbf{b}|\bar{\mathbf{a}}]\mathbf{Pr}[\bar{\mathbf{a}}]}$$

$$\approx \frac{\mathbf{Pr}[\mathbf{b}|\mathbf{a}]\left(1 - \frac{2}{\ln n}\right)}{\mathbf{Pr}[\mathbf{b}|\mathbf{a}]\left(1 - \frac{2}{\ln n}\right) + \frac{2}{\ln n}}$$

$$= \frac{\mathbf{Pr}[\mathbf{b}|\mathbf{a}](\ln n - 2)}{\mathbf{Pr}[\mathbf{b}|\mathbf{a}](\ln n - 2) + 2}$$

$$\leq \frac{2^{-m}(\ln n - 2)}{2^{-m}(\ln n - 2) + 2}$$

$$= \frac{\ln n - 2}{\ln n - 2 + 2^{m+1}}.$$

Note that in this computation, $\bar{\mathbf{a}}$ denotes the event

"a random odd integer n is prime."

FIGURE 5.2
Error probabilities for the SOLOVAY-STRASSEN ALGORITHM

m	2^{-m}	bound on error probability
1	.500	.989
2	.250	.978
5	$.312 \times 10^{-1}$	.847
10	$.977 \times 10^{-3}$	.147
20	$.954 \times 10^{-6}$	$.168 \times 10^{-3}$
30	$.931 \times 10^{-9}$	$.164 \times 10^{-6}$
50	$.888 \times 10^{-15}$	$.157 \times 10^{-12}$
100	$.789 \times 10^{-30}$	$.139 \times 10^{-27}$

It is interesting to compare the two quantities $(\ln n - 2)/(\ln n - 2 + 2^{m+1})$ and 2^{-m} as a function of m. Suppose that $n \approx 2^{512} \approx e^{355}$, since these are the sizes of primes p and q used to construct an RSA modulus. Then the first function is roughly $353/(353 + 2^{m+1})$. We tabulate the two functions for some values of m in Figure 5.2.

Although $353/(353 + 2^{m+1})$ approaches zero exponentially quickly, it does not do so as quickly as 2^{-m}. In practice, however, one would take m to be something like 50 or 100, which will reduce the probability of error to a very small quantity.

5.4.3   The Miller-Rabin Algorithm

We now present another Monte Carlo algorithm for **Composites** which is called the MILLER-RABIN ALGORITHM (this is also known as the "strong pseudo-prime test"). This algorithm is presented as Algorithm 5.7.

Algorithm 5.7 is clearly a polynomial-time algorithm: an elementary analysis shows that its complexity is $O((\log n)^3)$, as is the SOLOVAY-STRASSEN ALGORITHM. In fact, the MILLER-RABIN ALGORITHM performs better in practice than the SOLOVAY-STRASSEN ALGORITHM.

We show now that this algorithm cannot answer "n is composite" if n is prime, i.e., the algorithm is yes-biased.

THEOREM 5.11   *The* MILLER-RABIN *ALGORITHM for* **Composites** *is a yes-biased Monte Carlo algorithm.*

PROOF   We will prove this by assuming that Algorithm 5.7 answers "n is composite" for some prime integer n, and obtain a contradiction. Since the algorithm answers "n is composite," it must be the case that $a^m \not\equiv 1 \pmod{n}$. Now consider the sequence of values b tested in the algorithm. Since b is squared in each iteration of the **for** loop, we are testing the values $a^m, a^{2m}, \ldots, a^{2^{k-1}m}$. Since the algorithm answers "n is composite," we conclude that

$$a^{2^i m} \not\equiv -1 \pmod{n}$$

for $0 \le i \le k - 1$.

Now, using the assumption that n is prime, Fermat's theorem (Corollary 5.6) tells us that

$$a^{2^k m} \equiv 1 \pmod{n}$$

since $n - 1 = 2^k m$. Then $a^{2^{k-1} m}$ is a square root of 1 modulo n. Because n is prime, there are only two square roots of 1 modulo n, namely, $\pm 1 \bmod n$. We have that

$$a^{2^{k-1} m} \not\equiv -1 \pmod{n},$$

so it follows that

$$a^{2^{k-1} m} \equiv 1 \pmod{n}.$$

Then $a^{2^{k-2} m}$ must be a square root of 1. By the same argument,

$$a^{2^{k-2} m} \equiv 1 \pmod{n}.$$

Repeating this argument, we eventually obtain

$$a^m \equiv 1 \pmod{n},$$

which is a contradiction, since the algorithm would have answered "n is prime" in this case. ∎

It remains to consider the error probability of the MILLER-RABIN ALGO-RITHM. Although we will not prove it here, the error probability can be shown to be at most $1/4$.

5.5 Square Roots Modulo n

In this section, we briefly discuss several useful results related to the existence of square roots modulo n. Throughout this section, we will suppose that n is odd and $\gcd(n, a) = 1$. The first question we will consider is the number of solutions $y \in \mathbb{Z}_n$ to the congruence $y^2 \equiv a \pmod{n}$. We already know from Section 5.4 that this congruence has either zero or two solutions if n is prime, depending on whether $\left(\frac{a}{n}\right) = -1$ or $\left(\frac{a}{n}\right) = 1$.

Our next theorem extends this characterization to (odd) prime powers. A proof is outlined in the Exercises.

Algorithm 5.7: MILLER-RABIN(n)

write $n - 1 = 2^k m$, where m is odd
choose a random integer a, $1 \le a \le n - 1$
$b \leftarrow a^m \bmod n$
if $b \equiv 1 \pmod{n}$
  then return ("n is prime")
for $i \leftarrow 0$ **to** $k - 1$
  do $\begin{cases} \textbf{if } b \equiv -1 \pmod{n} \\ \quad \textbf{then return } (\text{"}n \text{ is prime"}) \\ \quad \textbf{else } b \leftarrow b^2 \bmod n \end{cases}$
return ("n is composite")

THEOREM 5.12 *Suppose that p is an odd prime, e is a positive integer, and $\gcd(a, p) = 1$. Then the congruence $y^2 \equiv a \pmod{p^e}$ has no solutions if $\left(\frac{a}{p}\right) = -1$, and two solutions (modulo p^e) if $\left(\frac{a}{p}\right) = 1$.*

Notice that Theorem 5.12 tells us that the existence of square roots of a modulo p^e can be determined by evaluating the Legendre symbol $\left(\frac{a}{p}\right)$.

It is not difficult to extend Theorem 5.12 to the case of an arbitrary odd integer n. The following result is basically an application of the Chinese remainder theorem.

THEOREM 5.13 *Suppose that $n > 1$ is an odd integer having factorization*

$$n = \prod_{i=1}^{\ell} p_i^{e_i},$$

where the p_i's are distinct primes and the e_i's are positive integers. Suppose further that $\gcd(a, n) = 1$. Then the congruence $y^2 \equiv a \pmod{n}$ has 2^ℓ solutions modulo n if $\left(\frac{a}{p_i}\right) = 1$ for all $i \in \{1, \ldots, \ell\}$, and no solutions, otherwise.

PROOF It is clear that $y^2 \equiv a \pmod{n}$ if and only if $y^2 \equiv a \pmod{p_i^{e_i}}$ for all $i \in \{1, \ldots, \ell\}$. If $\left(\frac{a}{p_i}\right) = -1$ for some i, then the congruence $y^2 \equiv a \pmod{p_i^{e_i}}$ has no solutions, and hence $y^2 \equiv a \pmod{n}$ has no solutions.

Now suppose that $\left(\frac{a}{p_i}\right) = 1$ for all $i \in \{1, \ldots, \ell\}$. It follows from Theorem 5.12 that each congruence $y^2 \equiv a \pmod{p_i^{e_i}}$ has two solutions modulo $p_i^{e_i}$, say $y \equiv b_{i,1}$ or $b_{i,2} \pmod{p_i^{e_i}}$. For $1 \le i \le \ell$, let $b_i \in \{b_{i,1}, b_{i,2}\}$. Then the system of congruences $y \equiv b_i \pmod{p_i^{e_i}}$ $(1 \le i \le \ell)$ has a unique solution modulo n, which can be found using the Chinese remainder theorem. There are 2^ℓ ways to choose the ℓ-tuple $(b_1, \ldots, b_\ell)$, and therefore there are 2^ℓ solutions modulo n to the congruence $y^2 \equiv a \pmod{n}$. $\blacksquare$

Suppose that $x^2 \equiv y^2 \equiv a \pmod{n}$, where $\gcd(a, n) = 1$. Let $z = xy^{-1} \bmod n$. It follows that $z^2 \equiv 1 \pmod{n}$. Conversely, if $z^2 \equiv 1 \pmod{n}$, then $(xz)^2 \equiv x^2 \pmod{n}$ for any x. It is therefore possible to obtain all 2^ℓ square roots of an element $a \in \mathbb{Z}_n{}^*$ by taking all 2^ℓ products of one given square root of a with the 2^ℓ square roots of 1. We will make use of this observation later in this chapter.

5.6 Factoring Algorithms

The most obvious way to attack the *RSA Cryptosystem* is to attempt to factor the public modulus. There is a huge amount of literature on factoring algorithms, and a thorough treatment would require more pages than we have in this book. We will just try to give a brief overview here, including an informal discussion of the best current factoring algorithms and their use in practice. The three algorithms that are most effective on very large numbers are the QUADRATIC SIEVE, the ELLIPTIC CURVE FACTORING ALGORITHM and the NUMBER FIELD SIEVE. Other well-known algorithms that were precursors include Pollard's rho-method and $p - 1$ algorithm, Williams' $p + 1$ algorithm, the continued fraction algorithm, and, of course, trial division.

Throughout this section, we suppose that the integer n that we wish to factor is odd. If n is composite, then it is easy to see that n has a prime factor $p \leq \lfloor \sqrt{n} \rfloor$. Therefore, the simple method of *trial division*, which consists of dividing n by every odd integer up to $\lfloor \sqrt{n} \rfloor$, suffices to determine if n is prime or composite. If $n < 10^{12}$, say, this is a perfectly reasonable factorization method, but for larger n we generally need to use more sophisticated techniques.

When we say that we want to factor n, we could ask for a complete factorization into primes, or we might be content with finding any non-trivial factor. In most of the algorithms we study, we are just searching for an arbitrary non-trivial factor. In general, we obtain factorizations of the form $n = n_1 n_2$, where $1 < n_1 < n$ and $1 < n_2 < n$. If we desire a complete factorization of n into primes, we could test n_1 and n_2 for primality using a randomized primality test, and then factor one or both of them further if they are not prime.

5.6.1 The Pollard $p - 1$ Algorithm

As an example of a simple algorithm that can sometimes be applied to larger integers, we describe the POLLARD $p - 1$ ALGORITHM, which dates from 1974. This algorithm, presented as Algorithm 5.8, has two inputs: the (odd) integer n to be factored, and a prespecified "bound," B.

Here is what is taking place in the POLLARD $p - 1$ ALGORITHM: Suppose p is a prime divisor of n, and suppose that $q \leq B$ for every prime power $q \mid (p - 1)$.

Algorithm 5.8: POLLARD $p - 1$ FACTORING ALGORITHM(n, B)

$a \leftarrow 2$
for $j \leftarrow 2$ **to** B
  do $a \leftarrow a^j \bmod n$
$d \leftarrow \gcd(a - 1, n)$
if $1 < d < n$
  then return (d)
  else return ("failure")

Then it must be the case that
$$(p - 1) \mid B!$$
At the end of the **for** loop, we have that
$$a \equiv 2^{B!} \pmod{n}.$$
Since $p \mid n$, it must be the case that
$$a \equiv 2^{B!} \pmod{p}.$$
Now,
$$2^{p-1} \equiv 1 \pmod{p}$$
by Fermat's theorem. Since $(p - 1) \mid B!$, it follows that
$$a \equiv 1 \pmod{p},$$
and hence $p \mid (a - 1)$. Since we also have that $p \mid n$, we see that $p \mid d$, where $d = \gcd(a - 1, n)$. The integer d will be a non-trivial divisor of n (unless $a = 1$). Having found a non-trivial factor d, we would then proceed to attempt to factor d and n/d if they are expected to be composite.

Here is an example to illustrate.

Example 5.9   Suppose $n = 15770708441$. If we apply Algorithm 5.8 with $B = 180$, then we find that $a = 11620221425$ and d is computed to be 135979. In fact, the complete factorization of n into primes is

$$15770708441 = 135979 \times 115979.$$

In this example, the factorization succeeds because 135978 has only "small" prime factors:
$$135978 = 2 \times 3 \times 131 \times 173.$$
Hence, by taking $B \geq 173$, it will be the case that $135978 \mid B!$, as desired.   □

In the POLLARD $p-1$ ALGORITHM, there are $B-1$ modular exponentiations, each requiring at most $2 \log_2 B$ modular multiplications using the SQUARE-AND-MULTIPLY ALGORITHM. The gcd can be computed in time $O((\log n)^3)$ using the EXTENDED EUCLIDEAN ALGORITHM. Hence, the complexity of the algorithm is $O(B \log B (\log n)^2 + (\log n)^3)$. If the integer B is $O((\log n)^i)$ for some fixed integer i, then the algorithm is indeed a polynomial-time algorithm (as a function of $\log n$); however, for such a choice of B the probability of success will be very small. On the other hand, if we increase the size of B drastically, say to $\sqrt{n}$, then the algorithm is guaranteed to be successful, but it will be no faster than trial division.

Thus, the drawback of this method is that it requires n to have a prime factor p such that $p-1$ has only "small" prime factors. It would be very easy to construct an RSA modulus $n = pq$ which would resist factorization by this method. One would start by finding a large prime p_1 such that $p = 2p_1 + 1$ is also prime, and a large prime q_1 such that $q = 2q_1 + 1$ is also prime (using one of the Monte Carlo primality testing algorithms discussed in Section 5.4). Then the RSA modulus $n = pq$ will be resistant to factorization using the $p-1$ method.

The more powerful elliptic curve algorithm, developed by Lenstra in the mid-1980s, is in fact a generalization of the POLLARD $p-1$ ALGORITHM. The success of the elliptic curve method depends on the more likely situation that an integer "close to" p has only "small" prime factors. Whereas the $p-1$ method depends on a relation that holds in the group $\mathbb{Z}_p$, the elliptic curve method involves groups defined on elliptic curves modulo p.

5.6.2 The Pollard Rho Algorithm

Let p be the smallest prime divisor of n. Suppose there exist two integers $x, x' \in \mathbb{Z}_n$, such that $x \neq x'$ and $x \equiv x' \pmod{p}$. Then $p \leq \gcd(x - x', n) < n$, so we obtain a non-trivial factor of n by computing a greatest common divisor. (Note that the value of p does not need to be known ahead of time in order for this method to work.)

Suppose we try to factor n by first choosing a random subset $X \subseteq \mathbb{Z}_n$, and then computing $\gcd(x - x', n)$ for all distinct values $x, x' \in X$. This method will be successful if and only if the mapping $x \mapsto x \bmod p$ yields at least one collision for $x \in X$. This situation can be analyzed using the birthday paradox described in Section 4.2.2: if $|X| \approx 1.17\sqrt{p}$, then there is a 50% probability that there is at least one collision, and hence a non-trivial factor of n will be found. However, in order to find a collision of the form $x \bmod p = x' \bmod p$, we need to compute $\gcd(x - x', n)$. (We cannot explicitly compute the values $x \bmod p$ for $x \in X$, and sort the resulting list, as suggested in Section 4.2.2, because the value of p is not known.) This means that we would expect to compute more than $\binom{|X|}{2} > p/2$ gcd's before finding a factor of n.

The POLLARD RHO ALGORITHM incorporates a variation of this technique that requires fewer gcd computations and less memory. Suppose that the function

f is a polynomial with integer coefficients, e.g., $f(x) = x^2 + a$, where a is a small constant ($a = 1$ is a commonly used value). Let's assume that the mapping $x \mapsto f(x) \bmod p$ behaves like a random mapping. (It is of course not "random," which means that what we are presenting is a heuristic analysis rather than a rigorous proof.) Let $x_1 \in \mathbb{Z}_n$, and consider the sequence $x_1, x_2, \ldots$, where

$$x_j = f(x_{j-1}) \bmod n,$$

for all $j \geq 2$. Let m be an integer, and define $X = \{x_1, \ldots, x_m\}$. To simplify matters, suppose that X consists of m distinct residues modulo n. Hopefully it will be the case that X is a random subset of m elements of $\mathbb{Z}_n$.

We are looking for two distinct values $x_i, x_j \in X$ such that $\gcd(x_j - x_i, n) > 1$. Each time we compute a new term x_j in the sequence, we could compute $\gcd(x_j - x_i, n)$ for all $i < j$. However, it turns out that we can reduce the number of gcd computations greatly. We describe how this can be done.

Suppose that $x_i \equiv x_j \pmod{p}$. Using the fact that f is a polynomial with integer coefficients, we have that $f(x_i) \equiv f(x_j) \pmod{p}$. Recall that $x_{i+1} = f(x_i) \bmod n$ and $x_{j+1} = f(x_j) \bmod n$. Then

$$x_{i+1} \bmod p = (f(x_i) \bmod n) \bmod p = f(x_i) \bmod p,$$

because $p \mid n$. Similarly,

$$x_{j+1} \bmod p = f(x_j) \bmod p.$$

Therefore, $x_{i+1} \equiv x_{j+1} \pmod{p}$. Repeating this argument, we obtain the following important result:

If $x_i \equiv x_j \pmod{p}$, then $x_{i+\delta} \equiv x_{j+\delta} \pmod{p}$ for all integers $\delta \geq 0$.

Denoting $\ell = j - i$, it follows that $x_{i'} \equiv x_{j'} \pmod{p}$ if $j' > i' \geq i$ and $j' - i' \equiv 0 \pmod{\ell}$.

Suppose that we construct a graph G on vertex set $\mathbb{Z}_p$, where for all $i \geq 1$, we have a directed edge from $x_i \bmod p$ to $x_{i+1} \bmod p$. There must exist a first pair x_i, x_j with $i < j$ such that $x_i \equiv x_j \pmod{p}$. By the observation made above, it is easily seen that the graph G consists of a "tail"

$$x_1 \bmod p \to x_2 \bmod p \to \cdots \to x_i \bmod p\,,$$

and an infinitely repeated cycle of length ℓ, having vertices

$$x_i \bmod p \to x_{i+1} \bmod p \to \cdots \to x_j \bmod p = x_i \bmod p.$$

Thus G looks like the Greek letter ρ, which is the reason for the name "rho algorithm."

We illustrate the above with an example.

Algorithm 5.9: POLLARD RHO FACTORING ALGORITHM(n, x_1)

external f
$x \leftarrow x_1$
$x' \leftarrow f(x) \bmod n$
$p \leftarrow \gcd(x - x', n)$
while $p = 1$
$\quad$ **comment:** in the ith iteration, $x = x_i$ and $x' = x_{2i}$

$\quad$ **do** $\begin{cases} x \leftarrow f(x) \bmod n \\ x' \leftarrow f(x') \bmod n \\ x' \leftarrow f(x') \bmod n \\ p \leftarrow \gcd(x - x', n) \end{cases}$

if $p = n$
$\quad$ **then return** ("failure")
$\quad$ **else return** (p)

Example 5.10 Suppose that $n = 7171 = 71 \times 101$, $f(x) = x^2 + 1$ and $x_1 = 1$. The sequence of x_i's begins as follows:

$$\begin{array}{ccccccc} 1 & 2 & 5 & 26 & 677 & 6557 & 4105 \\ 6347 & 4903 & 2218 & 219 & 4936 & 4210 & 4560 \\ 4872 & 375 & 4377 & 4389 & 2016 & 5471 & 88 \end{array}$$

The above values, when reduced modulo 71, are as follows:

$$\begin{array}{ccccccc} 1 & 2 & 5 & 26 & 38 & 25 & 58 \\ 28 & 4 & 17 & 6 & 37 & 21 & 16 \\ 44 & 20 & 46 & 58 & 28 & 4 & 17 \end{array}$$

The first collision in the above list is

$$x_7 \bmod 71 = x_{18} \bmod 71 = 58.$$

Therefore the graph G consists of a tail of length seven and a cycle of length 11.
$\square$

We have already mentioned that our goal is to discover two terms $x_i \equiv x_j$ (mod p) with $i < j$, by computing a greatest common divisor. It is not necessary that we discover the first occurrence of a collision of this type. In order to simplify and improve the algorithm, we restrict our search for collisions by taking $j = 2i$. The resulting algorithm is presented as Algorithm 5.9.

This algorithm is not hard to analyze. If $x_i \equiv x_j$ (mod p), then it is also the case that $x_{i'} \equiv x_{2i'}$ (mod p) for all i' such that $i' \equiv 0$ (mod ℓ) and $i' \geq i$.

Among the ℓ consecutive integers $i, \ldots, j - 1$, there must be one that is divisible by ℓ. Therefore the smallest value i' that satisfies the two conditions above is at most $j - 1$. Hence, the number of iterations required to find a factor p is at most j, which is expected to be at most $\sqrt{p}$.

In Example 5.10, the first collision modulo 71 occurs for $i = 7, j = 18$. The smallest integer $i' \geq 7$ that is divisible by 11 is $i' = 11$. Therefore Algorithm 5.9 will discover the factor 71 of n when it computes $\gcd(x_{11} - x_{22}, n) = 71$.

In general, since $p < \sqrt{n}$, the expected complexity of the algorithm is $O(n^{1/4})$ (ignoring logarithmic factors). However, we again emphasize that this is a heuristic analysis, and not a mathematical proof. On the other hand, the actual performance of the algorithm in practice is similar to this estimate.

It is possible that Algorithm 5.9 could fail to find a nontrivial factor of n. This happens if and only if the first values x and x' which satisfy $x \equiv x' \pmod{p}$ actually satisfy $x \equiv x' \pmod{n}$ (this is equivalent to $x = x'$, because x and x' are reduced modulo n). We would estimate heuristically that the probability of this situation occurring is roughly p/n, which is quite small when n is large, because $p < \sqrt{n}$. If the algorithm does fail in this way, it is a simple matter to run it again with a different initial value or a different choice for the function f.

The reader might wish to run Algorithm 5.9 on a larger value of n. When $n = 15770708441$ (the same value of n considered in Example 5.9), $x_1 = 1$ and $f(x) = x^2 + 1$, it can be verified that $x_{422} = 2261992698$, $x_{211} = 7149213937$, and

$$\gcd(x_{422} - x_{211}, n) = 135979.$$

5.6.3   Dixon's Random Squares Algorithm

Many factoring algorithms are based on the following very simple idea. Suppose we can find $x \not\equiv \pm y \pmod{n}$ such that $x^2 \equiv y^2 \pmod{n}$. Then

$$n \mid (x - y)(x + y)$$

but neither of $x - y$ or $x + y$ is divisible by n. It therefore follows that $\gcd(x + y, n)$ is a non-trivial factor of n (and similarly, $\gcd(x - y, n)$ is also a non-trivial factor of n).

As an example, it is easy to verify that $10^2 \equiv 32^2 \pmod{77}$. By computing $\gcd(10 + 32, 77) = 7$, we discover the factor 7 of 77.

The RANDOM SQUARES ALGORITHM uses a *factor base*, which is a set $\mathcal{B}$ of the b smallest primes, for an appropriate value b. We first obtain several integers z such that all the prime factors of $z^2 \bmod n$ occur in the factor base $\mathcal{B}$. (How this is done will be discussed a bit later.) The idea is to then take the product of a subset of these z's in such a way that every prime in the factor base is used an even number of times. This then gives us a congruence of the desired type $x^2 \equiv y^2 \pmod{n}$, which (we hope) will lead to a factorization of n.

We illustrate with a carefully contrived example.

Example 5.11 Suppose $n = 15770708441$ (this was the same n that we used in Example 5.9). Let $b = 6$; then $\mathcal{B} = \{2, 3, 5, 7, 11, 13\}$. Consider the three congruences:

$$8340934156^2 \equiv 3 \times 7 \pmod{n}$$

$$12044942944^2 \equiv 2 \times 7 \times 13 \pmod{n}$$

$$2773700011^2 \equiv 2 \times 3 \times 13 \pmod{n}.$$

If we take the product of these three congruences, then we have

$$(8340934156 \times 12044942944 \times 2773700011)^2 \equiv (2 \times 3 \times 7 \times 13)^2 \pmod{n}.$$

Reducing the expressions inside the parentheses modulo n, we have

$$9503435785^2 \equiv 546^2 \pmod{n}.$$

Then, using the EUCLIDEAN ALGORITHM, we compute

$$\gcd(9503435785 - 546, 15770708441) = 115759,$$

finding the factor 115759 of n. □

Suppose $\mathcal{B} = \{p_1, \ldots, p_b\}$ is the factor base. Let c be slightly larger than b (say $c = b + 4$), and suppose we have obtained c congruences:

$$z_j^2 \equiv p_1^{\alpha_{1j}} \times p_2^{\alpha_{2j}} \cdots \times p_b^{\alpha_{bj}} \pmod{n},$$

$1 \leq j \leq c$. For each j, consider the vector

$$a_j = (\alpha_{1j} \bmod 2, \ldots, \alpha_{bj} \bmod 2) \in (\mathbb{Z}_2)^b.$$

If we can find a subset of the a_j's that sum modulo 2 to the vector $(0, \ldots, 0)$, then the product of the corresponding z_j's will use each factor in $\mathcal{B}$ an even number of times.

We illustrate by returning to Example 5.11, where there exists a dependence even though $c < b$ in this case.

Example 5.11 **(Cont.)** The three vectors a_1, a_2, a_3 are as follows:

$$a_1 = (0, 1, 0, 1, 0, 0)$$

$$a_2 = (1, 0, 0, 1, 0, 1)$$

$$a_3 = (1, 1, 0, 0, 0, 1).$$

It is easy to see that

$$a_1 + a_2 + a_3 = (0, 0, 0, 0, 0, 0) \bmod 2.$$

This gives rise to the congruence we saw earlier that successfully factored n. □

Observe that finding a subset of the c vectors $a_1, \ldots, a_c$ that sums modulo 2 to the all-zero vector is nothing more than finding a linear dependence (over $\mathbb{Z}_2$) of these vectors. Provided $c > b$, such a linear dependence must exist, and it can be found easily using the standard method of Gaussian elimination. The reason why we take $c > b+1$ is that there is no guarantee that any given congruence $x^2 \equiv y^2$ (mod n) will yield the factorization of n. However, we argue heuristically that $x \equiv \pm y$ (mod n) at most 50% of the time, as follows. Suppose that $x^2 \equiv y^2 \equiv a$ (mod n), where $\gcd(a, n) = 1$. Theorem 5.13 tells us that a has 2^ℓ square roots modulo n, where ℓ is the number of prime divisors of n. If $\ell \geq 2$, then a has at least four square roots. Hence, if we assume that x and y are "random," we can then conclude that $x \equiv \pm y$ (mod n) with probability $2/2^\ell \leq 1/2$.

Now, if $c > b+1$, we can obtain several such congruences of the form $x^2 \equiv y^2$ (mod n) (arising from different linear dependencies among the a_j's). Hopefully, at least one of the resulting congruences will yield a congruence of the form $x^2 \equiv y^2$ mod n where $x \not\equiv \pm y$ (mod n), and a non-trivial factor of n will be obtained.

We now discuss how to obtain integers z such that the values $z^2 \bmod n$ factor completely over a given factor base $\mathcal{B}$. There are several methods of doing this. One way is simply to choose the z's at random; this approach yields the so-called RANDOM SQUARES ALGORITHM. However, it is particularly useful to try integers of the form $j + \lceil \sqrt{kn} \rceil$, $j = 0, 1, 2, \ldots$, $k = 1, 2, \ldots$. These integers tend to be small when squared and reduced modulo n, and hence they have a higher than average probability of factoring over $\mathcal{B}$. Another useful trick is to try integers of the form $z = \lfloor \sqrt{kn} \rfloor$. When squared and reduced modulo n, these integers are a bit less than n. This means that $-z^2 \bmod n$ is small and can perhaps be factored over $\mathcal{B}$. Therefore, if we include -1 in $\mathcal{B}$, we can factor $z^2 \bmod n$ over $\mathcal{B}$.

We illustrate these techniques with a small example.

Example 5.12   Suppose that $n = 1829$ and $\mathcal{B} = \{-1, 2, 3, 5, 7, 11, 13\}$. We compute $\sqrt{n} = 42.77$, $\sqrt{2n} = 60.48$, $\sqrt{3n} = 74.07$ and $\sqrt{4n} = 85.53$. Suppose we take $z = 42, 43, 60, 61, 74, 75, 85, 86$. We obtain several factorizations of $z^2 \bmod n$ over $\mathcal{B}$. In the following table, all congruences are modulo n:

$$
\begin{aligned}
z_1{}^2 &\equiv 42^2 &\equiv -65 &\equiv (-1) \times 5 \times 13 \\
z_2{}^2 &\equiv 43^2 &\equiv 20 &\equiv 2^2 \times 5 \\
z_3{}^2 &\equiv 61^2 &\equiv 63 &\equiv 3^2 \times 7 \\
z_4{}^2 &\equiv 74^2 &\equiv -11 &\equiv (-1) \times 11 \\
z_5{}^2 &\equiv 85^2 &\equiv -91 &\equiv (-1) \times 7 \times 13 \\
z_6{}^2 &\equiv 86^2 &\equiv 80 &\equiv 2^4 \times 5.
\end{aligned}
$$

We therefore have six factorizations, which yield six vectors in $(\mathbb{Z}_2)^7$. This is not enough to guarantee a dependence relation, but it turns out to be sufficient in this particular case. The six vectors are as follows:

$$a_1 = (1, 0, 0, 1, 0, 0, 1)$$
$$a_2 = (0, 0, 0, 1, 0, 0, 0)$$
$$a_3 = (0, 0, 0, 0, 1, 0, 0)$$
$$a_4 = (1, 0, 0, 0, 0, 1, 0)$$
$$a_5 = (1, 0, 0, 0, 1, 0, 1)$$
$$a_6 = (0, 0, 0, 1, 0, 0, 0).$$

Clearly $a_2 + a_6 = (0, 0, 0, 0, 0, 0, 0)$; however, the reader can check that this dependence relation does not yield a factorization of n. A dependence relation that does work is

$$a_1 + a_2 + a_3 + a_5 = (0, 0, 0, 0, 0, 0, 0).$$

The congruence that we obtain is

$$(42 \times 43 \times 61 \times 85)^2 \equiv (2 \times 3 \times 5 \times 7 \times 13)^2 \pmod{1829}.$$

This simplifies to give

$$1459^2 \equiv 901^2 \pmod{1829}.$$

It is then straightforward to compute

$$\gcd(1459 + 901, 1829) = 59,$$

and thus we have obtained a nontrivial factor of n. ▯

An important general question is how large the factor base should be (as a function of the integer n that we are attempting to factor) and what the complexity of the algorithm is. In general, there is a trade-off: if $b = |\mathcal{B}|$ is large, then it is more likely that an integer $z^2 \bmod n$ factors over $\mathcal{B}$. But the larger b is, the more congruences we need to accumulate before we are able to find a dependence relation. A good choice for b can be determined with the help of some results from number theory. We discuss some of the main ideas now. This will be a heuristic analysis in which we will be assuming that the integers z are chosen randomly.

Suppose that n and m are positive integers. We say that n is *m-smooth* provided that every prime factor of n is less than or equal to m. $\Psi(n, m)$ is defined to be the number of positive integers less than or equal to n that are m-smooth. An important result in number theory says that, if $n \gg m$, then

$$\frac{\Psi(n, m)}{n} \approx \frac{1}{u^u},$$

where $u = \log n / \log m$. Observe that $\Psi(n, m)/n$ represents the probability that a random integer in the set $\{1, \ldots, n\}$ is m-smooth.

Suppose that $n \approx 2^r$ and $m \approx 2^s$. Then

$$u = \frac{\log n}{\log m} \approx \frac{r}{s}.$$

Division of an r-bit integer by an s-bit integer can be done in time $O(rs)$. From this, it is possible to show that we can determine if an integer in the set $\{1, \ldots, n\}$ is m-smooth in time $O(rsm)$ if we assume that $r < m$ (see the Exercises).

Our factor base $\mathcal{B}$ consists of all the primes less than or equal to m. Therefore, applying the Prime number theorem, we have that

$$|\mathcal{B}| = b = \pi(m) \approx \frac{m}{\ln m}.$$

We need to find slightly more than b m-smooth squares modulo n in order for the algorithm to succeed. We expect to test bu^u integers in order to find b of them that are m-smooth. Therefore, the expected time to find the necessary m-smooth squares is $O(bu^u \times rsm)$. We have that b is $O(m/s)$, so the running time of the first part of the algorithm is $O(rm^2 u^u)$.

In the second part of the algorithm, we need to reduce the associated matrix modulo 2, construct our congruence of the form $x^2 \equiv y^2 \pmod{n}$, and apply the EUCLIDEAN ALGORITHM. It can be checked without too much difficulty that these steps can be done in time that is polynomial in r and m, say $O(r^i m^j)$, where i and j are positive integers. (On average, this second part of the algorithm must be done at most twice, because the probability that a congruence does not provide a factor of n is at most $1/2$. This contributes a constant factor of at most 2, which is absorbed into the big-oh.)

At this point, we know that the total running time of the algorithm can be written in the form $O(rm^2 u^u + r^i m^j)$. Recall that $n \approx 2^r$ is given, and we are trying to choose $m \approx 2^s$ to optimize the running time. It turns out that a good choice for m is to take $s \approx \sqrt{r \log_2 r}$. Then

$$u \approx \frac{r}{s} \approx \sqrt{\frac{r}{\log_2 r}}.$$

Now we can compute

$$\log_2 u^u = u \log_2 u$$

$$\approx \sqrt{\frac{r}{\log_2 r}} \log_2 \left(\sqrt{\frac{r}{\log_2 r}} \right)$$

$$< \sqrt{\frac{r}{\log_2 r}} \log_2 \sqrt{r}$$

$$= \sqrt{\frac{r}{\log_2 r}} \times \frac{\log_2 r}{2}$$

$$= \frac{\sqrt{r \log_2 r}}{2}.$$

It follows that

$$u^u \leq 2^{0.5 \sqrt{r \log_2 r}}.$$

We also have that

$$m \approx 2^{\sqrt{r \log_2 r}}$$

and

$$r = 2^{\log_2 r}.$$

Hence, the total running time can be expressed in the form

$$O \left(2^{\log_2 r + 2\sqrt{r \log_2 r} + 0.5\sqrt{r \log_2 r}} + 2^{i \log_2 r + j\sqrt{r \log_2 r}} \right),$$

which is easily seen to be

$$O \left(2^{c \sqrt{r \log_2 r}} \right)$$

for some constant c. Using the fact that $r \approx \log_2 n$, we obtain a running time of

$$O \left(2^{c \sqrt{\log_2 n \log_2 \log_2 n}} \right).$$

Often the running time is expressed in terms of logarithms and exponentials to the base e. A more precise analysis, using an optimal choice of m, leads to the following commonly stated expected running time:

$$O \left(e^{(1 + o(1)) \sqrt{\ln n \ln \ln n}} \right).$$

5.6.4   Factoring Algorithms in Practice

One specific, well-known algorithm that has been widely used in practice is the QUADRATIC SIEVE due to Pomerance. The name "quadratic sieve" comes from a sieving procedure (which we will not describe here) that is used to determine the values $z^2 \bmod n$ that factor over $\mathcal{B}$. The NUMBER FIELD SIEVE is a more recent factoring algorithm from the late 1980s. It also factors n by constructing a congruence $x^2 \equiv y^2 \pmod{n}$, but it does so by means of computations in rings of algebraic integers. In recent years, the number field sieve has become the algorithm of choice for factoring large integers.

The asymptotic running times of the QUADRATIC SIEVE, ELLIPTIC CURVE and NUMBER FIELD SIEVE factoring algorithms are as follows:

quadratic sieve	$O\left(e^{(1+o(1))\sqrt{\ln n \ln \ln n}}\right)$
elliptic curve	$O\left(e^{(1+o(1))\sqrt{2 \ln p \ln \ln p}}\right)$
number field sieve	$O\left(e^{(1.92+o(1))(\ln n)^{1/3}(\ln \ln n)^{2/3}}\right)$

The notation $o(1)$ denotes a function of n that approaches 0 as $n \to \infty$, and p denotes the smallest prime factor of n.

In the worst case, $p \approx \sqrt{n}$ and the asymptotic running times of the QUADRATIC SIEVE and ELLIPTIC CURVE algorithms are essentially the same. But in such a situation, QUADRATIC SIEVE is faster than ELLIPTIC CURVE. The ELLIPTIC CURVE ALGORITHM is more useful if the prime factors of n are of differing size. One very large number that was factored using the ELLIPTIC CURVE ALGORITHM was the Fermat number $2^{2^{11}} + 1$, which was factored in 1988 by Brent.

For factoring RSA moduli (where $n = pq$, p, q are prime, and p and q are roughly the same size), the QUADRATIC SIEVE was the most-used algorithm up until the mid-1990s. The number field sieve is the most recently developed of the three algorithms. Its advantage over the other algorithms is that its asymptotic running time is faster than either QUADRATIC SIEVE or ELLIPTIC CURVE. The NUMBER FIELD SIEVE has proven to be faster for numbers having more than about 125–130 digits. An early use of the NUMBER FIELD SIEVE was in 1990, when Lenstra, Lenstra, Manasse, and Pollard factored $2^{2^9} + 1$ into three primes having 7, 49 and 99 digits.

Some notable milestones in factoring have included the following factorizations. In 1983, the QUADRATIC SIEVE successfully factored a 69-digit number that was a (composite) factor of $2^{251} - 1$ (a computation which was done by Davis, Holdridge, and Simmons). Progress continued throughout the 1980s, and by 1989, numbers having up to 106 digits were factored by this method by Lenstra and Manasse, by distributing the computations to hundreds of widely separated workstations (they called this approach "factoring by electronic mail").

In the early 1990s, RSA published a series of "challenge" numbers for factoring algorithms on the Internet. The numbers were known as RSA-100, RSA-110, . . . , RSA-500, where each number RSA-d is a d-digit integer that is the product of two primes of approximately the same length. Several of the challenges were factored, culminating in the factorization of RSA-160 in April, 2003.

RSA put forward a new factoring challenge in 2001, where the size of the numbers involved were measured in bits rather than decimal digits. There are eight numbers in the new RSA challenge: RSA-576, RSA-640, RSA-704, RSA-768, RSA-896, RSA-1024, RSA-1536 and RSA-2048. There are prizes for factoring these numbers, ranging from $10,000 to $200,000. The factorization of RSA-576 was announced in December, 2003 by Jens Franke; the factorization was accomplished using the GENERAL NUMBER FIELD SIEVE.

Extrapolating current trends in factoring into the future, it has been suggested that 768-bit moduli may be factored by 2010, and 1024-bit moduli may be factored by the year 2018.

5.7   Other Attacks on RSA

In this section, we address the following question: are there possible attacks on the *RSA Cryptosystem* other than factoring n? For example, it is at least conceivable that there could exist a method of decrypting RSA ciphertexts that does not involve finding the factorization of the modulus n.

5.7.1   Computing $\phi(n)$

We first observe that computing $\phi(n)$ is no easier than factoring n. For, if n and $\phi(n)$ are known, and n is the product of two primes p, q, then n can be easily factored, by solving the two equations

$$n = pq$$
$$\phi(n) = (p-1)(q-1)$$

for the two "unknowns" p and q. This is easily accomplished, as follows. If we substitute $q = n/p$ into the second equation, we obtain a quadratic equation in the unknown value p:

$$p^2 - (n - \phi(n) + 1)p + n = 0. \tag{5.1}$$

The two roots of equation (5.1) will be p and q, the factors of n. Hence, if a cryptanalyst can learn the value of $\phi(n)$, then he can factor n and break the system. In other words, computing $\phi(n)$ is no easier than factoring n.

Here is an example to illustrate.

Example 5.13   Suppose $n = 84773093$, and the adversary has learned that $\phi(n) = 84754668$. This information gives rise to the following quadratic equation:

$$p^2 - 18426p + 84773093 = 0.$$

This can be solved by the quadratic formula, yielding the two roots 9539 and 8887. These are the two factors of n. ∎

5.7.2   The Decryption Exponent

We will now prove the very interesting result that, if the decryption exponent a is known, then n can be factored in polynomial time by means of a randomized algorithm. Therefore we can say that computing a is (essentially) no easier than factoring n. (However, this does not rule out the possibility of breaking the *RSA Cryptosystem* without computing a.) Notice that this result is of much more than theoretical interest. It tells us that if a is revealed (accidently or otherwise), then it is not sufficient for Bob to choose a new encryption exponent; he must also choose a new modulus n.

The algorithm we are going to describe is a randomized algorithm of the Las Vegas type (see Section 4.2.2 for the definition). Here, we consider Las Vegas algorithms having worst-case success probability at least $1 - \epsilon$. Therefore, for any problem instance, the algorithm may fail to give an answer with probability at most ϵ.

If we have such a Las Vegas algorithm, then we simply run the algorithm over and over again until it finds an answer. The probability that the algorithm will return "no answer" m times in succession is ϵ^m. The average (i.e., expected) number of times the algorithm must be run in order to obtain an answer is $1/(1-\epsilon)$ (see the Exercises).

We will describe a Las Vegas algorithm that will factor n with probability at least $1/2$ when given the values a, b and n as input. Hence, if the algorithm is run m times, then n will be factored with probability at least $1 - 1/2^m$.

The algorithm is based on certain facts concerning square roots of 1 modulo n, where $n = pq$ is the product of two distinct odd primes. $x^2 \equiv 1 \pmod{p}$ and Theorem 5.13 asserts that there are four square roots of 1 modulo n. Two of these square roots are $\pm 1 \bmod n$; these are called the *trivial* square roots of 1 modulo n. The other two square roots are called *non-trivial*; they are also negatives of each other modulo n.

Here is a small example to illustrate.

Example 5.14   Suppose $n = 403 = 13 \times 31$. The four square roots of 1 modulo 403 are $1, 92, 311$ and 402. The square root 92 is obtained by solving the system

$$x \equiv 1 \pmod{13},$$

$$x \equiv -1 \pmod{31}.$$

using the Chinese remainder theorem. The other non-trivial square root is $403 - 92 = 311$. It is the solution to the system

$$x \equiv -1 \pmod{13},$$

$$x \equiv 1 \pmod{31}.$$

□

Suppose x is a non-trivial square root of 1 modulo n. Then

$$x^2 \equiv 1^2 \pmod{n}$$

but

$$x \not\equiv \pm 1 \pmod{n}.$$

Then, as in the Random squares factoring algorithm, we can find the factors of n by computing $\gcd(x + 1, n)$ and $\gcd(x - 1, n)$. In Example 5.14 above,

$$\gcd(93, 403) = 31$$

and

$$\gcd(312, 403) = 13.$$

Algorithm 5.10 attempts to factor n by finding a non-trivial square root of 1 modulo n. Before analyzing the algorithm, we first do an example to illustrate its application.

Example 5.15 Suppose $n = 89855713$, $b = 34986517$ and $a = 82330933$, and the random value $w = 5$. We have

$$ab - 1 = 2^3 \times 360059073378795.$$

Then

$$w^r \bmod n = 85877701.$$

It happens that

$$85877701^2 \equiv 1 \pmod{n}.$$

Therefore the algorithm will return the value

$$x = \gcd(85877702, n) = 9103.$$

This is one factor of n; the other is $n/9103 = 9871.$ ☐

Let's now proceed to the analysis of Algorithm 5.10. First, observe that if we are lucky enough to choose w to be a multiple of p or q, then we can factor n immediately. If w is relatively prime to n, then we compute $w^r, w^{2r}, w^{4r}, \ldots$, by successive squaring, until

$$w^{2^t r} \equiv 1 \pmod{n}$$

for some t. Since

$$ab - 1 = 2^s r \equiv 0 \pmod{\phi(n)},$$

we know that $w^{2^s r} \equiv 1 \pmod{n}$. Hence, the **while** loop terminates after at most s iterations. At the end of the **while** loop, we have found a value v_0 such

---

Algorithm 5.10: RSA-FACTOR(n, a, b)

comment: we are assuming that $ab \equiv 1 \pmod{\phi(n)}$

write $ab - 1 = 2^s r, r$ odd
choose w at random such that $1 \le w \le n - 1$
$x \leftarrow \gcd(w, n)$
if $1 < x < n$
  then return (x)
comment: x is a factor of n

$v \leftarrow w^r \bmod n$
if $v \equiv 1 \pmod{n}$
  then return ("failure")
while $v \not\equiv 1 \pmod{n}$
  do $\begin{cases} v_0 \leftarrow v \\ v \leftarrow v^2 \bmod n \end{cases}$
if $v_0 \equiv -1 \pmod{n}$
  then return ("failure")
  else $\begin{cases} x \leftarrow \gcd(v_0 + 1, n) \\ \textbf{return } (x) \end{cases}$
comment: x is a factor of n

---

that $(v_0)^2 \equiv 1 \pmod{n}$ but $v_0 \not\equiv 1 \pmod{n}$. If $v_0 \equiv -1 \pmod{n}$, then the algorithm fails; otherwise, v_0 is a non-trivial square root of 1 modulo n and we are able to factor n.

The main task facing us now is to prove that the algorithm succeeds with probability at least $1/2$. There are two ways in which the algorithm can fail to factor n:

1. $w^r \equiv 1 \pmod{n}$, or
2. $w^{2^t r} \equiv -1 \pmod{n}$ for some $t, 0 \le t \le s - 1$.

We have $s + 1$ congruences to consider. If a random value w is a solution to at least one of these $s + 1$ congruences, then it is a "bad" choice, and the algorithm fails. So we proceed by counting the number of solutions to each of these congruences.

First, consider the congruence $w^r \equiv 1 \pmod{n}$. The way to analyze a congruence such as this is to consider solutions modulo p and modulo q separately, and then combine them using the Chinese remainder theorem. Observe that $x \equiv 1 \pmod{n}$ if and only if $x \equiv 1 \pmod{p}$ and $x \equiv 1 \pmod{q}$.

So, we first consider $w^r \equiv 1 \pmod{p}$. Since p is prime, $\mathbb{Z}_p^*$ is a cyclic group by Theorem 5.7. Let g be a primitive element modulo p. We can write $w = g^u$

for a unique integer u, $0 \leq u \leq p - 2$. Then we have

$$w^r \equiv 1 \pmod{p},$$

$$g^{ur} \equiv 1 \pmod{p}, \text{ and hence}$$

$$(p - 1) \mid ur.$$

Let us write

$$p - 1 = 2^i p_1$$

where p_1 is odd, and

$$q - 1 = 2^j q_1$$

where q_1 is odd. Since

$$\phi(n) = (p - 1)(q - 1) \mid (ab - 1) = 2^s r,$$

we have that

$$2^{i+j} p_1 q_1 \mid 2^s r.$$

Hence

$$i + j \leq s$$

and

$$p_1 q_1 \mid r.$$

Now, the condition $(p - 1) \mid ur$ becomes $2^i p_1 \mid ur$. Since $p_1 \mid r$ and r is odd, it is necessary and sufficient that $2^i \mid u$. Hence, $u = k2^i$, $0 \leq k \leq p_1 - 1$, and the number of solutions to the congruence $w^r \equiv 1 \pmod{p}$ is p_1.

By an identical argument, the congruence $w^r \equiv 1 \pmod{q}$ has exactly q_1 solutions. We can combine any solution modulo p with any solution modulo q to obtain a unique solution modulo n, using the Chinese remainder theorem. Consequently, the number of solutions to the congruence $w^r \equiv 1 \pmod{n}$ is $p_1 q_1$.

The next step is to consider a congruence $w^{2^t r} \equiv -1 \pmod{n}$ for a fixed value t (where $0 \leq t \leq s - 1$). Again, we first look at the congruence modulo p and then modulo q (note that $w^{2^t r} \equiv -1 \pmod{n}$ if and only if $w^{2^t r} \equiv -1 \pmod{p}$ and $w^{2^t r} \equiv -1 \pmod{q}$). First, consider $w^{2^t r} \equiv -1 \pmod{p}$. Writing $w = g^u$, as above, we get

$$g^{u 2^t r} \equiv -1 \pmod{p}.$$

Since $g^{(p-1)/2} \equiv -1 \pmod{p}$, we have that

$$u 2^t r \equiv \frac{p - 1}{2} \pmod{p - 1}$$

$$(p - 1) \mid \left(u 2^t r - \frac{p - 1}{2} \right)$$

$$2(p - 1) \mid (u 2^{t+1} r - (p - 1)).$$

Since $p - 1 = 2^i p_1$, we get

$$2^{i+1} p_1 \mid (u 2^{t+1} r - 2^i p_1).$$

Taking out a common factor of p_1, this becomes

$$2^{i+1} \mid \left(\frac{u 2^{t+1} r}{p_1} - 2^i \right).$$

Now, if $t \geq i$, then there can be no solutions since $2^{i+1} \mid 2^{t+1}$ but $2^{i+1} \nmid 2^i$. On the other hand, if $t \leq i - 1$, then u is a solution if and only if u is an odd multiple of 2^{i-t-1} (note that r/p_1 is an odd integer). So, the number of solutions in this case is

$$\frac{p-1}{2^{i-t-1}} \times \frac{1}{2} = 2^t p_1.$$

By similar reasoning, the congruence $w^{2^t r} \equiv -1 \pmod{q}$ has no solutions if $t \geq j$, and $2^t q_1$ solutions if $t \leq j - 1$. Using the Chinese remainder theorem, we see that the number of solutions of $w^{2^t r} \equiv -1 \pmod{n}$ is

$$\begin{array}{ll} 0 & \text{if } t \geq \min\{i, j\} \\ 2^{2t} p_1 q_1 & \text{if } t \leq \min\{i, j\} - 1. \end{array}$$

Now, t can range from 0 to $s - 1$. Without loss of generality, suppose $i \leq j$; then the number of solutions is 0 if $t \geq i$. The total number of "bad" choices for w is at most

$$p_1 q_1 + p_1 q_1 (1 + 2^2 + 2^4 + \cdots + 2^{2i-2}) = p_1 q_1 \left(1 + \frac{2^{2i} - 1}{3} \right)$$

$$= p_1 q_1 \left(\frac{2}{3} + \frac{2^{2i}}{3} \right).$$

Recall that $p - 1 = 2^i p_1$ and $q - 1 = 2^j q_1$. Now, $j \geq i \geq 1$, so $p_1 q_1 < n/4$. We also have that

$$2^{2i} p_1 q_1 \leq 2^{i+j} p_1 q_1 = (p-1)(q-1) < n.$$

Hence, we obtain

$$p_1 q_1 \left(\frac{2}{3} + \frac{2^{2i}}{3} \right) < \frac{n}{6} + \frac{n}{3}$$

$$= \frac{n}{2}.$$

Since at most $(n-1)/2$ choices for w are "bad," it follows that at least $(n-1)/2$ choices are "good" and hence the probability of success of the algorithm is at least $1/2$.

5.7.3 Wiener's Low Decryption Exponent Attack

As always, suppose that $n = pq$ where p and q are prime; then $\phi(n) = (p - 1)(q - 1)$. In this section, we present an attack, due to M. Wiener, that succeeds in computing the secret decryption exponent, a, whenever the following hypotheses are satisfied:

$$3a < n^{1/4} \quad \text{and} \quad q < p < 2q. \tag{5.2}$$

If n has ℓ bits in its binary representation, then the attack will work when a has fewer than $\ell/4 - 1$ bits in its binary representation and p and q are not too far apart.

Note that Bob might be tempted to choose his decryption exponent to be small in order to speed up decryption. If he uses Algorithm 5.5 to compute $y^a \bmod n$, then the running time of decryption will be reduced by roughly 75% if he chooses a value of a that satisfies (5.2). The results we prove in this section show that this method of reducing decryption time should be avoided.

Since $ab \equiv 1 \pmod{\phi(n)}$, it follows that there is an integer t such that

$$ab - t\phi(n) = 1.$$

Since $n = pq > q^2$, we have that $q < \sqrt{n}$. Hence,

$$0 < n - \phi(n) = p + q - 1 < 2q + q - 1 < 3q < 3\sqrt{n}.$$

Now, we see that

$$\left| \frac{b}{n} - \frac{t}{a} \right| = \left| \frac{ba - tn}{an} \right|$$

$$= \left| \frac{1 + t(\phi(n) - n)}{an} \right|$$

$$< \frac{3t\sqrt{n}}{an}$$

$$= \frac{3t}{a\sqrt{n}}.$$

Since $t < a$, we have that $3t < 3a < n^{1/4}$, and hence

$$\left| \frac{b}{n} - \frac{t}{a} \right| < \frac{1}{an^{1/4}}.$$

Finally, since $3a < n^{1/4}$, we have that

$$\left| \frac{b}{n} - \frac{t}{a} \right| < \frac{1}{3a^2}.$$

Therefore the fraction t/a is a very close approximation to the fraction b/n. From the theory of continued fractions, it is known that any approximation of b/n that is this close must be one of the convergents of the continued fraction expansion of b/n (see Theorem 5.14). This expansion can be obtained from the EUCLIDEAN ALGORITHM, as we describe now.

A (finite) *continued fraction* is an m-tuple of non-negative integers, say

$$[q_1, \ldots, q_m],$$

which is shorthand for the following expression:

$$q_1 + \cfrac{1}{q_2 + \cfrac{1}{q_3 + \cdots + \frac{1}{q_m}}}.$$

Suppose a and b are positive integers such that $\gcd(a, b) = 1$, and suppose that the output of Algorithm 5.1 is the m-tuple $(q_1, \ldots, q_m)$. Then it is not hard to see that $a/b = [q_1, \ldots, q_m]$. We say that $[q_1, \ldots, q_m]$ is the *continued fraction expansion* of a/b in this case. Now, for $1 \le j \le m$, define $C_j = [q_1, \ldots, q_j]$. C_j is said to be the jth *convergent* of $[q_1, \ldots, q_m]$. Each C_j can be written as a rational number c_j/d_j, where the c_j's and d_j's satisfy the following recurrences:

$$c_j = \begin{cases} 1 & \text{if } j = 0 \\ q_1 & \text{if } j = 1 \\ q_j c_{j-1} + c_{j-2} & \text{if } j \ge 2 \end{cases}$$

and

$$d_j = \begin{cases} 0 & \text{if } j = 0 \\ 1 & \text{if } j = 1 \\ q_j d_{j-1} + d_{j-2} & \text{if } j \ge 2. \end{cases}$$

Example 5.16   We compute the continued fraction expansion of $34/99$. The EUCLIDEAN ALGORITHM proceeds as follows:

$$\begin{aligned} 34 &= 0 \times 99 + 34 \\ 99 &= 2 \times 34 + 31 \\ 34 &= 1 \times 31 + 3 \\ 31 &= 10 \times 3 + 1 \\ 3 &= 3 \times 1. \end{aligned}$$

Hence, the continued fraction expansion of $34/99$ is $[0, 2, 1, 10, 3]$, i.e.,

$$\frac{34}{99} = 0 + \cfrac{1}{2 + \cfrac{1}{1 + \cfrac{1}{10 + \frac{1}{3}}}}.$$

The convergents of this continued fraction are as follows:

$$[0] = 0$$

$$[0, 2] = 1/2$$

$$[0, 2, 1] = 1/3$$

$$[0, 2, 1, 10] = 11/32, \quad \text{and}$$

$$[0, 2, 1, 10, 3] = 34/99.$$

The reader can verify that these convergents can be computed using the recurrence relations given above. ▯

The convergents of a continued fraction expansion of a rational number satisfy many interesting properties. For our purposes, the most important property is the following.

THEOREM 5.14 *Suppose that* $\gcd(a, b) = \gcd(c, d) = 1$ *and*

$$\left| \frac{a}{b} - \frac{c}{d} \right| < \frac{1}{2d^2}.$$

Then c/d *is one of the convergents of the continued fraction expansion of* a/b.

Now we can apply this result to the *RSA Cryptosystem*. We already observed that, if condition (5.2) holds, then the unknown fraction t/a is a close approximation to b/n. Theorem 5.14 tells us that t/a must be one of the convergents of the continued fraction expansion of b/n. Since the value of b/n is public information, it is a simple matter to compute its convergents. All we need is a method to test each convergent to see if it is the "right" one.

But this is also not difficult to do. If t/a is a convergent of b/n, then we can compute the value of $\phi(n)$ to be $\phi(n) = (ab - 1)/t$. Once n and $\phi(n)$ are known, we can factor n by solving the quadratic equation (5.1) for p. We do not know ahead of time which convergent of b/n will yield the factorization of n, so we try each one in turn, until the factorization of n is found. If we do not succeed in factoring n by this method, then it must be the case that the hypotheses (5.2) are not satisfied.

A pseudocode description of WIENER'S ALGORITHM is presented as Algorithm 5.11.

We present an example to illustrate.

Example 5.17 Suppose that $n = 160523347$ and $b = 60728973$. The continued fraction expansion of b/n is

$$[0, 2, 1, 1, 1, 4, 12, 102, 1, 1, 2, 3, 2, 2, 36].$$

Algorithm 5.11: WIENER'S ALGORITHM(n, b)

$(q_1, \ldots, q_m; r_m) \leftarrow$ EUCLIDEAN ALGORITHM(b, n)
$c_0 \leftarrow 1$
$c_1 \leftarrow q_1$
$d_0 \leftarrow 0$
$d_1 \leftarrow 1$
for $j \leftarrow 2$ **to** m

$\mathbf{do} \begin{cases} \begin{cases} c_j \leftarrow q_j c_{j-1} + c_{j-2} \\ d_j \leftarrow q_j d_{j-1} + d_{j-2} \\ n' \leftarrow (d_j b - 1)/c_j \\ \textbf{comment: } n' = \phi(n) \text{ if } c_j/d_j \text{ is the correct convergent} \\ \textbf{if } n' \text{ is an integer} \\ \quad \textbf{then} \begin{cases} \text{let } p \text{ and } q \text{ be the roots of the equation} \\ \quad x^2 - (n - n' + 1)x + n = 0 \\ \textbf{if } p \text{ and } q \text{ are positive integers less than } n \\ \quad \textbf{then return } (p, q) \end{cases} \end{cases} \end{cases}$

return ("failure")

The first few convergents are

$$0, \frac{1}{2}, \frac{1}{3}, \frac{2}{5}, \frac{3}{8}, \frac{14}{37}.$$

The reader can verify that the first five convergents do not produce a factorization of n. However, the convergent $14/37$ yields

$$n' = \frac{37 \times 60728973 - 1}{14} = 160498000.$$

Now, if we solve the equation

$$x^2 - 25348x + 160523347 = 0,$$

then we find the roots $x = 12347, 13001$. Therefore we have discovered the factorization

$$160523347 = 12347 \times 13001.$$

Notice that, for the modulus $n = 160523347$, WIENER'S ALGORITHM will work for

$$a < \frac{n^{1/4}}{3} \approx 37.52.$$

$\square$

Cryptosystem 5.2: *Rabin Cryptosystem*

Let $n = pq$, where p and q are primes and $p, q \equiv 3 \pmod{4}$. Let $\mathcal{P} = \mathcal{C} = \mathbb{Z}_n{}^*$, and define

$$\mathcal{K} = \{(n, p, q)\}.$$

For $K = (n, p, q)$, define

$$e_K(x) = x^2 \bmod n$$

and

$$d_K(y) = \sqrt{y} \bmod n.$$

The value n is the public key, while p and q are the private key.

5.8 The Rabin Cryptosystem

In this section, we describe the *Rabin Cryptosystem*, which is computationally secure against a chosen-plaintext attack provided that the modulus $n = pq$ cannot be factored. Therefore, the *Rabin Cryptosystem* provides an example of a provably secure cryptosystem: assuming that the problem of factoring is computationally infeasible, the *Rabin Cryptosystem* is secure. We present the *Rabin Cryptosystem* as Cryptosystem 5.2.

REMARK   The requirement that $p, q \equiv 3 \pmod{4}$ can be omitted. As well, the cryptosystem still "works" if we take $\mathcal{P} = \mathcal{C} = \mathbb{Z}_n$ instead of $\mathbb{Z}_n{}^*$. However, the more restrictive description we use simplifies some aspects of computation and analysis of the cryptosystem. ∎

One drawback of the *Rabin Cryptosystem* is that the encryption function e_K is not an injection, so decryption cannot be done in an unambiguous fashion. We prove this as follows. Suppose that y is a valid ciphertext; this means that $y = x^2 \bmod n$ for some $x \in \mathbb{Z}_n{}^*$. Theorem 5.13 proves that there are four square roots of y modulo n, which are the four possible plaintexts that encrypt to y. In general, there will be no way for Bob to distinguish which of these four possible plaintexts is the "right" plaintext, unless the plaintext contains sufficient redundancy to eliminate three of these four possible values.

Let us look at the decryption problem from Bob's point of view. He is given a ciphertext y and wants to determine x such that

$$x^2 \equiv y \pmod{n}.$$

This is a quadratic equation in $\mathbb{Z}_n$ in the unknown x, and decryption requires extracting square roots modulo n. This is equivalent to solving the two congruences

$$z^2 \equiv y \pmod{p}$$

and

$$z^2 \equiv y \pmod{q}.$$

We can use Euler's criterion to determine if y is a quadratic residue modulo p (and modulo q). In fact, y will be a quadratic residue modulo p and modulo q if encryption was performed correctly. Unfortunately, Euler's criterion does not help us find the square roots of y; it yields only an answer "yes" or "no."

When $p \equiv 3 \pmod{4}$, there is a simple formula to compute square roots of quadratic residues modulo p. Suppose y is a quadratic residue modulo p, where $p \equiv 3 \pmod{4}$. Then we have that

$$(\pm y^{(p+1)/4})^2 \equiv y^{(p+1)/2} \pmod{p}$$
$$\equiv y^{(p-1)/2} y \pmod{p}$$
$$\equiv y \pmod{p}.$$

Here we have again made use of Euler's criterion, which says that if y is a quadratic residue modulo p, then $y^{(p-1)/2} \equiv 1 \pmod{p}$. Hence, the two square roots of y modulo p are $\pm y^{(p+1)/4} \bmod p$. In a similar fashion, the two square roots of y modulo q are $\pm y^{(q+1)/4} \bmod q$. It is then straightforward to obtain the four square roots of y modulo n using the Chinese remainder theorem.

REMARK   For $p \equiv 1 \pmod{4}$, there is no known polynomial-time deterministic algorithm to compute square roots of quadratic residues modulo p. (There is a polynomial-time Las Vegas algorithm, however.) This is why we stipulated that $p, q \equiv 3 \pmod{4}$ in the definition of the *Rabin Cryptosystem*.   ∎

Example 5.18   Let's illustrate the encryption and decryption procedures for the *Rabin Cryptosystem* with a toy example. Suppose $n = 77 = 7 \times 11$. Then the encryption function is

$$e_K(x) = x^2 \bmod 77$$

and the decryption function is

$$d_K(y) = \sqrt{y} \bmod 77.$$

Suppose Bob wants to decrypt the ciphertext $y = 23$. It is first necessary to find the square roots of 23 modulo 7 and modulo 11. Since 7 and 11 are both congruent to 3 modulo 4, we use our formula:

$$23^{(7+1)/4} \equiv 2^2 \equiv 4 \pmod{7}$$

and
$$23^{(11+1)/4} \equiv 1^3 \equiv 1 \pmod{11}.$$

Using the Chinese remainder theorem, we compute the four square roots of 23 modulo 77 to be $\pm 10, \pm 32 \bmod 77$. Therefore, the four possible plaintexts are $x = 10, 32, 45$ and 67. It can be verified that each of these plaintexts yields the value 23 when squared and reduced modulo 77. This proves that 23 is indeed a valid ciphertext. □

5.8.1 Security of the Rabin Cryptosystem

We now discuss the (provable) security of the *Rabin Cryptosystem*. The security proof uses a Turing reduction, which is defined as follows.

Definition 5.5: Suppose that **G** and **H** are problems. A *Turing reduction* from **G** to **H** is an algorithm SOLVEG with the following properties:

1. SOLVEG assumes the existence of an arbitrary algorithm SOLVEH that solves the problem **H**.

2. SOLVEG can call the algorithm SOLVEH and make use of any values it outputs, but SOLVEG cannot make any assumption about the actual computations performed by SOLVEH (in other words, SOLVEH is treated as a "black box" and is termed an oracle).

3. SOLVEG is a polynomial-time algorithm.

4. SOLVEG correctly solves the problem **G**.

If there is a Turing reduction from **G** to **H**, we denote this by writing **G** $\propto_T$ **H**.

A Turing reduction **G** $\propto_T$ **H** does not necessarily yield a polynomial-time algorithm to solve **G**. It actually proves the truth of the following implication:

> If there exists a polynomial-time algorithm to solve **H**, then there exists a polynomial-time algorithm to solve **G**.

This is because any algorithm SOLVEH that solves **H** can be "plugged into" the algorithm SOLVEG, thereby producing an algorithm that solves **G**. Clearly this resulting algorithm will be a polynomial-time algorithm if SOLVEH is a polynomial-time algorithm.

We will provide an explicit example of a Turing reduction: We will prove that a decryption oracle RABIN DECRYPT can be incorporated into a Las Vegas algorithm, Algorithm 5.12, that factors the modulus n with probability at least $1/2$. In other words, we show that **Factoring** $\propto_T$ **Rabin decryption**, where the Turing reduction is itself a randomized algorithm. In Algorithm 5.12, we assume that n is the product of two distinct primes p and q; and RABIN DECRYPT is an oracle that performs Rabin decryption, returning one of the four possible plaintexts corresponding to a given ciphertext.

Algorithm 5.12: RABIN ORACLE FACTORING(n)

external RABIN DECRYPT
choose a random integer $r \in \mathbb{Z}_n^*$
$y \leftarrow r^2 \bmod n$
$x \leftarrow$ RABIN DECRYPT(y)
if $x \equiv \pm r \pmod{n}$
  then return ("failure")
  else $\begin{cases} p \leftarrow \gcd(x + r, n) \\ q \leftarrow n/p \\ \textbf{return } ("n = p \times q") \end{cases}$

There are several points of explanation needed. First, observe that y is a valid ciphertext and RABIN DECRYPT(y) will return one of four possible plaintexts as the value of x. In fact, it holds that $x \equiv \pm r \pmod{n}$ or $x \equiv \pm \omega r \pmod{n}$, where ω is one of the non-trivial square roots of 1 modulo n. In the second case, we have $x^2 \equiv r^2 \pmod{n}$, $x \not\equiv \pm r \pmod{n}$. Hence, computation of $\gcd(x + r, n)$ must yield either p or q, and the factorization of n is accomplished.

Let's compute the probability of success of this algorithm, over all choices for the random value $r \in \mathbb{Z}_n^*$. For a residue $r \in \mathbb{Z}_n^*$, define

$$[r] = \{\pm r \bmod n, \pm \omega r \bmod n\}.$$

Clearly any two residues in $[r]$ yield the same y-value in Algorithm 5.12, and the value of x that is output by the oracle RABIN DECRYPT is also in $[r]$. We have already observed that Algorithm 5.12 succeeds if and only if $x \equiv \pm \omega r \pmod{n}$. The oracle does not know which of four possible r-values was used to construct y, and r was chosen at random before the oracle RABIN DECRYPT is called. Hence, the probability that $x \equiv \pm \omega r \pmod{n}$ is $1/2$. We conclude that the probability of success of Algorithm 5.12 is $1/2$.

We have shown that the *Rabin Cryptosystem* is provably secure against a chosen plaintext attack. However, the system is completely insecure against a chosen ciphertext attack. In fact, Algorithm 5.12 can be used to break the *Rabin Cryptosystem* in a chosen ciphertext attack! In the chosen ciphertext attack, the (hypothetical) oracle RABIN DECRYPT is replaced by an actual decryption algorithm. (Informally, the security proof says that a decryption oracle can be used to factor n; and a chosen ciphertext attack assumes that a decryption oracle exists. Together, these break the cryptosystem!)

5.9 Semantic Security of RSA

To this point in the text, we have assumed that an adversary trying to break a cryptosystem is actually trying to determine the secret key (in the case of a symmetric-key cryptosystem) or the private key (in the case of a public-key cryptosystem). If Oscar can do this, then the system is completely broken. However, it is possible that the goal of an adversary is somewhat less ambitious. Even if Oscar cannot find the secret or private key, he still may be able to gain more information than we would like. If we want to be assured that the cryptosystem is "secure," we should take into account these more modest goals that an adversary might have.

Here is a short list of potential adversarial goals:

total break
>   The adversary is able to determine Bob's private key (in the case of a public-key cryptosystem) or the secret key (in the case of a symmetric-key cryptosystem). Therefore he can decrypt any ciphertext that has been encrypted using the given key.

partial break
>   With some non-negligible probability, the adversary is able to decrypt a previously unseen ciphertext (without knowing the key). Or, the adversary can determine some specific information about the plaintext, given the ciphertext.

distinguishability of ciphertexts
>   With some probability exceeding $1/2$, the adversary is able to distinguish between encryptions of two given plaintexts, or between an encryption of a given plaintext and a random string.

In the next sections, we will consider some possible attacks against RSA-like cryptosystems that achieve some of these types of goals. We also describe how to construct a public-key cryptosystem in which the adversary cannot (in polynomial time) distinguish ciphertexts, provided that certain computational assumptions hold. Such cryptosystems are said to achieve *semantic security*. Achieving semantic security is quite difficult, because we are providing protection against a very weak, and therefore easy to achieve, adversarial goal.

5.9.1 Partial Information Concerning Plaintext Bits

A weakness of some cryptosystems is the fact that partial information about the plaintext might be "leaked" by the ciphertext. This represents a type of partial break of the system, and it happens, in fact, in the *RSA Cryptosystem*. Suppose we are given a ciphertext, $y = x^b \bmod n$, where x is the plaintext. Since

$\gcd(b, \phi(n)) = 1$, it must be the case that b is odd. Therefore the Jacobi symbol

$$\left(\frac{y}{n}\right) = \left(\frac{x}{n}\right)^b = \left(\frac{x}{n}\right).$$

Hence, given the ciphertext y, anyone can efficiently compute $\left(\frac{x}{n}\right)$ without decrypting the ciphertext. In other words, an RSA encryption "leaks" some information concerning the plaintext x, namely, the value of the Jacobi symbol $\left(\frac{x}{n}\right)$.

In this section, we consider some other specific types of partial information that could be leaked by a cryptosystem:

1. given $y = e_K(x)$, compute $parity(y)$, where $parity(y)$ denotes the low-order bit of x (i.e., $parity(y) = 0$ if x is even and $parity(y) = 1$ if x is odd).

2. given $y = e_K(x)$, compute $half(y)$, where $half(y) = 0$ if $0 \leq x < n/2$ and $half(y) = 1$ if $n/2 < x \leq n - 1$.

We will prove that the *RSA Cryptosystem* does not leak these types of information provided that RSA encryption is secure. More precisely, we show that the problem of RSA decryption can be Turing reduced to the problem of computing $half(y)$. This means that the existence of a polynomial-time algorithm that computes $half(y)$ implies the existence of a polynomial-time algorithm for RSA decryption. In other words, computing certain partial information about the plaintext, namely $half(y)$, is no easier than decrypting the ciphertext to obtain the whole plaintext.

We will now show how to compute $x = d_K(y)$, given a hypothetical algorithm (oracle) HALF which computes $half(y)$. The algorithm is presented as Algorithm 5.13.

We explain what is happening in the above algorithm. First, we note that the RSA encryption function satisfies the following multiplicative property in $\mathbb{Z}_n$:

$$e_K(x_1)e_K(x_2) = e_K(x_1 x_2).$$

Now, using the fact that

$$y = e_K(x) = x^b \bmod n,$$

it is easily seen in the ith iteration of the first **for** loop that

$$h_i = half(y \times (e_K(2))^i) = half(e_K(x \times 2^i)),$$

for $0 \leq i \leq \lfloor \log_2 n \rfloor$. We observe that

$$half(e_K(x)) = 0 \Leftrightarrow x \in \left[0, \frac{n}{2}\right)$$

$$half(e_K(2x)) = 0 \Leftrightarrow x \in \left[0, \frac{n}{4}\right) \cup \left[\frac{n}{2}, \frac{3n}{4}\right)$$

$$half(e_K(4x)) = 0 \Leftrightarrow x \in \left[0, \frac{n}{8}\right) \cup \left[\frac{n}{4}, \frac{3n}{8}\right) \cup \left[\frac{n}{2}, \frac{5n}{8}\right) \cup \left[\frac{3n}{4}, \frac{7n}{8}\right),$$

Algorithm 5.13: ORACLE RSA DECRYPTION(n, b, y)

external HALF
$k \leftarrow \lfloor \log_2 n \rfloor$
for $i \leftarrow 0$ **to** k
$\quad$ **do** $\begin{cases} h_i \leftarrow \text{HALF}(n, b, y) \\ y \leftarrow (y \times 2^b) \bmod n \end{cases}$
$lo \leftarrow 0$
$hi \leftarrow n$
for $i \leftarrow 0$ **to** k
$\quad$ **do** $\begin{cases} mid \leftarrow (hi + lo)/2 \\ \textbf{if } h_i = 1 \\ \quad \textbf{then } lo \leftarrow mid \\ \quad \textbf{else } hi \leftarrow mid \end{cases}$
return $(\lfloor hi \rfloor)$

---

and so on. Hence, we can find x by a binary search technique, which is done in the second **for** loop. Here is a small example to illustrate.

Example 5.19 Suppose $n = 1457$, $b = 779$, and we have a ciphertext $y = 722$. Then suppose, using our oracle HALF, that we obtain the following values for h_i:

i	0	1	2	3	4	5	6	7	8	9	10
h_i	1	0	1	0	1	1	1	1	1	0	0

Then the binary search proceeds as shown in Figure 5.3. Hence, the plaintext is $x = \lfloor 999.55 \rfloor = 999$. $\qquad \square$

The complexity of Algorithm 5.13 is easily seen to be

$$O((\log n)^3) + O(\log n) \times \text{the complexity of HALF}.$$

Therefore we will obtain a polynomial-time algorithm for RSA decryption if HALF is a polynomial-time algorithm.

It is a simple matter to observe that computing $parity(y)$ is polynomially equivalent to computing $half(y)$. This follows from the following two easily proved identities involving RSA encryption (see the exercises):

$$half(y) = parity((y \times e_K(2)) \bmod n) \tag{5.3}$$

$$parity(y) = half((y \times e_K(2^{-1})) \bmod n), \tag{5.4}$$

FIGURE 5.3
Binary search for RSA decryption

i	lo	mid	hi
0	0.00	728.50	1457.00
1	728.50	1092.75	1457.00
2	728.50	910.62	1092.75
3	910.62	1001.69	1092.75
4	910.62	956.16	1001.69
5	956.16	978.92	1001.69
6	978.92	990.30	1001.69
7	990.30	996.00	1001.69
8	996.00	998.84	1001.69
9	998.84	1000.26	1001.69
10	998.84	999.55	1000.26
	998.84	999.55	999.55

and from the above-mentioned multiplicative rule, $e_K(x_1)e_K(x_2) = e_K(x_1x_2)$. Hence, from the results proved above, it follows that the existence of a polynomial-time algorithm to compute *parity* implies the existence of a polynomial-time algorithm for RSA decryption.

We have provided evidence that computing *parity* or *half* is difficult, provided that RSA decryption is difficult. However, the proofs we have presented do not rule out the possibility that it might be possible to find an efficient algorithm that computes *parity* or *half* with 75% accuracy, say. There are also many other types of plaintext information that could possibly be considered, and we certainly cannot deal with all possible types of information using separate proofs. Therefore the results of this section only provide evidence of security against certain types of attacks.

5.9.2 Optimal Asymmetric Encryption Padding

What we really want is to find a method of designing a cryptosystem that allows us to prove (assuming some plausible computational assumptions) that no information of any kind regarding the plaintext is revealed in polynomial time by examining the ciphertext. It can be shown that this is equivalent to showing that an adversary cannot distinguish ciphertexts. Therefore we consider the problem of **Ciphertext Distinguishability**, which is defined as follows:

Problem 5.3:   **Ciphertext Distinguishability**

Instance: An encryption function $f : X \rightarrow X$; two plaintexts $x_1, x_2 \in X$; and a ciphertext $y = f(x_i)$, where $i \in \{1, 2\}$.

Question: Is $i = 1$?

Problem 5.3 is of course trivial if the encryption function f is deterministic, since it suffices to compute $f(x_1)$ and $f(x_2)$ and see which one yields the ciphertext y. Hence, if **Ciphertext Distinguishability** is going to be computationally infeasible, then it will be necessary for the encryption process to be randomized. We now present some concrete methods to realize this objective. First, we consider Cryptosystem 5.3, which is based on an arbitrary *trapdoor one-way permutation*, which is a (bijective) trapdoor one-way function from a set X to itself. If $f : X \rightarrow X$ is a trapdoor one-way permutation, then the inverse permutation is denoted, as usual, by f^{-1}. f is the encryption function, and f^{-1} is the decryption function of the public-key cryptosystem.

In the case of the *RSA Cryptosystem*, we would take $n = pq$, $X = \mathbb{Z}_n$, $f(x) = x^b \bmod n$ and $f^{-1}(x) = x^a \bmod n$, where $ab \equiv 1 \pmod{\phi(n)}$. Cryptosystem 5.3 also employs a certain random function, G. Actually, G will be modeled by a random oracle, which was defined in Section 4.2.1.

We observe that Cryptosystem 5.3 is quite efficient: it requires little additional computation as compared to the underlying public-key cryptosystem based on f. In practice, the function G could be built from a secure hash function, such as *SHA-1*, in a very efficient manner. The main drawback of Cryptosystem 5.3 is the data expansion: m bits of plaintext are encrypted to form $k + m$ bits of ciphertext. If f is based on the RSA encryption function, for example, then it will be necessary to take $k \geq 1024$ in order for the system to be secure.

It is easy to see that there must be some data expansion in any semantically secure cryptosystem due to the fact that encryption is randomized. However, there are more efficient schemes that are still provably secure. The most important of these, *Optimal Asymmetric Encryption Padding*, will be discussed later in this section (it is presented as Cryptosystem 5.4). We begin with Cryptosystem 5.3, however, because it is conceptually simpler and easier to analyze.

An intuitive argument that Cryptosystem 5.3 is semantically secure in the random oracle model goes as follows: In order to determine any information about the plaintext x, we need to have some information about $G(r)$. Assuming that G is a random oracle, the only way to ascertain any information about the value of $G(r)$ is to first compute $r = f^{-1}(y_1)$. (It is not sufficient to compute some partial information about r; it is necessary to have complete information about r in order to obtain any information about $G(r)$.) However, if f is one-way, then r cannot be computed in a reasonable amount of time by an adversary who does not know the trapdoor, f^{-1}.

The preceding argument might be fairly convincing, but it is not a proof. If we are going to massage this argument into a proof, we need to describe a reduction,

---

Cryptosystem 5.3: *Semantically Secure Public-key Cryptosystem*

Let m, k be positive integers; let $\mathcal{F}$ be a family of trapdoor one-way permutations such that $f : \{0,1\}^k \to \{0,1\}^k$ for all $f \in \mathcal{F}$; and let $G : \{0,1\}^k \to \{0,1\}^m$ be a random oracle. Let $\mathcal{P} = \{0,1\}^m$ and $\mathcal{C} = \{0,1\}^k \times \{0,1\}^m$, and define

$$\mathcal{K} = \{(f, f^{-1}, G) : f \in \mathcal{F}\}.$$

For $K = (f, f^{-1}, G)$, let $r \in \{0,1\}^k$ be chosen randomly, and define

$$e_K(x) = (y_1, y_2) = (f(r), G(r) \oplus x),$$

where $y_1 \in \{0,1\}^k$, $x, y_2 \in \{0,1\}^m$. Further, define

$$d_K(y_1, y_2) = G(f^{-1}(y_1)) \oplus y_2$$

($y_1 \in \{0,1\}^k$, $y_2 \in \{0,1\}^m$). The functions f and G are the public key; the function f^{-1} is the private key.

---

from the problem of inverting the function f to the problem of **Ciphertext Distinguishability**. When f is randomized, as in Cryptosystem 5.3, it may not be feasible to solve Problem 5.3 if there are sufficiently many possible encryptions of a given plaintext.

We are going to describe a reduction that is more general than the Turing reductions considered previously. We will assume the existence of an algorithm DISTINGUISH that solves the problem of **Ciphertext Distinguishability** for two plaintexts x_1 and x_2, and then we will modify this algorithm in such a way that we obtain an algorithm to invert f. The algorithm DISTINGUISH need not be a "perfect" algorithm; we will only require that it gives the right answer with some probability $1/2 + \epsilon$, where $\epsilon > 0$ (i.e., it is more accurate than a random guess of "1" or "2"). DISTINGUISH is allowed to query the random oracle, and therefore it can compute encryptions of plaintexts. In other words, we are assuming it is a chosen plaintext attack.

As mentioned above, we will prove that Cryptosystem 5.3 is semantically secure in the random oracle model. The main features of this model (which we introduced in Section 4.2.1), and the reduction we describe, are as follows.

1. G is assumed to be a random oracle, so the only way to determine any information about a value $G(r)$ is to call the function G with input r.

2. We construct a new algorithm INVERT, by modifying the algorithm DISTINGUISH, which will invert randomly chosen elements y with probability bounded away from 0 (i.e., given a value $y = f(x)$ where x is chosen ran-

---

Algorithm 5.14: INVERT(y)

external f
global $RList, GList, \ell$
procedure SIMG(r)
  $i \leftarrow 1$
  found $\leftarrow$ **false**
  while $i \leq \ell$ **and not** *found*
    do $\begin{cases} \textbf{if } RList[i] = r \\ \quad \textbf{then } found \leftarrow \textbf{ true} \\ \quad \textbf{else } i \leftarrow i + 1 \end{cases}$
  if *found*
    then return ($GList[i]$)
  if $f(r) = y$
    then $\begin{cases} \text{let } j \in \{1, 2\} \text{ be chosen at random} \\ g \leftarrow y_2 \oplus x_j \end{cases}$
    else let g be chosen at random
  $\ell \leftarrow \ell + 1$
  $RList[\ell] \leftarrow r$
  $GList[\ell] \leftarrow g$
  return (g)

main
$y_1 \leftarrow y$
choose y_2 at random
$\ell \leftarrow 0$
insert the code for DISTINGUISH($x_1, x_2, (y_1, y_2)$) here
for $i \leftarrow 1$ **to** ℓ
  do $\begin{cases} \textbf{if } f(RList[i]) = y \\ \quad \textbf{then return } (RList[i]) \end{cases}$
return ("failure")

---

domly, the algorithm INVERT will find x with some specified probability).

3. The algorithm INVERT will replace the random oracle by a specific function that we will describe, SIMG, all of whose outputs are random numbers. SIMG is a perfect simulation of a random oracle.

The algorithm INVERT is presented as Algorithm 5.14.

Given two plaintexts x_1 and x_2, DISTINGUISH solves the **Ciphertext Distinguishability** problem with probability $1/2 + \epsilon$. The input to INVERT is the element y to be inverted; the objective is to output $f^{-1}(y)$. INVERT begins by constructing a ciphertext (y_1, y_2) in which $y_1 = y$ and y_2 is random. INVERT runs the algorithm DISTINGUISH on the ciphertext (y_1, y_2), attempting to determine if it is an encryption of x_1 or of x_2. DISTINGUISH will query the simulated oracle, SIMG, at various points during its execution. The following points summarize the operation of SIMG:

1. SIMG maintains a list, denoted $RList$, of all inputs r for which it is queried during the execution of DISTINGUISH; and the corresponding list, denoted $GList$, of outputs SIMG(r).

2. If an input r satisfies $f(r) = y$, then SIMG(r) is defined so that (y_1, y_2) is a valid encryption of one of x_1 or x_2 (chosen at random).

3. If the oracle was previously queried with input r, then SIMG(r) is already defined.

4. Otherwise, the value for SIMG(r) is chosen randomly.

Observe that, for any possible plaintext $x_0 \in X$, (y_1, y_2) is a valid encryption of x_0 if and only if

$$\text{SIMG}(f^{-1}(y_1)) = y_2 \oplus x_0.$$

In particular, (y_1, y_2) can be a valid encryption of either of x_1 or x_2, provided that SIMG($f^{-1}(y_1)$) is defined appropriately. The description of the algorithm SIMG ensures that (y_1, y_2) is a valid encryption of one of x_1 or x_2.

Eventually, the algorithm DISTINGUISH will terminate with an answer "1" or "2," which may or may not be correct. At this point, the algorithm INVERT examines the list $RList$ to see if any r in the $RList$ satisfies $f(r) = y$. If such a value r is found, then it is the desired value $f^{-1}(y)$, and the algorithm INVERT succeeds (INVERT fails if $f^{-1}(y)$ is not discovered in $RList$).

It is in fact possible to make algorithm INVERT more efficient by observing that the function SIMG checks to see if $y = f(r)$ for every r that it is queried with. Once it is discovered, within the function SIMG, that $y = f(r)$, we can terminate the algorithm INVERT immediately, returning the value r as its output. It is not necessary to keep running the algorithm DISTINGUISH to its conclusion. However, the analysis of the success probability, which we are going to do next, is a bit easier to understand for Algorithm 5.14 as we have presented it. (The reader might want to verify that the abovementioned modification of INVERT will not change its success probability.)

We now proceed to compute a lower bound on the success probability of the algorithm INVERT. We do this by examining the success probability of DISTINGUISH. We are assuming that the success probability of DISTINGUISH is at least $1/2 + \epsilon$ when it interacts with a random oracle. In the algorithm INVERT, DISTINGUISH interacts with the simulated random oracle, SIMG. Clearly SIMG is completely indistinguishable from a true random oracle for all inputs, except possibly for the input $r = f^{-1}(y)$. However, if $f(r) = y$ and (y, y_2) is a valid encryption of x_1 or x_2, then it must be the case that SIMG(r) $= y_2 \oplus x_1$ or SIMG(r) $= y_2 \oplus x_2$. SIMG is choosing randomly from these two possible alternatives. Therefore, the output it produces is indistinguishable from a true random oracle for the input $r = f^{-1}(y)$, as well. Consequently, the success probability of DISTINGUISH is at least $1/2 + \epsilon$ when it interacts with the simulated random oracle, SIMG.

We now calculate the success probability of DISTINGUISH, conditioned on whether (or not) $f^{-1}(y) \in RList$:

$$\mathbf{Pr}[\text{DISTINGUISH succeeds}] =$$
$$\mathbf{Pr}[\text{DISTINGUISH succeeds} \mid f^{-1}(y) \in RList]\,\mathbf{Pr}[f^{-1}(y) \in RList] +$$
$$\mathbf{Pr}[\text{DISTINGUISH succeeds} \mid f^{-1}(y) \notin RList]\,\mathbf{Pr}[f^{-1}(y) \notin RList].$$

It is clear that

$$\mathbf{Pr}[\text{DISTINGUISH succeeds} \mid f^{-1}(y) \notin RList] = 1/2,$$

because there is no way to distinguish an encryption of x_1 from an encryption of x_2 if the value of $\text{SIMG}(f^{-1}(y))$ is not determined. Now, using the fact that

$$\mathbf{Pr}[\text{DISTINGUISH succeeds} \mid f^{-1}(y) \in RList] \leq 1,$$

we obtain the following:

$$\frac{1}{2} + \epsilon \leq \mathbf{Pr}[\text{DISTINGUISH succeeds}]$$

$$\leq \mathbf{Pr}[f^{-1}(y) \in RList] + \frac{1}{2}\mathbf{Pr}[f^{-1}(y) \notin RList]$$

$$\leq \mathbf{Pr}[f^{-1}(y) \in RList] + \frac{1}{2}.$$

Therefore, it follows that

$$\mathbf{Pr}[f^{-1}(y) \in RList] \geq \epsilon.$$

Since

$$\mathbf{Pr}[\text{INVERSE succeeds}] = \mathbf{Pr}[f^{-1}(y) \in RList],$$

it follows that

$$\mathbf{Pr}[\text{INVERSE succeeds}] \geq \epsilon.$$

It is straightforward to consider the running time of INVERT as compared to that of DISTINGUISH. Suppose that t_1 is the running time of DISTINGUISH, t_2 is the time required to evaluate the function f, and q denotes the number of oracle queries made by DISTINGUISH. Then it is not difficult to see that the running time of INVERT is $t_1 + O(q^2 + qt_2)$.

We close this section by presenting a more efficient provably secure cryptosystem known as *Optimal Asymmetric Encryption Padding* (or *OAEP*); see Cryptosystem 5.4.

In Cryptosystem 5.4, it suffices to take k_0 to be large enough that 2^{k_0} is an infeasibly large running time; $k_0 = 128$ should suffice for most applications. The

---

Cryptosystem 5.4: *Optimal Asymmetric Encryption Padding*

Let m, k be positive integers with $m < k$, and let $k_0 = k - m$. Let $\mathcal{F}$ be a family of trapdoor one-way permutations such that $f : \{0,1\}^k \to \{0,1\}^k$ for all $f \in \mathcal{F}$. Let $G : \{0,1\}^{k_0} \to \{0,1\}^m$ and let $H : \{0,1\}^m \to \{0,1\}^{k_0}$ be "random" functions. Define $\mathcal{P} = \{0,1\}^m$, $\mathcal{C} = \{0,1\}^k$, and define

$$\mathcal{K} = \{(f, f^{-1}, G, H) : f \in \mathcal{F}\}.$$

For $K = (f, f^{-1}, G, H)$, let $r \in \{0,1\}^{k_0}$ be chosen randomly, and define

$$e_K(x) = f(y_1 \| y_2),$$

where

$$y_1 = x \oplus G(r)$$

and

$$y_2 = r \oplus H(x \oplus G(r)),$$

$x, y_1 \in \{0,1\}^m$, $y_2 \in \{0,1\}^{k_0}$, and "$\|$" denotes concatenation of vectors. Further, define

$$f^{-1}(y) = x_1 \| x_2,$$

where $x_1 \in \{0,1\}^m$ and $x_2 \in \{0,1\}^{k_0}$. Then define

$$r = x_2 \oplus H(x_1)$$

and

$$d_K(y) = G(r) \oplus x_1.$$

The functions f, G and H are the public key; the function f^{-1} is the private key.

---

length of a ciphertext in Cryptosystem 5.4 exceeds the length of a plaintext by k_0 bits, so the data expansion is considerably less than Cryptosystem 5.3. The security proof for Cryptosystem 5.4 is more complicated, however.

The adjective "optimal" in Cryptosystem 5.4 refers to the message expansion. Observe that each plaintext has 2^{k_0} possible valid encryptions. One way of solving the problem of **Ciphertext Distinguishability** would be simply to compute all the possible encryptions of one of the two given plaintexts, say x_1, and check to see if the given ciphertext y is obtained. The complexity of this algorithm is 2^{k_0}. It is therefore clear that the message expansion of the cryptosystem must be at least as big as the logarithm to the base 2 of the amount of computation time of an algorithm that solves the **Ciphertext Distinguishability** problem.

5.10 Notes and References

The idea of public-key cryptography was introduced in the open literature by Diffie and Hellman in 1976. Although [117] is the most cited reference, the conference paper [116] actually appeared a bit earlier. The *RSA Cryptosystem* was discovered by Rivest, Shamir and Adleman [283]. For a general survey article on public-key cryptography, we Koblitz and Menezes [234]. A specialized survey on *RSA* was written by Boneh [57]. Mollin [245] and Wagstaff [335] are textbooks that emphasize topics discussed in this chapter.

The Solovay-Strassen test was first described in [313]. The Miller-Rabin test was given in [240] and [279]. Our discussion of error probabilities is motivated by observations of Brassard and Bratley [70] (see also [11]). The best current bounds on the error probability of the Miller-Rabin algorithm can be found in [104]. An expository article on polynomial-time deterministic primality testing has been written by Bornemann [63].

There are many sources of information on factoring algorithms. Lenstra [216] is a good survey on factoring, and Lenstra and Lenstra [218] is a good article on number-theoretic algorithms in general. Bressoud and Wagon [72] is an elementary textbook on factoring and primality testing. One recommended cryptography textbook that emphasizes number theory is Koblitz [197] (note that Example 5.12 is taken from Koblitz's book). Recommended number theory books that are useful for the study of cryptography include Bach and Shallit [5], von zur Gathen and Gerhard [154] and Yan [349]. Lenstra and Lenstra [217] is a monograph on the number field sieve. For information about factoring the RSA challenge numbers, see the Mathworld web site: http://mathworld.wolfram.com/.

The material in Sections 5.7.2 and 5.9.1 is based on the treatment by Salomaa [289, pp. 143–154] (the factorization of n given the decryption exponent was proved in [106]; the results on partial information revealed by RSA ciphertexts is from [164]). Wiener's attack can be found in [343]; a recent strengthening of the attack, due to Boneh and Durfee, was published in [60].

The *Rabin Cryptosystem* was described in Rabin [278]. Provably secure systems in which decryption is unambiguous have been found by Williams [346] and Kurosawa, Ito and Takeuchi [207].

Partial information leaked by RSA ciphertexts is studied in Alexi, Chor, Goldreich and Schnorr [2]. The concept of semantic security is due to Goldwasser and Micali [163]. The *Blum-Goldwasser Cryptosystem* [51] is an example of an early probabilistic public-key cryptosystem that is provably (semantically) secure; we discuss this system in Section 8.4.

For a recent survey on the topic of provably secure cryptosystems, see Bellare [14]. The random oracle model was first described by Bellare and Rogaway in [22]; Cryptosystem 5.3 was presented in that paper. *Optimal Asymmetric Encryption Padding* is presented in [24]; it is incorporated into the IEEE P1363 standard specifications for public-key cryptography. There have been several recent works discussing the security of *Optimal Asymmetric Encryption Padding* and related cryptosystems against chosen-ciphertext attacks: Shoup [302], Fujisaki, Okamoto, Pointcheval and Stern [149] and Boneh [58].

Exercises 5.15–5.17 give some examples of protocol failures. For a nice article on this subject, see Moore [246].

Exercises

5.1 In Algorithm 5.1, prove that

$$\gcd(r_0, r_1) = \gcd(r_1, r_2) = \cdots = \gcd(r_{m-1}, r_m) = r_m$$

and, hence, $r_m = \gcd(a, b)$.

5.2 Suppose that $a > b$ in Algorithm 5.1.
   (a) Prove that $r_i \geq 2r_{i+2}$ for all i such that $0 \leq i \leq m - 2$.
   (b) Prove that m is $O(\log a)$.
   (c) Prove that m is $O(\log b)$.

5.3 Use the EXTENDED EUCLIDEAN ALGORITHM to compute the following multiplicative inverses:
   (a) $17^{-1} \bmod 101$
   (b) $357^{-1} \bmod 1234$
   (c) $3125^{-1} \bmod 9987$.

5.4 Compute $\gcd(57, 93)$, and find integers s and t such that $57s + 93t = \gcd(57, 93)$.

5.5 Suppose $\chi : \mathbb{Z}_{105} \to \mathbb{Z}_3 \times \mathbb{Z}_5 \times \mathbb{Z}_7$ is defined as

$$\chi(x) = (x \bmod 3, x \bmod 5, x \bmod 7).$$

Give an explicit formula for the function χ^{-1}, and use it to compute $\chi^{-1}(2, 2, 3)$.

5.6 Solve the following system of congruences:

$$x \equiv 12 \pmod{25}$$

$$x \equiv 9 \pmod{26}$$

$$x \equiv 23 \pmod{27}.$$

5.7 Solve the following system of congruences:

$$13x \equiv 4 \pmod{99}$$

$$15x \equiv 56 \pmod{101}.$$

HINT First use the EXTENDED EUCLIDEAN ALGORITHM, and then apply the Chinese remainder theorem.

5.8 Use Theorem 5.8 to find the smallest primitive element modulo 97.

5.9 Suppose that $p = 2q + 1$, where p and q are odd primes. Suppose further that $\alpha \in \mathbb{Z}_p^*$, $\alpha \not\equiv \pm 1 \pmod{p}$. Prove that α is a primitive element modulo p if and only if $\alpha^q \equiv -1 \pmod{p}$.

5.10 Suppose that $n = pq$, where p and q are distinct odd primes and $ab \equiv 1 \pmod{(p-1)(q-1)}$. The RSA encryption operation is $e(x) = x^b \bmod n$ and the decryption operation is $d(y) = y^a \bmod n$. We proved that $d(e(x)) = x$ if $x \in \mathbb{Z}_n^*$. Prove that the same statement is true for any $x \in \mathbb{Z}_n$.

HINT   Use the fact that $x_1 \equiv x_2 \pmod{pq}$ if and only if $x_1 \equiv x_2 \pmod{p}$ and $x_1 \equiv x_2 \pmod{q}$. This follows from the Chinese remainder theorem.

5.11 For $n = pq$, where p and q are distinct odd primes, define
$$\lambda(n) = \frac{(p-1)(q-1)}{\gcd(p-1, q-1)}.$$
Suppose that we modify the *RSA Cryptosystem* by requiring that $ab \equiv 1 \pmod{\lambda(n)}$.
  (a) Prove that encryption and decryption are still inverse operations in this modified cryptosystem.
  (b) If $p = 37$, $q = 79$, and $b = 7$, compute a in this modified cryptosystem, as well as in the original *RSA Cryptosystem*.

5.12 Two samples of RSA ciphertext are presented in Tables 5.1 and 5.2. Your task is to decrypt them. The public parameters of the system are $n = 18923$ and $b = 1261$ (for Table 5.1) and $n = 31313$ and $b = 4913$ (for Table 5.2). This can be accomplished as follows. First, factor n (which is easy because it is so small). Then compute the exponent a from $\phi(n)$, and, finally, decrypt the ciphertext. Use the SQUARE-AND-MULTIPLY ALGORITHM to exponentiate modulo n.
   In order to translate the plaintext back into ordinary English text, you need to know how alphabetic characters are "encoded" as elements in $\mathbb{Z}_n$. Each element of $\mathbb{Z}_n$ represents three alphabetic characters as in the following examples:
$$
\begin{array}{rcl}
DOG & \to & 3 \times 26^2 + 14 \times 26 + 6 \quad = \quad 2398 \\
CAT & \to & 2 \times 26^2 + 0 \times 26 + 19 \quad = \quad 1371 \\
ZZZ & \to & 25 \times 26^2 + 25 \times 26 + 25 \quad = \quad 17575.
\end{array}
$$
You will have to invert this process as the final step in your program.
   The first plaintext was taken from "The Diary of Samuel Marchbanks," by Robertson Davies, 1947, and the second was taken from "Lake Wobegon Days," by Garrison Keillor, 1985.

5.13 A common way to speed up RSA decryption incorporates the Chinese remainder theorem, as follows. Suppose that $d_K(y) = y^d \bmod n$ and $n = pq$. Define $d_p = d \bmod (p-1)$ and $d_q = d \bmod (q-1)$; and let $M_p = q^{-1} \bmod p$ and $M_q = p^{-1} \bmod q$. Then consider the following algorithm:

Algorithm 5.15: CRT-OPTIMIZED RSA DECRYPTION$(n, d_p, d_q, M_p, M_q, y)$

$x_p \leftarrow y^{d_p} \bmod p$
$x_q \leftarrow y^{d_q} \bmod q$
$x \leftarrow M_p q x_p + M_q p x_q \bmod n$
return (x)

Algorithm 5.15 replaces an exponentiation modulo n by modular exponentiations modulo p and q. If p and q are ℓ-bit integers and exponentiation modulo an ℓ-bit integer takes time $c\ell^3$, then the time to perform the required exponentiation(s) is reduced from $c(2\ell)^3$ to $2c\ell^3$, a savings of 75%. The final step, involving the Chinese remainder theorem, requires time $O(\ell^2)$ if d_p, d_q, M_p and M_q have been pre-computed.

TABLE 5.1
RSA ciphertext

12423	11524	7243	7459	14303	6127	10964	16399
9792	13629	14407	18817	18830	13556	3159	16647
5300	13951	81	8986	8007	13167	10022	17213
2264	961	17459	4101	2999	14569	17183	15827
12693	9553	18194	3830	2664	13998	12501	18873
12161	13071	16900	7233	8270	17086	9792	14266
13236	5300	13951	8850	12129	6091	18110	3332
15061	12347	7817	7946	11675	13924	13892	18031
2620	6276	8500	201	8850	11178	16477	10161
3533	13842	7537	12259	18110	44	2364	15570
3460	9886	8687	4481	11231	7547	11383	17910
12867	13203	5102	4742	5053	15407	2976	9330
12192	56	2471	15334	841	13995	17592	13297
2430	9741	11675	424	6686	738	13874	8168
7913	6246	14301	1144	9056	15967	7328	13203
796	195	9872	16979	15404	14130	9105	2001
9792	14251	1498	11296	1105	4502	16979	1105
56	4118	11302	5988	3363	15827	6928	4191
4277	10617	874	13211	11821	3090	18110	44
2364	15570	3460	9886	9988	3798	1158	9872
16979	15404	6127	9872	3652	14838	7437	2540
1367	2512	14407	5053	1521	297	10935	17137
2186	9433	13293	7555	13618	13000	6490	5310
18676	4782	11374	446	4165	11634	3846	14611
2364	6789	11634	4493	4063	4576	17955	7965
11748	14616	11453	17666	925	56	4118	18031
9522	14838	7437	3880	11476	8305	5102	2999
18628	14326	9175	9061	650	18110	8720	15404
2951	722	15334	841	15610	2443	11056	2186

(a) Prove that the value x returned by Algorithm 5.15 is, in fact, $y^d \bmod n$.

(b) Given that $p = 1511$, $q = 2003$ and $d = 1234577$, compute d_p, d_q, M_p and M_q.

(c) Given the above values of p, q and d, decrypt the ciphertext $y = 152702$ using Algorithm 5.15.

5.14 Prove that the *RSA Cryptosystem* is insecure against a chosen ciphertext attack. In particular, given a ciphertext y, describe how to choose a ciphertext $\hat{y} \neq y$, such that knowledge of the plaintext $\hat{x} = d_K(\hat{y})$ allows $x = d_K(y)$ to be computed.

HINT   Use the multiplicative property of the *RSA Cryptosystem*, i.e., that

$$e_K(x_1)e_K(x_2) \bmod n = e_K(x_1 x_2 \bmod n).$$

5.15 This exercise exhibits what is called a *protocol failure*. It provides an example where ciphertext can be decrypted by an opponent, without determining the key, if a cryptosystem is used in a careless way. The moral is that it is not sufficient to use a "secure" cryptosystem in order to guarantee "secure" communication.

TABLE 5.2
RSA ciphertext

6340	8309	14010	8936	27358	25023	16481	25809
23614	7135	24996	30590	27570	26486	30388	9395
27584	14999	4517	12146	29421	26439	1606	17881
25774	7647	23901	7372	25774	18436	12056	13547
7908	8635	2149	1908	22076	7372	8686	1304
4082	11803	5314	107	7359	22470	7372	22827
15698	30317	4685	14696	30388	8671	29956	15705
1417	26905	25809	28347	26277	7897	20240	21519
12437	1108	27106	18743	24144	10685	25234	30155
23005	8267	9917	7994	9694	2149	10042	27705
15930	29748	8635	23645	11738	24591	20240	27212
27486	9741	2149	29329	2149	5501	14015	30155
18154	22319	27705	20321	23254	13624	3249	5443
2149	16975	16087	14600	27705	19386	7325	26277
19554	23614	7553	4734	8091	23973	14015	107
3183	17347	25234	4595	21498	6360	19837	8463
6000	31280	29413	2066	369	23204	8425	7792
25973	4477	30989					

Suppose Bob has an *RSA Cryptosystem* with a large modulus n for which the factorization cannot be found in a reasonable amount of time. Suppose Alice sends a message to Bob by representing each alphabetic character as an integer between 0 and 25 (i.e., $A \leftrightarrow 0$, $B \leftrightarrow 1$, etc.), and then encrypting each residue modulo 26 as a separate plaintext character.

(a) Describe how Oscar can easily decrypt a message which is encrypted in this way.

(b) Illustrate this attack by decrypting the following ciphertext (which was encrypted using an *RSA Cryptosystem* with $n = 18721$ and $b = 25$) without factoring the modulus:

$$365, 0, 4845, 14930, 2608, 2608, 0.$$

5.16 This exercise illustrates another example of a protocol failure (due to Simmons) involving the *RSA Cryptosystem*; it is called the "common modulus protocol failure." Suppose Bob has an *RSA Cryptosystem* with modulus n and encryption exponent b_1, and Charlie has an *RSA Cryptosystem* with (the same) modulus n and encryption exponent b_2. Suppose also that $\gcd(b_1, b_2) = 1$. Now, consider the situation that arises if Alice encrypts the same plaintext x to send to both Bob and Charlie. Thus, she computes $y_1 = x^{b_1} \bmod n$ and $y_2 = x^{b_2} \bmod n$, and then she sends y_1 to Bob and y_2 to Charlie. Suppose Oscar intercepts y_1 and y_2, and performs the computations indicated in Algorithm 5.16.

Algorithm 5.16: RSA COMMON MODULUS DECRYPTION(n, b_1, b_2, y_1, y_2)

$c_1 \leftarrow b_1^{-1} \bmod b_2$
$c_2 \leftarrow (c_1 b_1 - 1)/b_2$
$x_1 \leftarrow y_1^{c_1} (y_2^{c_2})^{-1} \bmod n$
return (x_1)

(a) Prove that the value x_1 computed in Algorithm 5.16 is in fact Alice's plain-text, x. Thus, Oscar can decrypt the message Alice sent, even though the cryptosystem may be "secure."

(b) Illustrate the attack by computing x by this method if $n = 18721$, $b_1 = 43$, $b_2 = 7717$, $y_1 = 12677$ and $y_2 = 14702$.

5.17 We give yet another protocol failure involving the *RSA Cryptosystem*. Suppose that three users in a network, say Bob, Bart and Bert, all have public encryption exponents $b = 3$. Let their moduli be denoted by n_1, n_2, n_3, and assume that n_1, n_2 and n_3, are pairwise relatively prime. Now suppose Alice encrypts the same plaintext x to send to Bob, Bart and Bert. That is, Alice computes $y_i = x^3 \bmod n_i$, $1 \leq i \leq 3$. Describe how Oscar can compute x, given y_1, y_2 and y_3, without factoring any of the moduli.

5.18 A plaintext x is said to be *fixed* if $e_K(x) = x$. Show that, for the *RSA Cryptosystem*, the number of fixed plaintexts $x \in \mathbb{Z}_n^*$ is equal to

$$\gcd(b - 1, p - 1) \times \gcd(b - 1, q - 1).$$

HINT   Consider the following system of two congruences:

$$e_K(x) \equiv x \pmod{p},$$

$$e_K(x) \equiv x \pmod{q}.$$

5.19 Suppose **A** is a deterministic algorithm which is given as input an RSA modulus n, an encryption exponent b, and a ciphertext y. **A** will either decrypt y or return no answer. Supposing that there are $\epsilon(n - 1)$ nonzero ciphertexts which **A** is able to decrypt, show how to use **A** as an oracle in a Las Vegas decryption algorithm having success probability ϵ.

5.20 Write a program to evaluate Jacobi symbols using the four properties presented in Section 5.4. The program should not do any factoring, other than dividing out powers of two. Test your program by computing the following Jacobi symbols:

$$\left(\frac{610}{987}\right), \left(\frac{20964}{1987}\right), \left(\frac{1234567}{11111111}\right).$$

5.21 For $n = 837$, 851 and 1189, find the number of bases b such that n is an Euler pseudo-prime to the base b.

5.22 The purpose of this question is to prove that the error probability of the Solovay-Strassen primality test is at most $1/2$. Let $\mathbb{Z}_n^*$ denote the group of units modulo n. Define

$$G(n) = \left\{ a : a \in \mathbb{Z}_n^*, \left(\frac{a}{n}\right) \equiv a^{(n-1)/2} \pmod{n} \right\}.$$

(a) Prove that $G(n)$ is a subgroup of $\mathbb{Z}_n^*$. Hence, by Lagrange's theorem, if $G(n) \neq \mathbb{Z}_n^*$, then

$$|G(n)| \leq \frac{|\mathbb{Z}_n^*|}{2} \leq \frac{n-1}{2}.$$

(b) Suppose $n = p^k q$, where p and q are odd, p is prime, $k \geq 2$, and $\gcd(p, q) = 1$. Let $a = 1 + p^{k-1}q$. Prove that

$$\left(\frac{a}{n}\right) \not\equiv a^{(n-1)/2} \pmod{n}.$$

HINT   Use the binomial theorem to compute $a^{(n-1)/2}$.

(c) Suppose $n = p_1 \ldots p_s$, where the p_i's are distinct odd primes. Suppose $a \equiv u \pmod{p_1}$ and $a \equiv 1 \pmod{p_2 p_3 \ldots p_s}$, where u is a quadratic non-residue modulo p_1 (note that such an a exists by the Chinese remainder theorem). Prove that

$$\left(\frac{a}{n}\right) \equiv -1 \pmod{n},$$

but

$$a^{(n-1)/2} \equiv 1 \pmod{p_2 p_3 \ldots p_s},$$

so

$$a^{(n-1)/2} \not\equiv -1 \pmod{n}.$$

(d) If n is odd and composite, prove that $|G(n)| \leq (n-1)/2$.
(e) Summarize the above: prove that the error probability of the Solovay-Strassen primality test is at most $1/2$.

5.23 Suppose we have a Las Vegas algorithm with failure probability ϵ.
    (a) Prove that the probability of first achieving success on the nth trial is $p_n = \epsilon^{n-1}(1 - \epsilon)$.
    (b) The average (expected) number of trials to achieve success is

$$\sum_{n=1}^{\infty} (n \times p_n).$$

Show that this average is equal to $1/(1 - \epsilon)$.
    (c) Let δ be a positive real number less than 1. Show that the number of iterations required in order to reduce the probability of failure to at most δ is

$$\left\lceil \frac{\log_2 \delta}{\log_2 \epsilon} \right\rceil.$$

5.24 Suppose throughout this question that p is an odd prime and $\gcd(a, p) = 1$.
    (a) Suppose that $i \geq 2$ and $b^2 \equiv a \pmod{p^{i-1}}$. Prove that there is a unique $x \in \mathbb{Z}_{p^i}$ such that $x^2 \equiv a \pmod{p^i}$ and $x \equiv b \pmod{p^{i-1}}$. Describe how this x can be computed efficiently.
    (b) Illustrate your method in the following situation: starting with the congruence $6^2 \equiv 17 \pmod{19}$, find square roots of 17 modulo 19^2 and modulo 19^3.
    (c) For all $i \geq 1$, prove that the number of solutions to the congruence $x^2 \equiv a \pmod{p^i}$ is either 0 or 2.

5.25 Using various choices for the bound, B, attempt to factor 262063 and 9420457 using the $p - 1$ method. How big does B have to be in each case to be successful?

5.26 Factor 262063, 9420457 and 181937053 using the POLLARD RHO ALGORITHM, if the function f is defined to be $f(x) = x^2 + 1$. How many iterations are needed to factor each of these three integers?

5.27 Suppose we want to factor the integer $n = 256961$ using the RANDOM SQUARES ALGORITHM. Using the factor base

$$\{-1, 2, 3, 5, 7, 11, 13, 17, 19, 23, 29, 31\},$$

test the integers $z^2 \bmod n$ for $z = 500, 501, \ldots$, until a congruence of the form $x^2 \equiv y^2 \pmod{n}$ is obtained and the factorization of n is found.

5.28 In the RANDOM SQUARES ALGORITHM, we need to test a positive integer $w \leq n - 1$ to see if it factors completely over the factor base $\mathcal{B} = \{p_1, \ldots, p_B\}$ consisting of the B smallest prime numbers. Recall that $p_B = m \approx 2^s$ and $n \approx 2^r$.

    (a) Prove that this can be done using at most $B + r$ divisions of an integer having at most r bits by an integer having at most s bits.

    (b) Assuming that $r < m$, prove that the complexity of this test is $O(rsm)$.

5.29 In this exercise, we show that parameter generation for the *RSA Cryptosystem* should take care to ensure that $q - p$ is not too small, where $n = pq$ and $q > p$.

    (a) Suppose that $q - p = 2d > 0$, and $n = pq$. Prove that $n + d^2$ is a perfect square.

    (b) Given an integer n which is the product of two odd primes, and given a small positive integer d such that $n + d^2$ is a perfect square, show how this information can be used to factor n.

    (c) Use this technique to factor $n = 2189284635403183$.

5.30 Suppose Bob has carelessly revealed his decryption exponent to be $a = 14039$ in an *RSA Cryptosystem* with public key $n = 36581$ and $b = 4679$. Implement the randomized algorithm to factor n given this information. Test your algorithm with the "random" choices $w = 9983$ and $w = 13461$. Show all computations.

5.31 If $q_1, \ldots, q_m$ is the sequence of quotients obtained in the applying the EUCLIDEAN ALGORITHM with input r_0, r_1, prove that the continued fraction $[q_1, \ldots, q_m] = r_0/r_1$.

5.32 Suppose that $n = 317940011$ and $b = 77537081$ in the *RSA Cryptosystem*. Using WIENER'S ALGORITHM, attempt to factor n.

5.33 Consider the modification of the *Rabin Cryptosystem* in which $e_K(x) = x(x + B) \bmod n$, where $B \in \mathbb{Z}_n$ is part of the public key. Supposing that $p = 199$, $q = 211$, $n = pq$ and $B = 1357$, perform the following computations.

    (a) Compute the encryption $y = e_K(32767)$.

    (b) Determine the four possible decryptions of this given ciphertext y.

5.34 Prove Equations (5.3) and (5.4) relating the functions *half* and *parity*.

5.35 Prove that Cryptosystem 5.3 is not semantically secure against a chosen ciphertext attack. Given x_1, x_2, a ciphertext (y_1, y_2) that is an encryption of x_i ($i = 1$ or 2), and given a decryption oracle DECRYPT for Cryptosystem 5.3, describe an algorithm to determine whether $i = 1$ or $i = 2$. You are allowed to call the algorithm DECRYPT with any input except for the given ciphertext (y_1, y_2), and it will output the corresponding plaintext.

6

Public-key Cryptography and Discrete Logarithms

The theme of this chapter concerns public-key cryptosystems based on the **Discrete Logarithm** problem. The first and best-known of these is the *ElGamal Cryptosystem*. The **Discrete Logarithm** problem forms the basis of numerous cryptographic protocols that we will study throughout the rest of the text. Thus we devote a considerable amount of time to discussion of this important problem. In later sections of this chapter, we give treatments of some other ElGamal-type systems based on finite fields and elliptic curves.

6.1 The ElGamal Cryptosystem

The *ElGamal Cryptosystem* is based on the **Discrete Logarithm** problem. We begin by describing this problem in the setting of a finite multiplicative group $(G, \cdot)$. For an element $\alpha \in G$ having order n, define

$$\langle \alpha \rangle = \{\alpha^i : 0 \leq i \leq n - 1\}.$$

It is easy to see that $\langle \alpha \rangle$ is a subgroup of G, and that $\langle \alpha \rangle$ is cyclic of order n.

An often-used example is to take G to be the multiplicative group of a finite field $\mathbb{Z}_p$ (where p is prime), and to let α be a primitive element modulo p. In this situation, we have that $n = |\langle \alpha \rangle| = p - 1$. Another frequently used setting is to take α to be an element having prime order q in the multiplicative group $\mathbb{Z}_p^*$ (where p is prime and $p - 1 \equiv 0 \pmod{q}$). Such an α can be obtained by raising a primitive element in $\mathbb{Z}_p^*$ to the $(p - 1)/q$th power.

We now define the **Discrete Logarithm** problem in a subgroup $\langle \alpha \rangle$ of a group $(G, \cdot)$.

Problem 6.1:   **Discrete Logarithm**

Instance: A multiplicative group $(G, \cdot)$, an element $\alpha \in G$ having order n, and an element $\beta \in \langle \alpha \rangle$.

Question: Find the unique integer a, $0 \leq a \leq n - 1$, such that

$$\alpha^a = \beta.$$

We will denote this integer a by $\log_\alpha \beta$; it is called the *discrete logarithm* of β.

The utility of the **Discrete Logarithm** problem in a cryptographic setting is that finding discrete logarithms is (probably) difficult, but the inverse operation of exponentiation can be computed efficiently by using the square-and-multiply method (Algorithm 5.5). Stated another way, exponentiation is a one-way function in suitable groups G.

ElGamal proposed a public-key cryptosystem which is based on the **Discrete Logarithm** problem in $(\mathbb{Z}_p^*, \cdot)$. This system is presented as Cryptosystem 6.1.

The encryption operation in the *ElGamal Cryptosystem* is randomized, since the ciphertext depends on both the plaintext x and on the random value k chosen by Alice. Hence, there will be many ciphertexts ($p-1$, in fact) that are encryptions of the same plaintext.

Informally, this is how the *ElGamal Cryptosystem* works: The plaintext x is "masked" by multiplying it by β^k, yielding y_2. The value α^k is also transmitted as part of the ciphertext. Bob, who knows the private key, a, can compute β^k from α^k. Then he can "remove the mask" by dividing y_2 by β^k to obtain x.

A small example will illustrate the computations performed in the *ElGamal Cryptosystem*.

Example 6.1   Suppose $p = 2579$ and $\alpha = 2$. α is a primitive element modulo p. Let $a = 765$, so

$$\beta = 2^{765} \bmod 2579 = 949.$$

Now, suppose that Alice wishes to send the message $x = 1299$ to Bob. Say $k = 853$ is the random integer she chooses. Then she computes

$$y_1 = 2^{853} \bmod 2579$$
$$= 435,$$

and

$$y_2 = 1299 \times 949^{853} \bmod 2579$$
$$= 2396.$$

Cryptosystem 6.1: *ElGamal Public-key Cryptosystem in $\mathbb{Z}_p{}^*$*

Let p be a prime such that the **Discrete Logarithm** problem in $(\mathbb{Z}_p{}^*, \cdot)$ is infeasible, and let $\alpha \in \mathbb{Z}_p{}^*$ be a primitive element. Let $\mathcal{P} = \mathbb{Z}_p{}^*$, $\mathcal{C} = \mathbb{Z}_p{}^* \times \mathbb{Z}_p{}^*$, and define

$$\mathcal{K} = \{(p, \alpha, a, \beta) : \beta \equiv \alpha^a \pmod{p}\}.$$

The values p, α and β are the public key, and a is the private key.

For $K = (p, \alpha, a, \beta)$, and for a (secret) random number $k \in \mathbb{Z}_{p-1}$, define

$$e_K(x, k) = (y_1, y_2),$$

where

$$y_1 = \alpha^k \bmod p$$

and

$$y_2 = x\beta^k \bmod p.$$

For $y_1, y_2 \in \mathbb{Z}_p{}^*$, define

$$d_K(y_1, y_2) = y_2(y_1{}^a)^{-1} \bmod p.$$

When Bob receives the ciphertext $y = (435, 2396)$, he computes

$$x = 2396 \times (435^{765})^{-1} \bmod 2579$$
$$= 1299,$$

which was the plaintext that Alice encrypted. $\square$

Clearly the *ElGamal Cryptosystem* will be insecure if Oscar can compute the value $a = \log_\alpha \beta$, for then Oscar can decrypt ciphertexts exactly as Bob does. Hence, a necessary condition for the *ElGamal Cryptosystem* to be secure is that the **Discrete Logarithm** problem in $\mathbb{Z}_p{}^*$ is infeasible. This is generally regarded as being the case if p is carefully chosen and α is a primitive element modulo p. In particular, there is no known polynomial-time algorithm for this version of the **Discrete Logarithm** problem. To thwart known attacks, p should have at least 300 digits, and $p - 1$ should have at least one "large" prime factor.

6.2 Algorithms for the Discrete Logarithm Problem

Throughout this section, we assume that $(G, \cdot)$ is a multiplicative group and $\alpha \in G$ has order n. Hence the **Discrete Logarithm** problem can be phrased in the following form: Given $\beta \in \langle \alpha \rangle$, find the unique exponent a, $0 \leq a \leq n-1$, such that $\alpha^a = \beta$.

We begin by analyzing some elementary algorithms which can be used to solve the **Discrete Logarithm** problem. In our analyses, we will assume that computing a product of two elements in the group G requires constant (i.e., $O(1)$) time.

First, we observe that the **Discrete Logarithm** problem can be solved by exhaustive search in $O(n)$ time and $O(1)$ space, simply by computing $\alpha, \alpha^2, \alpha^3, \ldots$, until $\beta = \alpha^a$ is found. (Each term α^i in the above list is computed by multiplying the previous term α^{i-1} by α, and hence the total time required is $O(n)$.)

Another approach is to precompute all possible values α^i, and then sort the list of ordered pairs (i, α^i) with respect to their second coordinates. Then, given β, we can perform a binary search of the sorted list in order to find the value a such that $\alpha^a = \beta$. This requires precomputation time $O(n)$ to compute the n powers of α, and time $O(n \log n)$ to sort the list of size n. (The sorting step can be done in time $O(n \log n)$ if an efficient sorting algorithm, such as the QUICKSORT algorithm, is used.) If we neglect logarithmic factors, as is usually done in the analysis of these algorithms, the precomputation time is $O(n)$. The time for a binary search of a sorted list of size n is $O(\log n)$. If we (again) ignore the logarithmic term, then we see that we can solve the **Discrete Logarithm** problem in $O(1)$ time with $O(n)$ precomputation and $O(n)$ memory.

6.2.1 Shanks' Algorithm

The first non-trivial algorithm we describe is a time-memory trade-off due to Shanks. SHANKS' ALGORITHM is presented in Algorithm 6.1.

Some comments are in order. First, steps 2 and 3 can be precomputed, if desired (this will not affect the asymptotic running time, however). Next, observe that if $(j, y) \in L_1$ and $(i, y) \in L_2$, then

$$\alpha^{mj} = y = \beta \alpha^{-i},$$

so

$$\alpha^{mj+i} = \beta,$$

as desired. Conversely, for any $\beta \in \langle \alpha \rangle$, we have that $0 \leq \log_\alpha \beta \leq n-1$. If we divide $\log_\alpha \beta$ by the integer m, then we can express $\log_\alpha \beta$ in the form

$$\log_\alpha \beta = mj + i,$$

where $0 \leq j, i \leq m-1$. The fact that $j \leq m-1$ follows because

$$\log_\alpha \beta \leq n - 1 \leq m^2 - 1 = m(m-1) + m - 1.$$

Algorithm 6.1: SHANKS(G, n, α, β)

1. $m \leftarrow \lceil \sqrt{n} \rceil$
2.   **for** $j \leftarrow 0$ **to** $m - 1$
         do compute α^{mj}
3.   Sort the m ordered pairs (j, α^{mj}) with respect to their second coordinates, obtaining a list L_1
4.   **for** $i \leftarrow 0$ **to** $m - 1$
         do compute $\beta \alpha^{-i}$
5.   Sort the m ordered pairs $(i, \beta \alpha^{-i})$ with respect to their second coordinates, obtaining a list L_2
6.   Find a pair $(j, y) \in L_1$ and a pair $(i, y) \in L_2$ (i.e., find two pairs having identical second coordinates)
7. $\log_\alpha \beta \leftarrow (mj + i) \bmod n$

Hence, the search in step 6 will be successful. (However, if it happened that $\beta \notin \langle \alpha \rangle$, then step 6 will not be successful.)

It is not difficult to implement the algorithm to run in $O(m)$ time with $O(m)$ memory (neglecting logarithmic factors). Here are a few details: Step 2 can be performed by first computing α^m, and then computing its powers by successively multiplying by α^m. The total time for this step is $O(m)$. In a similar fashion, step 4 also takes time $O(m)$. Steps 3 and 5 can be done in time $O(m \log m)$ using an efficient sorting algorithm. Finally, step 6 can be done with one (simultaneous) pass through each of the two lists L_1 and L_2; so it requires time $O(m)$.

Here is a small example to illustrate SHANKS' ALGORITHM.

Example 6.2 Suppose we wish to find $\log_3 525$ in $(\mathbb{Z}_{809}{}^*, \cdot)$. Note that 809 is prime and 3 is a primitive element in $\mathbb{Z}_{809}{}^*$, so we have $\alpha = 3$, $n = 808$, $\beta = 525$ and $m = \lceil \sqrt{808} \rceil = 29$. Then

$$\alpha^{29} \bmod 809 = 99.$$

First, we compute the ordered pairs $(j, 99^j \bmod 809)$ for $0 \leq j \leq 28$. We obtain the list

$$
\begin{array}{ccccc}
(0, 1) & (1, 99) & (2, 93) & (3, 308) & (4, 559) \\
(5, 329) & (6, 211) & (7, 664) & (8, 207) & (9, 268) \\
(10, 644) & (11, 654) & (12, 26) & (13, 147) & (14, 800) \\
(15, 727) & (16, 781) & (17, 464) & (18, 632) & (19, 275) \\
(20, 528) & (21, 496) & (22, 564) & (23, 15) & (24, 676) \\
(25, 586) & (26, 575) & (27, 295) & (28, 81) &
\end{array}
$$

which is then sorted to produce L_1.

The second list contains the ordered pairs $(i, 525 \times (3^i)^{-1} \bmod 809), 0 \leq j \leq 28$. It is as follows:

$$
\begin{array}{ccccc}
(0, 525) & (1, 175) & (2, 328) & (3, 379) & (4, 396) \\
(5, 132) & (6, 44) & (7, 554) & (8, 724) & (9, 511) \\
(10, 440) & (11, 686) & (12, 768) & (13, 256) & (14, 355) \\
(15, 388) & (16, 399) & (17, 133) & (18, 314) & (19, 644) \\
(20, 754) & (21, 521) & (22, 713) & (23, 777) & (24, 259) \\
(25, 356) & (26, 658) & (27, 489) & (28, 163) &
\end{array}
$$

After sorting this list, we get L_2.

Now, if we proceed simultaneously through the two sorted lists, we find that $(10, 644)$ is in L_1 and $(19, 644)$ is in L_2. Hence, we can compute

$$\log_3 525 = (29 \times 10 + 19) \bmod 808$$

$$= 309.$$

As a check, it can be verified that $3^{309} \equiv 525 \pmod{809}$. ⛛

6.2.2   The Pollard Rho Discrete Logarithm Algorithm

We previously discussed the POLLARD RHO ALGORITHM for factoring in Section 5.6.2. There is a corresponding algorithm for finding discrete logarithms, which we describe now. As before, let $(G, \cdot)$ be a group and let $\alpha \in G$ be an element having order n. Let $\beta \in \langle \alpha \rangle$ be the element whose discrete logarithm we want to find. Since $\langle \alpha \rangle$ is cyclic of order n, we can treat $\log_\alpha \beta$ as an element of $\mathbb{Z}_n$.

As with the rho algorithm for factoring, we form a sequence $x_1, x_2, \ldots$ by iteratively applying a random-looking function, f. Once we obtain two elements x_i and x_j in the sequence such that $x_i = x_j$ and $i < j$, we can (hopefully) compute $\log_\alpha \beta$. Just as we did in the case of the factoring algorithm, we will seek a collision of the form $x_i = x_{2i}$, in order to save time and memory.

Let $S_1 \cup S_2 \cup S_3$ be a partition of G into three subsets of roughly equal size. We define a function $f : \langle \alpha \rangle \times \mathbb{Z}_n \times \mathbb{Z}_n \to \langle \alpha \rangle \times \mathbb{Z}_n \times \mathbb{Z}_n$ as follows:

$$
f(x, a, b) = \begin{cases}
(\beta x, a, b + 1) & \text{if } x \in S_1 \\
(x^2, 2a, 2b) & \text{if } x \in S_2 \\
(\alpha x, a + 1, b) & \text{if } x \in S_3.
\end{cases}
$$

Further, each of the triples (x, a, b) that we form are required to have the property that $x = \alpha^a \beta^b$. We begin with an initial triple having this property, say $(1, 0, 0)$. Observe that $f(x, a, b)$ satisfies the desired property if (x, a, b) does. So we define

$$
(x_i, a_i, b_i) = \begin{cases}
(1, 0, 0) & \text{if } i = 0 \\
f(x_{i-1}, a_{i-1}, b_{i-1}) & \text{if } i \geq 1.
\end{cases}
$$

We compare the triples (x_{2i}, a_{2i}, b_{2i}) and (x_i, a_i, b_i) until we find a value of $i \geq 1$ such that $x_{2i} = x_i$. When this occurs, we have that

$$\alpha^{a_{2i}} \beta^{b_{2i}} = \alpha^{a_i} \beta^{b_i}.$$

If we denote $c = \log_\alpha \beta$, then it must be the case that

$$\alpha^{a_{2i} + cb_{2i}} = \alpha^{a_i + cb_i}.$$

Since α has order n, it follows that

$$a_{2i} + cb_{2i} \equiv a_i + cb_i \pmod{n}.$$

This can be rewritten as

$$c(b_{2i} - b_i) \equiv a_i - a_{2i} \pmod{n}.$$

If $\gcd(b_{2i} - b_i, n) = 1$, then we can solve for c as follows:

$$c = (a_i - a_{2i})(b_{2i} - b_i)^{-1} \bmod n.$$

We illustrate the application of the above algorithm with an example. Notice that we take care to ensure that $1 \notin S_2$ (since we would obtain $x_i = (1, 0, 0)$ for all integers $i \geq 0$ if $1 \in S_2$).

Example 6.3 The integer $p = 809$ is prime, and it can be verified that the element $\alpha = 89$ has order $n = 101$ in $\mathbb{Z}_{809}^*$. The element $\beta = 618$ is in the subgroup $\langle \alpha \rangle$; we will compute $\log_\alpha \beta$.

Suppose we define the sets S_1, S_2 and S_3 as follows:

$$S_1 = \{x \in \mathbb{Z}_{809} : x \equiv 1 \pmod{3}\}$$

$$S_2 = \{x \in \mathbb{Z}_{809} : x \equiv 0 \pmod{3}\}$$

$$S_3 = \{x \in \mathbb{Z}_{809} : x \equiv 2 \pmod{3}\}.$$

For $i = 1, 2, \ldots$, we obtain triples (x_{2i}, a_{2i}, b_{2i}) and (x_i, a_i, b_i) as follows:

i	(x_i, a_i, b_i)	(x_{2i}, a_{2i}, b_{2i})
1	$(618, 0, 1)$	$(76, 0, 2)$
2	$(76, 0, 2)$	$(113, 0, 4)$
3	$(46, 0, 3)$	$(488, 1, 5)$
4	$(113, 0, 4)$	$(605, 4, 10)$
5	$(349, 1, 4)$	$(422, 5, 11)$
6	$(488, 1, 5)$	$(683, 7, 11)$
7	$(555, 2, 5)$	$(451, 8, 12)$
8	$(605, 4, 10)$	$(344, 9, 13)$
9	$(451, 5, 10)$	$(112, 11, 13)$
10	$(422, 5, 11)$	$(422, 11, 15)$

Algorithm 6.2: POLLARD RHO DISCRETE LOG ALGORITHM(G, n, α, β)

procedure $f(x, a, b)$
  if $x \in S_1$
    then $f \leftarrow (\beta \cdot x, a, (b+1) \bmod n)$
    else if $x \in S_2$
    then $f \leftarrow (x^2, 2a \bmod n, 2b \bmod n)$
    else $f \leftarrow (\alpha \cdot x, (a+1) \bmod n, b)$
  return (f)

main
  define the partition $G = S_1 \cup S_2 \cup S_3$
  $(x, a, b) \leftarrow f(1, 0, 0)$
  $(x', a', b') \leftarrow f(x, a, b)$
  while $x \neq x'$
  do $\begin{cases} (x, a, b) \leftarrow f(x, a, b) \\ (x', a', b') \leftarrow f(x', a', b') \\ (x', a', b') \leftarrow f(x', a', b') \end{cases}$
  if $\gcd(b' - b, n) \neq 1$
    then return ("failure")
    else return $((a - a')(b' - b)^{-1} \bmod n)$

The first collision in the above list is $x_{10} = x_{20} = 422$. The equation to be solved is

$$c = (5 - 11)(15 - 11)^{-1} \bmod 101 = (-6 \times 4^{-1}) \bmod 101 = 49.$$

Therefore, $\log_{89} 618 = 49$ in the group $\mathbb{Z}_{809}^*$.   ◻

The POLLARD RHO ALGORITHM for discrete logarithms is presented as Algorithm 6.2. In this algorithm, we assume, as usual, that $\alpha \in G$ has order n and $\beta \in \langle \alpha \rangle$.

In the situation where $\gcd(b' - b, n) > 1$, Algorithm 6.2 terminates with the output "failure." The situation may not be so bleak, however. If $\gcd(b'-b, n) = d$, then it is not hard to show that the congruence $c(b' - b) \equiv a - a' \pmod{n}$ has exactly d possible solutions. If d is not too large, then it is relatively straightforward to find the d solutions to the congruence and check to see which one is the correct one.

Algorithm 6.2 can be analyzed in a similar fashion as the Pollard rho factoring algorithm. Under reasonable assumptions concerning the randomness of the function f, we expect to be able to compute discrete logarithms in cyclic groups of order n in $O(\sqrt{n})$ iterations of the algorithm.

6.2.3 The Pohlig-Hellman Algorithm

The next algorithm we study is the POHLIG-HELLMAN ALGORITHM. Suppose that

$$n = \prod_{i=1}^{k} p_i^{c_i},$$

where the p_i's are distinct primes. The value $a = \log_\alpha \beta$ is determined (uniquely) modulo n. We first observe that if we can compute $a \bmod p_i^{c_i}$ for each i, $1 \le i \le k$, then we can compute $a \bmod n$ by the Chinese remainder theorem. So, let's suppose that q is prime,

$$n \equiv 0 \pmod{q^c}$$

and

$$n \not\equiv 0 \pmod{q^{c+1}}.$$

We will show how to compute the value

$$x = a \bmod q^c,$$

where $0 \le x \le q^c - 1$. We can express x in radix q representation as

$$x = \sum_{i=0}^{c-1} a_i q^i,$$

where $0 \le a_i \le q - 1$ for $0 \le i \le c - 1$. Also, observe that we can express a as

$$a = x + sq^c$$

for some integer s. Hence, we have that

$$a = \sum_{i=0}^{c-1} a_i q^i + sq^c.$$

The first step of the algorithm is to compute a_0. The main observation used in the algorithm is the following:

$$\beta^{n/q} = \alpha^{a_0 n/q}. \tag{6.1}$$

We prove that equation (6.1) holds as follows:

$$
\begin{aligned}
\beta^{n/q} &= (\alpha^a)^{n/q} \\
&= (\alpha^{a_0 + a_1 q + \cdots + a_{c-1} q^{c-1} + sq^c})^{n/q} \\
&= (\alpha^{a_0 + Kq})^{n/q} \qquad \text{where } K \text{ is an integer} \\
&= \alpha^{a_0 n/q} \alpha^{Kn} \\
&= \alpha^{a_0 n/q}.
\end{aligned}
$$

Using equation (6.1), it is a simple matter to determine a_0. This can be done, for example, by computing

$$\gamma = \alpha^{n/q}, \gamma^2, \ldots,$$

until

$$\gamma^i = \beta^{n/q}$$

for some $i \leq q - 1$. When this happens, we know that $a_0 = i$.

Now, if $c = 1$, we're done. Otherwise $c > 1$, and we proceed to determine $a_1, \ldots, a_{c-1}$. This is done in a similar fashion as the computation of a_0. Denote $\beta_0 = \beta$, and define

$$\beta_j = \beta\alpha^{-(a_0 + a_1 q + \cdots + a_{j-1}q^{j-1})}$$

for $1 \leq j \leq c-1$. We make use of the following generalization of equation (6.1):

$$\beta_j^{n/q^{j+1}} = \alpha^{a_j n/q}. \tag{6.2}$$

Observe that equation (6.2) reduces to equation (6.1) when $j = 0$.

The proof of equation (6.2) is similar to that of equation (6.1):

$$\beta_j^{n/q^{j+1}} = \left(\alpha^{a - (a_0 + a_1 q + \cdots + a_{j-1}q^{j-1})}\right)^{n/q^{j+1}}$$

$$= \left(\alpha^{a_j q^j + \cdots + a_{c-1}q^{c-1} + sq^c}\right)^{n/q^{j+1}}$$

$$= \left(\alpha^{a_j q^j + K_j q^{j+1}}\right)^{n/q^{j+1}} \qquad \text{where } K_j \text{ is an integer}$$

$$= \alpha^{a_j n/q}\alpha^{K_j n}$$

$$= \alpha^{a_j n/q}.$$

Hence, given β_j, it is straightforward to compute a_j from equation (6.2).

To complete the description of the algorithm, it suffices to observe that β_{j+1} can be computed from β_j by means of a simple recurrence relation, once a_j is known:

$$\beta_{j+1} = \beta_j\alpha^{-a_j q^j}. \tag{6.3}$$

Therefore, we can compute $a_0, \beta_1, a_1, \beta_2, \ldots, \beta_{c-1}, a_{c-1}$ by alternately applying equations (6.2) and (6.3).

A pseudo-code description of the POHLIG-HELLMAN ALGORITHM is given as Algorithm 6.3. To summarize the operation of this algorithm, α is an element of order n in a multiplicative group G, q is prime,

$$n \equiv 0 \pmod{q^c}$$

and

$$n \not\equiv 0 \pmod{q^{c+1}}.$$

The algorithm calculates $a_0, \ldots, a_{c-1}$, where

$$\log_\alpha \beta \bmod q^c = \sum_{i=0}^{c-1} a_i q^i.$$

We illustrate the Pohlig-Hellman algorithm with a small example.

Algorithm 6.3: POHLIG-HELLMAN($G, n, \alpha, \beta, q, c$)

$j \leftarrow 0$
$\beta_j \leftarrow \beta$
while $j \leq c - 1$

$$\mathbf{do} \begin{cases} \delta \leftarrow \beta_j^{n/q^{j+1}} \\ \text{find } i \text{ such that } \delta = \alpha^{in/q} \\ a_j \leftarrow i \\ \beta_{j+1} \leftarrow \beta_j \alpha^{-a_j q^j} \\ j \leftarrow j + 1 \end{cases}$$

return $(a_0, \ldots, a_{c-1})$

Example 6.4 Suppose $p = 29$ and $\alpha = 2$. p is prime and α is a primitive element modulo p, and we have that

$$n = p - 1 = 28 = 2^2 7^1.$$

Suppose $\beta = 18$, so we want to determine $a = \log_2 18$. We proceed by first computing $a \bmod 4$ and then computing $a \bmod 7$.

We start by setting $q = 2$ and $c = 2$ and applying Algorithm 6.3. We find that $a_0 = 1$ and $a_1 = 1$. Hence, $a \equiv 3 \pmod 4$.

Next, we apply Algorithm 6.3 with $q = 7$ and $c = 1$. We find that $a_0 = 4$, so $a \equiv 4 \pmod 7$.

Finally, solving the system

$$a \equiv 3 \pmod 4$$

$$a \equiv 4 \pmod 7$$

using the Chinese remainder theorem, we get $a \equiv 11 \pmod{28}$. That is, we have computed $\log_2 18 = 11$ in $\mathbb{Z}_{29}$. ⬜

Let's consider the complexity of Algorithm 6.3. It is not difficult to see that a straightforward implementation of this algorithm runs in time $O(cq)$. This can be improved, however, by observing that each computation of a value i such that $\delta = \alpha^{in/q}$ can be viewed as the solution of a particular instance of the **Discrete Logarithm** problem. To be specific, we have that $\delta = \alpha^{in/q}$ if and only if

$$i = \log_{\alpha^{n/q}} \delta.$$

The element $\alpha^{n/q}$ has order q, and therefore each i can be computed (using SHANKS' ALGORITHM, for example) in time $O(\sqrt{q})$. The complexity of Algorithm 6.3 can therefore be reduced to $O(c\sqrt{q})$.

6.2.4   The Index Calculus Method

The algorithms in the three previous sections can be applied to any group. The algorithm we describe in this section, the INDEX CALCULUS ALGORITHM, is more specialized: it applies to the particular situation of finding discrete logarithms in $\mathbb{Z}_p^*$ when p is prime and α is a primitive element modulo p. In this situation, the INDEX CALCULUS ALGORITHM is faster than the algorithms previously considered.

The INDEX CALCULUS ALGORITHM for computing discrete logarithms bears considerable resemblance to many of the best factoring algorithms. The method uses a *factor base*, which, as in Section 5.6.3, is a set $\mathcal{B}$ of "small" primes. Suppose $\mathcal{B} = \{p_1, p_2, \ldots, p_B\}$. The first step (a preprocessing step) is to find the logarithms of the B primes in the factor base. The second step is to compute the discrete logarithm of a desired element β, using the knowledge of the discrete logarithms of the elements in the factor base.

Let C be a bit bigger than B; say $C = B + 10$. In the precomputation phase, we will construct C congruences modulo p, which have the following form:

$$\alpha^{x_j} \equiv p_1{}^{a_{1j}} p_2{}^{a_{2j}} \ldots p_B{}^{a_{Bj}} \pmod{p},$$

for $1 \leq j \leq C$. Notice that these congruences can be written equivalently as

$$x_j \equiv a_{1j} \log_\alpha p_1 + \cdots + a_{Bj} \log_\alpha p_B \pmod{p-1},$$

$1 \leq j \leq C$. Given C congruences in the B "unknowns" $\log_\alpha p_i$ $(1 \leq i \leq B)$, we hope that there is a unique solution modulo $p - 1$. If this is the case, then we can compute the logarithms of the elements in the factor base.

How do we generate the C congruences of the desired form? One elementary way is to take a random value x, compute $\alpha^x \bmod p$, and then determine if $\alpha^x \bmod p$ has all its factors in $\mathcal{B}$ (using trial division, for example).

Now, supposing that we have already successfully carried out the precomputation step, we compute a desired logarithm $\log_\alpha \beta$ by means of a Las Vegas type randomized algorithm. Choose a random integer s $(1 \leq s \leq p - 2)$ and compute

$$\gamma = \beta \alpha^s \bmod p.$$

Now attempt to factor γ over the factor base $\mathcal{B}$. If this can be done, then we obtain a congruence of the form

$$\beta \alpha^s \equiv p_1{}^{c_1} p_2{}^{c_2} \ldots p_B{}^{c_B} \pmod{p}.$$

This can be written equivalently as

$$\log_\alpha \beta + s \equiv c_1 \log_\alpha p_1 + \cdots + c_B \log_\alpha p_B \pmod{p-1}.$$

Since all terms in the above congruence are now known, except for $\log_\alpha \beta$, we can easily solve for $\log_\alpha \beta$.

Here is a small, very artificial, example to illustrate the two steps in the algorithm.

Example 6.5 The integer $p = 10007$ is prime. Suppose that $\alpha = 5$ is the primitive element used as the base of logarithms modulo p. Suppose we take $\mathcal{B} = \{2, 3, 5, 7\}$ as the factor base. Of course $\log_5 5 = 1$, so there are three logs of factor base elements to be determined.

Some examples of "lucky" exponents that might be chosen are $4063, 5136$ and 9865.

With $x = 4063$, we compute

$$5^{4063} \bmod 10007 = 42 = 2 \times 3 \times 7.$$

This yields the congruence

$$\log_5 2 + \log_5 3 + \log_5 7 \equiv 4063 \pmod{10006}.$$

Similarly, since

$$5^{5136} \bmod 10007 = 54 = 2 \times 3^3$$

and

$$5^{9865} \bmod 10007 = 189 = 3^3 \times 7,$$

we obtain two more congruences:

$$\log_5 2 + 3\log_5 3 \equiv 5136 \pmod{10006}$$

and

$$3\log_5 3 + \log_5 7 \equiv 9865 \pmod{10006}.$$

We now have three congruences in three unknowns, and there happens to be a unique solution modulo 10006, namely $\log_5 2 = 6578$, $\log_5 3 = 6190$ and $\log_5 7 = 1301$.

Now, let's suppose that we wish to find $\log_5 9451$. Suppose we choose the "random" exponent $s = 7736$, and compute

$$9451 \times 5^{7736} \bmod 10007 = 8400.$$

Since $8400 = 2^4 3^1 5^2 7^1$ factors over $\mathcal{B}$, we obtain

$$\log_5 9451 = (4\log_5 2 + \log_5 3 + 2\log_5 5 + \log_5 7 - s) \bmod 10006$$

$$= (4 \times 6578 + 6190 + 2 \times 1 + 1301 - 7736) \bmod 10006$$

$$= 6057.$$

To verify, we can check that $5^{6057} \equiv 9451 \pmod{10007}$. $\quad\square$

Heuristic analyses of various versions of the INDEX CALCULUS ALGORITHM have been done. Under reasonable assumptions, such as those considered in the analysis of DIXON'S ALGORITHM in Section 5.6.3, the asymptotic running time of the precomputation phase is

$$O\left(e^{(1+o(1))\sqrt{\ln p \ln \ln p}}\right),$$

and the time to find a particular discrete logarithm is

$$O\left(e^{(1/2+o(1))\sqrt{\ln p \ln \ln p}}\right).$$

6.3   Lower Bounds on the Complexity of Generic Algorithms

In this section, we turn our attention to an interesting lower bound on the complexity of the **Discrete Logarithm** problem. Several of the algorithms we have described for the discrete logarithm problem can be applied in any group. An algorithm of this type is called a *generic algorithm*, because it does not depend on any property of the representation of the group. Examples of generic algorithms for the discrete logarithm problem include SHANKS' ALGORITHM, the POLLARD RHO ALGORITHM and the POHLIG-HELLMAN ALGORITHM. On the other hand, the INDEX CALCULUS ALGORITHM studied in the previous section is not generic. This algorithm involves treating elements of $\mathbb{Z}_p^*$ as integers, and then computing their factorizations into primes. Clearly this is something that cannot be done in an arbitrary group.

Another example of a non-generic algorithm for a particular group is provided by studying the **Discrete Logarithm** problem in the additive group $(\mathbb{Z}_n, +)$. (We defined the **Discrete Logarithm** problem in a multiplicative group, but this was done solely to establish a consistent notation for the algorithms we presented.) Suppose that $\gcd(\alpha, n) = 1$, so α is a generator of $\mathbb{Z}_n$. Since the group operation is addition modulo n, an "exponentiation" operation, α^a, corresponds to multiplication by a modulo n. Hence, in this setting, the **Discrete Logarithm** problem is to find the integer a such that

$$\alpha a \equiv \beta \pmod{n}.$$

Since $\gcd(\alpha, n) = 1$, α has a multiplicative inverse modulo n, and we can compute $\alpha^{-1} \bmod n$ easily using the EXTENDED EUCLIDEAN ALGORITHM. Then we can solve for a, obtaining

$$\log_\alpha \beta = \beta \alpha^{-1} \bmod n.$$

This algorithm is of course very fast; its complexity is polynomial in $\log n$.

An even more trivial algorithm can be used to solve the **Discrete Logarithm** problem in $(\mathbb{Z}_n, +)$ when $\alpha = 1$. In this situation, we have that $\log_1 \beta = \beta$ for all $\beta \in \mathbb{Z}_n$.

The **Discrete Logarithm** problem, by definition, takes place in a cyclic (sub)group of order n. It is well known, and almost trivial to prove, that all cyclic groups of order n are isomorphic. By the discussion above, we know how to compute discrete logarithms quickly in the additive group $(\mathbb{Z}_n, +)$. This suggests that we might be able to solve the **Discrete Logarithm** problem in any subgroup $\langle \alpha \rangle$ of order n of any group G by "reducing" the problem to the the easily solved formulation in $(\mathbb{Z}_n, +)$.

Let us think about how (in theory, at least) this could be done. The statement that $\langle \alpha \rangle$ is isomorphic to $(\mathbb{Z}_n, +)$ means that there is a bijection

$$\phi : \langle \alpha \rangle \to \mathbb{Z}_n$$

such that

$$\phi(xy) = (\phi(x) + \phi(y)) \bmod n$$

for all $x, y \in \langle \alpha \rangle$. It follows easily that

$$\phi(\alpha^a) = a\phi(\alpha) \bmod n,$$

so we have that

$$\beta = \alpha^a \Leftrightarrow a\phi(\alpha) \equiv \phi(\beta) \pmod{n}.$$

Hence, solving for a as described above (using the EXTENDED EUCLIDEAN ALGORITHM), we have that

$$\log_\alpha \beta = \phi(\beta)(\phi(\alpha))^{-1} \bmod n.$$

Consequently, if we have an efficient method of computing the isomorphism ϕ, then we would have an efficient algorithm to compute discrete logarithms in $\langle \alpha \rangle$. The catch is that there is no known general method to efficiently compute the isomorphism ϕ for an arbitrary subgroup $\langle \alpha \rangle$ of an arbitrary group G, even though we know the two groups in question are isomorphic. In fact, it is not hard to see that computing discrete logarithms in $\langle \alpha \rangle$ is equivalent to finding an explicit isomorphism between $\langle \alpha \rangle$ and $(\mathbb{Z}_n, +)$. Hence, this approach seems to lead to a dead end.

In view of the fact that an extremely efficient algorithm exists for the **Discrete Logarithm** problem in $(\mathbb{Z}_n, +)$, it is perhaps surprising that there is a nontrivial lower bound on the complexity of the general problem. However, a result of Shoup provides a lower bound on the complexity of generic algorithms for the **Discrete Logarithm** problem. Recall that Shanks' and the rho algorithms have the property that their complexity (in terms of the number of group operations required to run the algorithm) is roughly $\sqrt{n}$, where n is the order of the (sub)group in which the discrete logarithm is being sought. Shoup's result establishes that these algorithms are essentially optimal within the class of generic algorithms.

We begin by giving a precise description of what we mean by a generic algorithm. We consider a cyclic group or subgroup of order n, which is therefore isomorphic to $(\mathbb{Z}_n, +)$. We will study generic algorithms for the **Discrete Logarithm** problem in $(\mathbb{Z}_n, +)$. (As we shall see, the particular group that is used is irrelevant in the context of generic algorithms; the choice of $(\mathbb{Z}_n, +)$ is arbitrary.)

An *encoding* of $(\mathbb{Z}_n, +)$ is any injective mapping $\sigma : \mathbb{Z}_n \to S$, where S is a finite set. The encoding function specifies how group elements are represented. Any discrete logarithm problem in a (sub)group of cardinality n of an arbitrary group G can be specified by defining a suitable encoding function. For example, consider the multiplicative group $(\mathbb{Z}_p^*, \cdot)$, and let α be a primitive element in $\mathbb{Z}_p^*$. Let $n = p - 1$, and define the encoding function σ as follows: $\sigma(i) = \alpha^i \bmod p$, $0 \leq i \leq n - 1$. Then it should be clear that solving the **Discrete Logarithm** problem in $(\mathbb{Z}_p^*, \cdot)$ with respect to the primitive element α is equivalent to solving the **Discrete Logarithm** problem in $(\mathbb{Z}_n, +)$ with generator 1 under the encoding σ.

A generic algorithm is one that works for any encoding. In particular, a generic algorithm must work correctly when the encoding function σ is a random injective function; for example, when $S = \mathbb{Z}_n$ and σ is a random permutation of $\mathbb{Z}_n$. This is very similar to the random oracle model, in which a hash function is regarded as a random function in order to define an idealized model in which a formal security proof can be given.

We suppose that we have a random encoding, σ, for the group $(\mathbb{Z}_n, +)$. In this group, the discrete logarithm of any element a to the base 1 is just a, of course. Given the encoding function σ, the encoding $\sigma(1)$ of the generator, and an encoding of an arbitrary group element $\sigma(a)$, a generic algorithm is trying to compute the value of a. In order to perform operations in this group when group elements are encoded using the function σ, we hypothesize the existence of an oracle (or subroutine) to perform this task.

Given encodings of two group elements, say $\sigma(i)$ and $\sigma(j)$, it should be possible to compute the encodings $\sigma((i + j) \bmod n)$ and $\sigma((i - j) \bmod n)$. This is necessary if we are going to add and subtract group elements, and we assume that our oracle will do this for us. By combining operations of the above type, it is possible to compute arbitrary linear combinations of the form $\sigma((ci \pm dj) \bmod n)$, where $c, d \in \mathbb{Z}_n$. However, using the fact that $-j \equiv n - j \pmod{n}$, we observe that we only need to be able to compute linear combinations of the form $\sigma((ci + dj) \bmod n)$. We will assume that the oracle can directly compute linear combinations of this form in one unit of time.

Group operations of the type described above are the only ones allowed in a generic algorithm. That is, we assume that we have some method of performing group operations on encoded elements, but we cannot do any more than that. Now let us consider how a generic algorithm, say GENLOG, can go about trying to compute a discrete logarithm. The input to the algorithm GENLOG consists of $\sigma_1 = \sigma(1)$ and $\sigma_2 = \sigma(a)$, where $a \in \mathbb{Z}_n$ is chosen randomly. GENLOG will be successful if and only if it outputs the value a. (We will assume that n is prime, in order to simplify the analysis.)

GENLOG will use the oracle to generate a sequence of m, say, encodings of linear combinations of 1 and a. The execution of GENLOG can be specified by a list of ordered pairs $(c_i, d_i) \in \mathbb{Z}_n \times \mathbb{Z}_n$, $1 \le m$. (We can assume that these m ordered pairs are distinct.) For each ordered pair (c_i, d_i), the oracle computes the encoding $\sigma_i = \sigma((c_i + d_i a) \bmod n)$. Note that we can define $(c_1, d_1) = (1, 0)$ and $(c_2, d_2) = (0, 1)$ so our notation is consistent with the input to the algorithm.

In this way, the algorithm GENLOG obtains a list of encoded group elements, $(\sigma_1, \dots, \sigma_m)$. Because the encoding function σ is injective, it follows immediately that $c_i + d_i a \equiv c_j + d_j a \pmod{n}$ if and only if $\sigma_i = \sigma_j$. This provides a method to possibly compute the value of the unknown a: Suppose that $\sigma_i = \sigma_j$ for two integers $i \ne j$. If $d_i = d_j$, then $c_i = c_j$ and the two ordered pairs (c_i, d_i) and (c_j, d_j) are the same. Since we are assuming the ordered pairs are distinct, it follows that $d_i \ne d_j$. Because n is prime, we can compute a as follows:

$$a = (c_i - c_j)(d_j - d_i)^{-1} \bmod n.$$

(Recall that we used a similar method of computing the value of a discrete logarithm in the POLLARD RHO ALGORITHM.)

Suppose first that the algorithm GENLOG chooses a set

$$\mathcal{C} = \{(c_i, d_i) : 1 \le i \le m\} \subseteq \mathbb{Z}_n \times \mathbb{Z}_n$$

of m distinct ordered pairs all at once, at the beginning of the algorithm. Such an algorithm is called a *non-adaptive algorithm* (SHANKS' ALGORITHM is an example of a non-adaptive algorithm). Then the list of m corresponding encodings is obtained from the oracle. Define $\mathsf{Good}(\mathcal{C})$ to consist of all elements $a \in \mathbb{Z}_n$ that are the solution of an equation $a = (c_i - c_j)(d_j - d_i)^{-1} \bmod n$ with $i \ne j$, $i, j \in \{1, \dots, m\}$. By what we have said above, we know that the value of a can be computed by GENLOG if and only if $a \in \mathsf{Good}(\mathcal{C})$. It is clear that $|\mathsf{Good}(\mathcal{C})| \le \binom{m}{2}$, so there are at most $\binom{m}{2}$ elements for which GENLOG can compute the discrete logarithm after having obtained a sequence of m encoded group elements corresponding to the ordered pairs in $\mathcal{C}$. The probability that $a \in \mathsf{Good}(\mathcal{C})$ is at most $\binom{m}{2}/n$.

If $a \notin \mathsf{Good}(\mathcal{C})$, then the best strategy for the algorithm GENLOG is to guess the value of a by choosing a random value in $\mathbb{Z}_n \backslash \mathsf{Good}(\mathcal{C})$. Denote $g = |\mathsf{Good}(\mathcal{C})|$. Then, by conditioning on whether or not $a \in \mathsf{Good}(\mathcal{C})$, we can compute a bound on the success probability of the algorithm. Suppose we define A to be the event $a \in \mathsf{Good}(\mathcal{C})$ and we let B denote the event "the algorithm returns the correct value of a." Then we have that

$$\mathbf{Pr}[B] = \mathbf{Pr}[B|A] \times \mathbf{Pr}[A] + \mathbf{Pr}[B|\overline{A}] \times \mathbf{Pr}[\overline{A}]$$

$$= 1 \times \frac{g}{n} + \frac{1}{n-g} \times \frac{n-g}{n}$$

$$= \frac{g+1}{n}$$

$$\le \frac{\binom{m}{2}+1}{n}.$$

If the algorithm always gives the correct answer, then $\mathbf{Pr}[B] = 1$. In this case, it is easy to see that m is $\Omega(\sqrt{n})$.

A generic discrete logarithm algorithm is not required to choose all the ordered pairs in $\mathcal{C}$ at the beginning of the algorithm, of course. It can choose later pairs after seeing what encodings of previous linear combinations look like (i.e., we allow the algorithm to be an *adaptive algorithm*). However, it can be shown that this does not improve the success probability of the algorithm.

Let GENLOG be an adaptive generic algorithm for the discrete logarithm problem. For $1 \leq i \leq m$, let $\mathcal{C}_i$ consist of the first i ordered pairs, for which the oracle computes the corresponding encodings $\sigma_1, \ldots, \sigma_i$. The set $\mathcal{C}_i$ and the list $\sigma_1, \ldots, \sigma_i$ represent all the information available to GENLOG at time i of its execution.

Now, it can be proven that the value of a can be computed at time i if $a \in$ Good$(\mathcal{C}_i)$. Furthermore, if $a \notin$ Good$(\mathcal{C}_i)$, then a is equally likely to take on any given value in the set $\mathbb{Z}_n \backslash$ Good$(\mathcal{C}_i)$.

From these facts, it can be shown that adaptive generic algorithms have the same success probability as non-adaptive ones. It follows that $\Omega(\sqrt{n})$ is a lower bound on the complexity of any generic algorithm for the **Discrete Logarithm** problem in a (sub)group of prime order n.

6.4   Finite Fields

The *ElGamal Cryptosystem* can be implemented in any group where the **Discrete Logarithm** problem is infeasible. We used the multiplicative group $\mathbb{Z}_p^*$ in the description of Cryptosystem 6.1, but other groups are also suitable candidates. Two such classes of groups are

1. the multiplicative group of the finite field $\mathbb{F}_{p^n}$
2. the group of an elliptic curve defined over a finite field.

We will discuss these two classes of groups in the next sections.

We have already discussed the fact that $\mathbb{Z}_p$ is a field if p is prime. However, there are other examples of finite fields not of this form. In fact, there is a finite field with q elements if $q = p^n$ where p is prime and $n \geq 1$ is an integer. We will now describe very briefly how to construct such a field. First, we need several definitions.

Definition 6.1:   Suppose p is prime. Define $\mathbb{Z}_p[x]$ to be the set of all polynomials in the indeterminate x. By defining addition and multiplication of polynomials in the usual way (and reducing coefficients modulo p), we construct a ring.

For $f(x), g(x) \in \mathbb{Z}_p[x]$, we say that $f(x)$ *divides* $g(x)$ (notation: $f(x) \mid g(x)$) if there exists $q(x) \in \mathbb{Z}_p[x]$ such that

$$g(x) = q(x)f(x).$$

For $f(x) \in \mathbb{Z}_p[x]$, define $\deg(f)$, the *degree* of f, to be the highest exponent in a term of f.

Suppose $f(x), g(x), h(x) \in \mathbb{Z}_p[x]$, and $\deg(f) = n \geq 1$. We define

$$g(x) \equiv h(x) \pmod{f(x)}$$

if

$$f(x) \mid (g(x) - h(x)).$$

Notice the resemblance of the definition of congruence of polynomials to that of congruence of integers.

We are now going to define a ring of polynomials "modulo $f(x)$" which we denote by $\mathbb{Z}_p[x]/(f(x))$. The construction of $\mathbb{Z}_p[x]/(f(x))$ from $\mathbb{Z}_p[x]$ is based on the idea of congruences modulo $f(x)$ and is analogous to the construction of $\mathbb{Z}_m$ from $\mathbb{Z}$.

Suppose $\deg(f) = n$. If we divide $g(x)$ by $f(x)$, we obtain a (unique) *quotient* $q(x)$ and *remainder* $r(x)$, where

$$g(x) = q(x)f(x) + r(x)$$

and

$$\deg(r) < n.$$

This can be done by usual long division of polynomials. Hence any polynomial in $\mathbb{Z}_p[x]$ is congruent modulo $f(x)$ to a unique polynomial of degree at most $n - 1$.

Now we define the elements of $\mathbb{Z}_p[x]/(f(x))$ to be the p^n polynomials in $\mathbb{Z}_p[x]$ of degree at most $n-1$. Addition and multiplication in $\mathbb{Z}_p[x]/(f(x))$ is defined as in $\mathbb{Z}_p[x]$, followed by a reduction modulo $f(x)$. Equipped with these operations, $\mathbb{Z}_p[x]/(f(x))$ is a ring.

Recall that $\mathbb{Z}_m$ is a field if and only if m is prime, and multiplicative inverses can be found using the Euclidean algorithm. A similar situation holds for $\mathbb{Z}_p[x]/(f(x))$. The analog of primality for polynomials is irreducibility, which we define as follows:

Definition 6.2:   A polynomial $f(x) \in \mathbb{Z}_p[x]$ is said to be *irreducible* if there do not exist polynomials $f_1(x), f_2(x) \in \mathbb{Z}_p[x]$ such that

$$f(x) = f_1(x)f_2(x),$$

where $\deg(f_1) > 0$ and $\deg(f_2) > 0$.

A very important fact is that $\mathbb{Z}_p[x]/(f(x))$ is a field if and only if $f(x)$ is irreducible. Further, multiplicative inverses in $\mathbb{Z}_p[x]/(f(x))$ can be computed using a straightforward modification of the (extended) Euclidean algorithm.

Here is an example to illustrate the concepts described above.

Example 6.6   Let's attempt to construct a field having eight elements. This can be done by finding an irreducible polynomial of degree three in $\mathbb{Z}_2[x]$. It is sufficient to consider the polynomials having constant term equal to 1, since any polynomial with constant term 0 is divisible by x and hence is reducible. There are four such polynomials:

$$f_1(x) = x^3 + 1$$
$$f_2(x) = x^3 + x + 1$$
$$f_3(x) = x^3 + x^2 + 1$$
$$f_4(x) = x^3 + x^2 + x + 1.$$

Now, $f_1(x)$ is reducible because

$$x^3 + 1 = (x + 1)(x^2 + x + 1)$$

(remember that all coefficients are to be reduced modulo 2). Also, f_4 is reducible because

$$x^3 + x^2 + x + 1 = (x + 1)(x^2 + 1).$$

However, $f_2(x)$ and $f_3(x)$ are both irreducible, and either one can be used to construct a field having eight elements.

Let us use $f_2(x)$, and thus construct the field $\mathbb{Z}_2[x]/(x^3 + x + 1)$. The eight field elements are the eight polynomials 0, 1, x, $x + 1$, x^2, $x^2 + 1$, $x^2 + x$ and $x^2 + x + 1$.

To compute a product of two field elements, we multiply the two polynomials together, and reduce modulo $x^3 + x + 1$ (i.e., divide by $x^3 + x + 1$ and find the remainder polynomial). Since we are dividing by a polynomial of degree three, the remainder will have degree at most two and hence is an element of the field.

For example, to compute $(x^2 + 1)(x^2 + x + 1)$ in $\mathbb{Z}_2[x]/(x^3 + x + 1)$, we first compute the product in $\mathbb{Z}_2[x]$, which is $x^4 + x^3 + x + 1$. Then we divide by $x^3 + x + 1$, obtaining the expression

$$x^4 + x^3 + x + 1 = (x + 1)(x^3 + x + 1) + x^2 + x.$$

Hence, in the field $\mathbb{Z}_2[x]/(x^3 + x + 1)$, we have that

$$(x^2 + 1)(x^2 + x + 1) = x^2 + x.$$

Below, we present a complete multiplication table for the non-zero field elements. To save space, we write a polynomial $a_2x^2 + a_1x + a_0$ as the ordered triple $a_2a_1a_0$.

	001	010	011	100	101	110	111
001	001	010	011	100	101	110	111
010	010	100	110	011	001	111	101
011	011	110	101	111	100	001	010
100	100	011	111	110	010	101	001
101	101	001	100	010	111	011	110
110	110	111	001	101	011	010	100
111	111	101	010	001	110	100	011

Computation of inverses can be done by using a straightforward adaptation of the extended Euclidean algorithm.

Finally, the multiplicative group of the non-zero polynomials in the field is a cyclic group of order seven. Since 7 is prime, it follows that any field element other than 0 or 1 is a generator of this group, i.e., a primitive element of the field.

For example, if we compute the powers of x, we obtain

$$x^1 = x$$

$$x^2 = x^2$$

$$x^3 = x + 1$$

$$x^4 = x^2 + x$$

$$x^5 = x^2 + x + 1$$

$$x^6 = x^2 + 1$$

$$x^7 = 1,$$

which comprise all the non-zero field elements.

It remains to discuss existence and uniqueness of fields of this type. It can be shown that there is at least one irreducible polynomial of any given degree $n \geq 1$ in $\mathbb{Z}_p[x]$. Hence, there is a finite field with p^n elements for all primes p and all integers $n \geq 1$. There are usually many irreducible polynomials of degree n in $\mathbb{Z}_p[x]$. But the finite fields constructed from any two irreducible polynomials of degree n can be shown to be isomorphic. Thus there is a unique finite field of any size p^n (p prime, $n \geq 1$), which is denoted by $\mathbb{F}_{p^n}$. In the case $n = 1$, the resulting field $\mathbb{F}_p$ is the same thing as $\mathbb{Z}_p$. Finally, it can be shown that there does not exist a finite field with r elements unless $r = p^n$ for some prime p and some integer $n \geq 1$.

We have already noted that the multiplicative group $\mathbb{Z}_p^*$ (p prime) is a cyclic group of order $p - 1$. In fact, the multiplicative group of any finite field is cyclic: $\mathbb{F}_{p^n} \backslash \{0\}$ is a cyclic group of order $p^n - 1$. This provides further examples of cyclic groups in which the **Discrete Logarithm** problem can be studied.

In practice, the finite fields $\mathbb{F}_{2^n}$ have been most studied. Any generic algorithm works in a field $\mathbb{F}_{2^n}$, of course. More importantly, however, the INDEX CALCULUS ALGORITHM can be modified in a straightforward manner to work in these fields. Recall that the main steps in the INDEX CALCULUS ALGORITHM involve trying to factor elements in $\mathbb{Z}_p$ over a given factor base which consists of small primes. The analog of a factor base in $\mathbb{Z}_2[x]$ is a set of irreducible polynomials of low degree. The idea then is to try to factor elements in $\mathbb{F}_{2^n}$ into polynomials in the given factor base. The reader can easily fill in the details.

Once the appropriate modifications have been made, the precomputation time of the INDEX CALCULUS ALGORITHM in $\mathbb{F}_{2^n}$ turns out to be

$$O\left(e^{(1.405+o(1))n^{1/3}(\ln n)^{2/3}}\right),$$

and the time to find an individual discrete logarithm is

$$O\left(e^{(1.098+o(1))n^{1/3}(\ln n)^{2/3}}\right).$$

This algorithm was successfully used by Thomé in 2001 to compute discrete logarithms in $\mathbb{F}_{2^{607}}$. For large values of n (say $n > 1024$), the **Discrete Logarithm problem** in $\mathbb{F}_{2^n}$ is thought to be infeasible at present, provided that $2^n - 1$ has at least one "large" prime factor (in order to thwart a Pohlig-Hellman attack).

6.5   Elliptic Curves

Elliptic curves are described by the set of solutions to certain equations in two variables. Elliptic curves defined modulo a prime p are of central importance in public-key cryptography. We begin by looking briefly at elliptic curves defined over the real numbers, because some of the basic concepts are easier to motivate in this setting.

6.5.1 Elliptic Curves over the Reals

> **Definition 6.3:** Let $a, b \in \mathbb{R}$ be constants such that $4a^3 + 27b^2 \neq 0$. A *non-singular elliptic curve* is the set E of solutions $(x, y) \in \mathbb{R} \times \mathbb{R}$ to the equation
>
> $$y^2 = x^3 + ax + b, \tag{6.4}$$
>
> together with a special point $\mathcal{O}$ called the *point at infinity*.

In Figure 6.1, we depict the elliptic curve $y^2 = x^3 - 4x$.

It can be shown that the condition $4a^3 + 27b^2 \neq 0$ is necessary and sufficient to ensure that the equation $x^3 + ax + b = 0$ has three distinct roots (which may be real or complex numbers). If $4a^3 + 27b^2 = 0$, then the corresponding elliptic curve is called a *singular elliptic curve*.

Suppose E is a non-singular elliptic curve. We will define a binary operation over E which makes E into an abelian group. This operation is usually denoted by addition. The point at infinity, $\mathcal{O}$, will be the identity element, so $P + \mathcal{O} = \mathcal{O} + P = P$ for all $P \in E$.

Suppose $P, Q \in E$, where $P = (x_1, y_1)$ and $Q = (x_2, y_2)$. We consider three cases:

1. $x_1 \neq x_2$
2. $x_1 = x_2$ and $y_1 = -y_2$
3. $x_1 = x_2$ and $y_1 = y_2$

In case 1, we define L to be the line through P and Q. L intersects E in the two points P and Q, and it is easy to see that L will intersect E in one further point, which we call R'. If we reflect R' in the x-axis, then we get a point which we name R. We define $P + Q = R$.

Let's work out an algebraic formula to compute R. First, the equation of L is $y = \lambda x + \nu$, where the slope of L is

$$\lambda = \frac{y_2 - y_1}{x_2 - x_1},$$

and

$$\nu = y_1 - \lambda x_1 = y_2 - \lambda x_2.$$

In order to find the points in $E \cap L$, we substitute $y = \lambda x + \nu$ into the equation for E, obtaining the following:

$$(\lambda x + \nu)^2 = x^3 + ax + b,$$

which is the same as

$$x^3 - \lambda^2 x^2 + (a - 2\lambda\nu)x + b - \nu^2 = 0. \tag{6.5}$$

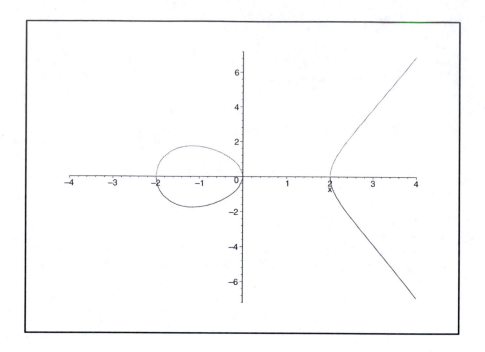

FIGURE 6.1
An elliptic curve over the reals

The roots of equation (6.5) are the x-co-ordinates of the points in $E \cap L$. We already know two points in $E \cap L$, namely, P and Q. Hence x_1 and x_2 are two roots of equation (6.5).

Since equation (6.5) is a cubic equation over the reals having two real roots, the third root, say x_3, must also be real. The sum of the three roots must be the negative of the coefficient of the quadratic term, or λ^2. Therefore

$$x_3 = \lambda^2 - x_1 - x_2.$$

x_3 is the x-co-ordinate of the point R'. We will denote the y-co-ordinate of R' by $-y_3$, so the y-co-ordinate of R will be y_3. An easy way to compute y_3 is to use the fact that the slope of L, namely λ, is determined by any two points on L. If we use the points (x_1, y_1) and $(x_3, -y_3)$ to compute this slope, we get

$$\lambda = \frac{-y_3 - y_1}{x_3 - x_1},$$

or

$$y_3 = \lambda(x_1 - x_3) - y_1.$$

Therefore we have derived a formula for $P + Q$ in case 1: if $x_1 \neq x_2$, then $(x_1, y_1) + (x_2, y_2) = (x_3, y_3)$, where

$$x_3 = \lambda^2 - x_1 - x_2,$$

$$y_3 = \lambda(x_1 - x_3) - y_1, \quad \text{and}$$

$$\lambda = \frac{y_2 - y_1}{x_2 - x_1}.$$

Case 2, where $x_1 = x_2$ and $y_1 = -y_2$, is simple: we define $(x, y) + (x, -y) = \mathcal{O}$ for all $(x, y) \in E$. Therefore (x, y) and $(x, -y)$ are inverses with respect to the elliptic curve addition operation.

Case 3 remains to be considered. Here we are adding a point $P = (x_1, y_1)$ to itself. We can assume that $y_1 \neq 0$, for then we would be in case 2. Case 3 is handled much like case 1, except that we define L to be the tangent to E at the point P. A little bit of calculus makes the computation quite simple. The slope of L can be computed using implicit differentiation of the equation of E:

$$2y \frac{dy}{dx} = 3x^2 + a.$$

Substituting $x = x_1, y = y_1$, we see that the slope of the tangent is

$$\lambda = \frac{3x_1{}^2 + a}{2y_1}.$$

The rest of the analysis in this case is the same as in case 1. The formula obtained is identical, except that λ is computed differently.

At this point the following properties of the addition operation, as defined above, should be clear:

1. addition is closed on the set E,

2. addition is commutative,

3. $\mathcal{O}$ is an identity with respect to addition, and

4. every point on E has an inverse with respect to addition.

In order to show that $(E, +)$ is an abelian group, it still must be proven that addition is associative. This is quite messy to prove by algebraic methods. The proof of associativity can be made simpler by using some results from geometry; however, we will not discuss the proof here.

6.5.2 Elliptic Curves Modulo a Prime

Let $p > 3$ be prime. Elliptic curves over $\mathbb{Z}_p$ can be defined exactly as they were over the reals (and the addition operation is also defined in an identical fashion) provided that all operations over $\mathbb{R}$ are replaced by analogous operations in $\mathbb{Z}_p$.

Definition 6.4:   Let $p > 3$ be prime. The *elliptic curve* $y^2 = x^3 + ax + b$ over $\mathbb{Z}_p$ is the set of solutions $(x, y) \in \mathbb{Z}_p \times \mathbb{Z}_p$ to the congruence

$$y^2 \equiv x^3 + ax + b \ (\text{mod } p), \qquad\qquad (6.6)$$

where $a, b \in \mathbb{Z}_p$ are constants such that $4a^3 + 27b^2 \not\equiv 0 \ (\text{mod } p)$, together with a special point $\mathcal{O}$ called the *point at infinity*.

The addition operation on E is defined as follows (where all arithmetic operations are performed in $\mathbb{Z}_p$): Suppose

$$P = (x_1, y_1)$$

and

$$Q = (x_2, y_2)$$

are points on E. If $x_2 = x_1$ and $y_2 = -y_1$, then $P + Q = \mathcal{O}$; otherwise $P + Q = (x_3, y_3)$, where

$$x_3 = \lambda^2 - x_1 - x_2$$

$$y_3 = \lambda(x_1 - x_3) - y_1,$$

and

$$\lambda = \begin{cases} (y_2 - y_1)(x_2 - x_1)^{-1}, & \text{if } P \neq Q \\ (3x_1{}^2 + a)(2y_1)^{-1}, & \text{if } P = Q. \end{cases}$$

Finally, define

$$P + \mathcal{O} = \mathcal{O} + P = P$$

for all $P \in E$.

Note that the addition of points on an elliptic curve over $\mathbb{Z}_p$ does not have the nice geometric interpretation that it does on an elliptic curve over the reals. However, the same formulas can be used to define addition, and the resulting pair $(E, +)$ still forms an abelian group.

Let us look at a small example.

Example 6.7   Let E be the elliptic curve $y^2 = x^3 + x + 6$ over $\mathbb{Z}_{11}$. Let's first determine the points on E. This can be done by looking at each possible $x \in \mathbb{Z}_{11}$, computing $x^3 + x + 6 \bmod 11$, and then trying to solve equation (6.6) for y. For a given x, we can test to see if $z = x^3 + x + 6 \bmod 11$ is a quadratic residue by applying Euler's criterion. Recall that there is an explicit formula to compute square roots of quadratic residues modulo p for primes $p \equiv 3 \ (\text{mod } 4)$. Applying this formula, we have that the square roots of a quadratic residue z are

$$\pm z^{(11+1)/4} \bmod 11 = \pm z^3 \bmod 11.$$

TABLE 6.1
Points on the elliptic curve $y^2 = x^3 + x + 6$ over $\mathbb{Z}_{11}$

x	$x^3 + x + 6 \bmod 11$	quadratic residue?	y
0	6	no	
1	8	no	
2	5	yes	4, 7
3	3	yes	5, 6
4	8	no	
5	4	yes	2, 9
6	8	no	
7	4	yes	2, 9
8	9	yes	3, 8
9	7	no	
10	4	yes	2, 9

The results of these computations are tabulated in Table 6.1.

E has 13 points on it. Since any group of prime order is cyclic, it follows that E is isomorphic to $\mathbb{Z}_{13}$, and any point other than the point at infinity is a generator of E. Suppose we take the generator $\alpha = (2,7)$. Then we can compute the "powers" of α (which we will write as multiples of α, since the group operation is additive). To compute $2\alpha = (2,7) + (2,7)$, we first compute

$$\lambda = (3 \times 2^2 + 1)(2 \times 7)^{-1} \bmod 11$$

$$= 2 \times 3^{-1} \bmod 11$$

$$= 2 \times 4 \bmod 11$$

$$= 8.$$

Then we have

$$x_3 = 8^2 - 2 - 2 \bmod 11$$

$$= 5$$

and

$$y_3 = 8(2 - 5) - 7 \bmod 11$$

$$= 2,$$

so $2\alpha = (5,2)$.

The next multiple would be $3\alpha = 2\alpha + \alpha = (5,2) + (2,7)$. Again, we begin by computing λ, which in this situation is done as follows:

$$\lambda = (7 - 2)(2 - 5)^{-1} \bmod 11$$
$$= 5 \times 8^{-1} \bmod 11$$
$$= 5 \times 7 \bmod 11$$
$$= 2.$$

Then we have

$$x_3 = 2^2 - 5 - 2 \bmod 11$$
$$= 8$$

and

$$y_3 = 2(5 - 8) - 2 \bmod 11$$
$$= 3,$$

so $3\alpha = (8, 3)$.

Continuing in this fashion, the remaining multiples can be computed to be the following:

α	$=$	$(2, 7)$	2α	$=$	$(5, 2)$	3α	$=$	$(8, 3)$
4α	$=$	$(10, 2)$	5α	$=$	$(3, 6)$	6α	$=$	$(7, 9)$
7α	$=$	$(7, 2)$	8α	$=$	$(3, 5)$	9α	$=$	$(10, 9)$
10α	$=$	$(8, 8)$	11α	$=$	$(5, 9)$	12α	$=$	$(2, 4)$

Hence, as we already knew, $\alpha = (2, 7)$ is indeed a primitive element. ◻

Let's now look at an example of ElGamal encryption and decryption using the elliptic curve of Example 6.7. Since the group operation of an elliptic curve is written additively, we will need to translate the operations in Cryptosystem 6.1 into additive notation.

Example 6.8   Suppose that $\alpha = (2, 7)$ and Bob's private key is 7, so

$$\beta = 7\alpha = (7, 2).$$

Thus the encryption operation is

$$e_K(x, k) = (k(2, 7), x + k(7, 2)),$$

where $x \in E$ and $0 \le k \le 12$, and the decryption operation is

$$d_K(y_1, y_2) = y_2 - 7y_1.$$

Suppose that Alice wishes to encrypt the plaintext $x = (10, 9)$ (which is a point on E). If she chooses the random value $k = 3$, then she will compute

$$y_1 = 3\,(2, 7)$$
$$= (8, 3)$$

and

$$y_2 = (10, 9) + 3\,(7, 2)$$
$$= (10, 9) + (3, 5)$$
$$= (10, 2).$$

Hence, $y = ((8, 3), (10, 2))$. Now, if Bob receives the ciphertext y, he decrypts it as follows:

$$x = (10, 2) - 7(8, 3)$$
$$= (10, 2) - (3, 5)$$
$$= (10, 2) + (3, 6)$$
$$= (10, 9).$$

Hence, the decryption yields the correct plaintext. $\square$

6.5.3   Properties of Elliptic Curves

An elliptic curve E defined over $\mathbb{Z}_p$ (p prime, $p > 3$) will have roughly p points on it. More precisely, a well-known theorem due to Hasse asserts that the number of points on E, which we denote by $\#E$, satisfies the following inequality

$$p + 1 - 2\sqrt{p} \leq \#E \leq p + 1 + 2\sqrt{p}.$$

Computing the exact value of $\#E$ is more difficult, but there is an efficient algorithm to do this, due to Schoof. (By "efficient" we mean that it has a running time that is polynomial in $\log p$. Schoof's algorithm has a running time of $O((\log p)^8)$ bit operations ($O((\log p)^6)$ operations in $\mathbb{Z}_p$) and is practical for primes p having several hundred digits.)

Now, given that we can compute $\#E$, we further want to find a cyclic subgroup of E in which the **Discrete Logarithm** problem is intractable. So we would like to know something about the structure of the group E. The following theorem gives a considerable amount of information on the group structure of E.

THEOREM 6.1   *Let E be an elliptic curve defined over $\mathbb{Z}_p$, where p is prime and $p > 3$. Then there exist positive integers n_1 and n_2 such that $(E, +)$ is isomorphic to $\mathbb{Z}_{n_1} \times \mathbb{Z}_{n_2}$. Further, $n_2 \mid n_1$ and $n_2 \mid (p - 1)$.*

Note that $n_2 = 1$ is allowed in the above theorem. In fact, $n_2 = 1$ if and only if E is a cyclic group. Also, if $\#E$ is a prime, or the product of distinct primes, then E must be a cyclic group.

In any event, if the integers n_1 and n_2 are computed, then we know that $(E, +)$ has a cyclic subgroup isomorphic to $\mathbb{Z}_{n_1}$ that can potentially be used as a setting for an *ElGamal Cryptosystem*.

Generic algorithms apply to the elliptic curve **Discrete Logarithm** problem, but there is no known adaptation of the INDEX CALCULUS ALGORITHM to the setting of elliptic curves. However, there is a method of exploiting an explicit isomorphism between elliptic curves and finite fields that leads to efficient algorithms for certain classes of elliptic curves. This technique, due to Menezes, Okamoto and Vanstone, can be applied to some particular examples within a special class of elliptic curves called supersingular curves that were suggested for use in cryptosystems.

Another class of weak elliptic curves are the so-called curves of "trace one." These are elliptic curves defined over $\mathbb{Z}_p$ (where p is prime) having exactly p points on them. The elliptic curve **Discrete Logarithm** problem can easily be solved on these elliptic curves.

If the classes of curves described above are avoided, however, then it appears that an elliptic curve having a cyclic subgroup of size about 2^{160} will provide a secure setting for a cryptosystem, provided that the order of the subgroup is divisible by at least one large prime factor (again, to guard against a Pohlig-Hellman attack).

6.5.4   Point Compression and the ECIES

There are some practical difficulties in implementing an *ElGamal Cryptosystem* on an elliptic curve. When implemented in $\mathbb{Z}_p$, the *ElGamal Cryptosystem* has a message expansion factor of two. An elliptic curve implementation has a message expansion factor of (about) four. This happens since there are approximately p plaintexts, but each ciphertext consists of four field elements. However, a more serious problem is that the plaintext space consists of the points on the curve E, and there is no convenient method known of deterministically generating points on E.

A more efficient ElGamal-type system is used in the so-called *ECIES* (*Elliptic Curve Integrated Encryption Scheme*). *ECIES* has a fairly complicated description that incorporates a symmetric-key encryption as well as a message authentication code; we present a simplification which basically implements the elliptic curve based ElGamal public-key encryption scheme used in *ECIES*. In this variation, the x-co-ordinate of a point on an elliptic curve is used for "masking," and a plaintext is allowed to be an arbitrary (nonzero) field element (i.e., it is not required to be a point on E).

We also use one other standard trick, called *point compression*, which reduces the storage requirement for points on elliptic curves. A (non-infinite) point on an

---

Algorithm 6.4: POINT-DECOMPRESS(x, i)

$z \leftarrow x^3 + ax + b \bmod p$
if z is a quadratic non-residue modulo p
  then return ("failure")
  else $\begin{cases} y \leftarrow \sqrt{z} \bmod p \\ \textbf{if } y \equiv i \pmod{2} \\ \quad \textbf{then return } (x, y) \\ \quad \textbf{else return } (x, p - y) \end{cases}$

---

elliptic curve E is a pair (x, y), where $y^2 \equiv x^3 + ax + b \pmod{p}$. Given a value for x, there are two possible values for y (unless $x^3 + ax + b \equiv 0 \pmod{p}$). These two possible y-values are negatives of each other modulo p. Since p is odd, one of the two possible values of $y \bmod p$ is even and the other is odd. Therefore we can determine a unique point $P = (x, y)$ on E by specifying the value of x, together with the single bit $y \bmod 2$. This reduces the storage by (almost) 50%, at the expense of requiring additional computations to reconstruct the y-co-ordinate of P.

The operation of point compression can be expressed as a function

$$\text{POINT-COMPRESS} : E \backslash \{\mathcal{O}\} \to \mathbb{Z}_p \times \mathbb{Z}_2,$$

which is defined as follows:

$$\text{POINT-COMPRESS}(P) = (x, y \bmod 2), \text{ where } P = (x, y) \in E.$$

The inverse operation, POINT-DECOMPRESS, reconstructs the elliptic curve point $P = (x, y)$ from $(x, y \bmod 2)$. It can be implemented as shown in Algorithm 6.4.

As previously mentioned, $\sqrt{z}$ can be computed as $z^{(p+1)/4} \bmod p$ provided that $p \equiv 3 \pmod{4}$ and z is a quadratic residue modulo p (or $z = 0$).

The cryptosystem that we call *Simplified ECIES* is presented as Cryptosystem 6.2.

Simplified ECIES has a message expansion (approximately) equal to two, which is similar to the *ElGamal Cryptosystem* over $\mathbb{Z}_p^*$. We illustrate encryption and decryption in *Simplified ECIES* using the curve $y^2 = x^3 + x + 6$ defined over $\mathbb{Z}_{11}$.

Example 6.9 As in the previous example, suppose that $P = (2, 7)$ and Bob's private key is 7, so

$$Q = 7P = (7, 2).$$

Cryptosystem 6.2: *Simplified ECIES*

Let E be an elliptic curve defined over $\mathbb{Z}_p$ ($p > 3$ prime) such that E contains a cyclic subgroup $H = \langle P \rangle$ of prime order n in which the **Discrete Logarithm** problem is infeasible.

Let $\mathcal{P} = \mathbb{Z}_p{}^*$, $\mathcal{C} = (\mathbb{Z}_p \times \mathbb{Z}_2) \times \mathbb{Z}_p{}^*$, and define

$$\mathcal{K} = \{(E, P, m, Q, n) : Q = mP\}.$$

The values P, Q and n are the public key, and $m \in \mathbb{Z}_n{}^*$ is the private key.

For $K = (E, P, m, Q, n)$, for a (secret) random number $k \in \mathbb{Z}_n{}^*$, and for $x \in \mathbb{Z}_p{}^*$, define

$$e_K(x, k) = (\text{POINT-COMPRESS}(kP), xx_0 \bmod p),$$

where $kQ = (x_0, y_0)$ and $x_0 \neq 0$.

For a ciphertext $y = (y_1, y_2)$, where $y_1 \in \mathbb{Z}_p \times \mathbb{Z}_2$ and $y_2 \in \mathbb{Z}_p{}^*$, define

$$d_K(y) = y_2(x_0)^{-1} \bmod p,$$

where

$$(x_0, y_0) = m\,\text{POINT-DECOMPRESS}(y_1).$$

Suppose Alice wants to encrypt the plaintext $x = 9$, and she chooses the random value $k = 6$. First, she computes

$$kP = 6\,(2, 7) = (7, 9)$$

and

$$kQ = 6\,(7, 2) = (8, 3),$$

so $x_0 = 8$.

Next, she calculates

$$y_1 = \text{POINT-COMPRESS}(7, 9) = (7, 1)$$

and

$$y_2 = 8 \times 9 \bmod 11 = 6.$$

The ciphertext she sends to Bob is

$$y = (y_1, y_2) = ((7, 1), 6).$$

When Bob receives the ciphertext y, he computes

$$\text{POINT-DECOMPRESS}(7, 1) = (7, 9),$$

$$7(7, 9) = (8, 3), \quad \text{and}$$

$$6 \times 8^{-1} \bmod 11 = 9.$$

Hence, the decryption yields the correct plaintext, 9. ⬜

6.5.5 Computing Point Multiples on Elliptic Curves

We can compute powers α^a in a multiplicative group efficiently using the SQUARE-AND-MULTIPLY ALGORITHM (Algorithm 5.5). In an elliptic curve setting, where the group operation is written additively, we would compute a multiple aP of an elliptic curve point P using an analogous DOUBLE-AND-ADD ALGORITHM. (The squaring operation $\alpha \mapsto \alpha^2$ would be replaced by the doubling operation $P \mapsto 2P$, and the multiplication of two group elements would be replaced by the addition of two elliptic curve points.)

The addition operation on an elliptic curve has the property that additive inverses are very easy to compute. This fact can be exploited in a generalization of the DOUBLE-AND-ADD ALGORITHM which we might call the DOUBLE-AND-(ADD OR SUBTRACT) ALGORITHM. We describe this technique now.

Let c be an integer. A *signed binary representation* of c is an equation of the form

$$c = \sum_{i=0}^{\ell-1} c_i 2^i,$$

where $c_i \in \{-1, 0, 1\}$ for all i. In general, there will be more than one signed binary representation of an integer c. For example, we have that

$$11 = 8 + 2 + 1 = 16 - 4 - 1,$$

so

$$(c_4, c_3, c_2, c_1, c_0) = (0, 1, 0, 1, 1) \quad \text{or} \quad (1, 0, -1, 0, -1)$$

are both signed binary representations of 11.

Let P be a point of order n on an elliptic curve. Given any signed binary representation $(c_{\ell-1}, \ldots, c_0)$ of an integer c, where $0 \leq c \leq n - 1$, it is possible to compute the multiple cP of the elliptic curve point P by a series of doublings, additions and subtractions, using the following algorithm.

In Algorithm 6.5, the subtraction operation $Q - P$ would be performed by first computing the additive inverse $-P$ of P, and then adding the result to Q.

A signed binary representation $(c_{\ell-1}, \ldots, c_0)$ of an integer c is said to be in *non-adjacent form* provided that no two consecutive c_i's are non-zero. Such a representation is denoted as a *NAF* representation. It is a simple matter to transform a binary representation of a positive integer c into a NAF representation. The

---

Algorithm 6.5: DOUBLE-AND-(ADD OR SUBTRACT)$(P, (c_{\ell-1}, \ldots, c_0))$

$Q \leftarrow \mathcal{O}$
for $i \leftarrow \ell - 1$ **downto** 0

$\quad$**do** $\begin{cases} Q \leftarrow 2Q \\ \textbf{if } c_i = 1 \\ \quad \textbf{then } Q \leftarrow Q + P \\ \quad \textbf{else if } c_i = -1 \\ \quad \quad \textbf{then } Q \leftarrow Q - P \end{cases}$

return (Q)

---

basis of this transformation is to replace substrings of the form $(0, 1, \cdots, 1, 1)$ in the binary representation by $(1, 0, \cdots, 0, -1)$. Substitutions of this type do not change the value of c, due to the identity

$$2^i + 2^{i-1} + \cdots + 2^j = 2^{i+1} - 2^j,$$

where $i > j$. This process is repeated as often as needed, starting with the right-most (i.e., low-order) bits and proceeding to the left.

We illustrate the above-described process with an example:

	1	1	1	1	0	0	1	1	0	1	1	1
						↓						
	1	1	1	1	0	0	1	1	1	0	0	-1
						↓						
	1	1	1	1	0	1	0	0	-1	0	0	-1
						↓						
1	0	0	0	-1	0	1	0	0	-1	0	0	-1

Hence the NAF representation of

$$(1, 1, 1, 1, 0, 0, 1, 1, 0, 1, 1, 1)$$

is

$$(1, 0, 0, 0, -1, 0, 1, 0, 0, -1, 0, 0, -1).$$

This discussion establishes that every non-negative integer has a NAF representation. It is also possible to prove that the NAF representation of an integer is unique (see the Exercises). Therefore we can speak of the NAF representation of an integer without ambiguity.

In a NAF representation, there do not exist two consecutive non-zero coefficients. We might expect that, on average, a NAF representation contains more zeroes than the traditional binary representation of a positive integer. This is indeed the case: it can be shown that, on average, an ℓ-bit integer contains $\ell/2$ zeroes in its binary representation and $2\ell/3$ zeroes in its NAF representation.

These results make it easy to compare the average efficiency of the DOUBLE-AND-ADD ALGORITHM using a binary representation to the DOUBLE-AND-(ADD OR SUBTRACT) ALGORITHM using the NAF representation. Each algorithm requires ℓ doublings, but the number of additions (or subtractions) is $\ell/2$ in the first case, and $\ell/3$ in the second case. If we assume that a doubling takes roughly the same amount of time as an addition (or subtraction), then the ratio of the average times required by the two algorithms is approximately

$$\frac{\ell + \frac{\ell}{2}}{\ell + \frac{\ell}{3}} = \frac{9}{8}.$$

We have therefore obtained a (roughly) 11% speedup, on average, by this simple technique.

6.6   Discrete Logarithm Algorithms in Practice

The most important settings (G, α) for the **Discrete Logarithm** problem in cryptographic applications are the following:

1. $G = (\mathbb{Z}_p^{\,*}, \cdot)$, p prime, α a primitive element modulo p;

2. $G = (\mathbb{Z}_p^{\,*}, \cdot)$, p, q prime, $p \equiv 1 \bmod q$, α an element in $\mathbb{Z}_p$ having order q;

3. $G = (\mathbb{F}_{2^n}^{\,*}, \cdot)$, α a primitive element in $\mathbb{F}_{2^n}^{\,*}$;

4. $G = (E, +)$, where E is an elliptic curve modulo a prime p, $\alpha \in E$ is a point having prime order $q = \#E/h$, where (typically) $h = 1, 2$ or 4; and

5. $G = (E, +)$, where E is an elliptic curve over a finite field $\mathbb{F}_{2^n}$, $\alpha \in E$ is a point having prime order $q = \#E/h$, where (typically) $h = 2$ or 4. (Note that we have defined ellipic curve over finite fields $\mathbb{F}_p$ only when p is a prime exceeding 3. Elliptic curves can be defined over any finite field, though a different equation is required if the field has characteristic 2 or 3.)

Cases 1, 2 and 3 can be attacked using the appropriate form of the INDEX CALCULUS ALGORITHM in $(\mathbb{Z}_p^{\,*}, \cdot)$ or $(\mathbb{F}_{2^n}^{\,*}, \cdot)$. Cases 2, 4 and 5 can be attacked using POLLARD RHO ALGORITHMS in subgroups of order q.

We briefly report some comparative security estimates due to Lenstra and Verheul. In order for an elliptic curve discrete logarithm based cryptosystem to be secure until the year 2020, it has been suggested that one should take $p \approx 2^{160}$ in case 4 (or $n \approx 160$ in case 5). In contrast, p needs to be at least 2^{1880} in cases 1 and 2 to achieve the same (predicted) level of security. The reason for this significant difference is the lack of a known index calculus attack on elliptic curve discrete logarithms. As a consequence, elliptic curve cryptography has become increasingly popular for practical applications, especially for applications on constrained platforms such as wireless devices and smart cards. On platforms such as these, available memory is very small, and a secure implementation of discrete logarithm based cryptography in $(\mathbb{Z}_p^*, \cdot)$, for example, would require too much space to be practical. The smaller space required for elliptic curve based cryptography is therefore very desirable.

The above estimates are conjectures that are based on the best currently implemented algorithms, as well as some reasonable hypotheses concerning possible progress in algorithm development and computing speed in the coming years. It is also of interest to look at the current state-of-the-art of algorithms for the **Discrete Logarithm** problem. Due to the practical importance of elliptic curve cryptography, algorithms for the **Discrete Logarithm** problem on elliptic curves have received the most attention in recent years. *Certicom Corporation* has issued a series of "challenges" to encourage the development of efficient implementations of discrete logarithm algorithms. The most recent challenges to be solved were the 109-bit challenges known as ECCp-109, which was solved in November, 2002; and ECC2-109, which was solved in April, 2004. Both of these challenges were solved by a team led by C. Monico.

6.7   Security of ElGamal Systems

In this section, we study several aspects of the security of ElGamal-type cryptosystems. First, we look at the bit security of discrete logarithms. Then we consider the semantic security of ElGamal-type cryptosystems, and introduce the **Diffie-Hellman** problems.

6.7.1   Bit Security of Discrete Logarithms

In this section, we consider whether individual bits of a discrete logarithm are easy or hard to compute. To be precise, consider Problem 6.2, which we call the **Discrete Logarithm ith Bit** problem (the setting for the discrete logarithms considered in this section is $(\mathbb{Z}_p^*, \cdot)$, where p is prime).

Problem 6.2:   **Discrete Logarithm ith Bit**

Instance: $I = (p, \alpha, \beta, i)$, where p is prime, $\alpha \in \mathbb{Z}_p^*$ is a primitive element, $\beta \in \mathbb{Z}_p^*$, and i is an integer such that $1 \leq i \leq \lceil \log_2(p-1) \rceil$.

Question: Compute $L_i(\beta)$, which (for the specified α and p) denotes the ith least significant bit in the binary representation of $\log_\alpha \beta$.

We will first show that computing the least significant bit of a discrete logarithm is easy. In other words, if $i = 1$, then the **Discrete Logarithm** i**th Bit** problem can be solved efficiently. This follows from Euler's criterion concerning quadratic residues modulo p, where p is prime.

Consider the mapping $f : \mathbb{Z}_p^* \to \mathbb{Z}_p^*$ defined by

$$f(x) = x^2 \bmod p.$$

Denote by $\mathsf{QR}(p)$ the set of quadratic residues modulo p; thus

$$\mathsf{QR}(p) = \{x^2 \bmod p : x \in \mathbb{Z}_p^*\}.$$

First, observe that $f(x) = f(p - x)$. Next, note that

$$w^2 \equiv x^2 \pmod p$$

if and only if

$$p \mid (w - x)(w + x),$$

which happens if and only if

$$w \equiv \pm x \pmod p.$$

It follows that

$$|f^{-1}(y)| = 2$$

for every $y \in \mathsf{QR}(p)$, and hence

$$|\mathsf{QR}(p)| = \frac{p - 1}{2}.$$

That is, exactly half the residues in $\mathbb{Z}_p^*$ are quadratic residues and half are not.

Now, suppose α is a primitive element of $\mathbb{Z}_p$. Then $\alpha^a \in \mathsf{QR}(p)$ if a is even. Since the $(p - 1)/2$ elements $\alpha^0 \bmod p, \alpha^2 \bmod p, \ldots, \alpha^{p-3} \bmod p$ are all distinct, it follows that

$$\mathsf{QR}(p) = \{\alpha^{2i} \bmod p : 0 \le i \le (p - 3)/2\}.$$

Hence, β is a quadratic residue if and only if $\log_\alpha \beta$ is even, that is, if and only if $L_1(\beta) = 0$. But we already know, by Euler's criterion, that β is a quadratic residue if and only if

$$\beta^{(p-1)/2} \equiv 1 \pmod p.$$

So we have the following efficient formula to calculate $L_1(\beta)$:

$$L_1(\beta) = \begin{cases} 0 & \text{if } \beta^{(p-1)/2} = 1 \pmod p \\ 1 & \text{otherwise.} \end{cases}$$

Let's now consider the computation of $L_i(\beta)$ for values of i exceeding 1. Suppose $p - 1 = 2^s t$, where t is odd. It can be shown that it is easy to compute $L_i(\beta)$

if $i \leq s$. On the other hand, computing $L_{s+1}(\beta)$ is (probably) difficult, in the sense that any hypothetical algorithm (or oracle) to compute $L_{s+1}(\beta)$ could be used to find discrete logarithms in $\mathbb{Z}_p$.

We shall prove this result in the case $s = 1$. More precisely, if $p \equiv 3 \pmod 4$ is prime, then we show how any oracle for computing $L_2(\beta)$ can be used to solve the **Discrete Logarithm** problem in $\mathbb{Z}_p$.

Recall that, if β is a quadratic residue in $\mathbb{Z}_p$ and $p \equiv 3 \pmod 4$, then the two square roots of β modulo p are $\pm\beta^{(p+1)/4} \bmod p$. It is also important that, for any $\beta \neq 0$,

$$L_1(\beta) \neq L_1(p - \beta)$$

if $p \equiv 3 \pmod 4$. We see this as follows. Suppose

$$\alpha^a \equiv \beta \pmod p;$$

then

$$\alpha^{a+(p-1)/2} \equiv -\beta \pmod p.$$

Since $p \equiv 3 \pmod 4$, the integer $(p-1)/2$ is odd, and the result follows.

Now, suppose that $\beta = \alpha^a$ for some (unknown) even exponent a. Then either

$$\beta^{(p+1)/4} \equiv \alpha^{a/2} \pmod p$$

or

$$-\beta^{(p+1)/4} \equiv \alpha^{a/2} \pmod p.$$

We can determine which of these two possibilities is correct if we know the value $L_2(\beta)$, since

$$L_2(\beta) = L_1(\alpha^{a/2}).$$

This fact is exploited in our algorithm, which we present as Algorithm 6.6.

At the end of Algorithm 6.6, the x_i's comprise the bits in the binary representation of $\log_\alpha \beta$; that is,

$$\log_\alpha \beta = \sum_{i \geq 0} x_i 2^i.$$

We will work out a small example to illustrate the algorithm.

Example 6.10   Suppose $p = 19$, $\alpha = 2$ and $\beta = 6$. Since the example is so small, we can tabulate the values of $L_1(\gamma)$ and $L_2(\gamma)$ for all $\gamma \in \mathbb{Z}_{19}^*$. (In general, L_1 can be computed efficiently using Euler's criterion, and L_2 is is computed using the hypothetical algorithm ORACLEL2.) These values are given in Table 6.2. Algorithm 6.6 then proceeds as shown in Figure 6.2.

The result is that $\log_2 6 = 1110_2 = 14$, as can easily be verified.   □

Algorithm 6.6: L_2Oracle-Discrete-Logarithm(p, α, β)

external L_1, OracleL_2
$x_0 \leftarrow L_1(\beta)$
$\beta \leftarrow \beta/\alpha^{x_0} \bmod p$
$i \leftarrow 1$
while $\beta \neq 1$
$\quad$ **do** $\begin{cases} x_i \leftarrow \text{Oracle}L_2(\beta) \\ \gamma \leftarrow \beta^{(p+1)/4} \bmod p \\ \textbf{if } L_1(\gamma) = x_i \\ \quad \textbf{then } \beta \leftarrow \gamma \\ \quad \textbf{else } \beta \leftarrow p - \gamma \\ \beta \leftarrow \beta/\alpha^{x_i} \bmod p \\ i \leftarrow i+1 \end{cases}$
return $(x_{i-1}, x_{i-2}, \ldots, x_0)$

It is possible to give formal proof of the algorithm's correctness using mathematical induction. Denote

$$x = \log_\alpha \beta = \sum_{i \geq 0} x_i 2^i.$$

For $i \geq 0$, define

$$Y_i = \left\lfloor \frac{x}{2^{i+1}} \right\rfloor.$$

Also, define β_0 to be the value of β just before the start of the **while** loop; and, for $i \geq 1$, define β_i to be the value of β at the end of the ith iteration of the **while** loop. It can be proved by induction that

$$\beta_i \equiv \alpha^{2Y_i} \pmod{p}$$

TABLE 6.2
Values of L_1 and L_2 for $p = 19, \alpha = 2$

γ	$L_1(\gamma)$	$L_2(\gamma)$	γ	$L_1(\gamma)$	$L_2(\gamma)$	γ	$L_1(\gamma)$	$L_2(\gamma)$
1	0	0	7	0	1	13	1	0
2	1	0	8	1	1	14	1	1
3	1	0	9	0	0	15	1	1
4	0	1	10	1	0	16	0	0
5	0	0	11	0	0	17	0	1
6	0	1	12	1	1	18	1	0

$x_0 \leftarrow 0, \beta \leftarrow 6, i \leftarrow 1$
$x_1 \leftarrow L_2(6) = 1, \gamma \leftarrow 5, L_1(5) = 0 \neq x_1, \beta \leftarrow 14, \beta \leftarrow 7, i \leftarrow 2$
$x_2 \leftarrow L_2(7) = 1, \gamma \leftarrow 11, L_1(11) = 0 \neq x_2, \beta \leftarrow 8, \beta \leftarrow 4, i \leftarrow 3$
$x_3 \leftarrow L_2(4) = 1, \gamma \leftarrow 17, L_1(17) = 0 \neq x_3, \beta \leftarrow 2, \beta \leftarrow 1, i \leftarrow 4$
return $(1, 1, 1, 0)$

FIGURE 6.2
Computation of $\log_2 6$ in $\mathbb{Z}_{19}{}^*$ using an oracle for L_2

for all $i \geq 0$. Now, with the observation that

$$2Y_i = Y_{i-1} - x_i,$$

it follows that

$$x_{i+1} = L_2(\beta_i),$$

$i \geq 0$. Since

$$x_0 = L_1(\beta),$$

the algorithm is correct. The details are left to the reader.

6.7.2 Semantic Security of ElGamal Systems

We first observe that the basic *ElGamal Cryptosystem*, as described in Cryptosystem 6.1, is not semantically secure. Recall that $\alpha \in \mathbb{Z}_p{}^*$ is a primitive element and $\beta = \alpha^a \bmod p$ where a is the private key. Given a plaintext element x, a random number k is chosen, and then $e_K(x, k) = (y_1, y_2)$ is computed, where $y_1 = \alpha^k \bmod p$ and $y_2 = x\beta^k \bmod p$.

We make use of the fact that it is easy, using Euler's criterion, to test elements of $\mathbb{Z}_p$ to see if they are quadratic residues modulo p. Recall from Section 6.7.1 that β is a quadratic residue modulo p if and only if a is even. Similarly, y_1 is a quadratic residue modulo p if and only if k is even. We can determine the parity of both a and k, and hence we can compute the parity of ak. Therefore, we can determine if β^k ($= \alpha^{ak}$) is a quadratic residue.

Now, suppose that we wish to distinguish encryptions of x_1 from encryptions of x_2, where x_1 is a quadratic residue and x_2 is a quadratic non-residue modulo p. It is a simple matter to determine the quadratic residuosity of y_2, and we have already discussed how the quadratic residuosity of β^k can be determined. It follows that (y_1, y_2) is an encryption of x_1 if and only if β^k and y_2 are both quadratic residues or both quadratic non-residues.

The above attack does not work if β is a quadratic residue and every plaintext x is required to be a quadratic residue. In fact, if $p = 2q + 1$ where q is prime, then it can be shown that restricting β, y_1 and x to be quadratic residues is equivalent to implementing the *ElGamal Cryptosystem* in the subgroup of quadratic residues modulo p (which is a cyclic subgroup of $\mathbb{Z}_p^*$ of order q). This version of the *ElGamal Cryptosystem* is conjectured to be semantically secure if the **Discrete Logarithm** problem in $\mathbb{Z}_p^*$ is infeasible.

6.7.3 The Diffie-Hellman Problems

We introduce two variants of the so-called **Diffie-Hellman** problems, a computational version and a decision version. The reason for calling them "**Diffie-Hellman** problems" comes from the origin of these two problems in connection with Diffie-Hellman key agreement protocols. In this section, we discuss some interesting connections between these problems and security of ElGamal-type cryptosystems.

Here are descriptions of the two problems.

Problem 6.3:   **Computational Diffie-Hellman**

Instance: A multiplicative group $(G, \cdot)$, an element $\alpha \in G$ having order n, and two elements $\beta, \gamma \in \langle \alpha \rangle$.

Question: Find $\delta \in \langle \alpha \rangle$ such that $\log_\alpha \delta \equiv \log_\alpha \beta \times \log_\alpha \gamma \pmod{n}$. (Equivalently, given α^b and α^c, find α^{bc}.)

Problem 6.4:   **Decision Diffie-Hellman**

Instance: A multiplicative group $(G, \cdot)$, an element $\alpha \in G$ having order n, and three elements $\beta, \gamma, \delta \in \langle \alpha \rangle$.

Question: Is it the case that $\log_\alpha \delta \equiv \log_\alpha \beta \times \log_\alpha \gamma \pmod{n}$? (Equivalently, given α^b, α^c and α^d, determine if $d \equiv bc \pmod{n}$.)

We often denote these two problems by **CDH** and **DDH**, respectively. It is easy to see that there exist Turing reductions

$$\textbf{DDH} \propto_T \textbf{CDH}$$

and

$$\textbf{CDH} \propto_T \textbf{Discrete Logarithm}.$$

The first reduction is proven as follows: Let $\alpha, \beta, \gamma, \delta$ be given. Use an algorithm that solves **CDH** to find the value δ' such that

$$\log_\alpha \delta' \equiv \log_\alpha \beta \times \log_\alpha \gamma \pmod{n}.$$

Then check to see if $\delta' = \delta$.

The second reduction is also very simple. Let α, β, γ be given. Use an algorithm that solves **Discrete Logarithm** to find $b = \log_\alpha \beta$ and $c = \log_\alpha \gamma$. Then compute $d = bc \bmod n$ and $\delta = \alpha^d$.

These reductions show that the assumption that **DDH** is infeasible is at least as strong as the assumption that **CDH** is infeasible, which in turn is at least as strong as the assumption that **Discrete Logarithm** is infeasible.

It is not hard to show that the semantic security of the *ElGamal Cryptosystem* is equivalent to the infeasibility of **DDH**; and ElGamal decryption (without knowing the private key) is equivalent to solving **CDH**. The assumptions necessary to prove the security of the *ElGamal Cryptosystem* are therefore (potentially) stronger than assuming just that **Discrete Logarithm** is infeasible. Indeed, we already showed that the *ElGamal Cryptosystem* in $\mathbb{Z}_p^*$ is not semantically secure, whereas the **Discrete Logarithm** problem is conjectured to be infeasible in $\mathbb{Z}_p^*$ for appropriately chosen primes p. This suggests that the security of the three problems may not be equivalent.

Here, we give a proof that any algorithm that solves **CDH** can be used to decrypt ElGamal ciphertexts, and vice versa. Suppose first that ORACLECDH is an algorithm for **CDH**, and let (y_1, y_2) be a ciphertext for the *ElGamal Cryptosystem* with public key α and β. Compute

$$\delta = \text{ORACLECDH}(\alpha, \beta, y_1),$$

and then define

$$x = y_2 \delta^{-1}.$$

It is easy to see that x is the decryption of the ciphertext (y_1, y_2).

Conversely, suppose that ORACLE-ELGAMAL-DECRYPT is an algorithm that decrypts ElGamal ciphertexts. Let α, β, γ be given as in **CDH**. Define α and β to be the public key for the *ElGamal Cryptosystem*. Then define $y_1 = \gamma$ and let $y_2 \in \langle \alpha \rangle$ be chosen randomly. Compute

$$x = \text{ORACLE-ELGAMAL-DECRYPT}(\alpha, \beta, (y_1, y_2)),$$

which is the decryption of the ciphertext (y_1, y_2). Finally, compute

$$\delta = y_2 x^{-1}.$$

δ is the solution to the given instance of **CDH**.

6.8   Notes and References

The *ElGamal Cryptosystem* was presented in [125]. The POHLIG-HELLMAN ALGORITHM was published in [270]. For further information on the **Discrete Logarithm** problem in general, we recommend the recent survey article by Odlyzko [258].

The POLLARD RHO ALGORITHM was first described in [272]. Brent [71] described a more efficient method to detect cycles (and, therefore, collisions), which can also be used in the corresponding factoring algorithm. There are many ways of defining the "random walks" used in the algorithm; for a thorough treatment of these topics, see Teske [328].

The lower bound on generic algorithms for the **Discrete Logarithm** problem was proven independently by Nechaev [249] and Shoup [301]. Our discussion is based on the treatment of Chateauneuf, Ling and Stinson [87].

The main reference book for finite fields is Lidl and Niederreiter [220]. McEliece [231] is a good elementary textbook on this subject.

An informative article on the **Discrete Logarithm** problem in $\mathbb{F}_{2^n}$ is Gordon and McCurley [166]. Some more recent results are found in the article by Thomé [329], where a discrete logarithm computation in $\mathbb{F}_{2^{607}}$ is described. Lenstra and Verheul [219] provide predictions of the feasibility of the **Discrete Logarithm** problem in coming years, and the ramifications for choosing key sizes for cryptosystems based on discrete logarithms.

The idea of using elliptic curves for public-key cryptosystems is due to Koblitz [196] and Miller [241]. Koblitz, Menezes and Vanstone [200] have written a recent survey article on this topic. Recent monographs on elliptic curve cryptography include Blake, Seroussi and Smart [42, 43], Enge [127] and Hankerson, Menezes and Vanstone [171]. For an elementary treatment of elliptic curves, see Silverman and Tate [303]. Galbraith and Menezes [152] is a recommended survey on the use of algebraic curves in cryptography (including elliptic curves); Koblitz [199] includes a chapter on cryptographic applications of hyperelliptic curves. Other textbooks which place an emphasis on topics from this chapter include Wagstaff [335] and Washington [340].

For a recent article that presents a thorough treatment of fast arithmetic on elliptic curves, see Solinas [312]. The Menezes-Okamoto-Vanstone reduction of discrete logarithms from elliptic curves to finite fields is given in [235] (see also [233]). The attack on "trace one" curves is due to Smart, Satoh, Araki and Semaev; see, for example, Smart [308].

The material we presented concerning the **Discrete Logarithm ith Bit** problem is based on Peralta [263]. Other papers that discuss related questions are Håstad, Schrift and Shamir [172] and Long and Wigderson [221].

Boneh [56] is an interesting survey article on the **Decision Diffie-Hellman** problem. Maurer and Wolf [230] is an article which further develops topics considered in Section 6.7.

Exercises

6.1 Implement SHANKS' ALGORITHM for finding discrete logarithms in $\mathbb{Z}_p^*$, where p is prime and α is a primitive element modulo p. Use your program to find

$\log_{106} 12375$ in $\mathbb{Z}_{24691}{}^*$ and $\log_6 248388$ in $\mathbb{Z}_{458009}{}^*$.

6.2 Describe how to modify SHANKS' ALGORITHM to compute the logarithm of β to the base α in a group G if it is specified ahead of time that this logarithm lies in the interval $[s, t]$, where s and t are integers such that $0 \leq s < t < n$, where n is the order of α. Prove that your algorithm is correct, and show that its complexity is $O(\sqrt{t - s})$.

6.3 The integer $p = 458009$ is prime and $\alpha = 2$ has order 57251 in $\mathbb{Z}_p{}^*$. Use the POL-LARD RHO ALGORITHM to compute the discrete logarithm in $\mathbb{Z}_p{}^*$ of $\beta = 56851$ to the base α. Take the initial value $x_0 = 1$, and define the partition (S_1, S_2, S_3) as in Example 6.3. Find the smallest integer i such that $x_i = x_{2i}$, and then compute the desired discrete logarithm.

6.4 Suppose that p is an odd prime and k is a positive integer. The multiplicative group $\mathbb{Z}_{p^k}{}^*$ has order $p^{k-1}(p-1)$, and is known to be cyclic. A generator for this group is called a *primitive element modulo p^k*.

    (a) Suppose that α is a primitive element modulo p. Prove that at least one of α or $\alpha + p$ is a primitive element modulo p^2.

    (b) Describe how to efficiently verify that 3 is a primitive root modulo 29 and modulo 29^2. Note: It can be shown that if α is a primitive root modulo p and modulo p^2, then it is a primitive root modulo p^k for all positive integers k (you do not have to prove this fact). Therefore, it follows that 3 is a primitive root modulo 29^k for all positive integers k.

    (c) Find an integer α that is a primitive root modulo 29 but not a primitive root modulo 29^2.

    (d) Use the POHLIG-HELLMAN ALGORITHM to compute the discrete logarithm of 3344 to the base 3 in the multiplicative group $\mathbb{Z}_{24389}{}^*$.

6.5 Implement the POHLIG-HELLMAN ALGORITHM for finding discrete logarithms in $\mathbb{Z}_p$, where p is prime and α is a primitive element. Use your program to find $\log_5 8563$ in $\mathbb{Z}_{28703}$ and $\log_{10} 12611$ in $\mathbb{Z}_{31153}$.

6.6 Let $p = 227$. The element $\alpha = 2$ is primitive in $\mathbb{Z}_p{}^*$.

    (a) Compute $\alpha^{32}, \alpha^{40}, \alpha^{59}$ and α^{156} modulo p, and factor them over the factor base $\{2, 3, 5, 7, 11\}$.

    (b) Using the fact that $\log 2 = 1$, compute $\log 3$, $\log 5$, $\log 7$ and $\log 11$ from the factorizations obtained above (all logarithms are discrete logarithms in $\mathbb{Z}_p{}^*$ to the base α).

    (c) Now suppose we wish to compute $\log 173$. Multiply 173 by the "random" value $2^{177} \bmod p$. Factor the result over the factor base, and proceed to compute $\log 173$ using the previously computed logarithms of the numbers in the factor base.

6.7 Suppose that $n = pq$ is an RSA modulus (i.e., p and q are distinct odd primes), and let $\alpha \in \mathbb{Z}_n{}^*$. For a positive integer m and for any $\alpha \in \mathbb{Z}_m{}^*$, define $\mathrm{ord}_m(\alpha)$ to be the order of α in the group $\mathbb{Z}_m{}^*$.

    (a) Prove that

$$\mathrm{ord}_n(\alpha) = \mathrm{lcm}(\mathrm{ord}_p(\alpha), \mathrm{ord}_q(\alpha)).$$

    (b) Suppose that $\gcd(p - 1, q - 1) = d$. Show that there exists an element $\alpha \in \mathbb{Z}_n{}^*$ such that

$$\mathrm{ord}_n(\alpha) = \frac{\phi(n)}{d}.$$

(c) Suppose that $\gcd(p - 1, q - 1) = 2$, and we have an oracle that solves the **Discrete Logarithm** problem in the subgroup $\langle \alpha \rangle$, where $\alpha \in \mathbb{Z}_n^*$ has order $\phi(n)/2$. That is, given any $\beta \in \langle \alpha \rangle$, the oracle will find the discrete logarithm $a = \log_\alpha \beta$, where $0 \leq a \leq \phi(n)/2 - 1$. (The value $\phi(n)/2$ is secret however.) Suppose we compute the value $\beta = \alpha^n \bmod n$ and then we use the oracle to find $a = \log_\alpha \beta$. Assuming that $p > 3$ and $q > 3$, prove that $n - a = \phi(n)$.

(d) Describe how n can easily be factored, given the discrete logarithm $a = \log_\alpha \beta$ from (c).

6.8 In this question, we consider a generic algorithm for the **Discrete Logarithm** problem in $(\mathbb{Z}_{19}, +)$.

(a) Suppose that the set C is defined as follows:

$$C = \{(1 - i^2 \bmod 19, i \bmod 19) : i = 0, 1, 2, 4, 7, 12\}.$$

Compute $\mathsf{Good}(C)$.

(b) Suppose that the output of the group oracle, given the ordered pairs in C, is as follows:

$$(0, 1) \mapsto 10111$$
$$(1, 0) \mapsto 01100$$
$$(16, 2) \mapsto 00110$$
$$(4, 4) \mapsto 01010$$
$$(9, 7) \mapsto 00100$$
$$(9, 12) \mapsto 11001,$$

where group elements are encoded as (random) binary 5-tuples. What can you say about the value of "a"?

6.9 Decrypt the ElGamal ciphertext presented in Table 6.3. The parameters of the system are $p = 31847$, $\alpha = 5$, $a = 7899$ and $\beta = 18074$. Each element of $\mathbb{Z}_n$ represents three alphabetic characters as in Exercise 5.12.

The plaintext was taken from "The English Patient," by Michael Ondaatje, Alfred A. Knopf, Inc., New York, 1992.

6.10 Determine which of the following polynomials are irreducible over $\mathbb{Z}_2[x]$: $x^5 + x^4 + 1$, $x^5 + x^3 + 1$, $x^5 + x^4 + x^2 + 1$.

6.11 The field $\mathbb{F}_{2^5}$ can be constructed as $\mathbb{Z}_2[x]/(x^5 + x^2 + 1)$. Perform the following computations in this field.

(a) Compute $(x^4 + x^2) \times (x^3 + x + 1)$.

(b) Using the extended Euclidean algorithm, compute $(x^3 + x^2)^{-1}$.

(c) Using the square-and-multiply algorithm, compute x^{25}.

6.12 We give an example of the *ElGamal Cryptosystem* implemented in $\mathbb{F}_{3^3}$. The polynomial $x^3 + 2x^2 + 1$ is irreducible over $\mathbb{Z}_3[x]$ and hence $\mathbb{Z}_3[x]/(x^3 + 2x^2 + 1)$ is the field $\mathbb{F}_{3^3}$. We can associate the 26 letters of the alphabet with the 26 nonzero field elements, and thus encrypt ordinary text in a convenient way. We will use a lexicographic ordering of the (nonzero) polynomials to set up the correspondence.

TABLE 6.3
ElGamal Ciphertext

$(3781, 14409)$	$(31552, 3930)$	$(27214, 15442)$	$(5809, 30274)$
$(5400, 31486)$	$(19936, 721)$	$(27765, 29284)$	$(29820, 7710)$
$(31590, 26470)$	$(3781, 14409)$	$(15898, 30844)$	$(19048, 12914)$
$(16160, 3129)$	$(301, 17252)$	$(24689, 7776)$	$(28856, 15720)$
$(30555, 24611)$	$(20501, 2922)$	$(13659, 5015)$	$(5740, 31233)$
$(1616, 14170)$	$(4294, 2307)$	$(2320, 29174)$	$(3036, 20132)$
$(14130, 22010)$	$(25910, 19663)$	$(19557, 10145)$	$(18899, 27609)$
$(26004, 25056)$	$(5400, 31486)$	$(9526, 3019)$	$(12962, 15189)$
$(29538, 5408)$	$(3149, 7400)$	$(9396, 3058)$	$(27149, 20535)$
$(1777, 8737)$	$(26117, 14251)$	$(7129, 18195)$	$(25302, 10248)$
$(23258, 3468)$	$(26052, 20545)$	$(21958, 5713)$	$(346, 31194)$
$(8836, 25898)$	$(8794, 17358)$	$(1777, 8737)$	$(25038, 12483)$
$(10422, 5552)$	$(1777, 8737)$	$(3780, 16360)$	$(11685, 133)$
$(25115, 10840)$	$(14130, 22010)$	$(16081, 16414)$	$(28580, 20845)$
$(23418, 22058)$	$(24139, 9580)$	$(173, 17075)$	$(2016, 18131)$
$(19886, 22344)$	$(21600, 25505)$	$(27119, 19921)$	$(23312, 16906)$
$(21563, 7891)$	$(28250, 21321)$	$(28327, 19237)$	$(15313, 28649)$
$(24271, 8480)$	$(26592, 25457)$	$(9660, 7939)$	$(10267, 20623)$
$(30499, 14423)$	$(5839, 24179)$	$(12846, 6598)$	$(9284, 27858)$
$(24875, 17641)$	$(1777, 8737)$	$(18825, 19671)$	$(31306, 11929)$
$(3576, 4630)$	$(26664, 27572)$	$(27011, 29164)$	$(22763, 8992)$
$(3149, 7400)$	$(8951, 29435)$	$(2059, 3977)$	$(16258, 30341)$
$(21541, 19004)$	$(5865, 29526)$	$(10536, 6941)$	$(1777, 8737)$
$(17561, 11884)$	$(2209, 6107)$	$(10422, 5552)$	$(19371, 21005)$
$(26521, 5803)$	$(14884, 14280)$	$(4328, 8635)$	$(28250, 21321)$
$(28327, 19237)$	$(15313, 28649)$		

This correspondence is as follows:

A	$\leftrightarrow$	1	B	$\leftrightarrow$	2	C	$\leftrightarrow$	x
D	$\leftrightarrow$	$x+1$	E	$\leftrightarrow$	$x+2$	F	$\leftrightarrow$	$2x$
G	$\leftrightarrow$	$2x+1$	H	$\leftrightarrow$	$2x+2$	I	$\leftrightarrow$	x^2
J	$\leftrightarrow$	x^2+1	K	$\leftrightarrow$	x^2+2	L	$\leftrightarrow$	x^2+x
M	$\leftrightarrow$	x^2+x+1	N	$\leftrightarrow$	x^2+x+2	O	$\leftrightarrow$	x^2+2x
P	$\leftrightarrow$	x^2+2x+1	Q	$\leftrightarrow$	x^2+2x+2	R	$\leftrightarrow$	$2x^2$
S	$\leftrightarrow$	$2x^2+1$	T	$\leftrightarrow$	$2x^2+2$	U	$\leftrightarrow$	$2x^2+x$
V	$\leftrightarrow$	$2x^2+x+1$	W	$\leftrightarrow$	$2x^2+x+2$	X	$\leftrightarrow$	$2x^2+2x$
Y	$\leftrightarrow$	$2x^2+2x+1$	Z	$\leftrightarrow$	$2x^2+2x+2$			

Suppose Bob uses $\alpha = x$ and $a = 11$ in an *ElGamal Cryptosystem*; then $\beta = x+2$. Show how Bob will decrypt the following string of ciphertext:

(K,H) (P,X) (N,K) (H,R) (T,F) (V,Y) (E,H) (F,A) (T,W) (J,D) (U,J)

6.13   Let E be the elliptic curve $y^2 = x^3 + x + 28$ defined over $\mathbb{Z}_{71}$.
   (a) Determine the number of points on E.
   (b) Show that E is not a cyclic group.
   (c) What is the maximum order of an element in E? Find an element having this order.

6.14 Suppose that $p > 3$ is an odd prime, and $a, b \in \mathbb{Z}_p$. Further, suppose that the equation $x^3 + ax + b \equiv 0 \pmod{p}$ has three distinct roots in $\mathbb{Z}_p$. Prove that the corresponding elliptic curve group $(E, +)$ is not cyclic.

HINT   Show that the points of order two generate a subgroup of $(E, +)$ that is isomorphic to $\mathbb{Z}_2 \times \mathbb{Z}_2$.

6.15 Consider an elliptic curve E described by the formula $y^2 \equiv x^3 + ax + b \pmod{p}$, where $4a^3 + 27b^2 \not\equiv 0 \pmod{p}$ and $p > 3$ is prime.
   (a) It is clear that a point $P = (x_1, y_1) \in E$ has order 3 if and only if $2P = -P$. Use this fact to prove that, if $P = (x_1, y_1) \in E$ has order 3, then
   $$3x_1{}^4 + 6ax_1{}^2 + 12x_1 b - a^2 \equiv 0 \pmod{p}. \qquad (6.7)$$
   (b) Conclude from equation (6.7) that there are at most 8 points of order 3 on the elliptic curve E.
   (c) Using equation (6.7), determine all points of order 3 on the elliptic curve $y^2 \equiv x^3 + 34x \pmod{73}$.

6.16 Suppose that E is an elliptic curve defined over $\mathbb{Z}_p$, where $p > 3$ is prime. Suppose that $\#E$ is prime, $P \in E$, and $P \neq \mathcal{O}$.
   (a) Prove that the discrete logarithm $\log_P(-P) = \#E - 1$.
   (b) Describe how to compute $\#E$ in time $O(p^{1/4})$ by using Hasse's bound on $\#E$, together with a modification of SHANKS' ALGORITHM. Give a pseudocode description of the algorithm.

6.17 Let E be the elliptic curve $y^2 = x^3 + 2x + 7$ defined over $\mathbb{Z}_{31}$. It can be shown that $\#E = 39$ and $P = (2, 9)$ is an element of order 39 in E. The *Simplified ECIES* defined on E has $\mathbb{Z}_{31}{}^*$ as its plaintext space. Suppose the private key is $m = 8$.
   (a) Compute $Q = mP$.
   (b) Decrypt the following string of ciphertext:

   $$((18, 1), 21), ((3, 1), 18), ((17, 0), 19), ((28, 0), 8).$$

   (c) Assuming that each plaintext represents one alphabetic character, convert the plaintext into an English word. (Here we will use the correspondence $A \leftrightarrow 1$, $\ldots, Z \leftrightarrow 26$, because 0 is not allowed in a (plaintext) ordered pair.)

6.18   (a) Determine the NAF representation of the integer 87.
   (b) Using the NAF representation of 87, use Algorithm 6.5 to compute $87P$, where $P = (2, 6)$ is a point on the elliptic curve $y^2 = x^3 + x + 26$ defined over $\mathbb{Z}_{127}$. Show the partial results during each iteration of the algorithm.

6.19 Let $\mathcal{L}_i$ denote the set of positive integers that have exactly i coefficients in their NAF representation, such that the leading coefficient is 1. Denote $k_i = |\mathcal{L}_i|$.
   (a) By means of a suitable decomposition of $\mathcal{L}_i$, prove that the k_i's satisfy the following recurrence relation:
   $$k_1 = 1$$
   $$k_2 = 1$$
   $$k_{i+1} = 2(k_1 + k_2 + \ldots + k_{i-1}) + 1 \quad \text{(for } i \geq 2\text{)}.$$
   (b) Derive a second degree recurrence relation for the k_i's, and obtain an explicit solution of the recurrence relation.

6.20 Find $\log_5 896$ in $\mathbb{Z}_{1103}$ using Algorithm 6.6, given that $L_2(\beta) = 1$ for $\beta = 25, 219$ and 841, and $L_2(\beta) = 0$ for $\beta = 163, 532, 625$ and 656.

6.21 Throughout this question, suppose that $p \equiv 5 \pmod{8}$ is prime and suppose that a is a quadratic residue modulo p.

(a) Prove that $a^{(p-1)/4} \equiv \pm 1 \pmod{p}$.

(b) If $a^{(p-1)/4} \equiv 1 \pmod{p}$, prove that $a^{(p+3)/8} \bmod p$ is a square root of a modulo p.

(c) If $a^{(p-1)/4} \equiv -1 \pmod{p}$, prove that $2^{-1}(4a)^{(p+3)/8} \bmod p$ is a square root of a modulo p.

HINT   Use the fact that $\left(\frac{2}{p}\right) = -1$ when $p \equiv 5 \pmod{8}$ is prime.

(d) Given a primitive element $\alpha \in \mathbb{Z}_p{}^*$, and given any $\beta \in \mathbb{Z}_p{}^*$, show that $L_2(\beta)$ can be computed efficiently.

HINT   Use the fact that it is possible to compute square roots modulo p, as well as the fact that $L_1(\beta) = L_1(p - \beta)$ for all $\beta \in \mathbb{Z}_p{}^*$, when $p \equiv 5 \pmod{8}$ is prime.

6.22 The *ElGamal Cryptosystem* can be implemented in any subgroup $\langle \alpha \rangle$ of a finite multiplicative group $(G, \cdot)$, as follows: Let $\beta \in \langle \alpha \rangle$ and define (α, β) to be the public key. The plaintext space is $\mathcal{P} = \langle \alpha \rangle$, and the encryption operation is $e_K(x) = (y_1, y_2) = (\alpha^k, x \cdot \beta^k)$, where k is random.

Here we show that distinguishing ElGamal encryptions of two plaintexts can be Turing reduced to **Decision Diffie-Hellman**, and vice versa.

(a) Assume that ORACLEDDH is an oracle that solves **Decision Diffie-Hellman** in $(G, \cdot)$. Prove that ORACLEDDH can be used as a subroutine in an algorithm that distinguishes ElGamal encryptions of two given plaintexts, say x_1 and x_2. (That is, given $x_1, x_2 \in \mathcal{P}$, and given a ciphertext (y_1, y_2) which is an encryption of x_i for some $i \in \{1, 2\}$, the distinguishing algorithm will determine if $i = 1$ or $i = 2$.)

(b) Assume that ORACLE-DISTINGUISH is an oracle that distinguishes ElGamal encryptions of any two given plaintexts x_1 and x_2, for any *ElGamal Cryptosystem* implemented in the group $(G, \cdot)$ as described above. Suppose further that ORACLE-DISTINGUISH will determine if a ciphertext (y_1, y_2) is not a valid encryption of either of x_1 or x_2. Prove that ORACLE-DISTINGUISH can be used as a subroutine in an algorithm that solves **Decision Diffie-Hellman** in $(G, \cdot)$.

7

Signature Schemes

7.1 Introduction

In this chapter, we study signature schemes, which are also called digital signatures. A "conventional" handwritten signature attached to a document is used to specify the person responsible for it. A signature is used in everyday situations such as writing a letter, withdrawing money from a bank, signing a contract, etc.

A signature scheme is a method of signing a message stored in electronic form. As such, a signed message can be transmitted over a computer network. In this chapter, we will study several signature schemes, but first we discuss some fundamental differences between conventional and digital signatures.

First is the question of signing a document. With a conventional signature, a signature is part of the physical document being signed. However, a digital signature is not attached physically to the message that is signed, so the algorithm that is used must somehow "bind" the signature to the message.

Second is the question of verification. A conventional signature is verified by comparing it to other, authentic signatures. For example, when someone signs a credit card purchase, the salesperson is supposed to compare the signature on the sales slip to the signature on the back of the credit card in order to verify the signature. Of course, this is not a very secure method as it is relatively easy to forge someone else's signature. Digital signatures, on the other hand, can be verified using a publicly known verification algorithm. Thus, "anyone" can verify a digital signature. The use of a secure signature scheme will prevent the possibility of forgeries.

Another fundamental difference between conventional and digital signatures is that a "copy" of a signed digital message is identical to the original. On the other hand, a copy of a signed paper document can usually be distinguished from an original. This feature means that care must be taken to prevent a signed digital message from being reused. For example, if Alice signs a digital message authorizing Bob to withdraw $100 from her bank account (i.e., a check), she wants Bob to be able to do so only once. So the message itself should contain information, such as a date, that prevents it from being reused.

A signature scheme consists of two components: a signing algorithm and a verification algorithm. Alice can sign a message x using a (private) signing algorithm $\mathbf{sig}_K$ which depends on a private key K. The resulting signature $\mathbf{sig}_K(x)$ can subsequently be verified using a public verification algorithm $\mathbf{ver}_K$. Given a pair (x, y), where x is a message and y is a purported signature on x, the verification algorithm returns an answer "true" or "false" depending on whether y is a valid signature for the message x.

Here is a formal definition of a signature scheme.

Definition 7.1: A *signature scheme* is a five-tuple $(\mathcal{P}, \mathcal{A}, \mathcal{K}, \mathcal{S}, \mathcal{V})$, where the following conditions are satisfied:

1. $\mathcal{P}$ is a finite set of possible *messages*

2. $\mathcal{A}$ is a finite set of possible *signatures*

3. $\mathcal{K}$, the *keyspace*, is a finite set of possible *keys*

4. For each $K \in \mathcal{K}$, there is a *signing algorithm* $\mathbf{sig}_K \in \mathcal{S}$ and a corresponding *verification algorithm* $\mathbf{ver}_K \in \mathcal{V}$. Each $\mathbf{sig}_K : \mathcal{P} \to \mathcal{A}$ and $\mathbf{ver}_K : \mathcal{P} \times \mathcal{A} \to \{true, false\}$ are functions such that the following equation is satisfied for every message $x \in \mathcal{P}$ and for every signature $y \in \mathcal{A}$:

$$\mathbf{ver}_K(x, y) = \begin{cases} true & \text{if } y = \mathbf{sig}_K(x) \\ false & \text{if } y \neq \mathbf{sig}_K(x). \end{cases}$$

A pair (x, y) with $x \in \mathcal{P}$ and $y \in \mathcal{A}$ is called a *signed message*.

For every $K \in \mathcal{K}$, the functions $\mathbf{sig}_K$ and $\mathbf{ver}_K$ should be polynomial-time functions. $\mathbf{ver}_K$ will be a public function and $\mathbf{sig}_K$ will be private. Given a message x, it should be computationally infeasible for anyone other than Alice to compute a signature y such that $\mathbf{ver}_K(x, y) = true$ (and note that there might be more than one such y for a given x, depending on how the function $\mathbf{ver}$ is defined). If Oscar can compute a pair (x, y) such that $\mathbf{ver}_K(x, y) = true$ and x was not previously signed by Alice, then the signature y is called a *forgery*. Informally, a forged signature is a valid signature produced by someone other than Alice.

As our first example of a signature scheme, we observe that the *RSA Cryptosystem* can be used to provide digital signatures; in this context, it is known as the *RSA Signature Scheme*. See Cryptosystem 7.1.

Thus, Alice signs a message x using the RSA decryption rule d_K. Alice is the only person who can create the signature because $d_K = \mathbf{sig}_K$ is private. The verification algorithm uses the RSA encryption rule e_K. Anyone can verify a signature because e_K is public.

Note that anyone can forge Alice's RSA signature by choosing a random y and computing $x = e_K(y)$; then $y = \mathbf{sig}_K(x)$ is a valid signature on the message

---

Cryptosystem 7.1: *RSA Signature Scheme*

Let $n = pq$, where p and q are primes. Let $\mathcal{P} = \mathcal{A} = \mathbb{Z}_n$, and define

$$\mathcal{K} = \{(n, p, q, a, b) : n = pq, p, q \text{ prime}, ab \equiv 1 \pmod{\phi(n)}\}.$$

The values n and b are the public key, and the values p, q, a are the private key.

For $K = (n, p, q, a, b)$, define

$$\mathbf{sig}_K(x) = x^a \bmod n$$

and

$$\mathbf{ver}_K(x, y) = \text{true} \Leftrightarrow x \equiv y^b \pmod{n}$$

$(x, y \in \mathbb{Z}_n)$.

---

x. (Note, however, that there does not seem to be an obvious way to first choose x and then compute the corresponding signature y; if this could be done, then the *RSA Cryptosystem* would be insecure.) One way to prevent this attack is to require that messages contain sufficient redundancy that a forged signature of this type does not correspond to a "meaningful" message x except with a very small probability. Alternatively, the use of hash functions in conjunction with signature schemes will eliminate this method of forging (cryptographic hash functions were discussed in Chapter 4). We pursue this approach further in the next section.

Finally, let's look briefly at how we would combine signing and public-key encryption. Suppose Alice wishes to send a signed, encrypted message to Bob. Given a plaintext x, Alice would compute her signature $y = \mathbf{sig}_{\text{Alice}}(x)$, and then encrypt both x and y using Bob's public encryption function e_{Bob}, obtaining $z = e_{\text{Bob}}(x, y)$. The ciphertext z would be transmitted to Bob. When Bob receives z, he first decrypts it with his decryption function d_{Bob} to get (x, y). Then he uses Alice's public verification function to check that $\mathbf{ver}_{\text{Alice}}(x, y) = \text{true}$.

What if Alice first encrypted x, and then signed the result? Then she would compute

$$z = e_{\text{Bob}}(x) \text{ and } y = \mathbf{sig}_{\text{Alice}}(z).$$

Alice would transmit the pair (z, y) to Bob. Bob would decrypt z, obtaining x, and then verify the signature y on z using $\mathbf{ver}_{\text{Alice}}$. One potential problem with this approach is that if Oscar obtains a pair (z, y) of this type, he could replace Alice's signature y by his own signature

$$y' = \mathbf{sig}_{\text{Oscar}}(z).$$

(Note that Oscar can sign the ciphertext $z = e_{\text{Bob}}(x)$ even though he doesn't know the plaintext x.) Then, if Oscar transmits (z, y') to Bob, Oscar's signature will be verified by Bob using $\mathbf{ver}_{Oscar}$, and Bob may infer that the plaintext x originated with Oscar. Because of this potential difficulty, most people recommend signing before encrypting.

The rest of this chapter is organized as follows. Section 7.2 introduces the notion of security for signature schemes and how hash functions are used in conjunction with signature schemes. Section 7.3 introduces the *ElGamal Signature Scheme* and discusses its security. Section 7.4 deals with three important schemes that evolved from the *ElGamal Signature Scheme*, namely, the *Schnorr Signature Scheme*, the *Digital Signature Algorithm* and the *Elliptic Curve Digital Signature Algorithm*. Provably secure signature schemes are the topic of Section 7.5. Finally, in Sections 7.6 and 7.7, we consider certain signature schemes with additional properties.

7.2 Security Requirements for Signature Schemes

In this section, we discuss what it means for a signature scheme to be "secure." As was the case with a cryptosystem, we need to specify an attack model, the goal of the adversary, and the type of security provided by the scheme.

Recall from Section 1.2 that the attack model defines the information available to the adversary. In the case of signature schemes, the following types of attack models are commonly considered:

key-only attack
> Oscar possesses Alice's public key, i.e., the verification function, $\mathbf{ver}_K$.

known message attack
> Oscar possesses a list of messages previously signed by Alice, say

$$(x_1, y_1), (x_2, y_2), \ldots,$$

> where the x_i's are messages and the y_i's are Alice's signatures on these messages (so $y_i = \mathbf{sig}_K(x_i), i = 1, 2, \ldots$).

chosen message attack
> Oscar requests Alice's signatures on a list of messages. Therefore he chooses messages $x_1, x_2, \ldots$, and Alice supplies her signatures on these messages, namely, $y_i = \mathbf{sig}_K(x_i), i = 1, 2, \ldots$.

We consider several possible adversarial goals:

total break
> The adversary is able to determine Alice's private key, i.e., the signing function $\mathbf{sig}_K$. Therefore he can create valid signatures on any message.

selective forgery

    With some non-negligible probability, the adversary is able to create a valid signature on a message chosen by someone else. In other words, if the adversary is given a message x, then he can determine (with some probability) the signature y such that $\mathbf{ver}_K(x, y) = \text{true}$. The message x should not be one that has previously been signed by Alice.

existential forgery

    The adversary is able to create a valid signature for at least one message. In other words, the adversary can create a pair (x, y) where x is a message and $\mathbf{ver}_K(x, y) = \text{true}$. The message x should not be one that has previously been signed by Alice.

A signature scheme cannot be unconditionally secure, since Oscar can test all possible signatures $y \in \mathcal{A}$ for a given message x, using the public algorithm $\mathbf{ver}_K$, until he finds a valid signature. So, given sufficient time, Oscar can always forge Alice's signature on any message. Thus, as was the case with public-key cryptosystems, our goal is to find signature schemes that are computationally or provably secure.

Notice that the above definitions have some similarity to the attacks on MACs that we considered in Section 4.4. In the MAC setting, there is no such thing as a public key; so it does not make sense to speak of a key-only attack (and a MAC does not have separate signing and verifying functions, of course). The attacks in Section 4.4 were existential forgeries using chosen message attacks.

We illustrate the concepts described above with a couple of examples based on the *RSA Signature Scheme*. In Section 7.1, we observed that Oscar can construct a valid signed message by choosing a signature y and then computing x such that $\mathbf{ver}_K(x, y) = \text{true}$. This would be an existential forgery using a key-only attack.

Another type of attack is based on the multiplicative property of RSA. Suppose that $y_1 = \mathbf{sig}_K(x_1)$ and $y_2 = \mathbf{sig}_K(x_2)$ are any two messages previously signed by Alice. Then

$$\mathbf{ver}_K(x_1 x_2 \bmod n, y_1 y_2 \bmod n) = \text{true},$$

and therefore Oscar can create the valid signature $y_1 y_2 \bmod n$ on the message $x_1 x_2 \bmod n$. This is an example of an existential forgery using a known message attack.

Here is one more variation. Suppose Oscar wants to forge a signature on the message x, where x was possibly chosen by someone else. It is a simple matter for him to find $x_1, x_2 \in \mathbb{Z}_n$ such that $x \equiv x_1 x_2 \pmod{n}$. Now suppose he asks Alice for her signatures on messages x_1 and x_2, which we denote by y_1 and y_2 respectively. Then, as in the previous attack, $y_1 y_2 \bmod n$ is the signature for the message $x = x_1 x_2 \bmod n$. This is a selective forgery using a chosen message attack.

FIGURE 7.1
Signing a message digest

message	x	$x \in \{0,1\}^*$
	$\downarrow$	
message digest	$z = h(x)$	$z \in \mathcal{Z}$
	$\downarrow$	
signature	$y = sig_K(z)$	$y \in \mathcal{Y}$

7.2.1 Signatures and Hash Functions

Signature schemes are almost always used in conjunction with a very fast public cryptographic hash function. The hash function $h : \{0,1\}^* \to \mathcal{Z}$ will take a message of arbitrary length and produce a message digest of a specified size (160 bits is a popular choice). The message digest will then be signed using a signature scheme $(\mathcal{P}, \mathcal{A}, \mathcal{K}, \mathcal{S}, \mathcal{V})$, where $\mathcal{Z} \subseteq \mathcal{P}$. This use of a hash function and signature scheme is depicted diagramatically in Figure 7.1.

Suppose Alice wants to sign a message x, which is a bitstring of arbitrary length. She first constructs the message digest $z = h(x)$, and then computes the signature on z, namely, $y = sig_K(z)$. Then she transmits the ordered pair (x, y) over the channel. Now the verification can be performed (by anyone) by first reconstructing the message digest $z = h(x)$ using the public hash function h, and then checking that $ver_K(z, y) = true$.

We have to be careful that the use of a hash function h does not weaken the security of the signature scheme, for it is the message digest that is signed, not the message. It will be necessary for h to satisfy certain properties in order to prevent various attacks. The desired properties of hash functions were the ones that were already discussed in Section 4.2.

The most obvious type of attack is for Oscar to start with a valid signed message (x, y), where $y = sig_K(h(x))$. (The pair (x, y) could be any message previously signed by Alice.) Then he computes $z = h(x)$ and attempts to find $x' \neq x$ such that $h(x') = h(x)$. If Oscar can do this, (x', y) would be a valid signed message; so y is a forged signature for the message x'. This is an existential forgery using a known message attack. In order to prevent this type of attack, we require that h be second preimage resistant.

Another possible attack is the following: Oscar first finds two messages $x \neq x'$ such that $h(x) = h(x')$. Oscar then gives x to Alice and persuades her to sign the message digest $h(x)$, obtaining y. Then (x', y) is a valid signed message and y is a forged signature for the message x'. This is an existential forgery using a chosen message attack; it can be prevented if h is collision resistant.

Here is a third variety of attack. It is often possible with certain signature schemes to forge signatures on random message digests z (we observed already that this could be done with the *RSA Signature Scheme*). That is, we assume that the signature scheme (without the hash function) is subject to existential forgery using a key-only attack. Now, suppose Oscar computes a signature on some message digest z, and then he finds a message x such that $z = h(x)$. If he can do this, then (x, y) is a valid signed message and y is a forged signature for the message x. This is an existential forgery on the signature scheme using a key-only attack. In order to prevent this attack, we desire that h be a preimage resistant hash function.

7.3 The ElGamal Signature Scheme

In this section, we present the *ElGamal Signature Scheme*, which was described in a 1985 paper. A modification of this scheme has been adopted as the *Digital Signature Algorithm* (or *DSA*) by the National Institute of Standards and Technology. The *DSA* also incorporates some ideas used in a scheme known as the *Schnorr Signature Scheme*. All of these schemes are designed specifically for the purpose of signatures, as opposed to the *RSA Cryptosystem*, which can be used both as a public-key cryptosystem and a signature scheme.

The *ElGamal Signature Scheme* is non-deterministic (recall that the *ElGamal Public-key Cryptosystem* is also non-deterministic). This means that there are many valid signatures for any given message, and the verification algorithm must be able to accept any of these valid signatures as authentic. The description of the *ElGamal Signature Scheme* is given as Cryptosystem 7.2.

If the signature was constructed correctly, then the verification will succeed, because

$$\beta^\gamma \gamma^\delta \equiv \alpha^{a\gamma} \alpha^{k\delta} \pmod{p}$$

$$\equiv \alpha^x \pmod{p},$$

where we use the fact that

$$a\gamma + k\delta \equiv x \pmod{p-1}.$$

Actually, it is probably less mysterious to begin with the verification equation, and then derive the signing function. Suppose we start with the congruence

$$\alpha^x \equiv \beta^\gamma \gamma^\delta \pmod{p}. \tag{7.1}$$

Then we make the substitutions

$$\gamma \equiv \alpha^k \pmod{p}$$

and

$$\beta \equiv \alpha^a \pmod{p},$$

Cryptosystem 7.2: *ElGamal Signature Scheme*

Let p be a prime such that the discrete log problem in $\mathbb{Z}_p$ is intractable, and let $\alpha \in \mathbb{Z}_p^*$ be a primitive element. Let $\mathcal{P} = \mathbb{Z}_p^*$, $\mathcal{A} = \mathbb{Z}_p^* \times \mathbb{Z}_{p-1}$, and define

$$\mathcal{K} = \{(p, \alpha, a, \beta) : \beta \equiv \alpha^a \pmod{p}\}.$$

The values p, α and β are the public key, and a is the private key.

For $K = (p, \alpha, a, \beta)$, and for a (secret) random number $k \in \mathbb{Z}_{p-1}^*$, define

$$\mathbf{sig}_K(x, k) = (\gamma, \delta),$$

where

$$\gamma = \alpha^k \bmod p$$

and

$$\delta = (x - a\gamma)k^{-1} \bmod (p - 1).$$

For $x, \gamma \in \mathbb{Z}_p^*$ and $\delta \in \mathbb{Z}_{p-1}$, define

$$\mathbf{ver}_K(x, (\gamma, \delta)) = \text{true} \Leftrightarrow \beta^\gamma \gamma^\delta \equiv \alpha^x \pmod{p}.$$

but we do not substitute for γ in the exponent of (7.1). We obtain the following:

$$\alpha^x \equiv \alpha^{a\gamma + k\delta} \pmod{p}.$$

Now, α is a primitive element modulo p; so this congruence is true if and only if the exponents are congruent modulo $p - 1$, i.e., if and only if

$$x \equiv a\gamma + k\delta \pmod{p - 1}.$$

Given x, a, γ and k, this congruence can be solved for δ, yielding the formula used in the signing function of Cryptosystem 7.2.

Alice computes a signature using both the private key, a, and the secret random number, k (which is used to sign one message, x). The verification can be accomplished using only public information.

Let's do a small example to illustrate the arithmetic.

Example 7.1   Suppose we take $p = 467$, $\alpha = 2$, $a = 127$; then

$$\beta = \alpha^a \bmod p$$

$$= 2^{127} \bmod 467$$

$$= 132.$$

Suppose Alice wants to sign the message $x = 100$ and she chooses the random value $k = 213$ (note that $\gcd(213, 466) = 1$ and $213^{-1} \bmod 466 = 431$). Then

$$\gamma = 2^{213} \bmod 467 = 29$$

and

$$\delta = (100 - 127 \times 29)431 \bmod 466 = 51.$$

Anyone can verify this signature by checking that

$$132^{29}29^{51} \equiv 189 \pmod{467}$$

and

$$2^{100} \equiv 189 \pmod{467}.$$

Hence, the signature is valid.   ▯

7.3.1   Security of the ElGamal Signature Scheme

Let's look at the security of the *ElGamal Signature Scheme*. Suppose Oscar tries to forge a signature for a given message x, without knowing a. If Oscar chooses a value γ and then tries to find the corresponding δ, he must compute the discrete logarithm $\log_\gamma \alpha^x \beta^{-\gamma}$. On the other hand, if he first chooses δ and then tries to find γ, he is trying to "solve" the equation

$$\beta^\gamma \gamma^\delta \equiv \alpha^x \pmod{p}$$

for the "unknown" γ. This is a problem for which no feasible solution is known; however, it does not seem to be related to any well-studied problem such as the **Discrete Logarithm** problem. There also remains the possibility that there might be some way to compute γ and δ simultaneously in such a way that (γ, δ) will be a signature. No one has discovered a way to do this, but conversely, no one has proved that it cannot be done.

If Oscar chooses γ and δ and then tries to solve for x, he is again faced with an instance of the **Discrete Logarithm** problem, namely the computation of $\log_\alpha \beta^\gamma \gamma^\delta$. Hence, Oscar cannot sign a given message x using this approach.

However, there is a method by which Oscar can sign a random message by choosing γ, δ and x simultaneously. Thus an existential forgery is possible under a key-only attack (assuming a hash function is not used). We describe how to do this now.

Suppose i and j are integers such that $0 \leq i \leq p - 2$, $0 \leq j \leq p - 2$, and suppose we express γ in the form $\gamma = \alpha^i \beta^j \bmod p$. Then the verification condition is

$$\alpha^x \equiv \beta^\gamma (\alpha^i \beta^j)^\delta \pmod{p}.$$

This is equivalent to

$$\alpha^{x-i\delta} \equiv \beta^{\gamma+j\delta} \pmod{p}.$$

This latter congruence will be satisfied if

$$x - i\delta \equiv 0 \pmod{p-1}$$

and

$$\gamma + j\delta \equiv 0 \pmod{p-1}.$$

Given i and j, we can easily solve these two congruences modulo $p-1$ for δ and x, provided that $\gcd(j, p-1) = 1$. We obtain the following:

$$\gamma = \alpha^i \beta^j \bmod p,$$

$$\delta = -\gamma j^{-1} \bmod (p-1), \quad \text{and}$$

$$x = -\gamma i j^{-1} \bmod (p-1).$$

By the way in which we constructed (γ, δ), it is clear that it is a valid signature for the message x.

We illustrate with an example.

Example 7.2 As in the previous example, suppose $p = 467$, $\alpha = 2$ and $\beta = 132$. Suppose Oscar chooses $i = 99$ and $j = 179$; then $j^{-1} \bmod (p-1) = 151$. He would compute the following:

$$\begin{aligned}
\gamma &= 2^{99}132^{179} \bmod 467 &&= 117 \\
\delta &= -117 \times 151 \bmod 466 &&= 41 \\
x &= 99 \times 41 \bmod 466 &&= 331.
\end{aligned}$$

Then $(117, 41)$ is a valid signature for the message 331, as may be verified by checking that

$$132^{117}117^{41} \equiv 303 \pmod{467}$$

and

$$2^{331} \equiv 303 \pmod{467}.$$

$\square$

Here is a second type of forgery, in which Oscar begins with a message previously signed by Alice. This is an existential forgery under a known message attack. Suppose (γ, δ) is a valid signature for a message x. Then it is possible for Oscar to sign various other messages. Suppose h, i and j are integers, $0 \leq h, i, j \leq p-2$, and $\gcd(h\gamma - j\delta, p-1) = 1$. Compute the following:

$$\lambda = \gamma^h \alpha^i \beta^j \bmod p$$

$$\mu = \delta\lambda(h\gamma - j\delta)^{-1} \bmod (p-1), \quad \text{and}$$

$$x' = \lambda(hx + i\delta)(h\gamma - j\delta)^{-1} \bmod (p-1).$$

Then, it is tedious but straightforward to check that the verification condition

$$\beta^\lambda \lambda^\mu \equiv \alpha^{x'} \pmod{p}$$

holds. Hence (λ, μ) is a valid signature for x'.

Both of these methods are existential forgeries, but it does not appear that they can be modified to yield selective forgeries. Hence, they do not seem to represent a threat to the security of the *ElGamal Signature Scheme*, provided that a secure hash function is used as described in Section 7.2.1.

We also mention a couple of ways in which the *ElGamal Signature Scheme* can be broken if it is used carelessly (these are further examples of protocol failures, some of which were discussed in the Exercises of Chapter 5). First, the random value k used in computing a signature should not be revealed. For, if k is known and $gcd(\gamma, p - 1) = 1$, then it is a simple matter to compute

$$a = (x - k\delta)\gamma^{-1} \bmod (p - 1).$$

Once a is known, then the system is completely broken and Oscar can forge signatures at will.

Another misuse of the system is to use the same value k in signing two different messages. This also makes it easy for Oscar to compute a and hence break the system. This can be done as follows. Suppose (γ, δ_1) is a signature on x_1 and (γ, δ_2) is a signature on x_2. Then we have

$$\beta^\gamma \gamma^{\delta_1} \equiv \alpha^{x_1} \pmod{p}$$

and

$$\beta^\gamma \gamma^{\delta_2} \equiv \alpha^{x_2} \pmod{p}.$$

Thus

$$\alpha^{x_1 - x_2} \equiv \gamma^{\delta_1 - \delta_2} \pmod{p}.$$

Writing $\gamma = \alpha^k$, we obtain the following equation in the unknown k:

$$\alpha^{x_1 - x_2} \equiv \alpha^{k(\delta_1 - \delta_2)} \pmod{p},$$

which is equivalent to

$$x_1 - x_2 \equiv k(\delta_1 - \delta_2) \pmod{p - 1}.$$

Now let $d = gcd(\delta_1 - \delta_2, p - 1)$. Since $d \mid (p - 1)$ and $d \mid (\delta_1 - \delta_2)$, it follows that $d \mid (x_1 - x_2)$. Define

$$x' = \frac{x_1 - x_2}{d}$$

$$\delta' = \frac{\delta_1 - \delta_2}{d}$$

$$p' = \frac{p - 1}{d}.$$

Then the congruence becomes:

$$x' \equiv k\delta' \pmod{p'}.$$

Since $\gcd(\delta', p') = 1$, we can compute

$$\epsilon = (\delta')^{-1} \bmod p'.$$

Then value of k is determined modulo p' to be

$$k = x'\epsilon \bmod p'.$$

This yields d candidate values for k:

$$k = x'\epsilon + ip' \bmod (p-1)$$

for some i, $0 \leq i \leq d - 1$. Of these d candidate values, the (unique) correct one can be determined by testing the condition

$$\gamma \equiv \alpha^k \pmod{p}.$$

7.4   Variants of the ElGamal Signature Scheme

In many situations, a message might be encrypted and decrypted only once, so it suffices to use any cryptosystem which is known to be secure at the time the message is encrypted. On the other hand, a signed message could function as a legal document such as a contract or will; so it is very likely that it would be necessary to verify a signature many years after the message is signed. It is therefore important to take even more precautions regarding the security of a signature scheme as opposed to a cryptosystem. Since the *ElGamal Signature Scheme* is no more secure than the **Discrete Logarithm** problem, this necessitates the use of a large modulus p. Most people would argue that the length of p should be at least 1024 bits in order to provide present-day security, and even larger to provide security into the foreseeable future (this was already mentioned in Section 6.6).

A 1024 bit modulus leads to an ElGamal signature having 2048 bits. For potential applications, many of which involve the use of smart cards, a shorter signature is desirable. In 1989, Schnorr proposed a signature scheme that can be viewed as a variant of the *ElGamal Signature Scheme* in which the signature size is greatly reduced. The *Digital Signature Algorithm* (or *DSA*) is another modification of the *ElGamal Signature Scheme*, which incorporates some of the ideas used in the *Schnorr Signature Scheme*. The *DSA* was published in the Federal Register on May 19, 1994 and was adopted as a standard on December 1, 1994 (however, it was first proposed in August 1991). We describe the *Schnorr Signature Scheme*, the *DSA*, and a modification of the *DSA* to elliptic curves (called the *ECDSA*) in the next subsections.

7.4.1 The Schnorr Signature Scheme

Suppose that p and q are primes such that $p - 1 \equiv 0 \pmod{q}$. Typically we will take $p \approx 2^{1024}$ and $q \approx 2^{160}$. The *Schnorr Signature Scheme* modifies the *ElGamal Signature Scheme* in an ingenious way so that a $\log_2 q$-bit message digest is signed using a $2 \log_2 q$-bit signature, but the computations are done in $\mathbb{Z}_p$. The way that this is accomplished is to work in a subgroup of $\mathbb{Z}_p^*$ of size q. The assumed security of the scheme is based on the belief that finding discrete logarithms in this specified subgroup of $\mathbb{Z}_p^*$ is secure. (This setting for the **Discrete Logarithm** problem was previously discussed in Section 6.6.)

We will take α to be a qth root of 1 modulo p. (It is easy to construct such an α: Let α_0 be a primitive element of $\mathbb{Z}_p$, and define $\alpha = \alpha_0^{(p-1)/q} \bmod p$.) The key in the *Schnorr Signature Scheme* is similar to the key in the *ElGamal Signature Scheme* in other respects. However, the *Schnorr Signature Scheme* integrates a hash function directly into the signing algorithm (as opposed to the usual hash-then-sign method that we discussed in Section 7.2.1). We will assume that $h : \{0, 1\}^* \to \mathbb{Z}_q$ is a secure hash function. A complete description of the *Schnorr Signature Scheme* is given as Cryptosystem 7.3.

It is easy to check that $\alpha^\delta \beta^{-\gamma} \equiv \alpha^k \pmod{p}$, and hence a Schnorr signature will be verified. Here is a small example to illustrate.

Example 7.3 Suppose we take $q = 101$ and $p = 78q + 1 = 7879$. 3 is a primitive element in $\mathbb{Z}_{7879}^*$, so we can take

$$\alpha = 3^{78} \bmod 7879 = 170.$$

α is a qth root of 1 modulo p. Suppose $a = 75$; then

$$\beta = \alpha^a \bmod 7879 = 4567.$$

Now, suppose Alice wants to sign the message x, and she chooses the random value $k = 50$. She first computes

$$\alpha^k \bmod p = 170^{50} \bmod 7879 = 2518.$$

The next step is to compute $h(x \parallel 2518)$, where h is a given hash function and 2518 is represented in binary (as a bitstring). Suppose for purposes of illustration that $h(x \parallel 2518) = 96$. Then δ is computed as

$$\delta = 50 + 75 \times 96 \bmod 101 = 79,$$

and the signature is $(96, 79)$.

This signature is verified by computing

$$170^{79} 4567^{-96} \bmod 7879 = 2518,$$

and then checking that $h(x \parallel 2518) = 96$. □

Cryptosystem 7.3: *Schnorr Signature Scheme*

Let p be a prime such that the discrete log problem in $\mathbb{Z}_p^*$ is intractable, and let q be a prime that divides $p - 1$. Let $\alpha \in \mathbb{Z}_p^*$ be a qth root of 1 modulo p. Let $\mathcal{P} = \{0, 1\}^*$, $\mathcal{A} = \mathbb{Z}_q \times \mathbb{Z}_q$, and define

$$\mathcal{K} = \{(p, q, \alpha, a, \beta) : \beta \equiv \alpha^a \pmod{p}\},$$

where $0 \leq a \leq q - 1$. The values p, q, α and β are the public key, and a is the private key. Finally, let $h : \{0, 1\}^* \to \mathbb{Z}_q$ be a secure hash function.

For $K = (p, q, \alpha, a, \beta)$, and for a (secret) random number k, $1 \leq k \leq q - 1$, define

$$\mathbf{sig}_K(x, k) = (\gamma, \delta),$$

where

$$\gamma = h(x \,||\, \alpha^k \bmod p)$$

and

$$\delta = k + a\gamma \bmod q.$$

For $x \in \{0, 1\}^*$ and $\gamma, \delta \in \mathbb{Z}_q$, verification is done by performing the following computations:

$$\mathbf{ver}_K(x, (\gamma, \delta)) = \text{true} \Leftrightarrow h(x \,||\, \alpha^\delta \beta^{-\gamma} \bmod p) = \gamma.$$

7.4.2 The Digital Signature Algorithm

We will outline the changes that are made to the verification function of the *ElGamal Signature Scheme* in the specification of the *DSA*. The *DSA* uses an order q subgroup of $\mathbb{Z}_p^*$, as does the *Schnorr Signature Scheme*. In the *DSA*, it is required that q is a 160-bit prime and p is an L-bit prime, where $L \equiv 0 \pmod{64}$ and $512 \leq L \leq 1024$. The key in the *DSA* has the same form as in the *Schnorr Signature Scheme*. It is also specified in the *DSA* that the message will be hashed using *SHA-1* before it is signed. The result is that a 160-bit message digest is signed with a 320-bit signature, and the computations are done in $\mathbb{Z}_p$ and $\mathbb{Z}_q$.

In the *ElGamal Signature Scheme*, suppose we change the "$-$" to a "$+$" in the definition of δ, so

$$\delta = (x + a\gamma)k^{-1} \bmod (p - 1).$$

It is easy to see that this changes the verification condition to the following:

$$\alpha^x \beta^\gamma \equiv \gamma^\delta \pmod{p}. \tag{7.2}$$

Cryptosystem 7.4: *Digital Signature Algorithm*

Let p be a L-bit prime such that the discrete log problem in $\mathbb{Z}_p$ is intractable, where $L \equiv 0 \pmod{64}$ and $512 \leq L \leq 1024$, and let q be a 160-bit prime that divides $p - 1$. Let $\alpha \in \mathbb{Z}_p^*$ be a qth root of 1 modulo p. Let $\mathcal{P} = \{0, 1\}^*$, $\mathcal{A} = \mathbb{Z}_q^* \times \mathbb{Z}_q^*$, and define

$$\mathcal{K} = \{(p, q, \alpha, a, \beta) : \beta \equiv \alpha^a \pmod{p}\},$$

where $0 \leq a \leq q - 1$. The values p, q, α and β are the public key, and a is the private key.

For $K = (p, q, \alpha, a, \beta)$, and for a (secret) random number k, $1 \leq k \leq q - 1$, define

$$\mathbf{sig}_K(x, k) = (\gamma, \delta),$$

where

$$\gamma = (\alpha^k \bmod p) \bmod q \quad \text{and}$$

$$\delta = (\text{SHA-1}(x) + a\gamma)k^{-1} \bmod q.$$

(If $\gamma = 0$ or $\delta = 0$, a new random value of k should be chosen.)

For $x \in \{0, 1\}^*$ and $\gamma, \delta \in \mathbb{Z}_q^*$, verification is done by performing the following computations:

$$e_1 = \text{SHA-1}(x)\,\delta^{-1} \bmod q$$

$$e_2 = \gamma\,\delta^{-1} \bmod q$$

$$\mathbf{ver}_K(x, (\gamma, \delta)) = \text{true} \Leftrightarrow (\alpha^{e_1}\beta^{e_2} \bmod p) \bmod q = \gamma.$$

Now, α has order q, and β and γ are powers of α, so they also have order q. This means that all exponents in (7.2) can be reduced modulo q without affecting the validity of the congruence. Since x will be replaced by a 160-bit message digest in the *DSA*, we will assume that $x \in \mathbb{Z}_q$. Further, we will alter the definition of δ, so that $\delta \in \mathbb{Z}_q$, as follows:

$$\delta = (x + a\gamma)k^{-1} \bmod q,$$

It remains to consider $\gamma = \alpha^k \bmod p$. Suppose we temporarily define

$$\gamma' = \gamma \bmod q = (\alpha^k \bmod p) \bmod q.$$

Note that

$$\delta = (x + a\gamma')k^{-1} \bmod q,$$

so δ is unchanged. We can write the verification equation as

$$\alpha^x \beta^{\gamma'} \equiv \gamma^\delta \pmod p. \tag{7.3}$$

Notice that we cannot replace the remaining occurrence of γ by γ'.

Now we proceed to rewrite (7.3), by raising both sides to the power $\delta^{-1} \bmod q$ (this requires that $\delta \neq 0$). We obtain the following:

$$\alpha^{x\delta^{-1}} \beta^{\gamma'\delta^{-1}} \bmod p = \gamma. \tag{7.4}$$

Now we can reduce both sides of (7.4) modulo q, which produces the following:

$$(\alpha^{x\delta^{-1}} \beta^{\gamma'\delta^{-1}} \bmod p) \bmod q = \gamma'. \tag{7.5}$$

The complete description of the *DSA* is given as Cryptosystem 7.4, in which we rename γ' as γ and replace x by SHA-1(x).

In October 2001, NIST recommended that p be chosen to be a 1024-bit prime (i.e., 1024 is the only value allowed for L). This is "neither a standard nor a guideline," but it does indicate some concern about the security of the discrete logarithm problem.

Notice that if Alice computes a value $\delta \equiv 0 \pmod q$ in the DSA signing algorithm, she should reject it and construct a new signature with a new random k. We should point out that this is not likely to cause a problem in practice: the probability that $\delta \equiv 0 \pmod q$ is likely to be on the order of 2^{-160}; so for all intents and purposes it will almost never happen.

Here is an example (with p and q much smaller than they are required to be in the *DSA*) to illustrate.

Example 7.4 Suppose we take the same values of p, q, α, a, β and k as in Example 7.3, and suppose Alice wants to sign the message digest SHA-1$(x) = 22$. Then she computes

$$k^{-1} \bmod 101 = 50^{-1} \bmod 101 = 99,$$

$$\gamma = (170^{50} \bmod 7879) \bmod 101$$
$$= 2518 \bmod 101$$
$$= 94,$$

and

$$\delta = (22 + 75 \times 94)99 \bmod 101$$
$$= 97.$$

The signature $(94, 97)$ on the message digest 22 is verified by the following computations:

$$\delta^{-1} = 97^{-1} \bmod 101 = 25$$

$$e_1 = 22 \times 25 \bmod 101 = 45$$

$$e_2 = 94 \times 25 \bmod 101 = 27$$

$$(170^{45}4567^{27} \bmod 7879) \bmod 101 = 2518 \bmod 101 = 94.$$

$\square$

When the *DSA* was proposed in 1991, there were several criticisms put forward. One complaint was that the selection process by NIST was not public. The standard was developed by the National Security Agency (NSA) without the input of U.S. industry. Regardless of the merits of the resulting scheme, many people resented the "closed-door" approach.

Of the technical criticisms put forward, the most serious was that the size of the modulus p was fixed initially at 512 bits. Many people suggested that the modulus size not be fixed, so that larger modulus sizes could be used if desired. In reponse to these comments, NIST altered the description of the standard so that a variety of modulus sizes were allowed.

7.4.3 The Elliptic Curve DSA

In 2000, the *Elliptic Curve Digital Signature Algorithm* (*ECDSA*) was approved as FIPS 186-2. This signature scheme can be viewed as a modification of the *DSA* to the setting of elliptic curves. We have two points A and B on an elliptic curve defined over $\mathbb{Z}_p$ for some prime p.[1] The discrete logarithm $m = \log_A B$ is the private key. (This is analogous to the relation $\beta = \alpha^a \bmod p$ in the *DSA*, where a is the private key.) The order of A is a large prime number q. Computing a signature involves first choosing a random value k and computing kA (this is analogous to the computation of α^k in the *DSA*).

Now here is the main difference between the *DSA* and the *ECDSA*. In the *DSA*, the value $\alpha^k \bmod p$ is reduced modulo q to yield a value γ which is the first component of the signature (γ, δ). In the *ECDSA*, the analogous value is r, which is the x-co-ordinate of the elliptic curve point kA, reduced modulo q. This value r is the first component of the signature (r, s).

Finally, in the *ECDSA*, the value s is computed from r, m, k, and the message x in exactly the same way as δ is computed from γ, a, k and the message x in the *DSA*.

We now present the complete description of the *ECDSA* as Cryptosystem 7.5.

---

[1] We note that the *ECDSA* also permits the use of elliptic curves defined over finite fields $\mathbb{F}_{2^n}$, but we will not describe this variation here.

Cryptosystem 7.5: *Elliptic Curve Digital Signature Algorithm*

Let p be a large prime and let E be an elliptic curve defined over $\mathbb{F}_p$. Let A be a point on E having prime order q, such that the **Discrete Logarithm** problem in $\langle A \rangle$ is infeasible. Let $\mathcal{P} = \{0, 1\}^*$, $\mathcal{A} = \mathbb{Z}_q^* \times \mathbb{Z}_q^*$, and define

$$\mathcal{K} = \{(p, q, E, A, m, B) : B = mA\},$$

where $0 \leq m \leq q - 1$. The values p, q, E, A and B are the public key, and m is the private key.

For $K = (p, q, E, A, m, B)$, and for a (secret) random number k, $1 \leq k \leq q - 1$, define

$$\mathbf{sig}_K(x, k) = (r, s),$$

where

$$kA = (u, v)$$

$$r = u \bmod q, \quad \text{and}$$

$$s = k^{-1}(\text{SHA-1}(x) + mr) \bmod q.$$

(If either $r = 0$ or $s = 0$, a new random value of k should be chosen.)

For $x \in \{0, 1\}^*$ and $r, s \in \mathbb{Z}_q^*$, verification is done by performing the following computations:

$$w = s^{-1} \bmod q$$

$$i = w\,\text{SHA-1}(x) \bmod q$$

$$j = wr \bmod q$$

$$(u, v) = iA + jB$$

$$\mathbf{ver}_K(x, (r, s)) = \text{true} \Leftrightarrow u \bmod q = r.$$

We work through a tiny example to illustrate the computations in the *ECDSA*.

Example 7.5   We will base our example on the same elliptic curve that was used in Section 6.5.2, namely, $y^2 = x^3 + x + 6$, defined over $\mathbb{Z}_{11}$. The parameters of the signature scheme are $p = 11, q = 13, A = (2, 7), m = 7$ and $B = (7, 2)$.

Suppose we have a message x with SHA-1$(x) = 4$, and Alice wants to sign the message x using the random value $k = 3$. She will compute

$$(u, v) = 3\,(2, 7) = (8, 3)$$

$$r = u \bmod 13 = 8, \quad \text{and}$$

$$s = 3^{-1}(4 + 7 \times 8) \bmod 13 = 7.$$

Therefore $(8, 7)$ is the signature.

Bob verifies the signature by performing the following computations:

$$w = 7^{-1} \bmod 13 = 2$$

$$i = 2 \times 4 \bmod 13 = 8$$

$$j = 2 \times 8 \bmod 13 = 3$$

$$(u, v) = 8A + 3B = (8, 3), \quad \text{and}$$

$$u \bmod 13 = 8 = r.$$

Hence, the signature is verified. $\square$

7.5 Provably Secure Signature Schemes

We present some examples of provably secure signature schemes in this section. First, we describe a construction for a one-time signature scheme based on an arbitrary one-way (i.e., preimage resistant) function, say f. This scheme can be proven secure against a key-only attack provided that f is a bijective function. The second construction is for a signature scheme known as *Full Domain Hash*. This signature scheme is provably secure in the random oracle model provided that it is constructed from a trap-door one-way permutation.

7.5.1 One-time Signatures

In this section, we describe a conceptually simple way to construct a provably secure one-time signature scheme from a one-way function. (A signature scheme is a *one-time* signature scheme if it is secure when only one message is signed. The signature can be verified an arbitrary number of times, of course.) The description of the scheme, which is known as the *Lamport Signature Scheme*, is given in Cryptosystem 7.6.

Informally, this is how the system works. A message to be signed is a binary k-tuple. Each bit of the message is signed individually. If the ith bit of the message equals j (where $j \in \{0, 1\}$), then the ith element of the signature is the value $y_{i,j}$, which is a preimage of the public key value $z_{i,j}$. The verification consists simply of checking that each element in the signature is a preimage of the public

---

Cryptosystem 7.6: *Lamport Signature Scheme*

Let k be a positive integer and let $\mathcal{P} = \{0, 1\}^k$. Suppose $f : Y \to Z$ is a one-way function, and let $\mathcal{A} = Y^k$. Let $y_{i,j} \in Y$ be chosen at random, $1 \leq i \leq k$, $j = 0, 1$, and let $z_{i,j} = f(y_{i,j})$, $1 \leq i \leq k$, $j = 0, 1$. The key K consists of the $2k$ y's and the $2k$ z's. The y's are the private key while the z's are the public key.

For $K = (y_{i,j}, z_{i,j} : 1 \leq i \leq k, j = 0, 1)$, define

$$\mathbf{sig}_K(x_1, \ldots, x_k) = (y_{1,x_1}, \ldots, y_{k,x_k}).$$

A signature $(a_1, \ldots, a_k)$ on the message $(x_1, \ldots, x_k)$ is verified as follows:

$$\mathbf{ver}_K((x_1, \ldots, x_k), (a_1, \ldots, a_k)) = \text{true} \Leftrightarrow f(a_i) = z_{i,x_i}, 1 \leq i \leq k.$$

---

key element $z_{i,j}$ that corresponds to the ith bit of the message. This can be done using the public function f.

We illustrate the scheme by considering one possible implementation using the exponentiation function $f(x) = \alpha^x \bmod p$, where α is a primitive element modulo p. Here $f : \{0, \ldots, p-2\} \to \mathbb{Z}_p^*$. We present a toy example to demonstrate the computations that take place in the scheme.

Example 7.6   7879 is prime and 3 is a primitive element in $\mathbb{Z}_{7879}^*$. Define

$$f(x) = 3^x \bmod 7879.$$

Suppose $k = 3$, and Alice chooses the six (secret) random numbers

$$y_{1,0} = 5831$$
$$y_{1,1} = 735$$
$$y_{2,0} = 803$$
$$y_{2,1} = 2467$$
$$y_{3,0} = 4285$$
$$y_{3,1} = 6449.$$

Then Alice computes the images of these six y's under the function f:

$$z_{1,0} = 2009$$
$$z_{1,1} = 3810$$
$$z_{2,0} = 4672$$
$$z_{2,1} = 4721$$
$$z_{3,0} = 268$$
$$z_{3,1} = 5731.$$

These z's are published. Now, suppose Alice wants to sign the message

$$x = (1, 1, 0).$$

The signature for x is

$$(y_{1,1}, y_{2,1}, y_{3,0}) = (735, 2467, 4285).$$

To verify this signature, it suffices to compute the following:

$$3^{735} \bmod 7879 = 3810$$
$$3^{2467} \bmod 7879 = 4721$$
$$3^{4285} \bmod 7879 = 268.$$

Hence, the signature is verified. ◻

Oscar cannot forge a signature because he is unable to invert the one-way function f to obtain the secret y's. However, the signature scheme can be used to sign only one message securely. Given signatures on two different messages, it is an easy matter for Oscar to construct signatures for another message different from the first two (unless the first two messages differ in exactly one bit).

For example, suppose the messages $(0, 1, 1)$ and $(1, 0, 1)$ are both signed using the same key. The message $(0, 1, 1)$ has as its signature the triple $(y_{1,0}, y_{2,1}, y_{3,1})$, and the message $(1, 0, 1)$ is signed with $(y_{1,1}, y_{2,0}, y_{3,1})$. Given these two signatures, Oscar can manufacture signatures for the messages $(1, 1, 1)$ (namely, $(y_{1,1}, y_{2,1}, y_{3,1})$) and $(0, 0, 1)$ (namely, $(y_{1,0}, y_{2,0}, y_{3,1})$).

The security of the *Lamport Signature Scheme* can be proven if we assume that $f : Y \to Z$ is a bijective one-way function and a public key consists of $2k$ distinct elements of Z. We consider a key-only attack; so an adversary is given only the public key. We will assume that the adversary can carry out an existential forgery. In other words, given the public key, the adversary outputs a message, x, and a valid signature, y. (We are assuming that f, Y, Z and k are fixed.)

Algorithm 7.1: LAMPORT-PREIMAGE(z)

external f, LAMPORT-FORGE
comment: we assume $f : Y \to Z$ is a bijection

choose a random $i_0 \in \{1, \ldots, k\}$ and a random $j_0 \in \{0, 1\}$
construct a random public key $\mathcal{Z} = (z_{i,j} : 1 \le i \le k, j = 0, 1)$
    such that $z_{i_0, j_0} = z$
$((x_1, \ldots, x_k), (a_1, \ldots, a_k)) \leftarrow$ LAMPORT-FORGE($\mathcal{Z}$)
if $x_{i_0} = j_0$
    then return (a_{i_0})
    else return (fail)

The adversary is modeled by an algorithm, say LAMPORT-FORGE. For simplicity, we assume that LAMPORT-FORGE is deterministic: given any particular public key, it always outputs the same forgery. We will describe an algorithm, LAMPORT-PREIMAGE, to find preimages of the function f on randomly chosen elements $z \in Z$. This algorithm is a reduction which uses the algorithm LAMPORT-FORGE as an oracle. The existence of this reduction contradicts the assumed one-wayness of f. Hence, if we believe that f is one-way, then we conclude that a key-only existential forgery is computationally infeasible. LAMPORT-PREIMAGE is presented as Algorithm 7.1.

Let's consider the (average) success probability of Algorithm 7.1, where the average is computed over all $z \in Z$. We are assuming that LAMPORT-FORGE always succeeds in finding a forgery. If $x_{i_0} = j_0$ in the forgery, then it is the case that

$$f(a_{i_0}) = z_{i_0, x_{i_0}} = z_{i_0, j_0} = z,$$

and we have found $f^{-1}(z)$, as desired. Recall that each x_i has the value 0 or 1. We will prove that the probability that $x_{i_0} = j_0$, averaged over all possible runs of the algorithm, is $1/2$. Therefore, the average success probability of Algorithm 7.1 is equal to $1/2$. We give a proof of this in the following theorem.

THEOREM 7.1 *Suppose that $f : Y \to Z$ is a one-way bijection, and suppose there exists a deterministic algorithm, LAMPORT-FORGE, that will create an existential forgery for the Lamport Signature Scheme using a key-only attack, for any public key $\mathcal{Z}$ consisting of $2k$ distinct elements of Z. Then there exists an algorithm, LAMPORT-PREIMAGE, that will find preimages of random elements $z \in Z$ with average probability at least $1/2$.*

PROOF Let $\mathcal{S}$ denote the set of all possible public keys, and for any $z \in Z$, let $\mathcal{S}_z$ denote the set of all possible public keys that contain z. Denote $s = |\mathcal{S}|$ and, for all $z \in Z$, denote $s_z = |\mathcal{S}_z|$. Let $\mathcal{T}_z$ consist of all public keys in $\mathcal{Z} \in \mathcal{S}_z$ such that LAMPORT-PREIMAGE(z) succeeds when Z is the public key that is chosen by LAMPORT-PREIMAGE(z), and denote $t_z = |\mathcal{I}_z|$.

We will make use of two equations, which follow from elementary counting techniques. First, we have the following:

$$\sum_{z \in Z} t_z = ks. \tag{7.6}$$

This is seen as follows: There are s possible public keys. For each public key $Z \in S$, LAMPORT-FORGE finds inverses of k elements in Z. On the other hand, the total number of inverses computed by LAMPORT-FORGE, over all possible public keys, can be computed to be $\sum t_z$.

Second, for any $z \in Z$, the following equation holds:

$$2ks = s_z |Z|. \tag{7.7}$$

Again, this is easy to prove. Each of the s possible public keys contains $2k$ elements of Z. However, it is obvious that every element $z \in Z$ occurs in the same number of public keys. Therefore s_z is constant (i.e., independent of z) and $2ks = s_z |Z|$.

Now, let p_z denote the probability that LAMPORT-PREIMAGE(z) succeeds. It is clear that $p_z = t_z/s_z$. We compute the average value of p_z, denoted $\bar{p}$, as follows:

$$\bar{p} = \frac{1}{|Z|} \sum_{z \in Z} p_z$$

$$= \frac{1}{|Z|} \sum_{z \in Z} \frac{t_z}{s_z}$$

$$= \frac{1}{s_z |Z|} \sum_{z \in Z} t_z$$

$$= \frac{1}{2ks} \sum_{z \in Z} t_z \qquad \text{from (7.6)}$$

$$= \frac{ks}{2ks} \qquad \text{from (7.7)}$$

$$= \frac{1}{2}.$$

The *Lamport Signature Scheme* is quite elegant, but it is not of practical use. One problem is the size of the signatures it produces. For example, if we use the modular exponentiation function to construct f, as in Example 7.6, then a secure

---

Cryptosystem 7.7: *Full Domain Hash*

Let k be a positive integer; let $\mathcal{F}$ be a family of trapdoor one-way permutations such that $f : \{0, 1\}^k \to \{0, 1\}^k$ for all $f \in \mathcal{F}$; and let $G : \{0, 1\}^* \to \{0, 1\}^k$ be a "random" function. Let $\mathcal{P} = \{0, 1\}^*$ and $\mathcal{A} = \{0, 1\}^k$, and define

$$\mathcal{K} = \{(f, f^{-1}, G) : f \in \mathcal{F}\}.$$

Given a key $K = (f, f^{-1}, G)$, f^{-1} is the private key and (f, G) is the public key.

For $K = (f, f^{-1}, G)$ and $x \in \{0, 1\}^*$, define

$$\mathbf{sig}_K(x) = f^{-1}(G(x)).$$

A signature $y = (y_1, \dots, y_k) \in \{0, 1\}^k$ on the message x is verified as follows:

$$\mathbf{ver}_K(x, y) = \text{true} \Leftrightarrow f(y) = G(x).$$

---

implementation would require that p be at least 1024 bits in length. This means that each bit of the message is signed using 1024 bits. Consequently, the signature is 1024 times as long as the message! It might be more efficient to use a one-way bijection whose security is based on the infeasibility of the elliptic curve **Discrete Logarithm** problem, but the scheme still would not be very practical.

7.5.2   Full Domain Hash

In Section 5.9.2, we showed how to construct provably secure public-key cryptosystems from trapdoor one-way permutations (in the random oracle model). Practical implementations of these systems are based on the *RSA Cryptosystem* and they replace the random oracle by a hash function such as *SHA-1*. In this section, we use a trapdoor one-way permutation to construct a secure signature scheme in the random oracle model. The scheme we present is called *Full Domain Hash*. The name of this scheme comes from the requirement that the range of the random oracle be the same as the domain of the trapdoor one-way permutation used in the scheme. The scheme is presented as Cryptosystem 7.7.

Full Domain Hash uses the familiar hash-then-sign method. $G(x)$ is the message digest produced by the random oracle, G. f^{-1} is used to sign the message digest, and f is used to verify it.

Let's briefly consider an RSA-based implementation of this scheme. The function f^{-1} would be the RSA signing (i.e., decryption) function, and f would be the RSA verification (i.e., encryption) function. In order for this to be secure, we would have to take $k = 1024$, say. Now suppose that the random oracle G is replaced by the hash function *SHA-1*. *SHA-1* constructs a 160-bit message digest, so the range of the hash function, namely $\{0, 1\}^{160}$, is a very small subset of $\{0, 1\}^k = \{0, 1\}^{1024}$. In practice, it is necessary to specify some padding scheme in order to expand a 160-bit message to 1024 bits before applying f^{-1}. This is typically done using a fixed (deterministic) padding scheme.

We now proceed to our security proof, in which we assume that $\mathcal{F}$ is a family of trapdoor one-way permutations and G is a "full domain" random oracle. (Note that the security proofs we will present do not apply when the random oracle is replaced by a fully specified hash function such as *SHA-1*.) It can be proven that *Full Domain Hash* is secure against existential forgery using a chosen-message attack; however, we will only prove the easier result that *Full Domain Hash* is secure against existential forgery using a key-only attack.

As usual, the security proof is a type of reduction. We assume that there is an adversary (i.e., a randomized algorithm, which we denote by FDH-FORGE) which is able to forge signatures (with some specified probability) when it is given the public key and access to the random oracle (recall that it can query the random oracle for values $G(x)$, but there is no algorithm specified to evaluate the function G). FDH-FORGE makes some number of oracle queries, say q_h. Eventually, FDH-FORGE outputs a valid forgery with some probability, denoted by ϵ.

We construct an algorithm, FDH-INVERT, which attempts to invert randomly chosen elements $z_0 \in \{0, 1\}^k$. That is, given $z_0 \in \{0, 1\}^k$, our hope is that FDH-INVERT$(z_0) = f^{-1}(z_0)$. We now present FDH-INVERT as Algorithm 7.2.

Algorithm 7.2 is fairly simple. It basically consists of running the adversary, FDH-FORGE. Hash queries made by FDH-FORGE are handled by the function SIMG, which is a simulation of a random oracle. We have assumed that FDH-FORGE will make q_h hash queries, say $x_1, \ldots, x_{q_h}$. For simplicity, we assume that the x_i's are distinct. (If they are not, then we need to ensure that SIMG$(x_i) =$ SIMG(x_j) whenever $x_i = x_j$. This is not difficult to do; it just requires some bookkeeping, as was done in Algorithm 5.14.) We randomly choose one query, say the j_0th query, and define SIMG$(x_{j_0}) = z_0$ (z_0 is the value we are trying to invert). For all other queries, the value SIMG(x_j) is chosen to be a random number. Because z_0 is also random, it is easy to see that SIMG is indistinguishable from a true random oracle. It therefore follows that FDH-FORGE outputs a message and a valid forged signature, which we denote by (x, y), with probability ϵ. We then check to see if $f(y) = z_0$; if so, then $y = f^{-1}(z_0)$ and we have succeeded in inverting z_0.

Our main task is to analyze the success probability of the algorithm FDH-INVERT, as a function of the success probability, ϵ, of FDH-FORGE. We will assume that $\epsilon > 2^{-k}$, because a random choice of y will be a valid signature for a message x with probability 2^{-k}, and we are only interested in adversaries that have a higher success probability than a random guess. As we did above, we denote the hash queries made by FDH-FORGE by $x_1, \ldots, x_{q_h}$, where x_j is the jth hash query, $1 \leq j \leq q_h$.

Algorithm 7.2: FDH-INVERT(z_0, q_h)

  external f
  procedure SIMG(x)
    if $j > q_h$
      then return ("failure")
      else if $j = j_0$
      then $z \leftarrow z_0$
      else let $z \in \{0, 1\}^k$ be chosen at random
    $j \leftarrow j + 1$
    return (z)

  main
    choose $j_0 \in \{1, \ldots, q_h\}$ at random
    $j \leftarrow 1$
    insert the code for FDH-FORGE(f) here
    if FDH-FORGE(f) $= (x, y)$
            $\Big\{$ **if** $f(y) = z_0$
      then   **then return** (y)
                else return ("failure")

We begin by conditioning the success probability, ϵ, on whether or not $x \in \{x_1, \ldots, x_{q_h}\}$:

$$\epsilon = \mathbf{Pr}[\text{FDH-FORGE succeeds} \wedge (x \in \{x_1, \ldots, x_{q_h}\})]$$
$$+ \mathbf{Pr}[\text{FDH-FORGE succeeds} \wedge (x \notin \{x_1, \ldots, x_{q_h}\})]. \quad (7.8)$$

It is not hard to see that

$$\mathbf{Pr}[\text{FDH-FORGE succeeds} \wedge (x \notin \{x_1, \ldots, x_{q_h}\})] = 2^{-k}.$$

This is because the (undetermined) value SIMG(x) is equally likely to take on any given value in $\{0, 1\}^k$, and hence the probability that SIMG(x) $= f(y)$ is 2^{-k}. (This is where we use the assumption that the hash function is a "full domain" hash.) Substituting into (7.8), we obtain the following:

$$\mathbf{Pr}[\text{FDH-FORGE succeeds} \wedge (x \in \{x_1, \ldots, x_{q_h}\})] \geq \epsilon - 2^{-k}. \quad (7.9)$$

Now we turn to the success probability of FDH-INVERT. The next inequality is obvious:

$$\mathbf{Pr}[\text{FDH-INVERT succeeds}] \geq \mathbf{Pr}[\text{FDH-FORGE succeeds} \wedge (x = x_{j_0})].$$
$$(7.10)$$

Our final observation is that

$$\mathbf{Pr}[\text{FDH-Forge succeeds} \wedge (x = x_{j_0})]$$

$$= \frac{1}{q_h} \times \mathbf{Pr}[\text{FDH-Forge succeeds} \wedge (x \in \{x_1, \ldots, x_{q_h}\})]. \quad (7.11)$$

Note that equation (7.11) is true because there is a $1/q_h$ chance that $x = x_{j_0}$, given that $x \in \{x_1, \ldots, x_{q_h}\}$. Now, if we combine (7.9), (7.10) and (7.11), then we obtain the following bound:

$$\mathbf{Pr}[\text{FDH-Invert succeeds}\,] \geq \frac{\epsilon - 2^{-k}}{q_h}. \quad (7.12)$$

Therefore we have obtained a concrete lower bound on the success probability of FDH-Invert. We have proven the following result.

THEOREM 7.2 *Suppose there exists an algorithm* FDH-Forge *that will output an existential forgery for* **Full Domain Hash** *with probability* $\epsilon > 2^{-k}$ *after making* q_h *queries to the random oracle, using a key-only attack. Then there exists an algorithm* FDH-Invert *that will find inverses of random elements* $z_0 \in \{0, 1\}^k$ *with probability at least* $(\epsilon - 2^{-k})/q_h$.

Observe that the usefulness of the resulting inversion algorithm depends on the ability of FDH-Forge to find forgeries using as few hash queries as possible.

7.6 Undeniable Signatures

Undeniable signature schemes were introduced by Chaum and van Antwerpen in 1989. They have several novel features. Primary among these is that a signature cannot be verified without the cooperation of the signer, Alice. This protects Alice against the possibility that documents signed by her are duplicated and distributed electronically without her approval. The verification will be accomplished by means of a *challenge-and-response protocol*.

If Alice's cooperation is required to verify a signature, what is to prevent Alice from disavowing a signature she made at an earlier time? Alice might claim that a valid signature is a forgery, and either refuse to verify it, or carry out the protocol in such a way that the signature will not be verified. To prevent this from happening, an undeniable signature scheme incorporates a *disavowal protocol* by which Alice can prove that a signature is a forgery. Thus, Alice will be able to prove "in court" that a forged signature is, in fact, a forgery. (If she refuses to take part in the disavowal protocol, this would be regarded as evidence that the signature is, in fact, genuine.)

Cryptosystem 7.8: *Chaum-van Antwerpen Signature Scheme*

Let $p = 2q + 1$ be a prime such that q is prime and the discrete log problem in $\mathbb{Z}_p^*$ is intractable. Let $\alpha \in \mathbb{Z}_p^*$ be an element of order q. Let $1 \leq a \leq q - 1$ and define $\beta = \alpha^a \bmod p$. Let G denote the multiplicative subgroup of $\mathbb{Z}_p^*$ of order q (G consists of the quadratic residues modulo p). Let $\mathcal{P} = \mathcal{A} = G$, and define

$$\mathcal{K} = \{(p, \alpha, a, \beta) : \beta \equiv \alpha^a \pmod{p}\}.$$

The values p, α and β are the public key, and a is the private key.

For $K = (p, \alpha, a, \beta)$ and $x \in G$, define

$$y = \mathbf{sig}_K(x) = x^a \bmod p.$$

For $x, y \in G$, verification is done by executing the following protocol:

1. Bob chooses e_1, e_2 at random, $e_1, e_2 \in \mathbb{Z}_q$.
2. Bob computes $c = y^{e_1} \beta^{e_2} \bmod p$ and sends it to Alice.
3. Alice computes $d = c^{a^{-1} \bmod q} \bmod p$ and sends it to Bob.
4. Bob accepts y as a valid signature if and only if

$$d \equiv x^{e_1} \alpha^{e_2} \pmod{p}.$$

Thus, an undeniable signature scheme consists of three components: a signing algorithm, a verification protocol, and a disavowal protocol. First, we present the signing algorithm and verification protocol of the *Chaum-van Antwerpen Signature Scheme* as Cryptosystem 7.8.

We should explain the roles of p and q in this scheme. The scheme lives in $\mathbb{Z}_p$; however, we need to be able to do computations in a multiplicative subgroup G of $\mathbb{Z}_p^*$ of prime order. In particular, we need to be able to compute inverses modulo $|G|$, which is why $|G|$ should be prime. It is convenient to take $p = 2q + 1$ where q is prime. In this way, the subgroup G is as large as possible, which is desirable since messages and signatures are both elements of G.

We first prove that Bob will accept a valid signature. In the following computations, all exponents are to be reduced modulo q. First, we show that

$$d \equiv c^{a^{-1}} \pmod{p}$$

$$\equiv y^{e_1 a^{-1}} \beta^{e_2 a^{-1}} \pmod{p}.$$

Since

$$\beta \equiv \alpha^a \pmod{p},$$

we have that
$$\beta^{a^{-1}} \equiv \alpha \pmod{p}.$$

Similarly,
$$y \equiv x^a \pmod{p}$$

implies that
$$y^{a^{-1}} \equiv x \pmod{p}.$$

Hence,
$$d \equiv x^{e_1} \alpha^{e_2} \pmod{p},$$

as desired.

Here is a small example to illustrate.

Example 7.7 Suppose we take $p = 467$. Since 2 is a primitive element, $2^2 = 4$ is a generator of G, the quadratic residues modulo 467. So we can take $\alpha = 4$. Suppose $a = 101$; then

$$\beta = \alpha^a \bmod 467 = 449.$$

Alice will sign the message $x = 119$ with the signature

$$y = 119^{101} \bmod 467 = 129.$$

Now, suppose Bob wants to verify the signature y. Suppose he chooses the random values $e_1 = 38$, $e_2 = 397$. He will compute $c = 13$, whereupon Alice will respond with $d = 9$. Bob checks the response by verifying that

$$119^{38} 4^{397} \equiv 9 \pmod{467}.$$

Hence, Bob accepts the signature as valid. ▯

We next prove that Alice cannot fool Bob into accepting a fraudulent signature as valid, except with a very small probability. This result does not depend on any computational assumptions, i.e., the security is unconditional.

THEOREM 7.3 If $y \not\equiv x^a \pmod{p}$, then Bob will accept y as a valid signature for x with probability at most $1/q$.

PROOF We will assume that Alice actually sends a response d which is in the group G (if she does not do so, then Bob will reject) First, we observe that each possible challenge c corresponds to exactly q ordered pairs (e_1, e_2) (this is because y and β are both elements of the multiplicative group G of prime order q). Now, when Alice receives the challenge c, she has no way of knowing which of the q possible ordered pairs (e_1, e_2) Bob used to construct c. We claim that, if $y \not\equiv x^a \pmod{p}$, then any possible response $d \in G$ that Alice might make is consistent with exactly one of the q possible ordered pairs (e_1, e_2).

Since α generates G, we can write any element of G as a power of α, where the exponent is defined uniquely modulo q. So write $c = \alpha^i$, $d = \alpha^j$, $x = \alpha^k$, and $y = \alpha^\ell$, where $i, j, k, \ell \in \mathbb{Z}_q$ and all arithmetic is modulo p. Consider the following two congruences:

$$c \equiv y^{e_1} \beta^{e_2} \pmod{p}$$

$$d \equiv x^{e_1} \alpha^{e_2} \pmod{p}.$$

This system is equivalent to the following system:

$$i \equiv \ell e_1 + a e_2 \pmod{q}$$

$$j \equiv k e_1 + e_2 \pmod{q}.$$

Now, we are assuming that

$$y \not\equiv x^a \pmod{p},$$

so it follows that

$$\ell \not\equiv ak \pmod{q}.$$

Hence, the coefficient matrix of this system of congruences modulo q has non-zero determinant, and thus there is a unique solution to the system. That is, every $d \in G$ is the correct response for exactly one of the q possible ordered pairs (e_1, e_2). Consequently, the probability that Alice gives Bob a response d that will be verified is exactly $1/q$, and the theorem is proved. ∎

We now turn to the disavowal protocol. This protocol consists of two runs of the verification protocol and is presented as Algorithm 7.3.

Algorithm 7.3: DISAVOWAL

1. Bob chooses e_1, e_2 at random, $e_1, e_2 \in \mathbb{Z}_q^*$
2. Bob computes $c = y^{e_1} \beta^{e_2} \bmod p$ and sends it to Alice
3. Alice computes $d = c^{a^{-1} \bmod q} \bmod p$ and sends it to Bob
4. Bob verifies that $d \not\equiv x^{e_1} \alpha^{e_2} \pmod{p}$
5. Bob chooses f_1, f_2 at random, $f_1, f_2 \in \mathbb{Z}_q^*$
6. Bob computes $C = y^{f_1} \beta^{f_2} \bmod p$ and sends it to Alice
7. Alice computes $D = C^{a^{-1} \bmod q} \bmod p$ and sends it to Bob
8. Bob verifies that $D \not\equiv x^{f_1} \alpha^{f_2} \pmod{p}$
9. Bob concludes that y is a forgery if and only if

$$(d\alpha^{-e_2})^{f_1} \equiv (D\alpha^{-f_2})^{e_1} \pmod{p}.$$

Steps 1 to 4 and steps 5 to 8 comprise two unsuccessful runs of the verification protocol. Step 9 is a "consistency check" that enables Bob to determine if Alice is forming her responses in the manner specified by the protocol.

The following example illustrates the disavowal protocol.

Example 7.8   As before, suppose $p = 467$, $\alpha = 4$, $a = 101$ and $\beta = 449$. Suppose the message $x = 286$ is signed with the (bogus) signature $y = 83$, and Alice wants to convince Bob that the signature is invalid.

Suppose Bob begins by choosing the random values $e_1 = 45$, $e_2 = 237$. Bob computes $c = 305$ and Alice responds with $d = 109$. Then Bob computes

$$286^{45}4^{237} \bmod 467 = 149.$$

Since $149 \neq 109$, Bob proceeds to step 5 of the protocol.

Now suppose Bob chooses the random values $f_1 = 125$, $f_2 = 9$. Bob computes $C = 270$ and Alice responds with $D = 68$. Bob computes

$$286^{125}4^9 \bmod 467 = 25.$$

Since $25 \neq 68$, Bob proceeds to step 9 of the protocol and performs the consistency check. This check succeeds, since

$$(109 \times 4^{-237})^{125} \equiv 188 \pmod{467}$$

and

$$(68 \times 4^{-9})^{45} \equiv 188 \pmod{467}.$$

Hence, Bob is convinced that the signature is invalid. ▢

We have to prove two things at this point:

1. Alice can convince Bob that an invalid signature is a forgery.

2. Alice cannot make Bob believe that a valid signature is a forgery except with a very small probability.

THEOREM 7.4   *If $y \not\equiv x^a \pmod{p}$ and Bob and Alice follow the disavowal protocol, then*

$$(da^{-e_2})^{f_1} \equiv (Da^{-f_2})^{e_1} \pmod{p}.$$

PROOF   Using the facts that

$$d \equiv c^{a^{-1}} \pmod{p},$$

$$c \equiv y^{e_1}\beta^{e_2} \pmod{p}, \quad \text{and}$$

$$\beta \equiv \alpha^a \pmod{p},$$

we have the following:

$$(d\alpha^{-e_2})^{f_1} \equiv \left((y^{e_1}\beta^{e_2})^{a^{-1}}\alpha^{-e_2}\right)^{f_1} \pmod{p}$$

$$\equiv y^{e_1 a^{-1} f_1}\beta^{e_2 a^{-1} f_1}\alpha^{-e_2 f_1} \pmod{p}$$

$$\equiv y^{e_1 a^{-1} f_1}\alpha^{e_2 f_1}\alpha^{-e_2 f_1} \pmod{p}$$

$$\equiv y^{e_1 a^{-1} f_1} \pmod{p}.$$

A similar computation, using the facts that $D \equiv C^{a^{-1}} \pmod{p}$, $C \equiv y^{f_1}\beta^{f_2}$ $\pmod{p}$ and $\beta \equiv \alpha^a \pmod{p}$, establishes that

$$(D\alpha^{-f_2})^{e_1} \equiv y^{e_1 a^{-1} f_1} \pmod{p},$$

so the consistency check in step 9 succeeds. ∎

Now we look at the possibility that Alice might attempt to disavow a valid signature. In this situation, we do not assume that Alice follows the protocol. That is, Alice might not construct d and D as specified by the protocol. Hence, in the following theorem, we assume only that Alice is able to produce values d and D which satisfy the conditions in steps 4, 8, and 9 of Algorithm 7.3.

THEOREM 7.5 *Suppose* $y \equiv x^a \pmod{p}$ *and Bob follows the disavowal protocol. If* $d \not\equiv x^{e_1}\alpha^{e_2} \pmod{p}$ *and* $D \not\equiv x^{f_1}\alpha^{f_2} \pmod{p}$, *then the probability that* $(d\alpha^{-e_2})^{f_1} \not\equiv (D\alpha^{-f_2})^{e_1} \pmod{p}$ *is* $1 - 1/q$.

PROOF Suppose that the following congruences are satisfied:

$$y \equiv x^a \pmod{p}$$

$$d \not\equiv x^{e_1}\alpha^{e_2} \pmod{p}$$

$$D \not\equiv x^{f_1}\alpha^{f_2} \pmod{p}$$

$$(d\alpha^{-e_2})^{f_1} \equiv (D\alpha^{-f_2})^{e_1} \pmod{p}.$$

We will derive a contradiction.

The consistency check (step 9) can be rewritten in the following form: $D \equiv d_0^{f_1}\alpha^{f_2} \pmod{p}$, where $d_0 = d^{1/e_1}\alpha^{-e_2/e_1} \bmod p$ is a value that depends only on steps 1 to 4 of the protocol.

Applying Theorem 7.3, we conclude that y is a valid signature for d_0 with probability $1 - 1/q$. But we are assuming that y is a valid signature for x. That is, with high probability we have $x^a \equiv d_0^a \pmod{p}$, which implies that $x = d_0$. However, the fact that $d \not\equiv x^{e_1}\alpha^{e_2} \pmod{p}$ means that $x \not\equiv d^{1/e_1}\alpha^{-e_2/e_1}$ $\pmod{p}$. Since $d_0 \equiv d^{1/e_1}\alpha^{-e_2/e_1} \pmod{p}$, we conclude that $x \neq d_0$ and we have a contradiction. Hence, Alice can fool Bob in this way with probability $1/q$. ∎

7.7 Fail-stop Signatures

A *fail-stop signature scheme* provides enhanced security against the possibility that a very powerful adversary might be able to forge a signature. In the event that Oscar is able to forge Alice's signature on a message, Alice will (with high probability) subsequently be able to prove that Oscar's signature is a forgery.

In this section, we describe a fail-stop signature scheme constructed by van Heyst and Pedersen in 1992. Like the *Lamport Signature Scheme*, this is a one-time signature scheme. The system consists of signing and verification algorithms, as well as a "proof of forgery" algorithm. The description of the signing and verification algorithms of the *van Heyst and Pedersen Signature Scheme* are presented as Cryptosystem 7.9.

Cryptosystem 7.9: *van Heyst and Pedersen Signature Scheme*

Let $p = 2q + 1$ be a prime such that q is prime and the discrete log problem in $\mathbb{Z}_p^*$ is intractable. Let $\alpha \in \mathbb{Z}_p^*$ be an element of order q. Let $1 \leq a_0 \leq q - 1$ and define $\beta = \alpha^{a_0} \bmod p$. The values p, q, α, β, and a_0 are chosen by a central (trusted) authority. p, q, α, and β are public and will be regarded as fixed. The value of a_0 is kept secret from everyone (even Alice).

Let $\mathcal{P} = \mathbb{Z}_q$ and $\mathcal{A} = \mathbb{Z}_q \times \mathbb{Z}_q$. A key has the form

$$K = (\gamma_1, \gamma_2, a_1, a_2, b_1, b_2),$$

where $a_1, a_2, b_1, b_2 \in \mathbb{Z}_q$,

$$\gamma_1 = \alpha^{a_1} \beta^{a_2} \bmod p \quad \text{and} \quad \gamma_2 = \alpha^{b_1} \beta^{b_2} \bmod p.$$

(γ_1, γ_2) is the public key and (a_1, a_2, b_1, b_2) is the private key.

For $K = (\gamma_1, \gamma_2, a_1, a_2, b_1, b_2)$ and $x \in \mathbb{Z}_q$, define

$$\mathbf{sig}_K(x) = (y_1, y_2),$$

where

$$y_1 = a_1 + x b_1 \bmod q \quad \text{and} \quad y_2 = a_2 + x b_2 \bmod q.$$

For $y = (y_1, y_2) \in \mathbb{Z}_q \times \mathbb{Z}_q$, we have

$$\mathbf{ver}_K(x, y) = \text{true} \Leftrightarrow \gamma_1 \gamma_2^{x} \equiv \alpha^{y_1} \beta^{y_2} \pmod{p}.$$

It is straightforward to see that a signature produced by Alice will satisfy the verification condition, so let's turn to the security aspects of this scheme and how the fail-stop property works. First we establish some important facts concerning the keys of the scheme. We begin with a definition. Two keys $(\gamma_1, \gamma_2, a_1, a_2, b_1, b_2)$ and $(\gamma_1', \gamma_2', a_1', a_2', b_1', b_2')$ are said to be *equivalent* if $\gamma_1 = \gamma_1'$ and $\gamma_2 = \gamma_2'$. It is easy to see that there are exactly q^2 keys in any equivalence class.

We establish several lemmas.

LEMMA 7.6 *Suppose K and K' are equivalent keys and suppose that $\mathbf{ver}_K(x, y) =$ true. Then $\mathbf{ver}_{K'}(x, y) =$ true.*

PROOF Suppose $K = (\gamma_1, \gamma_2, a_1, a_2, b_1, b_2)$ and $K' = (\gamma_1, \gamma_2, a_1', a_2', b_1', b_2')$, where

$$\gamma_1 = \alpha^{a_1} \beta^{a_2} \bmod p = \alpha^{a_1'} \beta^{a_2'} \bmod p$$

and

$$\gamma_2 = \alpha^{b_1} \beta^{b_2} \bmod p = \alpha^{b_1'} \beta^{b_2'} \bmod p.$$

Suppose x is signed using K, producing the signature $y = (y_1, y_2)$. Observe that $\mathbf{ver}_K(x, y)$ only depends on the values of γ_1, γ_2, x, and y, and these values are the same when $\mathbf{ver}_{K'}(x, y)$ is computed. Thus, y will also be verified using K'. ∎

LEMMA 7.7 *Suppose K is a key and $y = \mathbf{sig}_K(x)$. Then there are exactly q keys K' equivalent to K such that $y = \mathbf{sig}_{K'}(x)$.*

PROOF Suppose (γ_1, γ_2) is the public key. We want to determine the number of 4-tuples (a_1, a_2, b_1, b_2) such that the following congruences are satisfied:

$$\gamma_1 \equiv \alpha^{a_1} \beta^{a_2} \pmod{p}$$
$$\gamma_2 \equiv \alpha^{b_1} \beta^{b_2} \pmod{p}$$
$$y_1 \equiv a_1 + x b_1 \pmod{q}$$
$$y_2 \equiv a_2 + x b_2 \pmod{q}.$$

Since α generates G, there exist unique exponents $c_1, c_2, a_0 \in \mathbb{Z}_q$ such that

$$\gamma_1 \equiv \alpha^{c_1} \pmod{p},$$
$$\gamma_2 \equiv \alpha^{c_2} \pmod{p}, \quad \text{and}$$
$$\beta \equiv \alpha^{a_0} \pmod{p}.$$

Hence, it is necessary and sufficient that the following system of congruences be satisfied:

$$c_1 \equiv a_1 + a_0 a_2 \pmod{q}$$

$$c_2 \equiv b_1 + a_0 b_2 \pmod{q}$$

$$y_1 \equiv a_1 + x b_1 \pmod{q}$$

$$y_2 \equiv a_2 + x b_2 \pmod{q}.$$

This system can be written as a matrix equation in $\mathbb{Z}_q$, as follows:

$$\begin{pmatrix} 1 & a_0 & 0 & 0 \\ 0 & 0 & 1 & a_0 \\ 1 & 0 & x & 0 \\ 0 & 1 & 0 & x \end{pmatrix} \begin{pmatrix} a_1 \\ a_2 \\ b_1 \\ b_2 \end{pmatrix} = \begin{pmatrix} c_1 \\ c_2 \\ y_1 \\ y_2 \end{pmatrix}.$$

Now, the coefficient matrix of this system can be seen to have rank[2] three: Clearly, the rank is at least three because rows 1, 2 and 4 are linearly independent over $\mathbb{Z}_q$. The rank is at most three because

$$r_1 + x r_2 - r_3 - a_0 r_4 = (0, 0, 0, 0),$$

where r_i denotes the ith row of the matrix.

This system of equations has at least one solution, obtained by using the key K. Since the rank of the coefficient matrix is three, it follows that the dimension of the solution space is $4 - 3 = 1$, and hence there are exactly q solutions. The result follows. ∎

By similar reasoning, the following result can be proved. We omit the proof.

LEMMA 7.8 *Suppose K is a key, $y = \text{sig}_K(x)$, and $\text{ver}_K(x', y') = \text{true}$, where $x' \neq x$. Then there is at most one key K' equivalent to K such that $y = \text{sig}_{K'}(x)$ and $y' = \text{sig}_{K'}(x')$.*

Let's interpret what the preceding two lemmas say about the security of the scheme. Given that y is a valid signature for message x, there are q possible keys that would have signed x with y. But for any message $x' \neq x$, these q keys will produce q different signatures on x'. Thus, the following theorem results.

THEOREM 7.9 *Given that $\text{sig}_K(x) = y$ and $x' \neq x$, Oscar can compute $\text{sig}_K(x')$ with probablity $1/q$.*

---

[2] The *rank* of a matrix is the maximum number of linearly independent rows it contains.

Note that this theorem does not depend on the computational power of Oscar: the stated level of security is obtained because Oscar cannot tell which of q possible keys is being used by Alice. So the security is unconditional.

We now go on to look at the fail-stop concept. What we have said so far is that, given a signature y on message x, Oscar cannot compute Alice's signature y' on a different message x'. It is still conceivable that Oscar can compute a forged signature $y'' \neq \mathbf{sig}_K(x')$ which will still be verified. However, if Alice is given a valid forged signature, then with probability $1 - 1/q$ she can produce a proof of forgery. The proof of forgery is the value $a_0 = \log_\alpha \beta$, which is known only to the central authority.

We assume that Alice possesses a pair (x', y'') such that $\mathbf{ver}_K(x', y'') = \text{true}$ and $y'' \neq \mathbf{sig}_K(x')$. That is,

$$\gamma_1 \gamma_2^{x'} \equiv \alpha^{y_1''} \beta^{y_2''} \pmod{p},$$

where $y'' = (y_1'', y_2'')$. Now, Alice can compute her own signature on x', namely $y' = (y_1', y_2')$, and it will be the case that

$$\gamma_1 \gamma_2^{x'} \equiv \alpha^{y_1'} \beta^{y_2'} \pmod{p}.$$

Hence,

$$\alpha^{y_1''} \beta^{y_2''} \equiv \alpha^{y_1'} \beta^{y_2'} \pmod{p}.$$

Writing $\beta = \alpha^{a_0} \bmod p$, we have that

$$\alpha^{y_1'' + a_0 y_2''} \equiv \alpha^{y_1' + a_0 y_2'} \pmod{p},$$

or

$$y_1'' + a_0 y_2'' \equiv y_1' + a_0 y_2' \pmod{q}.$$

This simplifies to give

$$y_1'' - y_1' \equiv a_0(y_2' - y_2'') \pmod{q}.$$

Now, $y_2' \not\equiv y_2'' \pmod{q}$ since y' is a forgery. Hence, $(y_2' - y_2'')^{-1} \bmod q$ exists, and

$$a_0 = \log_\alpha \beta = (y_1'' - y_1')(y_2' - y_2'')^{-1} \bmod q.$$

Of course, by accepting such a proof of forgery, we assume that Alice cannot compute the discrete logarithm $\log_\alpha \beta$ by herself. This is a computational assumption.

Finally, we remark that the scheme is a one-time scheme since Alice's key K can easily be computed if two messages are signed using K.

We close with an example illustrating how Alice can produce a proof of forgery.

Example 7.9   Suppose $p = 3467 = 2 \times 1733 + 1$. The element $\alpha = 4$ has order 1733 in $\mathbb{Z}_{3467}^*$. Suppose that $a_0 = 1567$, so

$$\beta = 4^{1567} \bmod 3467 = 514.$$

(Recall that Alice knows the values of α and β, but not a_0.) Suppose Alice forms her key using $a_1 = 888, a_2 = 1024, b_1 = 786$ and $b_2 = 999$, so

$$\gamma_1 = 4^{888} 514^{1024} \bmod 3467 = 3405$$

and

$$\gamma_2 = 4^{786} 514^{999} \bmod 3467 = 2281.$$

Now, suppose Alice is presented with the forged signature $(822, 55)$ on the message 3383. This is a valid signature because the verification condition is satisfied:

$$3405 \times 2281^{3383} \equiv 2282 \pmod{3467}$$

and

$$4^{822} 514^{55} \equiv 2282 \pmod{3467}.$$

On the other hand, this is not the signature Alice would have constructed. Alice can compute her own signature to be

$$(888 + 3383 \times 786 \bmod 1733, 1024 + 3383 \times 999 \bmod 1733) = (1504, 1291).$$

Then, she proceeds to calculate the secret discrete logarithm

$$a_0 = (822 - 1504)(1291 - 55)^{-1} \bmod 1733 = 1567.$$

This is the proof of forgery. $\Box$

7.8   Notes and References

For a nice survey of signature schemes, we recommend Mitchell, Piper and Wild [243]. This paper also contains the two methods of forging ElGamal signatures that we presented in Section 7.3. Pedersen [262] is a more recent survey that is also worth reading.

The *ElGamal Signature Scheme* was presented by ElGamal [125], and the *Schnorr Signature Scheme* is due to Schnorr [294]. Another popular scheme, which we have not discussed in this book, is the *Fiat-Shamir Signature Scheme* [135].

The *Digital Signature Algorithm* was first published by NIST in August 1991, and it was adopted as FIPS 186 in December 1994 [143]. There is a lengthy discussion of *DSA* and some controversies surrounding it in the July 1992 issue of the *Communications of the ACM*; for a response by NIST to some of the questions raised, see [311]. FIPS 186-2 [144] is a revised version of the standard, which now includes the *RSA Signature Scheme* as well as the *Elliptic Curve Digital Signature Algorithm*. A complete description of the *ECDSA* is found in Johnson, Menezes and Vanstone [183].

The *Lamport Signature Scheme* is described in the 1976 paper by Diffie and Hellman [117]. A more efficient modification, due to Lamport and (independently) Bos and Chaum, is described in [64]. A more general treatment of the construction of signature schemes from arbitrary one-way functions is given by Bleichenbacher and Maurer [48].

Full Domain Hash is due to Bellare and Rogaway [22, 26]. The paper [26] also includes a more efficient variant, known as the *Probabilistic Signature Scheme* (*PSS*). Provably secure ElGamal-type schemes have also been studied; see, for example, Pointcheval and Stern [271].

The undeniable signature scheme presented in Section 7.6 is due to Chaum and van Antwerpen [91]. The fail-stop signature scheme from Section 7.7 is due to van Heyst and Pedersen [332]; see Pfitzmann [265] for an expanded treatment of this topic.

Some of the Exercises point out some security problems with ElGamal type schemes if the "k" values are reused or generated in a predictable fashion. There are now several works that pursue this theme; see, for example, Bellare, Goldwasser and Micciancio [17] and Nguyen and Shparlinski [253].

Exercises

7.1 Suppose Alice is using the *ElGamal Signature Scheme* with $p = 31847$, $\alpha = 5$ and $\beta = 25703$. Compute the values of k and a (without solving an instance of the **Discrete Logarithm** problem), given the signature $(23972, 31396)$ for the message $x = 8990$ and the signature $(23972, 20481)$ for the message $x = 31415$.

7.2 Suppose I implement the *ElGamal Signature Scheme* with $p = 31847$, $\alpha = 5$ and $\beta = 26379$. Write a computer program which does the following:

    (a) Verify the signature $(20679, 11082)$ on the message $x = 20543$.

    (b) Determine my private key, a, by solving an instance of the **Discrete Logarithm** problem.

    (c) Then determine the random value k used in signing the message x, without solving an instance of the **Discrete Logarithm** problem.

7.3 Suppose that Alice is using the *ElGamal Signature Scheme*. In order to save time in generating the random numbers k that are used to sign messages, Alice chooses an initial random value k_0, and then signs the ith message using the value $k_i = k_0 + 2i \bmod (p - 1)$ (therefore $k_i = k_{i-1} + 2 \bmod (p - 1)$ for all $i \geq 1$).

    (a) Suppose that Bob observes two consecutive signed messages, say $(x_i, \mathbf{sig}(x_i, k_i))$ and $(x_{i+1}, \mathbf{sig}(x_{i+1}, k_{i+1}))$. Describe how Bob can easily compute Alice's secret key, a, given this information, without solving an instance of the **Discrete Logarithm** problem. (Note that the value of i does not have to be known for the attack to succeed.)

(b) Suppose that the parameters of the scheme are $p = 28703$, $\alpha = 5$ and $\beta = 11339$, and the two messages observed by Bob are

$$x_i = 12000 \qquad \textbf{sig}(x_i, k_i) = (26530, 19862)$$
$$x_{i+1} = 24567 \qquad \textbf{sig}(x_{i+1}, k_{i+1}) = (3081, 7604).$$

Find the value of a using the attack you described in part (a).

7.4   (a) Prove that the second method of forgery on the *ElGamal Signature Scheme*, described in Section 7.3, also yields a signature that satisfies the verification condition.

      (b) Suppose Alice is using the *ElGamal Signature Scheme* as implemented in Example 7.1: $p = 467$, $\alpha = 2$ and $\beta = 132$. Suppose Alice has signed the message $x = 100$ with the signature $(29, 51)$. Compute the forged signature that Oscar can then form by using $h = 102$, $i = 45$ and $j = 293$. Check that the resulting signature satisfies the verification condition.

7.5   (a) A signature in the *ElGamal Signature Scheme* or the *DSA* is not allowed to have $\delta = 0$. Show that if a message were signed with a "signature" in which $\delta = 0$, then it would be easy for an adversary to compute the secret key, a.

      (b) A signature in the *DSA* is not allowed to have $\gamma = 0$. Show that if a "signature" in which $\gamma = 0$ is known, then the value of k used in that "signature" can be determined. Given that value of k, show that it is now possible to forge a "signature" (with $\gamma = 0$) for any desired message (i.e., a selective forgery can be carried out).

      (c) Evaluate the consequences of allowing a signature in the *ECDSA* to have $r = 0$ or $s = 0$.

7.6 Here is a variation of the *ElGamal Signature Scheme*. The key is constructed in a similar manner as before: Alice chooses $\alpha \in \mathbb{Z}_p^*$ to be a primitive element, $0 \leq a \leq p - 2$ where $\gcd(a, p - 1) = 1$, and $\beta = \alpha^a \bmod p$. The key $K = (\alpha, a, \beta)$, where α and β are the public key and a is the private key. Let $x \in \mathbb{Z}_p$ be a message to be signed. Alice computes the signature $\textbf{sig}(x) = (\gamma, \delta)$, where

$$\gamma = \alpha^k \bmod p$$

and

$$\delta = (x - k\gamma)a^{-1} \bmod (p - 1).$$

The only difference from the original *ElGamal Signature Scheme* is in the computation of δ. Answer the following questions concerning this modified scheme.

      (a) Describe how a signature (γ, δ) on a message x would be verified using Alice's public key.

      (b) Describe a computational advantage of the modified scheme over the original scheme.

      (c) Briefly compare the security of the original and modified scheme.

7.7 Suppose Alice uses the *DSA* with $q = 101$, $p = 7879$, $\alpha = 170$, $a = 75$ and $\beta = 4567$, as in Example 7.4. Determine Alice's signature on a message x such that $\text{SHA-1}(x) = 52$, using the random value $k = 49$, and show how the resulting signature is verified.

7.8 We showed that using the same value k to sign two messages in the *ElGamal Signature Scheme* allows the scheme to be broken (i.e., an adversary can determine the

secret key without solving an instance of the **Discrete Logarithm** problem). Show
how similar attacks can be carried out for the *Schnorr Signature Scheme*, the *DSA*
and the *ECDSA*.

7.9 Suppose that $x_0 \in \{0, 1\}^*$ is a bitstring such that $\text{SHA-1}(x_0) = 0\,0\cdots 0$. Therefore, when used in *DSA* or *ECDSA*, we have that $\text{SHA-1}(x_0) \equiv 0 \pmod{q}$.

    (a) Show how it is possible to forge a *DSA* signature for the message x_0.

        HINT   Let $\delta = \gamma$, where γ is chosen appropriately.

    (b) Show how it is possible to forge an *ECDSA* signature for the message x_0.

7.10   (a) We describe a potential attack against the *DSA*. Suppose that x is given, let
$z = (\text{SHA-1}(x))^{-1} \bmod q$, and let $\epsilon = \beta^z \bmod p$. Now suppose it is
possible to find $\gamma, \lambda \in \mathbb{Z}_q^*$ such that

$$\left((\alpha\,\epsilon^\gamma)^{\lambda^{-1}\bmod q} \right) \bmod p \bmod q = \gamma.$$

        Define $\delta = \lambda\,\text{SHA-1}(x) \bmod q$. Prove that (γ, δ) is a valid signature for x.

    (b) Describe a similar (possible) attack against the *ECDSA*.

7.11 In a verification of a signature constructed using the *ElGamal Signature Scheme*
(or many of its variants), it is necessary to compute a value of the form $\alpha^c \beta^d$. If
c and d are random ℓ-bit exponents, then a straightforward use of the SQUARE-
AND-MULTIPLY algorithm would require (on average) $\ell/2$ multiplications and ℓ
squarings to compute each of α^c and β^d. The purpose of this exercise is to show
that the product $\alpha^c \beta^d$ can be computed much more efficiently.

    (a) Suppose that c and d are represented in binary, as in Algorithm 5.5. Suppose
also that the product $\alpha\beta$ is precomputed. Describe a modification of Algorithm 5.5, in which at most one multiplication is performed in each iteration
of the algorithm.

    (b) Suppose that $c = 26$ and $d = 17$. Show how your algorithm would compute
$\alpha^c \beta^d$, i.e., what are the values of the exponents i and j at the end of each
iteration of your algorithm (where $z = \alpha^i \beta^j$).

    (c) Explain why, on average, this algorithm requires ℓ squarings and $3\ell/4$ mul-
tiplications to compute $\alpha^c \beta^d$, if c and d are randomly chosen ℓ-bit integers.

    (d) Estimate the average speedup achieved, as compared to using the original
SQUARE-AND-MULTIPLY algorithm to compute α^c and β^d separately, assuming that a squaring operation takes roughly the same time as a multiplication operation.

7.12 Prove that a correctly constructed signature in the *ECDSA* will satisfy the verification condition.

7.13 Let E denote the elliptic curve $y^2 \equiv x^3 + x + 26 \bmod 127$. It can be shown that
$\#E = 131$, which is a prime number. Therefore any non-identity element in E is a
generator for $(E, +)$. Suppose the *ECDSA* is implemented in E, with $A = (2, 6)$
and $m = 54$.

    (a) Compute the public key $B = mA$.

    (b) Compute the signature on a message x if $\text{SHA-1}(x) = 10$, when $k = 75$.

    (c) Show the computations used to verify the signature constructed in part (b).

7.14 In the *Lamport Signature Scheme*, suppose that two k-tuples, x and x', were signed
by Alice using the same key. Let ℓ denote the number of coordinates in which x and
x' differ, i.e.,

$$\ell = |\{i : x_i \neq x'_i\}|.$$

Show that Oscar can now sign $2^\ell - 2$ new messages.

7.15 Suppose Alice is using the *Chaum-van Antwerpen Signature Scheme* as in Example 7.7. That is, $p = 467$, $\alpha = 4$, $a = 101$ and $\beta = 449$. Suppose Alice is presented with a signature $y = 25$ on the message $x = 157$ and she wishes to prove it is a forgery. Suppose Bob's random numbers are $e_1 = 46$, $e_2 = 123$, $f_1 = 198$ and $f_2 = 11$ in the disavowal protocol. Compute Bob's challenges, c and C, and Alice's responses, d and D, and show that Bob's consistency check will succeed.

7.16 Prove that each equivalence class of keys in the *Pedersen-van Heyst Signature Scheme* contains q^2 keys.

7.17 Suppose Alice is using the *Pedersen-van Heyst Signature Scheme*, where $p = 3467$, $\alpha = 4$, $a_0 = 1567$ and $\beta = 514$ (of course, the value of a_0 is not known to Alice).

(a) Using the fact that $a_0 = 1567$, determine all possible keys

$$K = (\gamma_1, \gamma_2, a_1, a_2, b_1, b_2)$$

such that $\mathbf{sig}_K(42) = (1118, 1449)$.

(b) Suppose that $\mathbf{sig}_K(42) = (1118, 1449)$ and $\mathbf{sig}_K(969) = (899, 471)$. Without using the fact that $a_0 = 1567$, determine the value of K (this shows that the scheme is a one-time scheme).

7.18 Suppose Alice is using the *Pedersen-van Heyst Signature Scheme* with $p = 5087$, $\alpha = 25$ and $\beta = 1866$. Suppose the key is

$$K = (5065, 5076, 144, 874, 1873, 2345).$$

Now, suppose Alice finds the signature $(2219, 458)$ has been forged on the message 4785.

(a) Prove that this forgery satisfies the verification condition, so it is a valid signature.

(b) Show how Alice will compute the proof of forgery, a_0, given this forged signature.

8

Pseudo-random Number Generation

8.1 Introduction and Examples

There are many situations in cryptography where it is important to be able to generate random numbers, bitstrings, etc. For example, cryptographic keys are normally generated at random from a specified keyspace, and many encryption schemes and signature schemes require random numbers to be generated during their execution. Generating random numbers by means of coin tosses or other physical processes is time-consuming and expensive, so in practice it is common to use a *pseudo-random bit generator*. A bit generator starts with a short random bitstring (a *seed*) and expands it into a much longer bitstring. Thus, a bit generator reduces the number of random bits that are required in a cryptographic application.

More formally, we have the following definition.

Definition 8.1: Let k, ℓ be positive integers such that $\ell \geq k+1$. A (k, ℓ)-*bit generator* is a function $f : (\mathbb{Z}_2)^k \to (\mathbb{Z}_2)^\ell$ that can be computed in polynomial time (as a function of k). The input $s_0 \in (\mathbb{Z}_2)^k$ is called the *seed*, and the output $f(s_0) \in (\mathbb{Z}_2)^\ell$ is called the *generated bitstring*. It will always be required that ℓ is a polynomial function of k.

The function f is deterministic, so the bitstring $f(s_0)$ is dependent only on the seed. Our goal is that the generated bitstring $f(s_0)$ should "look like" truly random bits, given that the seed is chosen at random. If this property holds, then the bit generator will be "secure" and it will be termed a *pseudo-random bit generator* (or *PRBG*). Giving a precise definition of security of a bit generator is quite difficult, but we will try to give an intuitive explanation of the concept as we progress through this chapter.

One specific motivating example for studying pseudo-random bit generators is as follows. Recall the concept of perfect secrecy that we studied in Chapter 2. One realization of perfect secrecy is the *One-time Pad*, where the plaintext and the key are both bitstrings of a specified length, and the ciphertext is constructed by taking the bitwise exclusive-or of the plaintext and the key. The practical difficulty of the *One-time Pad* is that the key, which must be randomly generated and communicated over a secure channel, must be as long as the plaintext in order to ensure perfect secrecy. Pseudo-random bit generators provide a possible way of alleviating this problem. Suppose Alice and Bob agree on a PRBG and communicate a seed over the secure channel. Alice and Bob can then both compute the same string of pseudo-random bits, which will be xor-ed with the plaintext. Thus the seed functions as a key, and the PBRG can be thought of as a keystream generator for a stream cipher. (Of course, when we replace the one-time pad with a string of pseudo-random bits, we lose the perfect secrecy property.)

Random number generators are used in numerous areas of computer science in addition to cryptography. Applications include simulations, Monte Carlo algorithms, sampling, testing, any many others. Often, it is sufficient for generated random numbers to have a relatively uniform distribution. Various characteristics of a sequence of pseudorandom numbers can be measured, such as frequencies, runs and gaps between the numbers in the sequence. The uniformity of these statistics are often evaluated using traditional tests such as a χ^2 test.

The security of a generator for cryptographic use is more difficult to achieve, however. Passing the above-mentioned tests would certainly be necessary for cryptographic security, but it is not sufficient. Cryptographic security relies on notations such as polynomial-time predictability, which is introduced in Section 8.2.1.

We now present some well-known bit generators to motivate and illustrate some of the concepts we will be studying. First, we observe that a linear feedback shift register, as described in Section 1.1.7, can be thought of as a bit generator. Given a k-bit seed, an LFSR of degree k can be used to produce as many as $2^k - k - 1$ further bits before repeating. The bit generator obtained from an LFSR is very insecure: we already observed in Section 1.2.5 that knowledge of any $2k$ consecutive generated bits suffice to allow the seed to be determined, and hence the entire sequence can then be reconstructed by an opponent. (Although we have not yet defined security of a bit generator, it should be clear that the existence of an attack of this type means that the generator is insecure!)

Another well-known (but insecure) bit generator, called the *Linear Congruential Generator*, is presented as Algorithm 8.1. The basic idea is to generate a sequence of residues modulo M, where each element in the sequence is a certain linear function modulo M of the previous element in the sequence. The seed will be a residue modulo M, and the least significant bits of the elements in the sequence form the generated bitstring.

Here is a very small example to illustrate the workings of the *Linear Congruential Generator*. This example also demonstrates the periodic nature of PRBGs, i.e., that PRBGs eventually repeat if a sufficient number of bits are generated.

Algorithm 8.1: *Linear Congruential Generator*

Suppose $M \geq 2$ is an integer and suppose $1 \leq a, b \leq M - 1$. Define $k = 1 + \lfloor \log_2 M \rfloor$ and let $k + 1 \leq \ell \leq M - 1$.

The seed is an integer s_0, where $0 \leq s_0 \leq M - 1$. (Observe that the binary representation of a seed is a bitstring of length not exceeding k; however, not all bitstrings of length k are permissible seeds.) Now, define

$$s_i = (a s_{i-1} + b) \bmod M$$

for $1 \leq i \leq \ell$, and then define

$$f(s_0) = (z_1, z_2, \ldots, z_\ell),$$

where

$$z_i = s_i \bmod 2,$$

$1 \leq i \leq \ell$. Then f is a (k, ℓ)-*Linear Congruential Generator*.

Example 8.1 Suppose we construct a $(5, 10)$-bit generator by taking $M = 31$, $a = 3$ and $b = 5$ in the *Linear Congruential Generator*. Suppose we consider the mapping $s \mapsto 3s + 5 \bmod 31$. Then $13 \mapsto 13$, and the other 30 residues are permuted in a cycle of length 30, namely

$$0, 5, 20, 3, 14, 16, 22, 9, 1, 8,$$
$$29, 30, 2, 11, 7, 26, 21, 6, 23, 12,$$
$$10, 4, 17, 25, 18, 28, 27, 24, 15, 19.$$

If the seed is anything other than 13, then the seed specifies a starting point in this cycle, and the next 10 elements, reduced modulo 2, form the pseudo-random sequence.

The 31 possible bitstrings produced by this generator are shown in Table 8.1. For example, the sequence constructed from the seed 0 is obtained by taking the ten integers following 0 in the above list, namely, $5, 20, 3, 14, 16, 22, 9, 1, 8, 29$, and reducing them modulo 2. ⬚

We can use some concepts developed in earlier chapters to construct bit generators. For example, the output feedback mode of a block cipher, as described in Section 3.7, can be thought of as a bit generator; moreover, it appears to be computationally secure if the underlying block cipher satisfies some reasonable security properties.

Another approach in constructing very fast bit generators is to combine LFSRs

TABLE 8.1
Bitstrings produced by the linear congruential generator

seed	sequence		seed	sequence
0	1010001101		16	0110100110
1	0100110101		17	1001011010
2	1101010001		18	0101101010
3	0001101001		19	0101000110
4	1100101101		20	1000110100
5	0100011010		21	0100011001
6	1000110010		22	1101001101
7	0101000110		23	0001100101
8	1001101010		24	1101010001
9	1010011010		25	0010110101
10	0110010110		26	1010001100
11	1010100011		27	0110101000
12	0011001011		28	1011010100
13	1111111111		29	0011010100
14	0011010011		30	0110101000
15	1010100011			

in some way so that the output looks non-linear. One such method, due to Coppersmith, Krawczyk and Mansour, is called the *Shrinking Generator*. Suppose we have two LFSRs, one of degree k_1 and one of degree k_2. We will require a total of $k_1 + k_2$ bits as our seed, in order to initialize both LFSRs. The first LFSR will produce a sequence of bits, say $a_1, a_2, \ldots$, and the second one produces a sequence of bits $b_1, b_2, \ldots$. Then we define a sequence of bits $z_1, z_2, \ldots$ by the rule

$$z_i = a_{i_k},$$

where i_k is the position of the kth "1" in the sequence $b_1, b_2, \ldots$. These bits comprise a subsequence of the bits generated by the first LFSR.

In the rest of this chapter, we will investigate bit generators that can be proved to be secure given some plausible computational assumption. There are PRBGs based on the fundamental problems of factoring (as it relates to the *RSA Cryptosystem*) and the **Discrete Logarithm** problem. A PRBG based on the RSA encryption function is shown in Algorithm 8.2, and a PRBG based on the **Discrete Logarithm** problem is discussed in the Exercises. Briefly, the *RSA Generator* chooses an initial element of $\mathbb{Z}_N$ to be the seed. A sequence of elements of $\mathbb{Z}_N$ is formed, in which each element in the sequence is the RSA encryption of the previous element. Then the least significant bits of the elements in the sequence form the generated bitstring.

We now give an example of the *RSA Generator*.

Example 8.2 Suppose $n = 91261 = 263 \times 347$, $b = 1547$, and $s_0 = 75634$.

Algorithm 8.2: *RSA Generator*

Suppose p, q are two $(k/2)$-bit primes, and define $n = pq$. Suppose b is chosen such that $\gcd(b, \phi(n)) = 1$. As always, n and b are public while p and q are secret.

A seed s_0 is any element of $\mathbb{Z}_n^*$, so s_0 has k bits. For $i \geq 1$, define

$$s_{i+1} = s_i^b \bmod n,$$

and then define

$$f(s_0) = (z_1, z_2, \ldots, z_\ell),$$

where

$$z_i = s_i \bmod 2,$$

$1 \leq i \leq \ell$. Then f is a (k, ℓ)-*RSA Generator*.

The first 20 bits produced by the *RSA Generator* are computed as shown in Table 8.2. The bitstring resulting from this seed is

$$10000111011110011000.$$

⬚

8.2 Indistinguishability of Probability Distributions

There are two main objectives of a pseudo-random number generator: it should be fast (i.e., computable in polynomial time as a function of k) and it should be secure. Of course, these two requirements are often conflicting. The bit generators based on linear congruences or linear feedback shift registers are indeed very fast. These bit generators are quite useful in simulations, but they are very insecure for cryptographic applications.

Let us now try to make precise the idea of what it means for a bit generator to be "secure." Intuitively, a string of ℓ bits produced by a bit generator should look "random." That is, it should be impossible in an amount of time that is polynomial in k (equivalently, polynomial in ℓ) to distinguish a string of ℓ bits produced by a PRBG from a string of ℓ truly random bits.

For example, if a bit generator produced "1"s with probability 2/3, say, then it would usually be easy to distinguish a generated bitstring from a truly random

TABLE 8.2
Bits produced by the *RSA Generator*

i	s_i	z_i
0	75634	
1	31483	1
2	31238	0
3	51968	0
4	39796	0
5	28716	0
6	14089	1
7	5923	1
8	44891	1
9	62284	0
10	11889	1
11	43467	1
12	71215	1
13	10401	1
14	77444	0
15	56794	0
16	78147	1
17	72137	1
18	89592	0
19	29022	0
20	13356	0

bitstring. To be specific, we might use the following distinguishing strategy. Suppose we were given a bitstring of length ℓ. Denote the number of "1"s in this bitstring by ℓ_1. On average, a random bitstring of length ℓ will contain $\ell/2$ "1"s, and a generated bitstring will contain (on average) $2\ell/3$ "1"s. Therefore, if

$$\ell_1 > \frac{\ell/2 + 2\ell/3}{2} = \frac{7\ell}{12},$$

then we would guess that this bitstring was more likely to be a generated bitstring than a random bitstring.

This example motivates the idea of distinguishability of probability distributions. We will now give a definition of this concept. In this definition and elsewhere, we will denote an i-tuple $(z_1, \ldots, z_i)$ by z^i for ease of notation.

Definition 8.2: Suppose p_0 and p_1 are two probability distributions on the set $(\mathbb{Z}_2)^\ell$ of all bitstrings of length ℓ. For $j = 0, 1$ and $z^\ell \in (\mathbb{Z}_2)^\ell$, $p_j(z^\ell)$ denotes the probability of the string z^ℓ occurring in the distribution p_j. Let $\mathbf{dst} : (\mathbb{Z}_2)^\ell \to \{0, 1\}$ be a function and let $\epsilon > 0$. For $j = 0, 1$, define

$$E_{\mathbf{dst}}(p_j) = \sum_{\{z^\ell \in (\mathbb{Z}_2)^\ell \,:\, \mathbf{dst}(z^\ell)=1\}} p_j(z^\ell).$$

We say that $\mathbf{dst}$ is an ϵ-*distinguisher* of p_0 and p_1 provided that

$$|E_{\mathbf{dst}}(p_0) - E_{\mathbf{dst}}(p_1)| \geq \epsilon,$$

and we say that p_0 and p_1 are ϵ-*distinguishable* if there exists an ϵ-distinguisher of p_0 and p_1.

A distinguisher, say $\mathbf{dst}$, is a *polynomial-time distinguisher* provided that $\mathbf{dst}(z^\ell)$ can be computed in polynomial time as a function of ℓ.

The intuition behind the definition of a distinguisher is as follows. The function (or algorithm) $\mathbf{dst}$ tries to decide if a given bitstring z^ℓ of length ℓ is more likely to have arisen from probability distribution p_0 or from probability distribution p_1. The output $\mathbf{dst}(z^\ell)$ represents the distinguisher's guess as to which of these two probability distributions is more likely to have produced z^ℓ. The quantity $E_{\mathbf{dst}}(p_j)$ represents the average (i.e., expected) value of the output of $\mathbf{dst}$ over the probability distribution p_j, for $j = 0, 1$. Then $\mathbf{dst}$ is an ϵ-distinguisher provided that the values of these two expectations are at least ϵ apart.

REMARK In the definition above, a distinguisher is a function, i.e., a deterministic algorithm. We could generalize the definition so that a distinguisher could be a randomized algorithm, if desired. That is, given an ℓ-tuple $z^\ell = (z_1, \ldots, z_\ell)$ a distinguisher might guess "0" with some probability p (depending on z^ℓ), and hence guess "1" with probability $1 - p$. In the case of a randomized distinguisher, it is not hard to see that

$$E_{\mathbf{dst}}(p_j) = \sum_{z^\ell \in (\mathbb{Z}_2)^\ell} \left(p_j(z^\ell) \times \mathbf{Pr}[\mathbf{dst}(z^\ell) = 1] \right).$$

All of the results we state and prove remain true for randomized distinguishers.

The relevance of distinguishers to PRBGs is as follows. Consider the sequence of ℓ bits produced by a bit generator. There are 2^ℓ possible sequences, and if these ℓ bits were chosen independently and uniformly at random, then each of these 2^ℓ sequences would occur with equal probability $1/2^\ell$. Thus a truly random sequence corresponds to an equiprobable (or uniform) distribution on the set of all bitstrings of length ℓ. Suppose we denote this uniform probability distribution by p_u.

Now, consider sequences produced by the bit generator f. Suppose a k-bit seed is chosen uniformly at random and the generator is used to construct a bitstring of length ℓ. Then we obtain a probability distribution on the set of all bitstrings of length ℓ, which we denote by p_f. For the purposes of illustration, suppose we make the simplifying assumption that no two seeds give rise to the same sequence of bits. Then, of the 2^ℓ possible sequences, 2^k sequences each occur with probability $1/2^k$, and the remaining $2^\ell - 2^k$ sequences never occur. Hence, the probability distribution p_f is very non-uniform.

Even though the two probability distributions p_u and p_f may be quite different, it is still conceivable that they might be ϵ-distinguishable in polynomial time only for small values of ϵ. This is our objective in constructing PRBGs. However, this goal can be quite difficult to achieve. As the next example shows, producing 0's and 1's with equal probability is not sufficient to ensure indistinguishability.

Example 8.3   Suppose that a bit generator f only produces sequences in which exactly $\ell/2$ bits have the value 0 and $\ell/2$ bits have the value 1. Define the function **dst** by

$$\mathbf{dst}(z_1, \ldots, z_\ell) = \begin{cases} 1 & \text{if } (z_1, \ldots, z_\ell) \text{ has exactly } \ell/2 \text{ bits equal to } 0 \\ 0 & \text{otherwise.} \end{cases}$$

It is not hard to see that

$$E_{\mathbf{dst}}(p_u) = \frac{\binom{\ell}{\ell/2}}{2^\ell}$$

and

$$E_{\mathbf{dst}}(p_f) = 1.$$

It can be shown that

$$\lim_{\ell \to \infty} \frac{\binom{\ell}{\ell/2}}{2^\ell} = 0.$$

Hence, for any fixed value of $\epsilon < 1$, p_u and p_f are ϵ-distinguishable if ℓ is sufficiently large.   ∎

8.2.1   Next Bit Predictors

Another useful concept in studying bit generators is that of a *next bit predictor*, which works as follows. Let f be a (k, ℓ)-bit generator. Suppose $1 \leq i \leq \ell - 1$, and we have a function **nbp** : $(\mathbb{Z}_2)^{i-1} \to \mathbb{Z}_2$, which takes as input an $(i-1)$-tuple $z^{i-1} = (z_1, \ldots, z_{i-1})$, which represents the first $i-1$ bits produced by f (given an unknown, random, k-bit seed). Then the function **nbp** attempts to predict the next bit produced by f, namely, z_i. We say that the function **nbp** is an

ϵ-*ith bit predictor* if **nbp** can predict the ith bit of the generated bitstring (given the first $i - 1$ bits) with probability at least $1/2 + \epsilon$, where $\epsilon > 0$.

We can give a more precise formulation of this concept in terms of probability distributions, as follows. We have already defined a probability distribution p_f on $(\mathbb{Z}_2)^\ell$ induced by the bit generator f. We can also look at the probability distributions induced by f on any one of the ℓ generated bits (or indeed on any subsequence of these ℓ generated bits). So, for $1 \le i \le \ell$, we will can think of the ith generated bit as a random variable that we will denote by z_i.

In view of these definitions, we have the following characterization of an ith bit predictor.

THEOREM 8.1   *Let f be a (k, ℓ)-bit generator. Then the function* **nbp** *is an ϵ-ith bit predictor for f if and only if*

$$\sum_{z^{i-1} \in (\mathbb{Z}_2)^{i-1}} \left(p_f(z^{i-1}) \times \mathbf{Pr}[z_i = \mathbf{nbp}(z^{i-1}) | z^{i-1}] \right) \ge \frac{1}{2} + \epsilon.$$

PROOF   The probability of correctly predicting the ith generated bit is computed by summing (over all possible $(i - 1)$-tuples $z^{i-1} = (z_1, \ldots, z_{i-1})$) the product of the probability that the $(i - 1)$-tuple z^{i-1} is produced by the bit generator f and the probability that the ith bit is predicted correctly, given the $(i - 1)$-tuple z^{i-1}. ∎

REMARK   The reason for the expression $1/2 + \epsilon$ in this definition is that any predicting algorithm will predict any bit of a truly random bitstring with probability $1/2$. If a generated bitstring is not truly random, then it may be possible to predict a given bit with higher probability. (Note that it is unnecessary to consider functions that predict a given bit with probability less than $1/2$, because in this case a function that replaces every prediction z by $1 - z$ will predict the bit with probability greater than $1/2$.)

Also, we could allow bit predictors to be randomized algorithms. The results we state and prove will remain true in this generalized setting. In fact, in one of the main results in this chapter (Theorem 8.3), we will be constructing a bit predictor that is randomized. ∎

We illustrate these ideas by producing a next-bit predictor for the *Linear Congruential Generator* of Example 8.1.

Example 8.1   *(Cont.)*   For any i such that $1 \le i \le 9$, define an ith bit predictor by the formula

$$\mathbf{nbp}(z^{i-1}) = 1 - z_{i-1}.$$

That is, the function **nbp** predicts that a 0 is most likely to be followed by a 1, and vice versa. It is not hard to compute from Table 8.1 that **nbp** is a $\frac{9}{62}$-next

---

Algorithm 8.3: DISTINGUISH(z^i)

external nbp
$z \leftarrow \mathbf{nbp}(z^{i-1})$
if $z = z_i$
   then return (1)
   else return (0)

---

bit predictor for any $i \geq 1$ (i.e., it predicts any bit correctly with probability $1/2 + 9/62 = 20/31$). $\square$

We can use a next bit predictor to construct a distinguishing algorithm. Suppose that **nbp** is an ϵ-ith bit predictor for a given integer $i \leq \ell$. The distinguisher is constructed as shown in Algorithm 8.3.

The input to the algorithm DISTINGUISH is a sequence of i bits, denoted by z^i. DISTINGUISH uses the function **nbp** to predict z_i, given the first $i-1$ bits in the sequence. If the predicted value is the same as the actual value of z_i, then the algorithm DISTINGUISH outputs "1," meaning that it deems the i-tuple z^i to have arisen from the bit generator f. Otherwise, DISTINGUISH outputs "0."

It is clear that DISTINGUISH (Algorithm 8.3) is a polynomial-time Turing reduction. The next theorem shows that DISTINGUISH is a good distinguisher provided that **nbp** is a good ith bit predictor.

THEOREM 8.2 *Suppose* **nbp** *is a (polynomial-time) ϵ-ith bit predictor for the (k, ℓ)-bit generator f. Let p_f be the probability distribution induced on $(\mathbb{Z}_2)^i$ by f, and let p_u be the uniform probability distribution on $(\mathbb{Z}_2)^i$. Then Algorithm 8.3 is a (polynomial-time) ϵ-distinguisher of p_f and p_u.*

PROOF First, observe that

$$\text{DISTINGUISH}(z^i) = 1 \Leftrightarrow \mathbf{nbp}(z^{i-1}) = z_i.$$

Thus, we can compute the expectation $E_{\text{DISTINGUISH}}(p_f)$ as follows:

$$E_{\text{DISTINGUISH}}(p_f) = \sum_{z^i \in (\mathbb{Z}_2)^i} \left(p_f(z^i) \times \mathbf{Pr}[\text{DISTINGUISH}(z^i) = 1] \right)$$

$$= \sum_{z^i \in (\mathbb{Z}_2)^i} \left(p_f(z^i) \times \mathbf{Pr}[\mathbf{nbp}(z^{i-1}) = z_i] \right).$$

Now, consider any $(i-1)$-tuple z^{i-1}. Define

$$z = (z_1, \ldots, z_{i-1}, 0)$$

and

$$z' = (z_1, \ldots, z_{i-1}, 1).$$

Consider the two terms in the above sum corresponding to z and z'. Then we have

$$p_f(z) \times \mathbf{Pr}[\mathbf{nbp}(z^{i-1}) = 0] + p_f(z') \times \mathbf{Pr}[\mathbf{nbp}(z^{i-1}) = 1]$$

$$= p_f(z^{i-1}) \times \sum_{j \in \{0,1\}} \left(\mathbf{Pr}[\mathbf{z_i} = j | z^{i-1}] \times \mathbf{Pr}[\mathbf{nbp}(z^{i-1}) = j] \right)$$

$$= p_f(z^{i-1}) \times \mathbf{Pr}[\mathbf{z_i} = \mathbf{nbp}(z^{i-1}) | z^{i-1}].$$

It therefore follows that

$$E_{\text{DISTINGUISH}}(p_f) = \sum_{z^{i-1} \in (\mathbb{Z}_2)^{i-1}} \left(p_f(z^{i-1}) \times \mathbf{Pr}[\mathbf{z_i} = \mathbf{nbp}(z^{i-1}) | z^{i-1}] \right)$$

$$\geq \frac{1}{2} + \epsilon,$$

because **nbp** is an ϵ-ith bit predictor.

On the other hand, any predictor will predict the ith bit of a truly random sequence with probability $1/2$. Therefore, it is easy to see that $E_{\text{DISTINGUISH}}(p_u) = 1/2$. Hence,

$$|E_{\text{DISTINGUISH}}(p_u) - E_{\text{DISTINGUISH}}(p_f)| \geq \epsilon,$$

as desired. ∎

One of the main results in the theory of pseudo-random bit generators, due to Yao, is that a next bit predictor is a *universal test*. That is, a bit generator is "secure" if and only if there does not exist any polynomial-time ϵ-ith bit predictor for the generator, except for very small values of ϵ. Theorem 8.2 proves the implication in one direction. To prove the converse, we need to show how the existence of a distinguisher implies the existence of a certain ith bit predictor. This is done in Theorem 8.3.

THEOREM 8.3 *Suppose* **dst** *is a (polynomial-time) ϵ-distinguisher of p_f and p_u, where p_f is the probability distribution induced on $(\mathbb{Z}_2)^\ell$ by the (k, ℓ)-bit generator f, and p_u is the uniform probability distribution on $(\mathbb{Z}_2)^\ell$. Then for some i, $1 \leq i \leq \ell - 1$, there exists a (polynomial-time) $\frac{\epsilon}{\ell}$-ith bit predictor for f.*

PROOF For $0 \leq i \leq \ell$, define q_i to be the probability distribution on $(\mathbb{Z}_2)^\ell$ where the first i bits are generated using the bit generator f (assuming that the k-bit seed is chosen uniformly at random), and the remaining $\ell - i$ bits are generated uniformly and independently at random. Thus $q_0 = p_u$ and $q_\ell = p_f$. We are given that

$$|E_{\mathbf{dst}}(q_0) - E_{\mathbf{dst}}(q_\ell)| \geq \epsilon.$$

Algorithm 8.4: NBP(z^{i-1})

> **external dst**
> choose $(z_i, \ldots, z_\ell) \in (\mathbb{Z}_2)^{\ell-i+1}$ at random
> $z \leftarrow \mathbf{dst}(z_1, \ldots, z_\ell)$
> **return** $(z + z_i \bmod 2)$

Applying the triangle inequality, we have that

$$\epsilon \leq |E_{\mathbf{dst}}(q_0) - E_{\mathbf{dst}}(q_\ell)| \leq \sum_{i=1}^{\ell} |E_{\mathbf{dst}}(q_{i-1}) - E_{\mathbf{dst}}(q_i)|.$$

Hence, it follows that there is at least one value i, $1 \leq i \leq \ell$, such that

$$|E_{\mathbf{dst}}(q_{i-1}) - E_{\mathbf{dst}}(q_i)| \geq \frac{\epsilon}{\ell}.$$

We will assume that

$$E_{\mathbf{dst}}(q_{i-1}) - E_{\mathbf{dst}}(q_i) \geq \frac{\epsilon}{\ell}$$

(alternatively, if

$$E_{\mathbf{dst}}(q_i) - E_{\mathbf{dst}}(q_{i-1}) \geq \frac{\epsilon}{\ell},$$

then straightforward modifications to our proof will yield the desired result).

We are going to construct an ith bit predictor (for this specified value of i). The predicting algorithm is a randomized algorithm and it is presented as Algorithm 8.4. It is clear that Algorithm 8.4 is a polynomial-time (randomized) Turing reduction.

Here is the idea behind Algorithm 8.4. NBP in fact produces an ℓ-tuple according to the probability distribution q_{i-1}, given that $z_1, \ldots, z_{i-1}$ are generated by the bit generator f. If **dst** answers "0," then it thinks that the ℓ-tuple was most likely generated according to the probability distribution q_i. Now q_{i-1} and q_i differ only in that the ith bit is generated at random in the distribution q_{i-1}, whereas it is generated using the bit generator f in the distribution q_i. Hence, when **dst** answers "0," it thinks that the ith bit, z_i, is what would be produced by the bit generator, f. Hence, in this case we take z_i as our prediction of the ith bit. On the other hand, if **dst** answers "1," it thinks that z_i is random, so we take $1 - z_i$ as our prediction of what the ith bit would be under the generator f.

We need to compute the probability that the ith bit is predicted correctly. Observe that if **dst** answers "0," then the prediction is correct with probability

$$\mathbf{Pr}[\mathbf{z_i} = z_i | z^{i-1}].$$

If **dst** answers "1," then the prediction is correct with probability

$$\mathbf{Pr}[\mathbf{z_i} \neq z_i | z^{i-1}].$$

In our computation, we will make use of a useful fact that relates the probability distribution q_{i-1} to the probability distribution q_i. Namely, the following holds:

$$q_{i-1}(z^\ell) \times \mathbf{Pr}[\mathbf{z_i} = z_i | z^{i-1}] = \frac{q_i(z^\ell)}{2}. \tag{8.1}$$

This can be proven fairly easily, as follows:

$$q_{i-1}(z^\ell) \times \mathbf{Pr}[\mathbf{z_i} = z_i | z^{i-1}] = q_{i-1}(z^{i-1}) \times \frac{1}{2^{\ell-i+1}} \times \mathbf{Pr}[\mathbf{z_i} = z_i | z^{i-1}]$$

$$= q_i(z^i) \times \frac{1}{2^{\ell-i+1}}$$

$$= \frac{q_i(z^\ell)}{2}.$$

Now we are ready to perform our main computation, which consists of a complicated series of sums of probabilities. The reader is encouraged to work through the proof, filling in all the details required to proceed from each line to the next. It is useful to note that the fact (8.1) proven above will be used in a couple of places later on in the proof.

$$\sum_{z^{i-1} \in (\mathbb{Z}_2)^{i-1}} \left(p_f(z^{i-1}) \times \mathbf{Pr}[\mathbf{z_i} = \mathbf{nbp}(z^{i-1})] \right)$$

$$= \sum_{z^\ell \in (\mathbb{Z}_2)^\ell} \left(q_{i-1}(z^\ell) \times \mathbf{Pr}[\mathbf{z_i} = \mathbf{nbp}(z^{i-1})] \right)$$

$$= \sum_{z^\ell \in (\mathbb{Z}_2)^\ell} \left(q_{i-1}(z^\ell)(\mathbf{Pr}[\mathbf{dst}(z^\ell) = 0 | z^\ell] \times \mathbf{Pr}[\mathbf{z_i} = z_i | z^{i-1}] + \right.$$

$$\left. \mathbf{Pr}[\mathbf{dst}(z^\ell) = 1 | z^\ell] \times \mathbf{Pr}[\mathbf{z_i} \neq z_i | z^{i-1}]) \right)$$

$$= \sum_{z^\ell \in (\mathbb{Z}_2)^\ell} \left(q_{i-1}(z^\ell)((1 - \mathbf{Pr}[\mathbf{dst}(z^\ell) = 1 | z^\ell]) \times \mathbf{Pr}[\mathbf{z_i} = z_i | z^{i-1}] + \right.$$

$$\left. \mathbf{Pr}[\mathbf{dst}(z^\ell) = 1 | z^\ell] \times (1 - \mathbf{Pr}[\mathbf{z_i} = z_i | z^{i-1}])) \right)$$

$$= \sum_{z^\ell \in (\mathbb{Z}_2)^\ell} \left(q_{i-1}(z^\ell) \times \mathbf{Pr}[\mathbf{z_i} = z_i | z^{i-1}] \right)$$

$$- 2 \sum_{z^\ell \in (\mathbb{Z}_2)^\ell} \left(q_{i-1}(z^\ell) \times \mathbf{Pr}[\mathbf{dst}(z^\ell) = 1 | z^\ell] \times \mathbf{Pr}[\mathbf{z_i} = z_i | z^{i-1}] \right)$$

$$+ \sum_{z^\ell \in (\mathbb{Z}_2)^\ell} \left(q_{i-1}(z^\ell) \times \mathbf{Pr}[\mathbf{dst}(z^\ell) = 1 | z^\ell] \right)$$

Algorithm 8.5: *Blum-Blum-Shub Generator*

Let p, q be two $(k/2)$-bit primes such that $p \equiv q \equiv 3 \bmod 4$, and define $n = pq$. Let $\mathbf{QR}(n)$ denote the set of quadratic residues modulo n.

A seed s_0 is any element of $\mathbf{QR}(n)$. For $0 \le i \le \ell - 1$, define

$$s_{i+1} = s_i^2 \bmod n,$$

and then define

$$f(s_0) = (z_1, z_2, \ldots, z_\ell),$$

where

$$z_i = s_i \bmod 2,$$

$1 \le i \le \ell$. Then f is a (k, ℓ)-PRBG, called the *Blum-Blum-Shub Generator*, which we abbreviate to *BBS Generator*.

One way to choose an appropriate seed is to select an element $s_{-1} \in \mathbb{Z}_n^*$ and compute $s_0 = s_{-1}^2 \bmod n$. This ensures that $s_0 \in \mathbf{QR}(n)$.

$$= \frac{1}{2} \sum_{z^\ell \in (\mathbb{Z}_2)^\ell} q_i(z^\ell) - \sum_{z^\ell \in (\mathbb{Z}_2)^\ell} \left(q_i(z^\ell) \times \mathbf{Pr}[\mathbf{dst}(z^\ell) = 1 | z^\ell] \right)$$

$$+ \sum_{z^\ell \in (\mathbb{Z}_2)^\ell} \left(q_{i-1}(z^\ell) \times \mathbf{Pr}[\mathbf{dst}(z^\ell) = 1 | z^\ell] \right)$$

$$= \frac{1}{2} + E_{\mathbf{dst}}(q_{i-1}) - E_{\mathbf{dst}}(q_i)$$

$$\ge \frac{1}{2} + \frac{\epsilon}{\ell},$$

which was what we wanted to prove. ∎

8.3   The Blum-Blum-Shub Generator

In this section we will describe and analyze one of the most popular PRBGs, due to Blum, Blum, and Shub. For any odd integer n, denote the quadratic residues modulo n by $\mathbf{QR}(n)$. That is, $\mathbf{QR}(n) = \{x^2 \bmod n : x \in \mathbb{Z}_n^*\}$. The *Blum-Blum-Shub Generator* is presented in Algorithm 8.5.

The generator works quite simply. Given a seed $s_0 \in \mathbf{QR}(n)$, we compute the sequence $s_1, s_2, \ldots, s_\ell$ by successive squarings modulo n, and then we reduce

TABLE 8.3
Bits produced by BBS generator

i	s_i	z_i
0	20749	
1	143135	1
2	177671	1
3	97048	0
4	89992	0
5	174051	1
6	80649	1
7	45663	1
8	69442	0
9	186894	0
10	177046	0
11	137922	0
12	123175	1
13	8630	0
14	114386	0
15	14863	1
16	133015	1
17	106065	1
18	45870	0
19	137171	1
20	48060	0

each s_i modulo 2 to obtain z_i. It follows that

$$z_i = \left(s_0^{2^i} \bmod n\right) \bmod 2,$$

$1 \le i \le \ell$.

We now give a toy example of the *BBS Generator*.

Example 8.4   Suppose $n = 192649 = 383 \times 503$ and $s_0 = 101355^2 \bmod n = 20749$. The first 20 bits produced by the *BBS Generator* are computed as shown in Table 8.3. Hence the bitstring resulting from this seed is

$$11001110000100111010.$$

$\square$

Next, we review some results on Jacobi symbols from Section 5.4 and some other number-theoretic facts from other parts of Chapter 5. Suppose p and q are two distinct odd primes, and let $n = pq$. From the definition of the Jacobi symbol, it is easily seen that

$$\left(\frac{x}{n}\right) = \begin{cases} 0 & \text{if } \gcd(x, n) > 1 \\ 1 & \text{if } \left(\frac{x}{p}\right) = \left(\frac{x}{q}\right) = 1 \text{ or if } \left(\frac{x}{p}\right) = \left(\frac{x}{q}\right) = -1 \\ -1 & \text{if one of } \left(\frac{x}{p}\right) \text{ and } \left(\frac{x}{q}\right) = 1 \text{ and the other } = -1. \end{cases}$$

Recall that x is a quadratic residue modulo n if and only if

$$\left(\frac{x}{p}\right) = \left(\frac{x}{q}\right) = 1.$$

Define

$$\widetilde{\mathbf{QR}}(n) = \left\{ x \in \mathbb{Z}_n^* \backslash \mathbf{QR}(n) : \left(\frac{x}{n}\right) = 1 \right\}.$$

Thus

$$\widetilde{\mathbf{QR}}(n) = \left\{ x \in \mathbb{Z}_n^* : \left(\frac{x}{p}\right) = \left(\frac{x}{q}\right) = -1 \right\}.$$

An element $x \in \widetilde{\mathbf{QR}}(n)$ is called a *pseudo-square* modulo n. It is not hard to see that $|\mathbf{QR}(n)| = |\widetilde{\mathbf{QR}}(n)| = (p-1)(q-1)/4$.

The security of the *Blum-Blum-Shub Generator* is based on the intractability of the **Composite Quadratic Residues** problem, which we define as Problem 8.1. (In Chapter 5, we defined the **Quadratic Residues** problem modulo a prime and showed that it is easy to solve; here we have a composite modulus.)

Problem 8.1:	**Composite Quadratic Residues**
Instance:	A positive integer n that is the product of two unknown distinct odd primes p and q, and an integer $x \in \mathbb{Z}_n^*$ such that $\left(\frac{x}{n}\right) = 1$.
Question:	Is $x \in \mathbf{QR}(n)$?

Basically, the **Composite Quadratic Residues** problem requires us to distinguish quadratic residues modulo n from pseudo-squares modulo n. This can be no more difficult than factoring n. For if the factorization $n = pq$ can be computed, then it is a simple matter to calculate $\left(\frac{x}{p}\right)$, say. Given that $\left(\frac{x}{n}\right) = 1$, it follows that x is a quadratic residue modulo n if and only if $\left(\frac{x}{p}\right) = 1$.

There does not seem to be any way to solve the **Composite Quadratic Residues** problem efficiently if the factorization of n is not known. So it is commonly conjectured that this problem is intractable if it is infeasible to factor n.

Here is a property of the *BBS Generator* that is important when we look at its security. Since $n = pq$ where $p \equiv q \equiv 3 \bmod 4$, it follows that, for any quadratic residue x, there is a unique square root of x that is also a quadratic residue. This particular square root is called the *principal square root* of x. As a consequence, the mapping $x \mapsto x^2 \bmod n$, which is used to define the *BBS Generator*, is a permutation on $\mathbf{QR}(n)$, the set of quadratic residues modulo n.

Example 8.5   Suppose that $n = 253 = 11 \times 23$. Then

$$|\mathbf{QR}(n)| = \frac{10 \times 22}{4} = 55.$$

It can be shown that *BBS Generator* in $\mathbb{Z}_{55}$ permutes the elements of $|\mathbf{QR}(55)|$ in one cycle of length 1, one cycle of length 4, one cycle of length 10 and two cycles of length 20. $\square$

8.3.1   Security of the BBS Generator

In this section, we look at the security of the *BBS Generator* in detail. We begin by supposing that the pseudo-random bits produced by the *BBS Generator* are ϵ-distinguishable from ℓ random bits, and then see where that leads us. We will present a series of polynomial-time Turing reductions. Throughout this section, $n = pq$, where p and q are distinct primes such that $p \equiv q \equiv 3 \pmod 4$, and the factorization $n = pq$ is unknown.

We have already discussed the idea of a next bit predictor. In this section, we consider a similar concept that we call a *previous bit predictor*. A previous bit predictor for a (k, ℓ)-*BBS Generator* will take as input ℓ pseudo-random bits produced by the generator (as determined by an unknown random seed $s_0 \in \mathbf{QR}(n)$), and attempt to compute the value $z_0 = s_0 \bmod 2$. A previous bit predictor can be a probabilistic algorithm, and we say that a previous bit predictor **pbp** is an ϵ-previous bit predictor if its probability of correctly guessing z_0 is at least $1/2 + \epsilon$, where this probability is computed over all possible seeds s_0.

We state the following theorem, which is similar to Theorem 8.3, without proof.

THEOREM 8.4   *Suppose there exists a (polynomial-time) ϵ-distinguisher of p_f and p_u, where p_f is the probability distribution induced on $(\mathbb{Z}_2)^\ell$ by the (k, ℓ)-BBS Generator, f, and p_u is the uniform probability distribution on $(\mathbb{Z}_2)^\ell$. Then there exists a (polynomial-time) $\frac{\epsilon}{\ell}$-previous bit predictor for f.*

Next, we show how to use a δ-previous bit predictor, **pbp**, to construct a probabilistic algorithm that distinguishes quadratic residues modulo n from pseudo-squares modulo n with probability $1/2 + \delta$. The algorithm QR-TEST, which is presented as Algorithm 8.6, uses **pbp** as a subroutine, or oracle. It is assumed that the input, x, is an element with Jacobi symbol equal to 1, i.e., $x \in \mathbf{QR}(n) \cup \widetilde{\mathbf{QR}}(n)$.

Briefly, Algorithm 8.6 constructs a sequence of ℓ bits by first squaring x and then using x^2 as a seed in the *BBS Generator*. If $x \in \mathbf{QR}(n)$, then x itself is a "legal" seed and the resulting sequence is identical to the one which would be obtained by running the *BBS Generator* with seed x. If $x \in \widetilde{\mathbf{QR}}(n)$, then the sequence is what would be obtained by running the *BBS Generator* with seed $-x$. The algorithm tests whether $x \in \mathbf{QR}(n)$ by appealing to the predictor **pbp**, as described in the following theorem and its proof.

Algorithm 8.6: QR-TEST(x, n)

external pbp
$s_1 \leftarrow x^2 \bmod n$
comment: s_1 is a quadratic residue modulo n

$z_1 \leftarrow s_1 \bmod 2$
use the *BBS Generator* to compute $z_2, \ldots, z_\ell$ from seed s_1
$z \leftarrow \mathbf{pbp}(z_1, \ldots, z_\ell)$
if $(x \bmod 2) = z$
  then return (yes)
  else return (no)

THEOREM 8.5 *Suppose* **pbp** *is a (polynomial-time) δ-previous bit predictor for the (k, ℓ)-BBS Generator f. Then the algorithm* QR-TEST, *determines quadratic residuosity correctly (in polynomial time) with probability at least $1/2 + \delta$, where this probability is averaged over all possible inputs $x \in \mathbf{QR}(n) \cup \widetilde{\mathbf{QR}}(n)$, which are chosen uniformly at random.*

PROOF Since $n = pq$ and $p \equiv q \equiv 3 \pmod 4$, it follows that $\left(\frac{-1}{n}\right) = 1$, so $-1 \in \widetilde{\mathbf{QR}}(n)$. Let s_0 denote the principal square root of s_1. We are assuming that $\left(\frac{x}{n}\right) = 1$. Therefore, $s_0 = x$ if $x \in \mathbf{QR}(n)$; whereas $s_0 = -x$ if $x \in \widetilde{\mathbf{QR}}(n)$.

The ℓ bits $z_1, \ldots, z_\ell$ are exactly those that would be generated by using the *BBS Generator* with s_0 as the seed. Now, the integer n is odd, so

$$(-x \bmod n) \bmod 2 \neq (x \bmod n) \bmod 2.$$

We have that

$$x \in \mathbf{QR}(n) \Leftrightarrow s_0 = x.$$

Therefore, it follows that QR-TEST gives the correct answer if and only if **pbp** correctly predicts z. The desired result then follows immediately. ∎

Example 8.6 Suppose that $n = 60485929729$ and $x = 349850938$. It can be verified that $\left(\frac{x}{n}\right) = 1$. QR-TEST will first compute $s_1 = 41061588913$ and $z_1 = 1$. Then we generate $\ell - 1$ additional bits using the *BBS Generator*. The sequence $s_1, s_2, \ldots$ begins with

$$41061588913, 24724816839, 40968882391, 2137662714, 32677932305.$$

Hence, the first five generated bits are

$$1, 1, 1, 0, 1.$$

Algorithm 8.7: MC-QR-TEST(x)

external QR-TEST
choose $r \in \mathbb{Z}_n^*$ randomly
$x' \leftarrow r^2 x \bmod n$
choose $s \in \{1, -1\}$ randomly
$x' \leftarrow sx' \bmod n$
$t \leftarrow$ QR-TEST(x')
if $((t = yes)$ **and** $(s = 1))$ **or** $((t = no)$ **and** $(s = -1))$
 then return (yes)
 else return (no)

The sequence of bits can be continued as far as desired.

We know that the seed for this sequence is either x or $-x$, depending on whether or not $x \in \mathbf{QR}(n)$. The constructed bit sequence is given to the predictor, which returns a value $z = 0$ or 1. Suppose that the value $z = 0$ is returned by **pbp**. Since $x \bmod 2 = 0 = z$, QR-TEST outputs "yes," meaning that it believes that $x \in \mathbf{QR}(n)$. ⬚

Theorem 8.5 shows how we can distinguish pseudo-squares from quadratic residues with probability at least $1/2 + \delta$. However, this success probability is an average over all possible inputs $x \in \mathbf{QR}(n) \cup \widetilde{\mathbf{QR}}(n)$. We now show that this result can be improved to yield a Monte Carlo algorithm that determines quadratic residuosity correctly with probability at least $1/2 + \delta$. In other words, for any $x \in \mathbf{QR}(n) \cup \widetilde{\mathbf{QR}}(n)$, the Monte Carlo algorithm we will describe gives the correct answer with probabilty at least $1/2 + \delta$. Note that this algorithm is an *unbiased* algorithm (i.e., it may give an incorrect answer for any input), in contrast to the Monte Carlo algorithms that we studied in Section 5.4 which were all biased algorithms.

The Monte Carlo algorithm MC-QR-TEST is presented in Algorithm 8.7. It calls the previous algorithm QR-TEST as a subroutine. Basically, Algorithm 8.7 is a randomization process. It constructs a residue x' by multiplying x by a random quadratic residue, and then randomly multiplying the result by ± 1. The value x' is tested for quadratic residuousity using the subroutine QR-TEST, and this information is used to make a decision on the quadratic residuousity of x.

THEOREM 8.6 *Suppose that* QR-TEST *determines quadratic residuosity correctly (in polynomial time) with average probability at least* $1/2 + \delta$. *Then the algorithm* MC-QR-TEST, *which is described in Algorithm 8.7, is a (polynomial-time) Monte Carlo algorithm for* **Composite Quadratic Residues** *having error* *probability at most* $1/2 - \delta$.

PROOF For any given input $x \in \mathbf{QR}(n) \cup \widetilde{\mathbf{QR}}(n)$, Algorithm 8.7 produces an element x' that is a random element of $\mathbf{QR}(n) \cup \widetilde{\mathbf{QR}}(n)$. Moreover, the status of x as a quadratic residue can be determined from the status of x' as a quadratic residue. ∎

The last algorithm in our series of reductions is a general result pertaining to unbiased Monte Carlo algorithms. It shows that any (unbiased) Monte Carlo algorithm having error probability at most $1/2 - \delta$ can be used to construct an unbiased Monte Carlo algorithm with error probability at most γ, for any $\gamma > 0$. In other words, we can make the probability of correctness arbitrarily close to 1. The idea is to run the given Monte Carlo algorithm $2m + 1$ times, for some integer m, and take the "majority vote" as the answer. By computing the error probability of this algorithm, we can also see how m depends on γ. This dependence is stated in the following theorem.

THEOREM 8.7 *Suppose* A *is an unbiased Monte Carlo algorithm with error probability at most* $1/2 - \delta$. *Suppose we define an algorithm* A^n *by running* A $n = 2m + 1$ *times on a given instance* I, *and outputing the most frequent answer. Then the error probability of the algorithm* A^n *is at most*

$$\frac{(1 - 4\delta^2)^m}{2}.$$

PROOF The probability of obtaining exactly i correct answers in the n trials is at most

$$\binom{n}{i} \left(\frac{1}{2} + \delta \right)^i \left(\frac{1}{2} - \delta \right)^{n-i}.$$

The probability that the most frequently occurring answer is incorrect is equal to the probability that the number of correct answers in the n trials is at most m.

Hence, we compute as follows:

$$\mathbf{Pr}[\text{error}] \le \sum_{i=0}^{m} \binom{n}{i} \left(\frac{1}{2} + \delta\right)^{i} \left(\frac{1}{2} - \delta\right)^{2m+1-i}$$

$$= \left(\frac{1}{2} + \delta\right)^{m} \left(\frac{1}{2} - \delta\right)^{m+1} \sum_{i=0}^{m} \binom{n}{i} \left(\frac{1/2 - \delta}{1/2 + \delta}\right)^{m-i}$$

$$\le \left(\frac{1}{2} + \delta\right)^{m} \left(\frac{1}{2} - \delta\right)^{m+1} \sum_{i=0}^{m} \binom{n}{i}$$

$$= \left(\frac{1}{2} + \delta\right)^{m} \left(\frac{1}{2} - \delta\right)^{m+1} 2^{2m}$$

$$= \left(\frac{1}{4} - \delta^2\right)^{m} \left(\frac{1}{2} - \delta\right) 2^{2m}$$

$$= (1 - 4\delta^2)^{m} \left(\frac{1}{2} - \delta\right)$$

$$\le \frac{(1 - 4\delta^2)^{m}}{2},$$

as required. ∎

Suppose we want to lower the probability of error to some value γ, where $0 < \gamma < 1/2 - \delta$. We need to choose m so that

$$\frac{(1 - 4\delta^2)^{m}}{2} \le \gamma.$$

Hence, it suffices to take

$$m = \left\lceil \frac{1 + \log_2 \gamma}{\log_2 (1 - 4\delta^2)} \right\rceil.$$

If algorithm A is run $2m + 1$ times, then the majority vote yields the correct answer with probability at least $1 - \gamma$. It is not hard to show that this value of m is at most $c/(\gamma\delta^2)$ for some constant c. Hence, the number of times that the algorithm must be run is polynomial in $1/\gamma$ and $1/\delta$.

Example 8.7   Suppose we start with a Monte Carlo algorithm that gives the correct answer with probability at least .55, so $\delta = .05$. If we desire a Monte Carlo algorithm in which the probability of error is at most .05, then it suffices to take $m = 230$, so $n = 461$. □

Let us combine all the reductions we have done. We have the following sequence of implications:

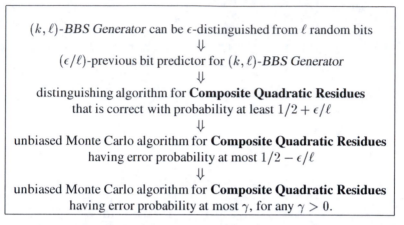

(k, ℓ)-**BBS Generator** can be ϵ-distinguished from ℓ random bits
⇓
(ϵ/ℓ)-previous bit predictor for (k, ℓ)-**BBS Generator**
⇓
distinguishing algorithm for **Composite Quadratic Residues**
that is correct with probability at least $1/2 + \epsilon/\ell$
⇓
unbiased Monte Carlo algorithm for **Composite Quadratic Residues**
having error probability at most $1/2 - \epsilon/\ell$
⇓
unbiased Monte Carlo algorithm for **Composite Quadratic Residues**
having error probability at most γ, for any $\gamma > 0$.

The reductions are all polynomial-time algorithms (i.e., polynomial time as a function of k, $1/\epsilon$, and $1/\gamma$). Since it is widely believed that there is no polynomial-time Monte Carlo algorithm for **Composite Quadratic Residues** with small error probability, we have some evidence that the *BBS Generator* is secure. This is yet another example of provable security.

We close this section by mentioning a way of improving the efficiency of the *BBS Generator*. The sequence of pseudo-random bits is constructed by taking the least significant bit of each s_i, where $s_i = s_0^{2^i} \bmod n$. Suppose instead that we extract the r least significant bits from each s_i, for some positive integer r. This will improve the efficiency of the PRBG by a factor of r, but we need to ask if the PRBG will remain secure (assuming the intractability of **Composite Quadratic Residues**). It has been shown that this approach is secure provided that $r \leq \log_2 \log_2 n$. So we can extract about $\log_2 \log_2 n$ pseudo-random bits per modular squaring. In a realistic implementation of the *BBS Generator*, say one in which $n \approx 10^{160}$, we can extract nine bits per squaring operation.

8.4   Probabilistic Encryption

In Section 5.9.2, we discussed notions such as semantic security of cryptosystems and ciphertext indistinguishability. An example of a semantically secure public-key cryptosystem was provided, based on a secure family of trapdoor one-way permutations and some appropriate hash functions. A proof of security was given, which applies when the hash functions are modeled as random oracles. The concrete system known as *Optimal Asymmetric Encryption Padding* is based on this type of construction.

In this section, we describe an alternative approach, known as probabilistic encryption, which is an interesting idea of Goldwasser and Micali. We begin with a definition of this concept. The definition is based on distinguishability of probability distributions.

Definition 8.3: A *probabilistic public-key cryptosystem* is defined to be a six-tuple $(\mathcal{P}, \mathcal{C}, \mathcal{K}, \mathcal{E}, \mathcal{D}, \mathcal{R})$, where $\mathcal{P}$ is the set of *plaintexts*, $\mathcal{C}$ is the set of *ciphertexts*, $\mathcal{K}$ is the *keyspace*, $\mathcal{R}$ is a set of *randomizers*, and for each *key* $K \in \mathcal{K}$, $e_K \in \mathcal{E}$ is a public *encryption rule* and $d_K \in \mathcal{D}$ is a secret *decryption rule*. The following properties should be satisfied:

1. Each $e_K : \mathcal{P} \times \mathcal{R} \to \mathcal{C}$ and $d_K : \mathcal{C} \to \mathcal{P}$ are functions such that

$$d_K(e_K(b, r)) = b$$

for every plaintext $b \in \mathcal{P}$ and every $r \in \mathcal{R}$. (In particular, this implies that $e_K(x, r) \neq e_K(x', r,)$ if $x \neq x'$.)

2. Security of the scheme is defined as follows. Let ϵ be a specified *security parameter*. For any fixed $K \in \mathcal{K}$ and for any $x \in \mathcal{P}$, define a probability distribution $p_{K,x}$ on $\mathcal{C}$, where $p_{K,x}(y)$ denotes the probability that y is the ciphertext given that K is the key and x is the plaintext (this probability is computed over all random choices $r \in \mathcal{R}$). Suppose $x, x' \in \mathcal{P}$, $x \neq x'$, and $K \in \mathcal{K}$. Then the probability distributions $p_{K,x}$ and $p_{K,x'}$ are not ϵ-distinguishable in polynomial time.

Here is how the scheme works. To encrypt a plaintext x, choose a (random) randomizer $r \in \mathcal{R}$ and compute $y = e_K(x, r)$. By property 1, any ciphertext $y = e_K(x, r)$ can be decrypted uniquely to the plaintext x. Property 2 is stating that the probability distribution of all encryptions of x cannot be distinguished (in polynomial time) from the probability distribution of all encryptions of x' if $x' \neq x$. Informally, an encryption of x "looks like" an encryption of x'. The security parameter ϵ should be small: in practice we would want to have $\epsilon = c/|\mathcal{R}|$ for some small $c > 0$.

It should be clear that a probabilistic public-key cryptosystem (as defined above) provides semantic security. Property 2 is stating that ciphertexts encrypting any two given plaintexts should be indistinguishable in polynomial time. This is the same requirement we had in Section 5.9.2.

We now present the *Goldwasser-Micali Public-key Cryptosystem* as Cryptosystem 8.1. Note that this cryptosystem encrypts every bit of plaintext independently, i.e., the plaintext space is $\mathcal{P} = \{0, 1\}$.

As noted above, this system encrypts one bit at a time. A 0 bit is encrypted to a random quadratic residue modulo n; a 1 bit is encrypted to a random pseudo-square modulo n. When Bob receives an element $y \in \mathbf{QR}(n) \cup \widetilde{\mathbf{QR}}(n)$, he can use his knowledge of the factorization of n to determine whether $y \in \mathbf{QR}(n)$ or

---

Cryptosystem 8.1: *Goldwasser-Micali Public-key Cryptosystem*

Let $n = pq$, where p and q are distinct odd primes, and let $m \in \widetilde{\mathbf{QR}}(n)$.[a]
The integers n and m are public; the factorization $n = pq$ is secret. Let
$\mathcal{P} = \{0, 1\}$, $\mathcal{C} = \mathcal{R} = \mathbb{Z}_n{}^*$, and define $\mathcal{K} = \{(n, p, q, m)\}$, where n, p, q and
m are as defined above.
For $K = (n, p, q, m)$, define

$$e_K(x, r) = m^x r^2 \bmod n$$

and

$$d_K(y) = \begin{cases} 0 & \text{if } y \in \mathbf{QR}(n) \\ 1 & \text{if } y \in \widetilde{\mathbf{QR}}(n), \end{cases}$$

where $x = 0$ or 1 and $r, y \in \mathbb{Z}_n{}^*$.

[a]If $p \equiv 3 \pmod 4$ and $q \equiv 3 \pmod 4$, then we can take $m = -1$. This makes encryption more efficient, because the computation of m^x would not require an exponentiation to be performed.

---

whether $y \in \widetilde{\mathbf{QR}}(n)$. He does this by computing

$$\left(\frac{y}{p}\right) = y^{(p-1)/2} \bmod p;$$

then

$$y \in \mathbf{QR}(n) \Leftrightarrow \left(\frac{y}{p}\right) = 1.$$

The *Goldwasser-Micali Cryptosystem* has a very high data expansion. This is because every plaintext bit is encrypted to a ciphertext of length $\log_2 n$ bits. In order for the system to be secure against factoring n, we should take n to be a 1024-bit integer. If this done, then the ciphertext is more than 1000 times as long as the plaintext.

A much more efficient probabilistic public-key cryptosystem (from the point of view of data expansion) was given by Blum and Goldwasser. The *Blum-Goldwasser Public-key Cryptosystem* is presented as Cryptosystem 8.2.

The *Blum-Goldwasser Public-key Cryptosystem* is a kind of public-key stream cipher. The basic idea is as follows. A random seed s_0 is used to generate a sequence of ℓ pseudorandom bits $z_1, \ldots, z_\ell$ using the *BBS Generator*. Then the z_i's are used exactly as if they were a keystream in a stream cipher: they are exclusive-ored with the ℓ plaintext bits to form the ciphertext. As well, the $(\ell+1)$st element, $s_{\ell+1} = s_0{}^{2^{\ell+1}} \bmod n$, is transmitted as part of the ciphertext.

Cryptosystem 8.2: *Blum-Goldwasser Public-key Cryptosystem*

Let $n = pq$, where p and q are primes, $p \equiv q \equiv 3 \pmod 4$. The integer n is public; the factorization $n = pq$ is secret. Let $\mathcal{P} = (\mathbb{Z}_2)^\ell$, $\mathcal{C} = (\mathbb{Z}_2)^\ell \times \mathbb{Z}_n^*$ and $\mathcal{R} = \mathbb{Z}_n^*$. Define $\mathcal{K} = \{(n, p, q)\}$, where n, p and q are as defined above. For $K = (n, p, q)$, $x \in (\mathbb{Z}_2)^\ell$ and $r \in \mathbb{Z}_n^*$, encrypt x as follows:

1. Compute $z_1, \ldots, z_\ell$ from seed $s_0 = r$ using the *BBS Generator*.
2. Compute $s_{\ell+1} = s_0^{2^{\ell+1}} \bmod n$.
3. Compute $y_i = (x_i + z_i) \bmod 2$ for $1 \le i \le \ell$.
4. Define $e_K(x, r) = (y_1, \ldots, y_\ell, s_{\ell+1})$.

To decrypt y, Bob performs the following steps:

1. Compute $a_1 = ((p+1)/4)^{\ell+1} \bmod (p-1)$.
2. Compute $a_2 = ((q+1)/4)^{\ell+1} \bmod (q-1)$.
3. Compute $b_1 = s_{\ell+1}^{a_1} \bmod p$.
4. Compute $b_2 = s_{\ell+1}^{a_2} \bmod q$.
5. Use the Chinese remainder theorem to find r such that

$$r \equiv b_1 \pmod p$$

and

$$r \equiv b_2 \pmod q.$$

6. Compute $z_1, \ldots, z_\ell$ from seed $s_0 = r$ using the *BBS Generator*.
7. Compute $x_i = (y_i + z_i) \bmod 2$ for $1 \le i \le \ell$.
8. The plaintext is $x = (x_1, \ldots, x_\ell)$.

When Bob receives the ciphertext, he can compute s_0 from $s_{\ell+1}$, then reconstruct the keystream, and finally exclusive-or the keystream with the ℓ ciphertext bits to obtain the plaintext. We should explain how Bob derives s_0 from $s_{\ell+1}$. Recall that each s_{i-1} is the principal square root of s_i. Now, $n = pq$ with $p \equiv q \equiv 3 \pmod 4$, so the square roots of any quadratic residue x modulo p are $\pm x^{(p+1)/4}$ (this was shown in Section 5.8). Using properties of Jacobi symbols, we have that

$$\left(\frac{x^{(p+1)/4}}{p}\right) = \left(\frac{x}{p}\right)^{(p+1)/4}$$
$$= 1.$$

It follows that $x^{(p+1)/4}$ is the principal square root of x modulo p. Similarly,

$x^{(q+1)/4}$ is the principal square root of x modulo q. Then, using the Chinese remainder theorem, Bob can find the principal square root of x modulo n.

More generally, $x^{((p+1)/4)^{\ell+1}}$ will be the principal $2^{\ell+1}$st root of x modulo p and $x^{((p+1)/4)^{\ell+1}}$ will be the principal $2^{\ell+1}$st root of x modulo q. Since $\mathbb{Z}_p^*$ has order $p-1$, we can reduce the exponent $((p+1)/4)^{\ell+1}$ modulo $p-1$ in the computation of $x^{((p+1)/4)^{\ell+1}}$ mod p. In a similar fashion, we can reduce the exponent $((q+1)/4)^{\ell+1}$ modulo $q-1$. In the decryption operations of Cryptosystem 8.2, we first compute the principal $2^{\ell+1}$st roots of $s_{\ell+1}$ modulo p and modulo q (see steps 1–4). Then the Chinese remainder theorem is used to compute the principal $2^{\ell+1}$st root of $s_{\ell+1}$ modulo n.

Here is an example to illustrate.

Example 8.8   Suppose $n = 192649$, as in Example 8.4. Suppose further that Alice chooses $r = 20749$ and she wants to encrypt the 20-bit plaintext string

$$x = 11010011010011101101.$$

Thus $\ell = 20$.

Alice will first compute the keystream

$$z = 11001110000100111010,$$

exactly as in Example 8.4, and then exclusive-or it with the plaintext, to obtain the ciphertext

$$y = 00011101010111010111$$

which she transmits to Bob. She also computes

$$s_{21} = s_{20}{}^2 \bmod n = 94739$$

and sends it to Bob.

Of course Bob knows the factorization $n = 383 \times 503$, so he can easily calculate $(p+1)/4 = 96$ and $(q+1)/4 = 126$. Then he computes

$$a_1 = ((p+1)/4)^{\ell+1} \bmod (p-1)$$
$$= 96^{21} \bmod 382$$
$$= 266$$

and

$$a_2 = ((q+1)/4)^{\ell+1} \bmod (q-1)$$
$$= 126^{21} \bmod 502$$
$$= 486.$$

Next, he calculates

$$b_1 = s_{21}{}^{a_1} \bmod p$$
$$= 94739^{266} \bmod 383$$
$$= 67$$

and

$$b_2 = s_{21}{}^{a_2} \bmod q$$
$$= 94739^{486} \bmod 503$$
$$= 126.$$

Now Bob proceeds to solve the system of congruences

$$r \equiv 67 \ (\mathrm{mod}\ 383)$$
$$r \equiv 126 \ (\mathrm{mod}\ 503)$$

to obtain Alice's seed $r = 20749$. Then he constructs Alice's keystream from r. Finally, he exclusive-ors the keystream with the ciphertext to get the plaintext.
□

The data expansion of the *Blum-Goldwasser Cryptosystem* is quite reasonable. Let's suppose for the purposes of illustration that we take $\ell = k^2$. That is, we use a k-bit seed to generate $\ell = k^2$ pseuodrandom bits. If n is a 1024-bit integer, then $k = 1024$, because the seed is an element of $\mathbb{Z}_n$. Therefore $\ell \approx 10^6$. For this choice of parameters, the encryption of 1,000,000 bits of plaintext yields a ciphertext whose length is 1,001,024. That is, the ciphertext is about .1% longer than the plaintext. This represents an enormous improvement over the data expansion of the *Goldwasser-Micali Cryptosystem*. Note, however, that there does not appear to be any concrete analysis that says that a particular value of ℓ is secure for a given modulus size (namely, k). Although $\ell = k^2$ is secure asymptotically (subject to our underlying computational assumptions), it is not known if it is secure for the fixed modulus size $n \approx 2^{1024}$.

8.5 Notes and References

A lengthy treatment of PRBGs can be found in the books by Kranakis [204] and Luby [223]. Also worth consulting is the survey paper by Lagarias [208] and two books by Goldreich [159, 160]. Knuth [195] discusses random number generation (mostly in a non-cryptographic context) in considerable detail.

The *Shrinking Generator* is due to Coppersmith, Krawczyk, and Mansour [97]; for information on attacks against this generator, see Golić [165]. Another practical method of constructing PBRGs using LFSRs has been given by Gunther [169]. For methods of breaking the *Linear Congruential Generator*, see Boyar [65].

The basic theory of secure PRBGs is due to Yao [350], who proved the universality of the next bit test. Further basic results can be found in Blum and Micali [52].

The *BBS Generator* is described by Blum, Blum and Shub in [50]. The security of the **Quadratic Residues** problem is studied by Goldwasser and Micali [163], on which we based much of Section 8.3.1. We have, however, used the approach of Brassard and Bratley [70, Section 10.6.4] to reduce the error probability of an unbiased Monte Carlo algorithm. We also note that it is known that the security of the *BBS Generator* can be proven assuming only that factoring the modulus n is intractable. A sufficient condition for the secure extraction of multiple bits per iteration of a PRBG was proved by Vazirani and Vazirani [334].

Properties of the *RSA Generator* are studied in Alexi, Chor, Goldreich, and Schnorr [2]. PRBGs based on the **Discrete Logarithm** problem are treated in Blum and Micali [52], Long and Wigderson [221], Håstad, Schrift, and Shamir [172], and Gennaro [155].

The idea of probabilistic encryption is due to Goldwasser and Micali [163]; the *Blum-Goldwasser Cryptosystem* is presented in [51].

Exercises

8.1 In this exercise, we consider some properties of the *Linear Congruential Generator* defined by $s_i = (as_{i-1} + b) \bmod M$. All operations are performed in $\mathbb{Z}_M$, and we assume that $a \neq 1$.

    (a) Prove that
$$s_i = s_0 a^i + \frac{b(a^i - 1)}{a - 1}$$
    for all $i \geq 0$.

    (b) The *period* of a *Linear Congruential Generator* is the smallest positive integer t such that $z_{i+t} = z_i$ for all $i \geq 0$. Prove that $t = 1$ if $s_0 = b/(a - 1)$.

    (c) Prove that the period t satisfies the inequality $t \leq n$, where n is the order of a in $\mathbb{Z}_M^*$.

8.2 Consider the *Linear Congruential Generator* defined by $s_i = (as_{i-1} + b) \bmod M$. Suppose that $M = qa + 1$ where a is odd and q is even, and suppose that $b = 1$. Show that the next bit predictor $\mathbf{nbp}(z) = 1 - z$ is an ϵ-next bit predictor for the ith bit, where
$$\frac{1}{2} + \epsilon = \frac{q(a + 1)}{2M}.$$

8.3 In the *Shrinking Generator*, suppose we take $k_1 = k_2$, the two LFSRs are identical, and we use the same seed for both LFSRs. Explain why this is a very bad idea.

8.4 Suppose we have an *RSA Generator* with $n = 36863$, $b = 229$ and seed $s_0 = 25$. Compute the first 100 bits produced by this generator.

8.5 A PRBG based on the **Discrete Logarithm** problem is given as Algorithm 8.8; this is called the *Discrete Logarithm Generator*. Suppose $p = 21383$, the primitive element $\alpha = 5$ and the seed $s_0 = 15886$. Compute the first 100 bits produced by this generator.

Algorithm 8.8: *Discrete Logarithm Generator*

Let p be a k-bit prime, and let α be a primitive element modulo p.

A seed x_0 is any element of $\mathbb{Z}_p^*$. For $i \geq 0$, define

$$x_{i+1} = \alpha^{x_i} \bmod p,$$

and then define

$$f(x_0) = (z_1, z_2, \ldots, z_\ell),$$

where

$$z_i = \begin{cases} 1 & \text{if } x_i > p/2 \\ 0 & \text{if } x_i < p/2. \end{cases}$$

Then f is called a (k, ℓ)-*Discrete Logarithm Generator*.

8.6 Suppose that Bob has knowledge of the factorization $n = pq$ in the *BBS Generator*.
  (a) Show how Bob can use this knowledge to compute any s_i from s_0 with $2k$ multiplications modulo $\phi(n)$ and $2k$ multiplications modulo n, where n has k bits in its binary representation. (If i is large compared to k, then this approach represents a substantial improvement over the i multiplications required to sequentially compute $s_0, \ldots, s_i$.)
  (b) Use this method to compute s_{10000} if $n = 59701 = 227 \times 263$ and $s_0 = 17995$.

8.7 In the *BBS Generator*, define $p_1 = (p - 1)/2$ and define $q_1 = (q - 1)/2$ (observe that p_1 and q_1 are both odd). Let u_1 denote the order of 2 modulo p_1 and let v_1 denote the order of 2 modulo q_1. Then, define t to be the least common multiple of u_1 and v_1.
  (a) Prove that the period of the *BBS Generator* does not exceed t.
  (b) Compute t when $p = 103$ and $q = 127$. Then, using 49 as a seed, verify that this instance of the *BBS Generator* has order t.

8.8 We proved that, in order to reduce the error probability of an unbiased Monte Carlo algorithm from $1/2 - \delta$ to γ, where $\gamma + \delta < 1/2$, it suffices to run the algorithm m times, where

$$m = \left\lceil \frac{1 + \log_2 \gamma}{\log_2(1 - 4\delta^2)} \right\rceil.$$

Prove that this value of m is $O(1/(\gamma \delta^2))$.

8.9 Assume that the *Goldwasser-Micali Probabilistic Public-key Cryptosystem* is implemented with $p = 1019$, $q = 1031$, $n = pq = 1050589$ and $m = 41$.
  (a) Verify that $m \in \mathbf{QR}(n)$.
  (b) Decrypt the five ciphertext elements

$$(y_1, y_2, y_3, y_4, y_5) = (734376, 721402, 133591, 824410, 757941).$$

TABLE 8.4
Blum-Goldwasser Ciphertext

```
E1866663F17FDBD1DC8C8FD2EEBC36AD7F53795DBA3C9CE22D
C9A9C7E2A56455501399CA6B98AED22C346A529A09C1936C61
ECDE10B43D226EC683A669929F2FFB912BFA96A8302188C083
46119E4F61AD8D0829BD1CDE1E37DBA9BCE65F40C0BCE48A80
0B3D087D76ECD1805C65D9DB730B8D0943266D942CF04D7D4D
76BFA891FA21BE76F767F1D5DCC7E3F1D86E39A9348B3
```

8.10 The purpose of this exercise is to decrypt some ciphertext which was encrypted with the *Blum-Goldwasser Probabilistic Public-key Cryptosystem*. The original plaintext consisted of English text. Each alphabetic character was converted to a bitstring of length five in the obvious way: $A \leftrightarrow 00000, B \leftrightarrow 00001, \ldots, Z \leftrightarrow 11001$. The plaintext consisted of 236 alphabetic characters, so a bitstring of length 1180 resulted. This bitstring was then encrypted. The resulting ciphertext bitstring was then converted to a hexadecimal representation, to save space. The final string of 295 hexadecimal characters is presented in Table 8.4.

Note that $s_{1181} = 20291$ is part of the ciphertext, and $n = 29893$ is the public key. The factorization of n is $n = pq$, where $p = 167$ and $q = 179$.

Your task is to decrypt the given ciphertext and restore the original English plaintext, which was taken from "Under the Hammer," by John Mortimer, Penguin Books, 1994.

9

Identification Schemes and Entity Authentication

9.1 Introduction

The topic of this chapter is *identification*, which is also known as *entity authentication*. Roughly speaking, the goal of an *identification scheme* is is to allow someone's identity to be confirmed. Normally this is done in "real time." In contrast, cryptographic tools such as signature schemes allow the authentication of data, which can be performed any time after the relevant message has been signed.

Suppose you want to prove your identity to someone else. It is sometimes said that this can be done in one of three ways, namely, based on what you are, what you have or what you know. "What you are" refers to behavioral and physical attributes; "what you have" refers to documents or credentials; and "what you know" encompasses passwords, personal information, etc. These techniques are described in more detail now.

physical attributes

People often identify other people already known to them by their appearance. This could include family and friends as well as famous celebrities. Specific features used for this purpose include sex, height, weight, racial origin, eye color, hair color, etc. Attributes that are unique to an individual are often more useful; these include fingerprints or retina scans. Sometimes automated identification schemes are based on biometrics such as these, and it seems likely that biometrics might frequently be used in the future.

credentials

A *credential* is defined, in the diplomatic usage of the word, as a letter of introduction. Trusted documents or cards such as driver's licences and passports function as credentials in many situations. Note that credentials often include photographs, which enables physical identification of the bearer of the credential.

knowledge

Knowledge is often used for identification when the person being identified is not in the same physical location as the person or entity performing the identification. In the context of identification, *knowledge* could be a password or *PIN* (*personal identification number*), or "your mother's maiden name" (a favorite of credit card companies). The difficulty with using knowledge for identification is that such knowledge may not be secret in the first place, and, moreover, it is usually revealed as part of the identification process. This allows for possible future impersonation of the person being identified, which is not a good thing! However, suitable cryptographic protocols will enable the construction of secure identification schemes, which will prevent these kinds of impersonation attacks from being carried out.

Let's consider some everyday situations where it is common to "prove" one's identity, either in person or electronically. Some typical scenarios are as follows:

telephone calling cards

To charge long-distance telephone calls (using a calling card), one requires only the knowledge of the telephone number that is being billed for the call, together with a four-digit PIN.

remote login

To do a remote login to a computer over a network via `telnet` or `ssh`, it suffices to know a valid user name and the corresponding password.

in-store credit card purchases

When a purchase is made at a store using a credit card, the sales clerk is supposed to verify that the customer's signature matches the signature on the back of the card. This provides a weak form of identification, as many signatures can be forged without too much difficulty. Some credit cards also have a photograph of the owner, which provides an additional level of authentication. Another way to enhance security is through the use of a personal identification number (or, PIN), which is sometimes used in conjunction with cards having embedded chips.

In contrast, consider the use of a credit card to purchase gasoline at the gas pump, without interacting with an attendant. Here, possession of the credit card is sufficient to allow it to be used; there is nothing at all to prevent the use of a stolen card.

credit card purchases without the credit card

In many situations, possession of the actual credit card is not required in order to use it. For example, to charge purchases made over the telephone or the internet to a credit card, all that usually is necessary is a valid credit card number (and perhaps the expiry date).

These uses of a credit card clearly do not provide any real security, because there is no meaningful authentication of identity being performed. Indeed, it is a simple matter for a dishonest person to collect valid credit card numbers and then use them in situations where the actual card is not required to carry out the transaction.

bank machine withdrawals

To withdraw money from an automated teller machine (or ATM), we use a bank card together with a four- or six-digit PIN. The card contains the owner's name and information about his or her bank accounts. The purpose of the PIN is to protect against fraudulent use of the card by someone else. The assumption is that the only person who knows the correct PIN is the owner of the card.

In practice, these types of schemes are not usually implemented in a secure way. In the protocols performed over the telephone, any eavesdropper can use the identifying information for their own purposes. This could include the person who is the recipient of the information; many credit card "scams" operate in this way. A bank card is somewhat more secure, but there are still weaknesses. For example, someone monitoring the communication line can obtain all the information encoded on the card's magnetic strip, as well as the PIN. This could perhaps allow an imposter to gain access to a bank account. Finally, remote computer login is a serious problem if user IDs and passwords are transmitted over the network in unencrypted form, because they can be read by anyone who is eavesdropping on the computer network.

The objective of an identification scheme would be that someone "listening in" as Alice identifies herself to Bob, say, should not subsequently be able to misrepresent herself as Alice. At the very least, the attack model allows the adversary to observe all the information being transmitted between Alice and Bob. The adversarial goal is to be able to impersonate Alice. Furthermore, we may even try to guard against the possibility that Bob himself might try to impersonate Alice after she has identified herself to him. Ultimately, we would like to devise "zero-knowledge" schemes whereby Alice can prove her identity electronically, without "giving away" the knowledge (or partial information about the knowledge) that is used as her identifying information.

Several practical and secure identification schemes have been discovered. One objective is to find a scheme that is simple enough that it can be implemented on a smart card, which is essentially a credit card equipped with a chip that can perform arithmetic computations. Hence, the amount of computation and the memory requirements should be kept as small as possible. Such a card would be a more secure alternative to many current bank cards. However, it is important to note that the "extra" security pertains to someone monitoring the communication line. Since it is the card that is "proving" its identity, we have no extra protection against a lost card. It would still be necessary to include a PIN in order to establish that it is the real owner of the card who is initiating the identification scheme.

A first observation is that any identification scheme should involve randomization in some way. If the information that Alice transmits to Bob to identify herself never changes, then the scheme is insecure in the model we introduced above.

Protocol 9.1: INSECURE CHALLENGE-AND-RESPONSE

1.   Bob chooses a random *challenge*, r, which he sends to Alice.
2.   Alice computes
$$y = MAC_K(r)$$
and sends y to Bob.
3.   Bob computes
$$y' = MAC_K(r).$$
If $y' = y$, then Bob "accepts"; otherwise, Bob "rejects."

Therefore, secure identification schemes usually include "random challenges" in them. This concept is explored more deeply in the next section.

We will take two approaches to the design of identification schemes. First, we explore the idea of building secure identification schemes from simpler crypto-graphic primitives, namely, message authentication codes or signature schemes. Schemes of this type are developed and analyzed in Sections 9.2 and 9.3. Then, in the remaining sections of this chapter, we discuss three identification schemes that are built "from scratch." These schemes due to Schnorr, Okamoto, and Guillou-Quisquater.

9.2   Challenge-and-Response in the Secret-key Setting

In later sections, we will describe some of the more popular zero-knowledge iden-tification schemes. First, we look at identification in the secret-key setting, where Alice and Bob both have the same secret key. We begin by examining a very simple (but insecure) scheme that can be based on any message authentication code, e.g., the MACs discussed in Chapter 4. The scheme, which is described as Protocol 9.1, is called *challenge-and-response*. In it, we assume that Alice is iden-tifying herself to Bob, and their common secret key is denoted by K. (Bob can also identify himself to Alice, by interchanging the roles of Alice and Bob in the scheme.) As usual, the message authentication code MAC_K is used to compute authentication tags.

We will often depict interactive protocols in diagrammatic fashion. Protocol 9.1 could be presented as shown in Figure 9.1.

Before analyzing the weaknesses of this scheme, let us define some basic ter-minology related to interactive protocols. In general, an *interactive protocol* will involve two or more parties that are communicating with each other. Each party

FIGURE 9.1
Information Flows in Protocol 9.1

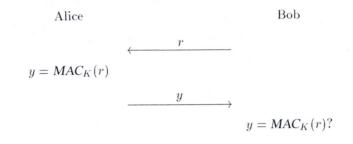

FIGURE 9.2
Attack on Protocol 9.1

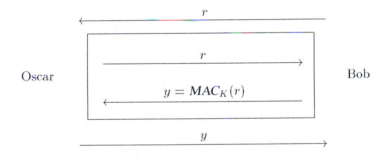

is modeled by an algorithm that alternately sends and receives information. Each run of a protocol will be called a *session*. Each step within a session of the protocol is called a *flow*; a flow consists of information transmitted from one party to another party. (Protocol 9.1 consists of two flows, the first one being from Bob to Alice, and the second one being from Alice to Bob.) At the end of a session, Bob (the initiator of the session) "accepts" or "rejects" (this is Bob's *internal state* at the end of the session). It may not be known to Alice whether Bob accepts or rejects.

It is not hard to see that Protocol 9.1 is insecure, even if the message authentication code used in it is secure. It is susceptible to a fairly standard type of attack known as a *parallel session attack*, wherein Oscar impersonates Alice. The attack is depicted in Figure 9.2.

Within the first session (in which it is supposed that Oscar is impersonating Alice to Bob), Oscar initiates a second session in which he asks Bob to identify himself. This second session is boxed in Figure 9.2. In this second session, Oscar

---

Protocol 9.2: (SECURE) CHALLENGE-AND-RESPONSE

1. Bob chooses a random challenge, r, which he sends to Alice.

2. Alice computes
$$y = MAC_K(ID(Alice) \parallel r)$$
and sends y to Bob.

3. Bob computes
$$y' = MAC_K(ID(Alice) \parallel r).$$

If $y' = y$, then Bob "accepts"; otherwise, Bob "rejects."

---

gives Bob the same challenge that he received from Bob in the first session. Once he receives Bob's response, Oscar resumes the first session, in which he relays Bob's response back to him. Thus Oscar is able to successfully complete the first session!

The reader might object to the premise that parallel sessions constitute a realistic threat. However, there are scenarios in which parallel sessions might be reasonable, or even desirable, and it would seem to be prudent to design an identification scheme to withstand such attacks. We present one easy way to rectify the problem in Protocol 9.2. The only change that is made is to include the identity of the person creating the MAC into the computation of the authentication tag.

In Protocol 9.2, we will assume that the random challenge is a bitstring of a specifed, predetermined length, say k bits (in practice, $k = 100$ will be a suitable choice). We also assume that the identity string ($ID(Alice)$ or $ID(Bob)$, depending on whose identity is being authenticated) is also a bitstring of a specified length, formatted in some standard, fixed manner. We assume that an identity string contains enough information to specify a unique individual in the network (so Bob does not have to worry about which "Alice" he is talking to).

We claim that a parallel session attack cannot be carried out against Protocol 9.2. If Oscar attempted to mount the same attack as before, he would receive the value $MAC_K(ID(Bob) \parallel r)$ from Bob in the second session. This is of no help in computing the value $MAC_K(ID(Alice) \parallel r)$ that is required in the first session to successfully respond to Bob's challenge.

The preceding discussion may convince the reader that the parallel session attack cannot be mounted against Protocol 9.2, but it does not present a proof of security against all possible attacks. We shortly will give a proof of security. First, however, we explicitly list all the assumptions we make regarding the cryptographic components used in the scheme. These assumptions are as follows:

secret key

> We assume that the secret key, K, is known only to Alice and Bob.

random challenges

> We assume that Alice and Bob both have perfect random number generators which they use to determine their challenges. Therefore, there is only a very small probability that the same challenge occurs by chance in two different sessions.

MAC security

> We assume that the message authentication code is secure. More precisely, there does not exist an (ϵ, Q)-forger for the MAC, for appropriate values of ϵ and Q. That is, the probability that Oscar can correctly compute $MAC_K(x)$ is at most ϵ, even when he is given Q other MACs, say $MAC_K(x_i)$, $i = 1, 2, \ldots, Q$, provided that $x \neq x_i$ for any i. As usual, Q is a specified security parmameter ($Q = 10000$ or 100000 might be a reasonable choice, depending on the application).

Oscar may observe several sessions between Alice and Bob. Oscar's goal is to deceive Alice or Bob, i.e., to cause Bob to "accept" in a session in which Alice is not taking part, or to cause Alice to "accept" in a session in which Bob is not taking part. We show that Oscar will not succeed in deceiving Alice or Bob in this way, except with small probability, when the above assumptions are valid. This is done fairly easily by analyzing the structure of the scheme.

Suppose that Bob "accepts." Then $y = MAC_K(ID(Alice) \| r)$, where y is the value he receives in the second flow and r was his challenge from the first flow of the scheme. We claim that, with high probability, this value y must have been constructed by Alice in response to the challenge r from the first flow of the scheme. To justify this claim, let's consider the possible sources of a response if it did not come directly from Alice. First, because the key K is assumed to be known only to Alice and Bob, we do not have to consider the possibility that $y = MAC_K(ID(Alice) \| r)$ was computed by some other party that knows the key K. So either Oscar (or someone else) computed y without knowing the key K, or the value y was computed by Alice or Bob in some previous session, copied and then reused by Oscar in the current session.

We now consider these possible cases in turn:

1. Suppose the value $y = MAC_K(ID(Alice) \| r)$ was previously constructed by Bob himself in some previous session. However, Bob only computes MACs of the form $MAC_K(ID(Bob) \| r)$, so he would not have created y himself. Therefore this case does not arise.

2. Suppose the value y was previously constructed by Alice in some earlier session. This can happen only if the challenge r is reused. However, the challenge r is assumed to be a challenge that is newly created by Bob using a perfect random number generator, so Bob would not have issued the same challenge in some other session, except with a very small probability.

3. Suppose the value y is a new MAC that is constructed by Oscar. Assuming that the message authentication code is secure and Oscar does not know the key K, Oscar cannot do this, except with a very small probability.

The informal proof given above can be made more precise. If we can prove an explicit, precise statement of the security of the underlying MAC, then we can give a precise security guarantee for the identification scheme. This is possible if the MAC is unconditionally secure. Alternatively, if we make an assumption about the MAC's security, then we can provide a security result for the identification scheme that depends on this assumption (this is the usual model of provable security). The security guarantees for the identification scheme quantify the probability that the adversary can fool Bob into accepting when the adversary is an active participant in the scheme.

A MAC is said to be *unconditionally* (ϵ, Q)-*secure* if the adversary cannot construct a valid MAC for any new message with probability greater than ϵ, given that the adversary has previously seen valid MACs for at most Q messages (i.e., there does not exist an (ϵ, Q)-forger). As usual, we assume a fixed key, K, whose value is not known to the adversary, is used to construct all Q of the MACs. An identification scheme is defined to be *unconditionally* (ϵ, Q)-*secure* if the adversary cannot fool Alice or Bob into accepting with probability greater then ϵ, given that the adversary has observed at most Q previous sessions between Alice and Bob.

Unconditionally secure (ϵ, Q)-secure MACs exist for any desired values of Q and ϵ (e.g., using almost strongly universal hash families, which are considered in the Exercises of Chapter 4). However, unconditionally secure MACs typically require fairly large keys (especially if Q is large). As a consequence, computationally secure MACs, such as *CBC-MAC*, are more often used in practice. In this situation, an assumption about the security of the MAC is necessary. This assumption would take a similar form, but could include time as an explicit parameter. A MAC would be said to be (ϵ, Q, T)-*secure* if the adversary cannot construct a valid MAC for any new message with probability greater than ϵ, given that his computation time is at most T and given that he has previously seen valid MACs for at most Q messages. An identification scheme is defined to be (ϵ, Q, T)-*secure* if the adversary cannot fool Alice or Bob into accepting with probability greater then ϵ, given that the adversary has observed at most Q previous sessions between Alice and Bob, and given that the adversary's computation time is at most T.

For simplicity of notation, we will usually omit an explicit specification of the time parameter. This allows us to use similar notations in both the computationally secure and unconditionally secure settings. Whether we are talking about unconditional or computational security should be clear from the context.

Suppose first that we base the identification scheme on an unconditionally secure MAC. Then the resulting identification scheme will also be unconditionally secure, provided that the adversary has access to at most Q valid MACs during some collection of sessions that all use the same MAC key. We need to recall one additional parameter, namely, the size (in bits) of the random challenge used in the scheme, which is denoted by k. Under these conditions, we can easily give an upper bound on the adversary's probability of deceiving Bob. We consider the same three cases as before:

1. As argued before, the value $y = MAC_K(ID(Alice) \parallel r)$ would not have been previously constructed by Bob himself in some other session. (So this case does not occur.)

2. Suppose the value y was previously constructed by Alice in some other session. The challenge r is assumed to be a random challenge newly created by Bob. The probability that Bob already used the challenge r in a specific previous session is $1/2^k$. There are at most Q previous sessions under consideration, so the probability that r was used as challenge in one of these previous sessions is at most $Q/2^k$. If this happens, then the adversary can re-use a MAC from a previous session.[1]

3. Suppose the value y is a new MAC that is constructed by Oscar. Then, Oscar will be successful in his deception with probability at most ϵ; this follows from the security of the message authentication code being used.

Summing up, Oscar's probability of deceiving Bob is at most $Q/2^k + \epsilon$. We therefore have established the security of the identification scheme as a function of the security of the underlying primitives.

The analysis is essentially identical if a computationally secure MAC is used. We summarize the results of this section in the following theorem.

THEOREM 9.1 *Suppose that MAC is an (ϵ, Q)-secure message authentication code, and suppose that random challenges are k bits in length. Then Protocol 9.2 is a $(Q/2^k + \epsilon, Q)$-secure identification scheme.*

9.2.1   Attack Model and Adversarial Goals

There are several subtleties associated with the attack model and the adversarial goals in an identification scheme. To illustrate, we depict an *intruder-in-the-middle* scenario in Figure 9.3.

At first glance, this might appear to be a parallel session attack. It could be argued that Oscar impersonates Alice to Bob in one session, and he impersonates Bob to Alice in a parallel session. When Oscar receives Bob's challenge, r, he sends it to Alice. Then Alice's response (namely, y) is sent by Oscar to Bob, and Bob will "accept." However, we do not consider this to be a real attack, because the "union" of the two "sessions" is a single session in which Alice has successfully identified herself to Bob. The overall result is that Bob issued a

---

[1] The exact probability that a challenge is repeated from a previous session is $1 - (1 - 2^{-k})^Q$, which is less than $Q/2^k$.

FIGURE 9.3
An Intruder-in-the-middle

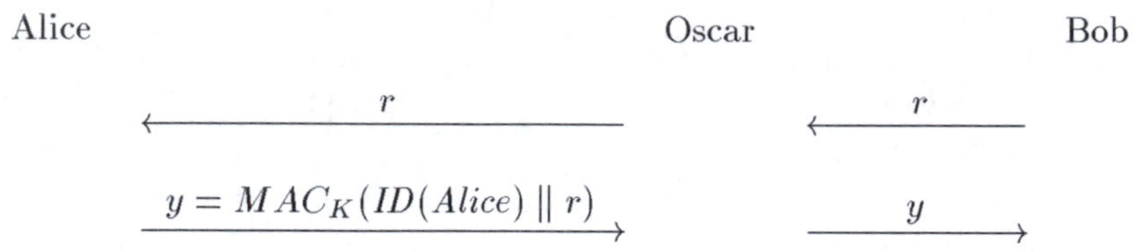

Alice   Oscar   Bob

$$\xleftarrow{\hspace{3cm} r \hspace{3cm}} \qquad \xleftarrow{\hspace{1.5cm} r \hspace{1.5cm}}$$

$$\xrightarrow{\hspace{0.5cm} y = MAC_K(ID(Alice) \,\|\, r) \hspace{0.5cm}} \qquad \xrightarrow{\hspace{1.5cm} y \hspace{1.5cm}}$$

challenge r and Alice computed the correct response y to the challenge. Oscar simply forwarded messages to their intended recipients without modifiying the messages, so Oscar was not an active participant in the scheme. The session executed exactly as it would have if Oscar had not been present.

A clear formulation of the adversarial goal should allow us to demonstrate that this is not an attack. We adopt the following approach. We will define the adversary (Oscar) to be *active* in a particular session if one of the following conditions holds:

1. Oscar creates a new message and places it in the channel

2. Oscar changes a message in the channel, or

3. Oscar diverts a message in the channel so it is sent to someone other than the intended receiver.

The goal of the adversary is to have the initiator of the scheme (e.g., Bob, who is assumed to be honest) "accept" in some session in which the adversary is active. According to this definition, Oscar is not active in the intruder-in-the-middle scenario considered above, and therefore the adversarial goal is not realized.

Another, esssentially equivalent, way to decide if an adversary is really active is to consider Alice and Bob's view of the scheme. Both Alice and Bob are interacting with an *intended peer*: Alice's intended peer is Bob and vice versa. Further, if there is no active adversary, then Alice and Bob will have compatible views of the session: every message sent by Alice is received by Bob, and vice versa. Moreover, no message will be received out of order. These features of a session are sometimes described as being *matching conversations*.

The above discussion of the model assumes that the legitimate participants in a session are honest. To be precise, a participant in a session of the scheme (e.g., Alice or Bob) is said to be an *honest participant* if she/he follows the scheme, performs correct computations, and does not reveal information to the adversary (Oscar). If a participant is not honest, then the scheme is completely broken, so statements of security generally require that participants are honest.

Let's now turn to a consideration of attack models. Before he actually tries to deceive Bob, say, Oscar carries out an *information-gathering phase*. Oscar is a *passive adversary* during this phase if he simply observes sessions between Alice and Bob. Alternatively, we might consider an attack model in which Oscar is active during the information-gathering phase. For example, Oscar might be given temporary access to an oracle that computes authentication tags $MAC_K(\cdot)$ for the (unknown) key K being used by Alice and Bob. During this time period, Oscar can successfully deceive Alice and Bob, of course, by using the oracle to respond to challenges. However, after the information-gathering phase, the MAC oracle is confiscated, and then Oscar carries out his attack, trying to get Alice or Bob to "accept" in a new session in which Oscar does not have a MAC oracle.

The security analysis that was performed in Section 9.2 applies to both of these attack models. The identification scheme is provably secure (more precisely, the adversary's success probability is at most $Q/2^k + \epsilon$) in the passive information-gathering model provided that the MAC is (ϵ, Q)-secure against a known message attack. Furthermore, the identification scheme is secure in the active information-gathering model provided that the MAC is (ϵ, Q)-secure against a chosen message attack.[2]

9.2.2 Mutual Authentication

A scheme in which Alice and Bob are both proving their identities to each other is called *mutual authentication* or *mutual identification*. Both participants are required to "accept" if a session of the scheme is to be considered a successfully completed session. The adversary could be trying to fool Alice, Bob, or both of them into accepting. The adversarial goal is to cause an honest participant to "accept" after a flow in which the adversary is active.

The following conditions specify what the outcome of a mutual identification scheme should be, if the scheme is to be considered secure:

1. Suppose Alice and Bob are the two participants in a session of the scheme and they are both honest. Suppose also that the adversary is passive. Then Alice and Bob will both "accept."

2. If the adversary is active during a given flow of the scheme, then no honest participant will "accept" after that flow.

Note that the adversary might be inactive in a particular session until after one participant accepts, and then become active. Therefore it is possible that one honest participant "accepts" and then the other honest participant "rejects." The adversary does not achieve his goal in this scenario, even though the session did not successfully complete, because the adversary was inactive before the first participant accepted. The outcome of the session is that Alice successfully identifies

---

[2]The attack model for MACs that we described in Chapter 4 is basically a chosen message attack. The notion of a known message attack for MACs is analogous to the corresponding notion for signature schemes that was introduced in Chapter 7.

Protocol 9.3: INSECURE MUTUAL CHALLENGE-AND-RESPONSE

1. Bob chooses a random challenge, r_1, which he sends to Alice.

2. Alice chooses a random challenge, r_2. She also computes

$$y_1 = MAC_K(ID(Alice) \parallel r_1)$$

and she sends r_2 and y_1 to Bob.

3. Bob computes

$$y_1' = MAC_K(ID(Alice) \parallel r_1).$$

If $y_1' = y_1$, then Bob "accepts"; otherwise, Bob "rejects." Bob also computes

$$y_2 = MAC_K(ID(Bob) \parallel r_2)$$

and he sends y_2 to Alice.

4. Alice computes

$$y_2' = MAC_K(ID(Bob) \parallel r_2).$$

If $y_2' = y_2$, then Alice "accepts"; otherwise, Alice "rejects."

---

herself to Bob (say), but Bob does not successfully identify himself to Alice. This could be considered a *disruption* of the scheme, but it is not a successful attack.

There are several ways in which the adversary could be active in a session of a scheme. We list some of these now:

1. The adversary impersonates Alice, hoping to cause Bob to accept.

2. The adversary impersonates Bob, hoping to cause Alice to accept.

3. The adversary is active in some session involving Alice and Bob, and he is trying to cause both Alice and Bob to accept.

We might try to achieve mutual authentication by running Protocol 9.2 twice (i.e., Alice verifies Bob's identity, and then Bob verifies Alice's identity in a separate session). However, it is generally more efficient to design a single scheme to accomplish both identifications at once.

What if Alice and Bob were to combine two sessions of one-way identification into a single scheme, in the obvious way? This is what is done in Protocol 9.3, and it reduces the number of flows required (compared to running the original one-way scheme twice) from four to three. However, it turns out that the resulting mutual identification scheme is flawed and can be attacked.

Protocol 9.3 is insecure because Oscar can fool Alice in a parallel session attack. Oscar, pretending to be Bob, initiates a session with Alice. When Oscar receives Alice's challenge, r_2, in the second flow, he "accepts," and then he ini-

FIGURE 9.4
Attack on Protocol 9.3

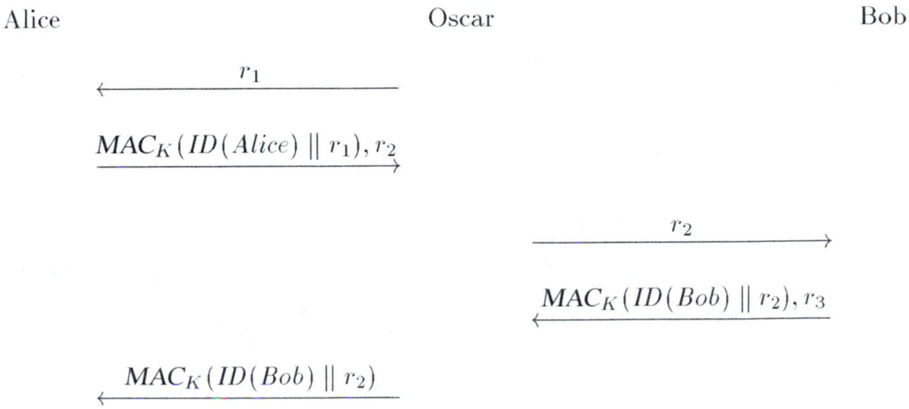

Alice   Oscar   Bob

$$\xleftarrow{\hspace{3cm} r_1 \hspace{3cm}}$$

$$\xrightarrow{\hspace{2cm} MAC_K(ID(Alice) \parallel r_1), r_2 \hspace{2cm}}$$

$$\xrightarrow{\hspace{3cm} r_2 \hspace{3cm}}$$

$$\xleftarrow{\hspace{2cm} MAC_K(ID(Bob) \parallel r_2), r_3 \hspace{2cm}}$$

$$\xleftarrow{\hspace{2cm} MAC_K(ID(Bob) \parallel r_2) \hspace{2cm}}$$

tiates a second session (pretending to be Alice) with Bob. In this second session, Oscar sends r_2 to Bob as his challenge in the first flow. When Oscar receives Bob's response (in the second flow in the second session), he forwards it to Alice as the third flow in the first session. Alice will "accept," and therefore Oscar has successfully impersonated Bob in this session. (The second session is dropped, i.e., it is never completed.) This constitutes a successful attack, because the honest participant in the first session (namely, Alice) accepted after a flow in which Oscar was active (namely, the initial flow of the session). An illustration of the attack is given in Figure 9.4.

Clearly, the attack is based on re-using a flow from one session in a different flow in another session. It is not difficult to rectify the problem, and there are in fact several ways to modify the scheme so that it is secure. Basically, what is required is to design the flows so that each flow contains information that is computed in a different manner. One solution along these lines is shown in Protocol 9.4.

The only change that was made in Protocol 9.4 is in the definition of y_1 in step 2. Now this MAC depends on two challenges, r_1 and r_2. This serves to distinguish the second flow from the third flow (in which the MAC depends only on the challenge r_2).

Protocol 9.4 can be analyzed in a similar fashion as Protocol 9.2. The analysis is a bit more complicated, however, because the adversary could try to play the role of Bob (fooling Alice) or Alice (fooling Bob). The probability that a value y_1 or y_2 can be "reused" from a previous session can be computed, as can the probability that the adversary can compute a new MAC from scratch.

Protocol 9.4: (SECURE) MUTUAL CHALLENGE-AND-RESPONSE

1. Bob chooses a random challenge, r_1, which he sends to Alice.
2. Alice chooses a random challenge, r_2. She also computes

$$y_1 = MAC_K(ID(Alice) \parallel r_1 \parallel r_2)$$

   and she sends r_2 and y_1 to Bob.
3. Bob computes

$$y_1' = MAC_K(ID(Alice) \parallel r_1 \parallel r_2).$$

   If $y_1' = y_1$, then Bob "accepts"; otherwise, Bob "rejects." Bob also computes

$$y_2 = MAC_K(ID(Bob) \parallel r_2)$$

   and he sends y_2 to Alice.
4. Alice computes

$$y_2' = MAC_K(ID(Bob) \parallel r_2).$$

   If $y_2' = y_2$, then Alice "accepts"; otherwise, Alice "rejects."

First, because any value y_1 is computed differently than any value y_2, it is impossible that a y_1 value from one session can be re-used as a y_2 value from another session (or vice versa). Oscar could try to play the role of Bob (fooling Alice) or Alice (fooling Bob), by determining y_2 or y_1, respectively. The probability that y_1 or y_2 can be reused from a previous session is at most $Q/2^k$, under the assumption that Oscar has seen at most Q MACs from previous sessions (this limits the number of previous sessions to $Q/2$, because there are two MACs per session). The probability that Oscar can compute a new y_1 is at most ϵ, and the probability that he can compute a new y_2 is at most ϵ. Therefore Oscar's probability of deceiving one of Alice or Bob is at most $Q/2^k + 2\epsilon$. Summarizing, we have the following theorem.

THEOREM 9.2 *Suppose that MAC is an (ϵ, Q)-secure message authentication code, and suppose that random challenges are k bits in length. Then Protocol 9.4 is a $(Q/2^k + 2\epsilon, Q/2)$-secure mutual identification scheme.*

9.3 Challenge-and-Response in the Public-key Setting

Now we turn to the public-key setting, where Alice and Bob might not have a prior shared secret key. However, we assume that Alice and Bob are members of a network, in which every participant has public and private keys for certain prespecified cryptosystems and/or signature schemes. In a setting such as this, it is always necessary to provide a mechanism to authenticate the public keys of other people in the network. This requires some kind of *public-key infrastructure* (also denoted as a *PKI*). In general, we assume that there is trusted authority, denoted by TA, who signs the public keys of all people in the network.[3] The (public) verification key of the TA, denoted ver_{TA}, is assumed to be known "by magic" to everyone in the network. This simplified setting is perhaps not completely realistic, but it allows us to concentrate on the design of the schemes. (The general topic of public key infrastructures will be studied in more detail in Chapter 12.)

9.3.1 Certificates

A *certificate* for someone in the network will consist of some identifying information for that person (e.g., their name, email address, etc.), their public key(s), and the signature of the TA on that information. A certificate allows network users to verify the authenticity of each other's keys.

Suppose, for example, that Alice wants to obtain a certificate from the TA which contains a copy of Alice's public verification key for a signature scheme. Then the steps in Protocol 9.5 would be executed.

We are not specifying exactly how Alice identifies herself to the TA, nor do we specify the precise format of $ID(Alice)$, or how the public and private keys of Alice are selected. In general, these implementation details could vary from one PKI to another.

It is possible for anyone who knows the TA's verification key, ver_{TA}, to verify anyone else's certificate. Suppose that Bob wants to be assured that Alice's public key is authentic. Alice can give her certificate to Bob. Bob can then verify the signature of the TA by checking that

$$ver_{TA}(ID(Alice) \parallel ver_{Alice}, s) = true.$$

The security of a certificate follows immediately from the security of the signature scheme used by the TA.

As mentioned above, the purpose of verifying a certificate is to authenticate someone's public key. The certificate itself does not provide any kind of proof of identity, because certificates contain only public information. Certificates can be distributed or redistributed to anyone, and possession of a certificate does not imply ownership of it.

---

[3] In the context of PKIs, the trusted authority is often called a *certification authority* and denoted CA. We will use the notation TA in this chapter, however.

Protocol 9.5: ISSUING A CERTIFICATE TO ALICE

1. The TA establishes Alice's identity by means of conventional forms of identification such as a birth certificate, passport, etc. Then the TA forms a string, denoted $ID(Alice)$, which contains Alice's identification information.

2. A private signing key for Alice, sig_{Alice}, and a corresponding public verification key, ver_{Alice}, are determined.

3. The TA generates its signature

$$s = sig_{TA}(ID(Alice) \parallel ver_{Alice})$$

on Alice's identity string and verification key. The certificate

$$\mathbf{Cert}(Alice) = (ID(Alice) \parallel ver_{Alice} \parallel s)$$

is given to Alice, along with Alice's private key, sig_{Alice}.

9.3.2 Public-key Identification Schemes

We now look at mutual identification schemes in the public-key setting. Our strategy is to modify Protocol 9.4 by replacing MACs by signatures. Another difference is that, in the secret-key setting, we included the name of the person who produced the MAC in each MAC (this was important because a secret key K, being known to two parties, allows either party to create MACs). In the public-key setting, only one person can create signatures using a specified private signing key, namely, the person possessing that key. Therefore we do not need to explicitly designate who created a particular signature.

As in the secret-key setting, at the beginning of a session, each participant has an intended peer (the person with whom each of them thinks they are communicating). Each participant will use the intended peer's verification key to verify signatures received in the forthcoming session. They will also include the name of the intended peer in all signatures that they create during the scheme.

Protocol 9.6 is a typical mutual identification scheme in the public-key setting. It can be proven to be secure if the signature scheme is secure and challenges are generated randomly. Figure 9.5 illustrates the scheme, omitting the transmission of the certificates of Alice and Bob. In this figure and elsewhere, "A" denotes "$ID(Alice)$" and "B" denotes "$ID(Bob)$."

Here is a theorem stating the security of Protocol 9.6 as a function of the security of the underlying signature scheme (where security of signature schemes is described using notation similar to that of MACs). The proof is left as an Exercise.

Protocol 9.6: PUBLIC-KEY MUTUAL AUTHENTICATION (VERSION 1)

1. Bob chooses a random challenge, r_1. He sends **Cert**(Bob) and r_1 to Alice.

2. Alice chooses a random challenge, r_2. She also computes $y_1 = sig_{Alice}(ID(Bob) \parallel r_1 \parallel r_2)$ and sends **Cert**($Alice$), r_2 and y_1 to Bob.

3. Bob verifies Alice's public key, ver_{Alice}, on the certificate **Cert**($Alice$). Then he checks that $ver_{Alice}(ID(Bob) \parallel r_1 \parallel r_2, y_1) = true$. If so, then Bob "accepts"; otherwise, Bob "rejects." Bob also computes $y_2 = sig_{Bob}(ID(Alice) \parallel r_2)$ and sends y_2 to Alice.

4. Alice verifies Bob's public key, ver_{Bob}, on the certificate **Cert**(Bob). Then she checks that $ver_{Bob}(ID(Alice) \parallel r_2, y_2) = true$. If so, then Alice "accepts"; otherwise, Alice "rejects."

FIGURE 9.5
Information Flows in Protocol 9.6

Alice   Bob

$$\xleftarrow{\quad r_1 \quad}$$

$$y_1 = sig_A(B \parallel r_1 \parallel r_2)$$

$$\xrightarrow{\quad r_2, y_1 \quad}$$

$$ver_A(B \parallel r_1 \parallel r_2, y_1) = true?$$
$$y_2 = sig_B(A \parallel r_2)$$

$$\xleftarrow{\quad y_2 \quad}$$

$$ver_B(A \parallel r_2, y_2) = true?$$

THEOREM 9.3 *Suppose that sig is an (ϵ, Q)-secure signature scheme, and suppose that random challenges are k bits in length. Then Protocol 9.6 is a $(Q/2^{k-1} + 2\epsilon, Q)$-secure mutual identification scheme.*

REMARK In Theorem 9.3, the number of previous sessions is Q, whereas in Theorem 9.2, the number of previous sessions was limited to $Q/2$. This is because the signatures created by Alice and Bob in Protocol 9.6 use different keys. The adversary is allowed to view Q signatures created by each of Alice and Bob. In

---

Protocol 9.7: (INSECURE) PUBLIC-KEY MUTUAL AUTHENTICATION

1. Bob chooses a random challenge, r_1. He sends **Cert**(Bob) and r_1 to Alice.

2. Alice chooses a random challenge, r_2. She also computes $y_1 = sig_{Alice}(ID(Bob) \parallel r_1 \parallel r_2)$ and sends **Cert**$(Alice)$, r_2 and y_1 to Bob.

3. Bob verifies Alice's public key, ver_{Alice}, on the certificate **Cert**$(Alice)$. Then he checks that $ver_{Alice}(ID(Bob) \parallel r_1 \parallel r_2, y_1) = true$. If so, then Bob "accepts"; otherwise, Bob "rejects." Bob also chooses a random number r_3, computes $y_2 = sig_{Bob}(ID(Alice) \parallel r_2 \parallel r_3)$ and sends r_3 and y_2 to Alice.

4. Alice verifies Bob's public key, ver_{Bob}, on the certificate **Cert**(Bob). Then she checks that $ver_{Bob}(ID(Alice) \parallel r_2 \parallel r_3, y_2) = true$. If so, then Alice "accepts"; otherwise, Alice "rejects."

---

contrast, in Protocol 9.4, both Alice and Bob use the same key to create MACs. Since we want to limit the adversary to seeing Q MACs created with any given key, this forces us to require that the adversary be allowed to eavesdrop in at most $Q/2$ previous sessions.   ∎

It is instructive to consider various modifications of this scheme. Some modifications turn out to be insecure, while others are secure. An example of an insecure (modified) scheme includes a third random number r_3 that is signed by Bob; with this modification, the scheme becomes vulnerable to a parallel session attack. The scheme is presented as Protocol 9.7.

In Protocol 9.7, the random value, r_3, is chosen by Bob and is signed by him (along with r_2) in the third flow of the scheme. Including this extra piece of information in the signature makes the scheme insecure because the signature in the third flow is now constructed in a similar fashion as the signature in the second flow. This allows the parallel session attack depicted in Figure 9.6 to be carried out. In this attack, Oscar initiates a sesssion with Alice, pretending to be Bob. Then he initiates a second session, with Bob, pretending to Alice. Bob's response in the second flow of the second session is forwarded to Alice in the third flow of the first session.

Finally, we note that another variation of Protocol 9.6 that is secure is discussed in the Exercises.

FIGURE 9.6
Attack on Protocol 9.7

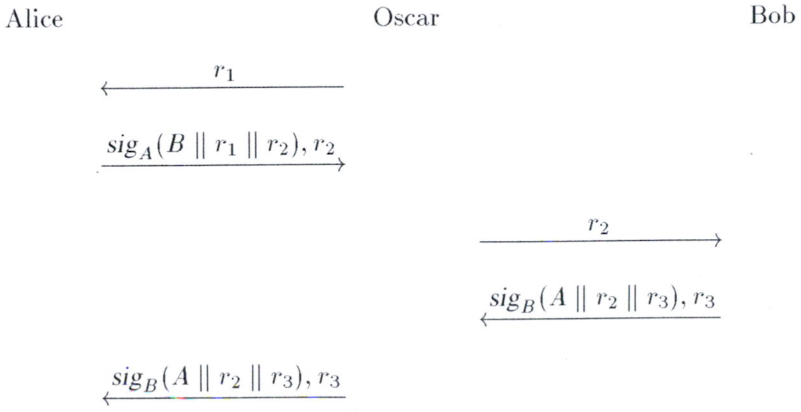

Alice　　　　　　　　　　　　　Oscar　　　　　　　　　　　　Bob

$$\xleftarrow{\qquad r_1 \qquad}$$

$$\xrightarrow{\quad sig_A(B \parallel r_1 \parallel r_2), r_2 \quad}$$

$$\xrightarrow{\qquad r_2 \qquad}$$

$$\xleftarrow{\quad sig_B(A \parallel r_2 \parallel r_3), r_3 \quad}$$

$$\xleftarrow{\quad sig_B(A \parallel r_2 \parallel r_3), r_3 \quad}$$

Note that "A" denotes "$ID(Alice)$" and "B" denotes "$ID(Bob)$."

9.4 The Schnorr Identification Scheme

Another approach to identification schemes is to design schemes "from scratch," without using any other cryptographic tools as building blocks. A potential advantage of schemes of this type is that they might be more efficient and have a lower communication complexity than the schemes considered in the previous sections. Such schemes typically involve having someone identify themselves by proving that they know the value of some secret quantity (i.e., a private key) without having to reveal its value.

The *Schnorr Identification Scheme* (Protocol 9.8) is an example of such a scheme. This scheme is based on the **Discrete Logarithm** problem, which we introduced as Problem 6.1. Here, we will take α to be an element having prime order q in the group $\mathbb{Z}_p^*$ (where p is prime and $p - 1 \equiv 0 \pmod{q}$). Then $\log_\alpha \beta$ is defined for any element $\beta \in \langle \alpha \rangle$, and $0 \leq \log_\alpha \beta \leq q - 1$. This is the same setting of the **Discrete Logarithm** problem that was used in the *Schnorr Signature Scheme* and the *Digital Signature Algorithm* (see Section 7.4). In order for this setting to be considered secure, we will specify that $p \approx 2^{1024}$ and $q \approx 2^{160}$.

The scheme requires a trusted authority, or *TA*, who chooses some common system parameters (*domain parameters*) for the scheme, as follows:

1. p is a large prime (i.e., $p \approx 2^{1024}$).

Protocol 9.8: SCHNORR IDENTIFICATION SCHEME

1. Alice chooses a random number, k, where $0 \leq k \leq q - 1$, and she computes $\gamma = \alpha^k \bmod p$. She sends **Cert**(*Alice*) and γ to Bob.

2. Bob verifies Alice's public key, v, on the certificate **Cert**(*Alice*). Bob chooses a random challenge r, $1 \leq r \leq 2^t$, and he sends r to Alice.

3. Alice computes $y = k + ar \bmod q$ and she sends the response y to Bob.

4. Bob verifies that $\gamma \equiv \alpha^y v^r \pmod{p}$. If so, then Bob "accepts"; otherwise, Bob "rejects."

---

2. q is a large prime divisor of $p - 1$ (i.e., $q \approx 2^{160}$).

3. $\alpha \in \mathbb{Z}_p^*$ has order q.

4. t is a *security parameter* such that $q > 2^t$. (The adversary's probability of deceiving Alice or Bob will be 2^{-t}, so $t = 40$ will provide adequate security for most practical applications.)

The domain parameters p, q, α and t are all made public, and they will be used by everyone in the network.

Every user in the network chooses their own private key, a, where $0 \leq a \leq q - 1$, and constructs a corresponding public key $v = \alpha^{-a} \bmod p$. Observe that v can be computed as $(\alpha^a)^{-1} \bmod p$, or (more efficiently) as $\alpha^{q-a} \bmod p$. The *TA* issues certificates for everyone in the network. Each user's certificate will contain their public key (and, perhaps, the public domain parameters). This information, as well as the user's identifying information, is signed by the *TA*, of course.

The following congruences demonstrate that Alice will be able to prove her identity to Bob, assuming that both parties are honest and perform correct computations:

$$\alpha^y v^r \equiv \alpha^{k+ar} v^r \pmod{p}$$

$$\equiv \alpha^{k+ar} \alpha^{-ar} \pmod{p}$$

$$\equiv \alpha^k \pmod{p}$$

$$\equiv \gamma \pmod{p}.$$

The fact that Bob will accept Alice's proof of identity (assuming that he and Alice are honest) is sometimes called the *completeness* property of the scheme.

Let's work out a small, toy example. The following example omits the authentication of Alice's public key by Bob.

Example 9.1 Suppose $p = 88667$, $q = 1031$ and $t = 10$. The element $\alpha = 70322$ has order q in $\mathbb{Z}_p^*$. Suppose Alice's private key is $a = 755$; then

$$v = \alpha^{-a} \bmod p$$
$$= 70322^{1031-755} \bmod 88667$$
$$= 13136.$$

Now suppose Alice chooses the random number $k = 543$. Then she computes

$$\gamma = \alpha^k \bmod p$$
$$= 70322^{543} \bmod 88667$$
$$= 84109,$$

and she sends γ, to Bob. Suppose Bob issues the challenge $r = 1000$. Then Alice computes

$$y = k + ar \bmod q$$
$$= 543 + 755 \times 1000 \bmod 1031$$
$$= 851,$$

and she sends y to Bob as her response. Bob then verifies that

$$84109 \equiv 70322^{851} 13136^{1000} \pmod{88667}.$$

Finally, Bob "accepts." $\square$

The *Schnorr Identification Scheme* was designed to be very fast and efficient, both from a computational point of view and in the amount of information that needs to be exchanged in the scheme. It is also designed to minimize the amount of computation performed by Alice, in particular. This is desirable because, in many practical applications, Alice's computations will be performed by a smart card with low computing power, while Bob's computations will be performed by a more powerful computer.

Let us consider Alice's computations. Step 1 requires an exponentiation (modulo p) to be performed; step 3 comprises one addition and one multiplication (modulo q) It is the modular exponentiation that is computationally intensive, but this can be precomputed offline, before the scheme is executed, if desired. The online computations to be performed by Alice are very modest.

It is also a simple matter to calculate the number of bits that are communicated during the scheme. We depict the information that is communicated (excluding Alice's certificate) in Figure 9.7. In that diagram, the notation $\in_R$ is used to denote a random choice made from a specified set.

FIGURE 9.7
Information Flows in Protocol 9.8

Alice   Bob
$k \in_R \{0, \ldots, q-1\}$
$\gamma = \alpha^k \bmod p$

$$\xrightarrow{\quad\gamma\quad (1024 \text{ bits})\quad}$$

$r \in_R \{1, \ldots, 2^t\}$

$$\xleftarrow{\quad r\quad (40 \text{ bits})\quad}$$

$y = k + ar \bmod q$

$$\xrightarrow{\quad y\quad (160 \text{ bits})\quad}$$

$\gamma \equiv \alpha^y v^r \pmod{p}$?

Alice gives Bob 1024 bits of information (excluding her certificate) in the first flow; Bob sends Alice 40 bits in the second flow; and Alice transmits 160 bits to Bob in the third flow. So the communication requirements are quite modest, as well.

The information transmitted in the second and third flows of the scheme have been reduced by the way in which the scheme is designed. In the second flow, a challenge could be taken to be any integer between 0 and $q - 1$; however, this would yield a 160-bit challenge. A 40-bit challenge provides sufficient security for many applications.

In the third flow, the value y is an exponent. This value is only 160 bits in length because the scheme is working inside a subgroup of $\mathbb{Z}_p^*$ of order $q \approx 2^{160}$. This permits the information transmitted in the third flow to be reduced significantly, as compared to an implementation of the scheme in the "whole group," $\mathbb{Z}_p^*$, in which an exponent would be 1024 bits in length.

The first flow clearly requires the most information to be transmitted. One possible way to reduce the amount of information is to replace the 1024-bit value γ by a 160-bit message digest, $\gamma' = SHA\text{-}1(\gamma)$. Then, in the last step of the scheme, Bob would verify that (the message digest) $\gamma' = SHA\text{-}1(\alpha^y v^r \pmod{p})$.

9.4.1 Security of the Schnorr Identification Scheme

Let's now study the security of the *Schnorr Identification Scheme*. As mentioned previously, t is a security parameter. It is sufficiently large to prevent an impostor posing as Alice, say Olga, from guessing Bob's challenge, r. (If Olga guessed the correct value of r, she could choose any value for y and precompute

$$\gamma = \alpha^y v^r \bmod p.$$

She would give Bob the value γ in the first flow of the scheme, and when she receives the challenge r, she would respond with the value y she has already chosen. Then the congruence involving γ would be verified by Bob, and he would "accept.") The probability that Olga will guess the value of r correctly is 2^{-t} if r is chosen at random by Bob.

Notice that Bob should choose a new, random challenge, r, every time Alice identifies herself to him. If Bob always used the same challenge r, then Olga could impersonate Alice by the method we just described.

Alice's computations in the scheme involve the use of her private key, a. The value a functions somewhat like a PIN, in that it convinces Bob that the person (or entity) carrying out the identification scheme is, indeed, Alice. But there is an important difference from a PIN: in this identification scheme, the value of a is not revealed. Instead, Alice (or more accurately, Alice's smart card) "proves" that she/it knows the value of a in the third flow of the scheme, by computing the correct response, y, to the challenge, r, issued by Bob. An adversary could attempt to compute a, because a is just a discrete logarithm of a known quantity: $a = -\log_\alpha v$ in $\mathbb{Z}_p^*$. However, we are assuming that this computation is infeasible.

We have argued that Olga can guess Bob's challenge, r, and thereby impersonate Alice, with probability 2^{-t}. Suppose that Olga can do better than this. If Olga can impersonate Alice successfully with a probability exceeding 2^{-t}, then it is plausible that Olga knows some γ (a value of her choosing), and two possible challenges, r_1 and r_2, such that she can compute responses y_1 and y_2, respectively, which would cause Bob to accept. (If Olga could only compute a correct response for one challenge for each γ, then her probability of success would be only 2^{-t}.)

So we assume that Olga knows (or she can compute) values r_1, r_2, y_1 and y_2 such that

$$\gamma \equiv \alpha^{y_1} v^{r_1} \equiv \alpha^{y_2} v^{r_2} \pmod{p}.$$

It follows that

$$\alpha^{y_1 - y_2} \equiv v^{r_2 - r_1} \pmod{p}.$$

It holds that $v \equiv \alpha^{-a} \pmod{p}$, where a is Alice's private key. Hence,

$$\alpha^{y_1 - y_2} \equiv \alpha^{-a(r_2 - r_1)} \pmod{p}.$$

The element α has order q, so it must be the case that

$$y_1 - y_2 \equiv a(r_1 - r_2) \pmod{q}.$$

Now, $0 < |r_2 - r_1| < 2^t$ and $q > 2^t$ is prime. Hence $\gcd(r_2 - r_1, q) = 1$, and therefore $(r_1 - r_2)^{-1} \bmod q$ exists. Hence, Olga can compute Alice's private key, a, as follows:

$$a = (y_1 - y_2)(r_1 - r_2)^{-1} \bmod q.$$

The above analysis suggests that anyone who is able to successfully impersonate Alice with a probability exceeding 2^{-t} must know (or be able to easily compute) Alice's private key, a. (We proved above that a "successful" impersonator can compute a. Conversely, it is obvious that anyone who knows the value of a can impersonate Alice, with probability equal to 1.) It therefore follows, roughly speaking, that being able to impersonate Alice is equivalent to knowing Alice's private key. This property is sometimes termed *soundness*.

An identification scheme that is both sound and complete is called a *proof of knowledge*. Our analysis so far has established that the *Schnorr Identification Scheme* is a proof of knowledge. We provide an example to illustrate the above discussion.

Example 9.2   Suppose we have the same parameters as in Example 9.1: $p = 88667$, $q = 1031$, $t = 10$, $\alpha = 70322$ and $v = 13136$. Suppose, for $\gamma = 84109$, that Olga is able somehow to determine two correct responses: $y_1 = 851$ is the correct repsonse for the challenge $r_1 = 1000$; and $y_1 = 454$ is the correct repsonse for the challenge $r_1 = 19$. In other words,

$$84109 \equiv \alpha^{851} v^{1000} \equiv \alpha^{454} v^{19} \pmod{p}.$$

Then Olga can compute

$$a = (851 - 454)(1000 - 19)^{-1} \bmod 1031 = 755,$$

and thus discover Alice's private key.   □

We have proved that the scheme is a proof of knowledge. But this is not sufficient to ensure that the scheme is "secure." We still need to consider the possibility that secret information (namely, Alice's private key) might be leaked to a verifier who takes part in the scheme, or an observer. (This could be thought of as the information gathering phase of an attack.) Our hope is that no information about a will be gained by Olga when Alice proves her identity. If this is true, then Olga will not be able subsequently to masquerade as Alice (assuming that the computation of the discrete logarithm a is infeasible).

In general, we could envision a situation whereby Alice proves her identity to Olga, say, on several different occasions. After several sessions of the scheme, Olga will try to determine the value of a so she can subsequently impersonate Alice. If Olga can determine no information about the value of a by taking part in a "reasonable" number of sessions of the scheme as the verifier, and then performing a "reasonable" amount of computation, then the scheme is termed a *zero-knowledge identification scheme*. This would prove that the scheme is secure, under the assumption that a is infeasible to compute. (Of course, it would be necessary to define, in a precise way, the term "reasonable," in order to have a meaningful statement of security.)

We will show that the *Schnorr Identification Scheme* is zero-knowledge for an honest verifier, where an *honest verifier* is defined to be one who chooses his or her challenges r at random, as specified by the scheme.

We require the notion of a *transcript* of a session, which consists of a triple $T = (\gamma, r, y)$ in which $\gamma \equiv \alpha^y v^r \pmod{p}$. The verifier (or an observer) can obtain a transcript $T(S)$ of each session S. The set of possible transcripts is

$$\mathcal{T} = \{(\gamma, r, y) : 1 \le r \le 2^t, 0 \le y \le q - 1, \gamma \equiv \alpha^y v^r \pmod{p}\}.$$

It is easy to see that $|\mathcal{T}| = q\, 2^t$. Further, it is not difficult to prove that the probability that any particular transcript occurs in any given session is $1/(q\, 2^t)$, assuming that the challenges r are generated at random. We argue this as follows: for any fixed value of r, there is a one-to-one correspondence between the value of $\gamma \in \langle \alpha \rangle$ and the value of $y \in \{0, \ldots, q - 1\}$ on a particular transcript. We are assuming that Alice chooses γ at random (namely, by choosing a random k and computing $\gamma = \alpha^k \bmod p$), and we also assume that Bob chooses r at random (because he is an honest verifier). These two values determine the value of y. Since there are q possible choices for γ and 2^t possible choices for r, it follows that every possible transcript occurs with the same probability, $1/(q\, 2^t)$, in sessions involving an honest verifier.

The key point of the zero-knowledge aspect of the scheme is a property called *simulatability*. It turns out that Olga (or anyone else, for that matter) can generate simulated transcripts, having exactly the same probability distribution as real transcripts, without taking part in the scheme. This is done by the following three simple steps:

1. choose $r \in_R \{1, \ldots, 2^t\}$
2. choose $y \in_R \{0, \ldots, q - 1\}$
3. compute $\gamma = \alpha^y v^r \bmod p$.

It is easy to see that the probability that any $T \in \mathcal{T}$ is generated by the above procedure is $1/(q\, 2^t)$. Therefore, it holds that

$$\mathbf{Pr}_{\text{real}}[T] = \mathbf{Pr}_{\text{sim}}[T] = \frac{1}{q\, 2^t}$$

for all $T \in \mathcal{T}$, where $\mathbf{Pr}_{\text{real}}[T]$ is the probability of generating the transcript T during a real session, and $\mathbf{Pr}_{\text{sim}}[T]$ is the probability of generating T as a simulated transcript.

What is the significance of the fact that transcripts can be simulated? We claim that, whatever an honest verifier can compute after taking part in several sessions of the scheme, the verifier can alternatively compute the same information without taking part in any sessions of the scheme. In particular, computing Alice's private key, a, which is necessary for Olga to be able to impersonate Alice, is not made easier for Olga if she plays the role of the verifier in one or more sessions in which he chooses his challenges randomly.

The above statements can be justified further, as follows. Suppose there exists an algorithm EXTRACT which, when given a set of transcripts, say $T_1, \ldots, T_\ell$, computes a private key, say a, with some probability, say ϵ. We assume that the transcripts are actual transcripts of sessions, in which the participants follow the scheme. Suppose that $T'_1, \ldots, T'_\ell$ are simulated transcripts. We have noted that the probability distribution on simulated transcripts is identical to the probability distribution on real transcripts. Therefore EXTRACT$(T'_1, \ldots, T'_\ell)$ will also compute a with probability ϵ. This establishes that executing the scheme does not make computing a easier, so the scheme is zero-knowledge.

Let's consider the possibility that Olga (a "dishonest verifier") might obtain some useful information by choosing his challenges r in a non-uniform way. To be specific, suppose that Olga chooses her challenge r using some function that depends, in a complicated way, on Alice's choice of γ. There does not seem to be any way to perfectly simulate the resulting probability distribution on transcripts, and therefore we cannot prove that the scheme is zero-knowledge in the way that we did for an honest verifier.

We should emphasize that there is no known attack on the scheme based on making non-random challenges; we are just saying that the proof technique we used previously does not seem to apply in this case. The only known security proofs of the scheme for arbitrary verifiers require additional assumptions.

To summarize, an interactive scheme is a proof of knowledge if it is impossible (except with a very small probability) to impersonate Alice without knowing the value of Alice's key. This means that the only way to "break" the scheme is to actually compute a. A scheme is termed zero-knowledge if it reveals no information about Alice's private key. Stated another way, computing Alice's private key is not made easier by taking part in the scheme (in Bob's role as the verifier) in some specified number of sessions. If a scheme is a zero-knowledge proof of knowledge, then it is "secure."

9.5 The Okamoto Identification Scheme

In this section, we present a modification of the *Schnorr Identification Scheme* due to Okamoto. This modification can be proven to be secure, assuming the intractibility of computing a particular discrete logarithm in $\mathbb{Z}_p^*$.

Protocol 9.9: OKAMOTO IDENTIFICATION SCHEME

1. Alice chooses random numbers k_1, k_2, where $0 \leq k_1, k_2 \leq q - 1$, and computes
$$\gamma = \alpha_1{}^{k_1}\alpha_2{}^{k_2} \bmod p.$$

2. Alice sends her certificate, **Cert**(*Alice*) $= (ID(Alice), v, s)$, and γ to Bob.

3. Bob verifies that $ver_{TA}(ID(Alice) \parallel v, s) = true$.

4. Bob chooses a random number r, $1 \leq r \leq 2^t$, and he gives it to Alice.

5. Alice computes
$$y_1 = k_1 + a_1 r \bmod q$$

and

$$y_2 = k_2 + a_2 r \bmod q$$

and she gives y_1 and y_2 to Bob.

6. Bob verifies that
$$\gamma \equiv \alpha_1{}^{y_1}\alpha_2{}^{y_2}v^r \pmod{p}.$$

To set up the scheme, the *TA* chooses p and q as in the *Schnorr Identification Scheme*. The *TA* also chooses two elements $\alpha_1, \alpha_2 \in \mathbb{Z}_p{}^*$ both having order q. Since $\mathbb{Z}_p{}^*$ is a cyclic group, it follows that it has a unique subgroup H of order q, which is cyclic, and any element of order q in $\mathbb{Z}_p{}^*$ is a generator of H. Therefore $\alpha_1 \in \langle \alpha_2 \rangle$ and $\alpha_2 \in \langle \alpha_1 \rangle$. Denote $c = \log_{\alpha_1} \alpha_2$. The value of c is kept secret from all the participants (including Alice). We will assume that it is infeasible for anyone (even a coalition of Alice and Olga, say) to compute the value c.

In addition, Alice's public key v is computed as

$$v = \alpha_1{}^{-a_1}\alpha_2{}^{-a_2} \pmod{p},$$

where a_1 and a_2 comprise Alice's private key. Now we can describe the *Okamoto Identification Scheme*; it is presented as Protocol 9.9.

Here is an example of the *Okamoto Identification Scheme*.

Example 9.3　As in previous examples, we will take $p = 88667, q = 1031$, and $t = 10$. Suppose $\alpha_1 = 58902$ and $\alpha_2 = 73611$ (both α_1 and α_2 have order q in $\mathbb{Z}_p{}^*$). Now, suppose $a_1 = 846$ and $a_2 = 515$; then $v = 13078$.

Suppose Alice chooses $k_1 = 899$ and $k_2 = 16$; then $\gamma = 14574$. If Bob issues the challenge $r = 489$ then Alice will respond with $y_1 = 131$ and $y_2 = 287$. Bob will verify that

$$58902^{131}73611^{287}13078^{489} \equiv 14574 \pmod{88667}.$$

So Bob will accept Alice's proof of identity.   ▯

The proof that the scheme is complete (i.e., that Bob will accept Alice's proof of identity) is straightforward. The main difference between Okamoto's and Schnorr's scheme is that we can prove that the *Okamoto Identification Scheme* is secure provided that the computation of the discrete logarithm $\log_{\alpha_1} \alpha_2$ is intractable.

The proof of security is quite subtle. Here is the general idea: Suppose that Alice identifies herself to Olga polynomially many times by executing the scheme. We then suppose (hoping to obtain a contradiction) that Olga is able to learn some information about the values of Alice's secret exponents a_1 and a_2. If this is so, then we will show that (with high probability) Alice and Olga together will be able to compute the discrete logarithm c in polynomial time. This contradicts the assumption made above, and proves that Olga must be unable to obtain any information about Alice's exponents by taking part in the scheme.

The first part of this procedure is similar to the proof given for the *Schnorr Identification Scheme*. Suppose Olga knows a value γ for which she can compute valid responses to two different challenges, say r and s. That is, suppose that Olga can compute values y_1, y_2, z_1, z_2, r and s with $r \neq s$ and

$$\gamma \equiv \alpha_1^{y_1} \alpha_2^{y_2} v^r \equiv \alpha_1^{z_1} \alpha_2^{z_2} v^s \pmod{p}.$$

Olga can set

$$b_1 = (y_1 - z_1)(r - s)^{-1} \bmod q$$

and

$$b_2 = (y_2 - z_2)(r - s)^{-1} \bmod q.$$

Then it is easy to check that

$$v \equiv \alpha_1^{-b_1} \alpha_2^{-b_2} \pmod{p}.$$

We now proceed to show how Alice and Olga together can compute the value of c (with high probability). We assume that Olga is able to determine values b_1 and b_2 such that

$$v \equiv \alpha_1^{-b_1} \alpha_2^{-b_2} \pmod{p},$$

as described above. Now suppose that Alice reveals the secret values a_1 and a_2 to Olga. Of course

$$v \equiv \alpha_1^{-a_1} \alpha_2^{-a_2} \pmod{p},$$

so it must be the case that

$$\alpha_1^{a_1 - b_1} \equiv \alpha_2^{b_2 - a_2} \pmod{p}.$$

Suppose that $(a_1, a_2) \neq (b_1, b_2)$. Then $(b_2 - a_2)^{-1} \bmod q$ exists, and the discrete log

$$c = \log_{\alpha_1} \alpha_2 = (a_1 - b_1)(b_2 - a_2)^{-1} \bmod q$$

can be computed in polynomial time.

There remains to be considered the possibility that $(a_1, a_2) = (b_1, b_2)$. If this happens, then the value of c cannot be computed as described above. However, we will argue that $(a_1, a_2) = (b_1, b_2)$ will happen only with very small probability $1/q$, so the procedure whereby Alice and Olga compute c will almost surely succeed.

Define

$$\mathcal{A} = \{(a_1', a_2') \in \mathbb{Z}_q \times \mathbb{Z}_q : \alpha_1^{-a_1'} \alpha_2^{-a_2'} \equiv \alpha_1^{-a_1} \alpha_2^{-a_2} \pmod{p}\}.$$

That is, $\mathcal{A}$ consists of all the possible ordered pairs that could be Alice's secret exponents. Observe that

$$\mathcal{A} = \{(a_1 - c\theta, a_2 + \theta) : \theta \in \mathbb{Z}_q\},$$

where $c = \log_{\alpha_1} \alpha_2$. Thus $\mathcal{A}$ consists of q ordered pairs.

The ordered pair (b_1, b_2) computed by Olga is certainly in the set $\mathcal{A}$. We will argue that the value of the pair (b_1, b_2) is independent of the value of the pair (a_1, a_2) that comprises Alice's secret exponents. Since (a_1, a_2) was originally chosen at random by Alice, it must be the case that the probability that $(a_1, a_2) = (b_1, b_2)$ is $1/q$.

So, we need to say what we mean by (b_1, b_2) being "independent" of (a_1, a_2). The idea is that Alice's pair (a_1, a_2) is one of the q possible ordered pairs in the set $\mathcal{A}$, and no information about which is the "correct" ordered pair is revealed by Alice identifying herself to Olga. (Stated informally, Olga knows that an ordered pair from $\mathcal{A}$ comprises Alice's exponents, but she has no way of telling which one.)

Let's look at the information that is exchanged during the identification scheme. Basically, in each execution of the scheme, Alice chooses a γ; Olga chooses an r; and Alice reveals y_1 and y_2 such that

$$\gamma \equiv \alpha_1^{y_1} \alpha_2^{y_2} v^r \pmod{p}.$$

Recall that Alice computes

$$y_1 = k_1 + a_1 r \bmod q$$

and

$$y_2 = k_2 + a_2 r \bmod q,$$

where

$$\gamma = \alpha_1^{k_1} \alpha_2^{k_2} \bmod p.$$

But note that k_1 and k_2 are not revealed (nor are a_1 and a_2).

The particular quadruple (γ, r, y_1, y_2) that is generated during one execution of the scheme appears to depend on Alice's ordered pair (a_1, a_2), since y_1 and y_2 are defined in terms of a_1 and a_2. But we will show that each such quadruple could equally well be generated from any other ordered pair $(a_1', a_2') \in \mathcal{A}$. To see this, suppose $(a_1', a_2') \in \mathcal{A}$, i.e., $a_1' = a_1 - c\theta$ and $a_2' = a_2 + \theta$, where $0 \le \theta \le q - 1$. We can express y_1 and y_2 as follows:

$$y_1 = k_1 + a_1 r$$
$$= k_1 + (a_1' + c\theta)r$$
$$= (k_1 + rc\theta) + a_1' r,$$

and

$$y_2 = k_2 + a_2 r$$
$$= k_2 + (a_2' - \theta)r$$
$$= (k_2 - r\theta) + a_2' r,$$

where all arithmetic is performed in $\mathbb{Z}_q$. That is, the quadruple (γ, r, y_1, y_2) is also consistent with the ordered pair (a_1', a_2') using the random choices $k_1' = k_1 + rc\theta$ and $k_2' = k_2 - r\theta$ to produce (the same) γ. We have already noted that the values of k_1 and k_2 are not revealed by Alice, so the quadruple (γ, r, y_1, y_2) yields no information regarding which ordered pair in $\mathcal{A}$ Alice is actually using for her secret exponents.

This security proof is certainly quite elegant and subtle. It would perhaps be useful to recap the features of the scheme that lead to the proof of security. The basic idea involves having Alice choose two secret exponents rather than one. There are a total of q pairs in the set $\mathcal{A}$ that are "equivalent" to Alice's pair (a_1, a_2). The fact that leads to the ultimate contradiction is that knowledge of two different pairs in $\mathcal{A}$ provides an efficient method of computing the discrete logarithm c. Alice, of course, knows one pair in $\mathcal{A}$; and we proved that if Olga can impersonate Alice, then Olga is able to compute a pair in $\mathcal{A}$ which (with high probability) is different from Alice's pair. Thus Alice and Olga together can find two pairs in $\mathcal{A}$ and compute c, which provides the desired contradiction.

Here is an example to illustrate the computation of $\log_{\alpha_1} \alpha_2$ by Alice and Olga.

Example 9.4   As in Example 9.3, we will take $p = 88667, q = 1031$ and $t = 10$, and assume that $v = 13078$. (Also, a_1, a_2, α_1 and α_2 have the same values as before.)

Suppose Olga has determined that

$$\alpha_1{}^{131} \alpha_2{}^{287} v^{489} \equiv \alpha_1{}^{890} \alpha_2{}^{303} v^{199} \pmod{p}.$$

Then she can compute

$$b_1 = (131 - 890)(489 - 199)^{-1} \bmod 1031 = 456$$

and
$$b_2 = (287 - 303)(489 - 199)^{-1} \bmod 1031 = 519.$$

Now, using the values of a_1 and a_2 supplied by Alice, the value

$$c = (846 - 456)(519 - 515)^{-1} \bmod 1031 = 613$$

is computed. This value c is in fact $\log_{\alpha_1} \alpha_2$, as can be verified by calculating

$$58902^{613} \bmod 88667 = 73611.$$

$\square$

9.6 The Guillou-Quisquater Identification Scheme

In this section, we describe another identification scheme, due to Guillou and Quisquater, that is based on the *RSA Cryptosystem*.

The set-up of the scheme is as follows: The TA chooses two primes p and q and forms the product $n = pq$. The values of p and q are secret, while n is public. As is usually the case, p and q should be chosen large enough so that factoring n is intractable. Also, the TA chooses a large prime integer b which will function as a security parameter as well as being a public RSA encryption exponent; to be specific, let us suppose that b is a 40-bit prime such that $\gcd(b, \phi(n)) = 1$.

Now Alice chooses an integer u, where $0 \leq u \leq n - 1$, and she computes

$$v = (u^{-1})^b \bmod n.$$

The value of v is given to the TA, and then the TA computes the signature

$$s = sig_{TA}(ID(Alice) \parallel v).$$

Then, $ID(Alice)$, v and s are placed on Alice's certificate. The integers n and b are public domain parameters, v is Alice's public key, and u is Alice's private key.

When Alice wants to prove her identity to Bob, Protocol 9.10 is executed. We illustrate this scheme in the following example.

Example 9.5   Suppose the TA chooses $p = 467$ and $q = 479$, so $n = 223693$. Suppose also that $b = 503$ and Alice's secret integer $u = 101576$. Then she will compute

$$v = (u^{-1})^b \bmod n$$
$$= (101576^{-1})^{503} \bmod 223693$$
$$= 89888.$$

Protocol 9.10: GUILLOU-QUISQUATER IDENTIFICATION SCHEME

1. Alice chooses a random number k, where $0 \leq k \leq n - 1$ and computes

$$\gamma = k^b \bmod n.$$

2. Alice gives Bob her certificate **Cert**$(Alice) = (ID(Alice), v, s)$ and γ.
3. Bob verifies that $ver_{TA}(ID(Alice) \parallel v, s) = true$.
4. Bob chooses a random number r, $0 \leq r \leq b - 1$ and gives it to Alice.
5. Alice computes
$$y = ku^r \bmod n$$

    and gives y to Bob.
6. Bob verifies that
$$\gamma \equiv v^r y^b \pmod{n}.$$

Now, let's assume that Alice is proving her identity to Bob and she chooses $k = 187485$; then she gives Bob the value

$$\gamma = k^b \bmod n$$
$$= 187485^{503} \bmod 223693$$
$$= 24412.$$

Suppose Bob responds with the challenge $r = 375$. Then Alice will compute

$$y = ku^r \bmod n$$
$$= 187485 \times 101576^{375} \bmod 223693$$
$$= 93725$$

and she gives y to Bob. Bob then verifies that

$$24412 \equiv 89888^{375} 93725^{503} \pmod{223693}.$$

Hence, Bob accepts Alice's proof of identity.   ⬜

We prove that the *Guillou-Quisquater Identification Scheme* is sound and complete. As is generally the case, proving completeness is quite simple:

$$v^r y^b \equiv (u^{-b})^r (ku^r)^b \pmod{n}$$
$$\equiv u^{-br} k^b u^{br} \pmod{n}$$
$$\equiv k^b \pmod{n}$$
$$\equiv \gamma \pmod{n}.$$

Now, let us consider soundness. We will prove that the scheme is sound provided that it is infeasible to compute u from v. Since v is formed from u by RSA encryption, this is a plausible assumption to make.

Suppose Olga knows a value γ for which she has probability $\epsilon \geq 2/b$ of successfully impersonating Alice in the verification scheme. For this γ, it is reasonable to assume that Olga can compute values y_1, y_2, r_1, r_2 with $r_1 \neq r_2$, such that

$$\gamma \equiv v^{r_1} y_1{}^b \equiv v^{r_2} y_2{}^b \pmod{n}.$$

Suppose, without loss of generality, that $r_1 > r_2$. Then we have

$$v^{r_1 - r_2} \equiv (y_2/y_1)^b \pmod{n}.$$

Since $0 < r_1 - r_2 < b$ and b is prime, $t = (r_1 - r_2)^{-1} \bmod b$ exists, and it can be computed in polynomial time by Olga using the Euclidean algorithm. Hence, we have that

$$v^{(r_1 - r_2)t} \equiv (y_2/y_1)^{bt} \pmod{n}.$$

Now,

$$(r_1 - r_2)t = \ell b + 1$$

for some positive integer ℓ, so

$$v^{\ell b + 1} \equiv (y_2/y_1)^{bt} \pmod{n},$$

or equivalently,

$$v \equiv (y_2/y_1)^{bt} (v^{-1})^{\ell b} \pmod{n}.$$

If both sides of the congruence are raised to the power $b^{-1} \bmod \phi(n)$, then we have the following:

$$u^{-1} \equiv (y_2/y_1)^t (v^{-1})^\ell \pmod{n}.$$

Finally, if we compute the inverse modulo n of both sides of this congruence, we obtain the following formula for u:

$$u = (y_1/y_2)^t v^\ell \bmod n.$$

Olga can use this formula to compute u in polynomial time.

Example 9.6 As in the previous example, suppose that $n = 223693$, $b = 503$, $u = 101576$ and $v = 89888$. Suppose Olga has learned that

$$v^{401}103386^b \equiv v^{375}93725^b \pmod{n}.$$

She will first compute

$$t = (r_1 - r_2)^{-1} \bmod b$$
$$= (401 - 375)^{-1} \bmod 503$$
$$= 445.$$

Next, she calculates

$$\ell = \frac{(r_1 - r_2)t - 1}{b}$$
$$= \frac{(401 - 375)445 - 1}{503}$$
$$= 23.$$

Finally, she can obtain the secret value u as follows:

$$u = (y_1/y_2)^t v^\ell \bmod n$$
$$= (103386/93725)^{445}89888^{23} \bmod 223693$$
$$= 101576.$$

Thus Alice's secret exponent has been compromised.   ☐

9.6.1 Identity-based Identification Schemes

The *Guillou-Quisquater Identification Scheme* can be tranformed into what is known as an *identity-based identification scheme*. This basically means that certificates are not necessary. However, we still require a *TA*, who will compute the value of u as a function of Alice's ID string.[4]

The computation of u is done as indicated in Protocol 9.11. In this scheme, h is a public hash function with range $\mathbb{Z}_n$ (h could be constructed as a suitable modification of *SHA-1*, say). The value u is an RSA ciphertext, which the *TA* computes using the secret encryption exponent a. As usual, $a = b^{-1} \bmod \phi(n)$. The identity-based identification scheme is described in Protocol 9.12.

The value v is computed from Alice's ID string via the public hash function h. In order to carry out the identification scheme, Alice needs to know the value of u, which can be computed only by the *TA* (assuming that the *RSA Cryptosystem* is secure). If Olga tries to identify herself as Alice, she will not succeed because she does not know the value of u.

---

[4]The *TA* would also be responsible for resolving any disputes that arise, e.g., concerning naming conventions, assignment of ID strings, etc.

Protocol 9.11: ISSUING A VALUE u TO ALICE

1. The TA establishes Alice's identity and issues an identification string $ID(Alice)$.

2. The TA computes

$$u = (h(ID(Alice))^{-1})^a \bmod n$$

and gives u to Alice.

Protocol 9.12: GUILLOU-QUISQUATER IDENTITY-BASED IDENTIFICATION SCHEME

1. Alice chooses a random number k, where $0 \leq k \leq n - 1$ and computes

$$\gamma = k^b \bmod n.$$

2. Alice gives $ID(Alice)$ and γ to Bob.

3. Bob computes
$$v = h(ID(Alice)).$$

4. Bob chooses a random number r, $0 \leq r \leq b - 1$ and he gives it to Alice.

5. Alice computes
$$y = ku^r \bmod n$$

and she gives y to Bob.

6. Bob verifies that
$$\gamma \equiv v^r y^b \pmod{n}.$$

9.7 Notes and References

The security model we use for identification schemes is adapted from the models described in Diffie, van Oorschot and Wiener [118], Bellare and Rogaway [23] and Blake-Wilson and Menezes [44]. Diffie [115] notes that cryptographic challenge-and-response protocols for identification date from the early 1950s. Research into identification schemes In the context of computer networks was initiated by Needham and Schroeder [251] in the late 1970s.

Protocol 9.6 is from [44]. Protocol 9.14 is very similar to one version of the mutual authentication scheme described in FIPS publication 196 [145]. The attack on Protocol 9.7 is from [115].

The *Schnorr Identification Scheme* is from [294] and the *Guillou-Quisquater Identification Scheme* can be found in [167]. Proofs of security of these schemes under certain reasonable computational assumptions were provided by Bellare and Palacio [20].

The *Okamoto Identification Scheme* was presented in [260], Another scheme that can be proved secure under a plausible computational assumption has been given by Brickell and McCurley in [76].

The *Feige-Fiat-Shamir Identification Scheme* (see [128, 135]) is another popular identification scheme. It can be proven secure using zero-knowledge techniques. The method of constructing signature schemes from identification schemes is due to Fiat and Shamir [135]. They also describe an identity-based version of their identification scheme.

Surveys on identification schemes have been published by Burmester, Desmedt and Beth [80] and de Waleffe and Quisquater [114].

Exercises

9.1 Prove that it is impossible to design a secure two-flow mutual identification scheme based on random challenges. (A two-flow scheme consists of one flow from Bob to Alice (say) followed by a flow from Alice to Bob. In a mutual identification scheme, both parties are required to "accept" in order for a session to terminate successfully.)

9.2 Consider the mutual identification scheme presented in Protocol 9.13. Prove that this scheme is insecure. (In particular, show that Olga can impersonate Bob by means of a certain type of parallel session attack, assuming that Olga has observed a previous session of the scheme between Alice and Bob.)

Protocol 9.13: INSECURE PUBLIC-KEY MUTUAL AUTHENTICATION

1. Bob chooses a random challenge, r_1. He also computes $y_1 = sig_{Bob}(r_1)$ and he sends **Cert**(Bob), r_1 and y_1 to Alice.

2. Alice verifies Bob's public key, ver_{Bob}, on the certificate **Cert**(Bob). Then she checks that $ver_{Bob}(r_1, y_1) = true$. If not, then Alice "rejects" and quits. Otherwise, Alice chooses a random challenge, r_2. She also computes $y_2 = sig_{Alice}(r_1)$ and $y_3 = sig_{Alice}(r_2)$ and she sends **Cert**($Alice$), r_2, y_2 and y_3 to Bob.

3. Bob verifies Alice's public key, ver_{Alice}, on the certificate **Cert**($Alice$). Then he checks that $ver_{Alice}(r_1, y_2) = true$ and $ver_{Alice}(r_2, y_3) = true$. If so, then Bob "accepts"; otherwise, Bob "rejects." Bob also computes $y_4 = sig_{Bob}(r_2)$ and he sends y_4 to Alice.

4. Alice checks that $ver_{Bob}(r_2, y_4) = true$. If so, then Alice "accepts"; otherwise, Alice "rejects."

9.3 Give a complete proof that Protocol 9.14 is secure. (This scheme is essentially identical to one of the schemes standardized in FIPS publication 196.)

Protocol 9.14: PUBLIC-KEY MUTUAL AUTHENTICATION (VERSION 2)

1. Bob chooses a random challenge, r_1. He sends **Cert**(Bob) and r_1 to Alice.
2. Alice chooses a random challenge, r_2. She also computes $y_1 = sig_{Alice}(ID(Bob) \parallel r_1 \parallel r_2)$ and she sends **Cert**($Alice$), r_2 and y_1 to Bob.
3. Bob verifies Alice's public key, ver_{Alice}, on the certificate **Cert**($Alice$). Then he checks that $ver_{Alice}(ID(Bob) \parallel r_1 \parallel r_2, y_1) = true$. If so, then Bob "accepts"; otherwise, Bob "rejects." Bob also computes $y_2 = sig_{Bob}(ID(Alice) \parallel r_2 \parallel r_1)$ and he sends y_2 to Alice.
4. Alice verifies Bob's public key, ver_{Bob}, on the certificate **Cert**(Bob). Then she checks that $ver_{Bob}(ID(Alice) \parallel r_2 \parallel r_1, y_2) = true$. If so, then Alice "accepts"; otherwise, Alice "rejects."

9.4 Discuss whether Protocol 9.15 is secure. (Certificates are omitted from its description, but they are assumed to be inlcuded in the scheme in the usual way.)

Protocol 9.15: UNKNOWN PROTOCOL

1. Bob chooses a random challenge, r_1, and he sends it to Alice.
2. Alice chooses a random challenge r_2, she computes $y_1 = sig_{Alice}(r_1)$, and she sends r_2 and y_1 to Bob.
3. Bob checks that $ver_{Alice}(r_1, y_1) = true$; if so, then Bob "accepts"; otherwise, Bob "rejects." Bob also computes $y_2 = sig_{Bob}(r_2)$ and he sends y_2 to Alice.
4. Alice checks that $ver_{Bob}(r_2, y_2) = true$. If so, then Alice "accepts"; otherwise, Alice "rejects."

9.5 Prove that Protocol 9.6 and Protocol 9.14 are both insecure if the identity of Alice (Bob, resp.) is omitted from the signature computed by Bob (Alice, resp.).

9.6 This question further investigates the soundness of the *Schnorr Identification Scheme*. Suppose that Olga has a "black box" $\mathcal{O}$ that will sometimes be able to compute correct responses to challenges. More precisely, let $\epsilon \geq 2^{-t+2}$, and suppose that $\mathcal{O}(\gamma, r)$ returns a valid response y with probability ϵ, when γ and r are chosen uniformly at random (and, with probability $1 - \epsilon$, $\mathcal{O}$ outputs "no repsonse"). Olga wants to use $\mathcal{O}$ to generate valid responses to two different challenges for the same γ. (Then, as described in Section 9.4.1, Olga can compute Alice's private key.)

Algorithm 9.1: FIND VALID SCHNORR RESPONSES

1. $N \leftarrow \frac{1}{\epsilon}$ (assume that N is an integer).
3.   Olga generates random pairs (γ_i, r_i), $1 \leq i \leq N$.
3.   **for** $1 \leq i \leq N$, Olga runs $\mathcal{O}(\gamma_i, r_i)$.
4.   If $\mathcal{O}(\gamma_i, r_i)$ yields a response y_i for some i, then Olga sets $(\gamma, r) \leftarrow (\gamma_i, r_i)$, and she proceeds to step 5. Otherwise, the attack fails.
5.   Olga generates random numbers $s_1, \ldots, s_N \in \{1, \ldots, 2^t\}$.
6.   **for** $1 \leq i \leq N$, Olga runs $\mathcal{O}(\gamma, s_i)$.
7.   If $\mathcal{O}(\gamma, s_i)$ yields a response z_i for some i, and $s_i \neq r$, then Olga sets $r' \leftarrow s_i$, she outputs the pairs (γ, r) and (γ, r'), and she quits (the attack succeeds). Otherwise, the attack fails.

Olga will use Algorithm 9.1 to try and generate the responses she needs. Prove the following assertions about this algorithm.

(a) The running time is $O(1/\epsilon)$.

(b) The probability that step 4 is successful is at least 0.63.

(c) For every fixed γ, define $p_\gamma = \mathbf{Pr}[\mathcal{O}(\gamma, r)$ responds$]$, where the probability is computed over a choice of r made uniformly at random. Further, define $\Gamma_0 = \{\gamma : p_\gamma \geq \epsilon/2\}$. Prove that $\mathbf{Pr}[\gamma \in \Gamma_0] \geq 1/2$, where γ is the value found in step 4 (assuming that this step is successful).

(d) Assuming that $\gamma \in \Gamma_0$, prove that the probability that step 7 is successful is bounded below by a positive constant.

(e) Prove that the attack succeeds with some probability that is bounded below by a positive constant.

HINT   You can use the following fact without proof: If $c > 0$, then $(1 - x)^{c/x} < e^{-c}$ for all $0 < x < 1$.

9.7   Consider the following possible identification scheme. Alice possesses a secret key $n = pq$, where p and q are prime and $p \equiv q \equiv 3 \pmod 4$. The value of n will be stored on Alice's certificate. When Alice wants to identify herself to Bob, say, Bob will present Alice with a random quadratic residue modulo n, say x. Then Alice will compute a square root y of x and give it to Bob. Bob then verifies that $y^2 \equiv x \pmod n$. Explain why this scheme is insecure.

9.8   Suppose Alice is using the *Schnorr Identification Scheme* where $q = 1201$, $p = 122503$, $t = 10$ and $\alpha = 11538$.

(a) Verify that α has order q in $\mathbb{Z}_p{}^*$.

(b) Suppose that Alice's secret exponent is $a = 357$. Compute v.

(c) Suppose that $k = 868$. Compute γ.

(d) Suppose that Bob issues the challenge $r = 501$. Compute Alice's response y.

(e) Perform Bob's calulations to verify y.

9.9   Suppose that Alice uses the *Schnorr Identification Scheme* with p, q, t and α as in the previous exercise. Now suppose that $v = 51131$, and Olga has learned that

$$\alpha^3 v^{148} \equiv \alpha^{151} v^{1077} \pmod p.$$

Show how Olga can compute Alice's secret exponent a.

9.10 Suppose that Alice is using the *Okamoto Identification Scheme* with $q = 1201$, $p = 122503$, $t = 10$, $\alpha_1 = 60497$ and $\alpha_2 = 17163$.

    (a) Suppose that Alice's secret exponents are $a_1 = 432$ and $a_2 = 423$. Compute v.

    (b) Suppose that $k_1 = 389$ and $k_2 = 191$. Compute γ.

    (c) Suppose that Bob issues the challenge $r = 21$. Compute Alice's response, y_1 and y_2.

    (d) Perform Bob's calculations to verify y_1 and y_2.

9.11 Suppose that Alice uses the *Okamoto Identification Scheme* with p, q, t, α_1 and α_2 as in the previous exercise. Suppose also that $v = 119504$.

    (a) Verify that

$$\alpha_1{}^{70} \alpha_2{}^{1033} v^{877} \equiv \alpha_1{}^{248} \alpha_2{}^{883} v^{992} \pmod{p}.$$

    (b) Use this information to compute b_1 and b_2 such that

$$\alpha_1{}^{-b_1} \alpha_2{}^{-b_2} \equiv v \pmod{p}.$$

    (c) Now suppose that Alice reveals that $a_1 = 484$ and $a_2 = 935$. Show how Alice and Olga together will be able to compute the qalue $\log_{\alpha_1} \alpha_2$.

9.12 Suppose that Alice is using the *Guillou-Quisquater Identification Scheme* with $p = 503$, $q = 379$ and $b = 509$.

    (a) Suppose that Alice's secret $u = 155863$. Compute v.

    (b) Suppose that $k = 123845$. Compute γ.

    (c) Suppose that Bob issues the challenge $r = 487$. Compute Alice's response, y.

    (d) Perform Bob's calculations to verify y.

9.13 Give a complete proof that the *Guillou-Quisquater Identification Scheme* is honest-verifier zero-knowledge. (Show that the set of real transcripts is identical to the set of simulated transcripts, and the two probability distributions are identical.) If you wish, you can assume in your proof that $\gcd(u, n) = 1$. This will be true with high probability for any secure implementation of the scheme.

9.14 Suppose that Alice is using the *Guillou-Quisquater Identification Scheme* with $n = 199543$, $b = 523$ and $v = 146152$. Suppose that Olga has somehow discovered that

$$v^{456} 101360^b \equiv v^{257} 36056^b \pmod{n}.$$

Show how Olga can compute u.

10

Key Distribution

10.1 Introduction

We have observed that public-key cryptosystems have the advantage over secret-key cryptosystems that a secure channel is not needed to exchange a key. But, unfortunately, most public-key cryptosystems (e.g., *RSA*) are much slower than secret-key systems (e.g., *AES*). So, in practice, secret-key systems are usually used to encrypt "long" messages. But then we come back to the problem of *key establishment*, the topic of this and the next chapter.

We will discuss several approaches to the problem of establishing secret keys. As our setting, we have an insecure network of n users. In some of our schemes, we will have a trusted authority (denoted by TA) who is reponsible for such things as verifying the identities of users, issuing certificates, choosing and transmitting keys to users, etc. There are many possible scenarios, including the following:

key predistribution

> In a *key predistribution scheme* (or *KPS*), a *TA* distributes keying information "ahead of time" in a secure fashion to everyone in the network. Note that a secure channel is required at the time that keys are distributed. Later, all network users can use these secret keys to encrypt messages they transmit over the network. Typically, every pair of users in the network will be able to determine a key, known only to them, as a result of the keying information they hold.

session key distribution

> In session key distribution, an online *TA* chooses session keys and distributes them to network users, when requested to do so, via an interactive protocol. Such a protocol is called a *session key distribution scheme* and denoted *SKDS*. Session keys are used to encrypt information for a specified, fairly short, period of time. The session keys will be encrypted by the *TA* using previously distributed secret keys (under the assumption that every network user possesses a secret key whose value is known to the *TA*).

key agreement

   Key agreement refers to the situation where network users employ an interactive protocol to construct a session key. Such a protocol is called a *key
   agreement scheme*, and it is denoted by *KAS*. These may be secret-key based
   or public-key based schemes, and usually they do not require an on-line *TA*.
   KAS will be studied in Chapter 11.

As always, we consider active and/or passive adversaries; various adversarial
goals, attack models, security levels, etc.

We now compare and contrast the above-mentioned methods of key establishment in more detail. First, we can distinguish between key distribution and key
agreement as follows. Key distribution is a mechanism whereby one party (usually a *TA*) chooses a secret key or keys and then transmits them to another party
or parties in encrypted form. Key agreement denotes a protocol whereby two (or
more) parties jointly establish a secret key by communicating over a public channel. In a key agreement scheme, the value of the key is determined as a function
of inputs provided by both parties and secret information of the two users. The
key is not "transported" from one party to another as in the case of a session key
distribution scheme.

It is important to distinguish between long-lived keys and session keys. Users
(or pairs of users) may have *long-lived keys* (*LL-keys*) which are pre-computed and
then stored securely. Alternatively, LL-keys might be computed non-interactively,
as needed, from securely stored secret information. LL-keys could be secret keys
known to a pair of users or to a user and the *TA*. On the other hand, they could be
private keys corresponding to public keys which are stored on users' certificates.

Pairs of users will often employ secret short-lived *session keys* in a particular
session, and then throw them away when the session has ended. Session keys are
usually secret keys, for use in a secret-key cryptosystem or MAC. LL-keys are
often used in protocols to transmit encrypted session keys (e.g., they may be used
as "key-encrypting keys" in an SKDS). LL-keys are also used to authenticate data
— using a message authentication code or signature scheme — that is sent in a
session of a scheme.

A key predistribution scheme provides one method to distribute secret LL-keys
ahead of time. It requires a secure channel between the *TA* and each network
user at the time that the keys are distributed. At a later time, a KAS might be
used by pairs of network users to generate session keys, as needed. One main
consideration in the study of KPS is the amount of secret information that must
be stored by each user in the network.

A session key distribution scheme is a three-party protocol, involving two users
U and *V* (say) and the *TA*. SKDS are usually based on long-lived secret keys
held by individual users and the *TA*. That is, *U* holds a secret key whose value is
known to the *TA*, and *V* holds a (different) secret key whose value is known to
the *TA*.

A key agreement scheme can be a secret-key based or a public/private-key based scheme. A KAS usually involves two users, say U and V, but it does not require an online TA. However, an offline TA might have distributed secret LL-keys in the past, say in the case of a secret-key based scheme. If the KAS is based on public keys, then a TA is implicitly required to issue certificates and (perhaps) to maintain a suitable public-key infrastructure. However, the TA does not take an active role in a session of the KAS.

There are several reasons why session keys are useful. First, they limit the amount of ciphertext (that is encrypted with one particular key) available to an attacker, because session keys are changed on a regular basis. Another advantage of session keys is that they limit exposure in the event of session key compromise, provided that the scheme is designed well (i.e., it is desirable that the compromise of a session key should not reveal information about the LL-key, or about other session keys). Session keys can therefore be used in "risky" environments where there is a higher possibility of exposure. Finally, the use of session keys often reduces the amount of long-term information that needs to be securely stored by each user, because keys for pairs of users are generated only when they are needed.

Long-lived keys should satisfy several requirements. The "type" of scheme used to construct session keys dictates the type of LL-keys required. As well, users' storage requirements depend on the type of keys used. We consider these requirements now, assuming that we have a network of n users. These users will typically be denoted by U, V, W, etc.

First, as mentioned above, if an SKDS is to be used for session key distribution, then each network user must have a secret LL-key in common with the TA. This entails a low storage requirement for network users, but the TA has a very high storage requirement.

A secret-key based KAS requires that every pair of network users has a secret LL-key known only to them. In a "naive" implementation, each user stores $n - 1$ long-lived keys, which necessitates a high storage requirement if n is large. Appropriate key predistribution schemes can reduce this storage requirement significantly, however.

Finally, in a public-key based KAS, we require that all network users have their own public/private LL-key pair. This yields a low storage requirement, because users only store their own private key and a certificate containing their public key.

Since the network is insecure, we need to protect against potential adversaries. Our opponent, Oscar, may be one of the users in the network. He may be active or passive during an information gathering phase. Later, when he carries out his attack, he might be a passive adversary, which means that his actions are restricted to eavesdropping on messages that are transmitted over the channel. On the other hand, we might want to guard against the possibility that Oscar is an active adversary. Recall that an active adversary can do various types of nasty things, such as:

1. alter messages that he observes being transmitted over the network,

2. save messages for reuse at a later time, or

3. attempt to masquerade as various users in the network.

The objective of an adversary might be:

1. to fool U and V into accepting an "invalid" key as valid (an invalid key could be an old key that has expired, or a key chosen by the adversary, to mention two possibilities),

2. to make U or V believe that they have exchanged a key with each other, when they have not done so, or

3. to determine some (partial) information about the key exchanged by U and V.

The first two of these adversarial goals involve active attacks, while the third goal could perhaps be accomplished within the context of a passive attack.

Summarizing, the objective of a session key distribution scheme or a key agreement scheme is that, at the end of a session of the scheme, the two parties involved in the session both have possession of the same key K, and the value of K is not known to any other party (except possibly the *TA*).

We sometimes desire *authenticated key agreement schemes*, which include (mutual) identification of U and V. Therefore, the schemes should be secure identification schemes (as defined in Chapter 9), and, in addition, U and V should possess a new secret key at the end of a session, whose value is not known to the adversary.

Extended attack models can also be considered. Suppose that the adversary learns the value of a particular session key (this is called a *known session key attack*). In this attack model, we would still want other session keys (as well as the LL-keys) to remain secure. As another possibility, suppose that the adversary learns the LL-keys of the participants (this is a *known LL-key attack*). This is a catastrophic attack, and consequently, a new scheme must be set up. However, can we limit the damage that is done in this type of attack? If the adverary cannot learn the values of previous session keys, then the scheme is said to possess the property of *perfect forward secrecy*. This is clearly a desirable attribute of a session key distribution scheme or a key agreement scheme.

In this chapter, we concentrate on key predistribution and session key distribution. For the problem of key predistribution, we study the classical *Diffie-Hellman Scheme* as well as some unconditionally secure schemes that use algebraic or combinatorial techniques. For session key distribution, we analyze some insecure schemes and then we present a secure scheme due to Bellare and Rogaway.

Protocol 10.1: DIFFIE-HELLMAN KPS

1. The public domain parameters consist of a group $(G, \cdot)$ and an element $\alpha \in G$ having order n.

2. V computes
$$K_{U,V} = \alpha^{a_U a_V} = b_U^{a_V},$$
using the public key b_U from U's certificate, together with her own private key a_V.

3. U computes
$$K_{U,V} = \alpha^{a_U a_V} = b_V^{a_U},$$
using the public key b_V from V's certificate, together with his own private key a_U.

10.2 Diffie-Hellman Key Predistribution

In this section, we describe a key predistribution scheme that is a modification of the well-known *Diffie-Hellman Key Agreement Scheme* which we will discuss in the next chapter. The scheme we describe now is the *Diffie-Hellman Key Predistribution Scheme*. This scheme is computationally secure provided that the **Decision Diffie-Hellman** problem (Problem 6.4) is intractable. Suppose that $(G, \cdot)$ is a group and suppose that $\alpha \in G$ is an element of order n such that the **Decision Diffie-Hellman** problem is intractible in the subgroup of G generated by α.

Every user U in the network has a private LL-key a_U (where $0 \leq a_U \leq n-1$) and a corresponding public key

$$b_U = \alpha^{a_U}.$$

The users' public keys are signed by the TA and stored on certificates, as usual. The common LL-key for any two users, say U and V, is defined to be

$$K_{U,V} = \alpha^{a_U a_V}.$$

The *Diffie-Hellman KPS* is summarized in Protocol 10.1.

We illustrate Protocol 10.1 with an insecure toy example.

Example 10.1 Suppose $p = 12987461$, $q = 1291$ and $\alpha = 3606738$ are the public domain parameters. Here, p and q are prime, $p - 1 \equiv 0 \pmod{q}$, and α has order q. We implement Protocol 10.1 in the subgroup of $(\mathbb{Z}_p^*, \cdot)$ generated by α. This subgroup has order q.

Suppose U chooses $a_U = 357$. Then he computes

$$b_U = \alpha^{a_U} \bmod p$$
$$= 3606738^{357} \bmod 12987461$$
$$= 7317197,$$

which is placed on his certificate. Suppose V chooses $a_V = 199$. Then she computes

$$b_V = \alpha^{a_V} \bmod p$$
$$= 3606738^{199} \bmod 12987461$$
$$= 138432,$$

which is placed on her certificate.

Now U can compute the key

$$K_{U,V} = b_V{}^{a_U} \bmod p$$
$$= 138432^{357} \bmod 12987461$$
$$= 11829605,$$

and V can compute the same key

$$K_{U,V} = b_U{}^{a_V} \bmod p$$
$$= 7317197^{199} \bmod 12987461$$
$$= 11829605.$$

$\square$

Let us think about the security of the *Diffie-Hellman KPS* in the presence of an adversary. Since there is no interaction in the scheme[1] and we assume that users' private keys are secure, we do not need to consider the possibility of an active adversary. Therefore, we need only to consider whether a bad user (say W) can compute $K_{U,V}$ if $W \neq U, V$. In other words, given public keys α^{a_U} and α^{a_V} (but not a_U or a_V), is it feasible to compute the secret key $K_{U,V} = \alpha^{a_U a_V}$? This is precisely the **Computational Diffie-Hellman** problem, which was defined as Problem 6.3. Therefore, the *Diffie-Hellman KPS* is secure against an adversary if and only if the **Computational Diffie-Hellman** problem in the subgroup $\langle \alpha \rangle$ is intractable.

---

[1] It might happen that two users exchange their IDs and/or their certificates, but this information is regarded as fixed, public information. Therefore we do not view the scheme as an interactive scheme.

Even if the adversary is unable to compute a Diffie-Hellman key, perhaps there is the possibility that he could determine some partial information about the key (in polynomial time). Therefore, we desire *semantic security* of the keys, which means that an adversary cannot compute any partial information about the key (in polynomial time). In other words, distinguishing Diffie-Hellman keys from random elements of the subgroup $\langle \alpha \rangle$ should be intractable. The semantic security of Diffie-Hellman keys is easily seen to be equivalent to the intractability of the **Decision Diffie-Hellman** problem (which was presented as Problem 6.4).

10.3 Unconditionally Secure Key Predistribution

In this section, we consider unconditionally secure key predistribution schemes. We begin by describing a "trivial" solution. For every pair of users $\{U, V\}$, the *TA* chooses a random key $K_{U,V} = K_{V,U}$ and transmits it "off-band" to U and V over a secure channel. (That is, the transmission of keys does not take place over the network, because the network is not secure.) Unfortunately, each user must store $n - 1$ keys, and the *TA* needs to transmit a total of $\binom{n}{2}$ keys securely (this is sometimes called the "n^2 problem"). Even for relatively small networks, this can become prohibitively expensive, and so it is not really a practical solution.

Thus, it is of interest to try to reduce the amount of information that needs to be transmitted and stored, while still allowing each pair of users U and V to be able to (independently) compute a secret key $K_{U,V}$. A particularly elegant scheme to accomplish this, called the *Blom Key Predistribution Scheme*, is discussed in the next subsection.

10.3.1 The Blom Key Predistribution Scheme

We begin by briefly discussing the security model used in the study of unconditionally secure KPS. We assume that the *TA* distributes secret information securely to the n network users. The adversary may corrupt a subset of at most k users, and obtain all their secret information, where k is a pre-specified *security parameter*. The adversary's goal is to determine the secret LL-key of a pair of uncorrupted users. The *Blom Key Predistribution Scheme* is a KPS that is unconditionally secure against adversaries of this type.

It is desired that each pair of users U and V will be able to compute a key $K_{U,V} = K_{V,U}$. Therefore, the security condition is as follows: any set of at most k users disjoint from $\{U, V\}$ must be unable to determine any information about $K_{U,V}$ (where we are speaking here about unconditional security).

In the *Blom Key Predistribution Scheme*, keys are chosen from a finite field $\mathbb{Z}_p$, where $p \geq n$ is prime. The *TA* will transmit $k + 1$ elements of $\mathbb{Z}_p$ to each user over a secure channel (as opposed to $n - 1$ elements in the trivial key predis-

---

Protocol 10.2: BLOM KPS ($k = 1$)

1. A prime number p is made public, and for each user U, an element $r_U \in \mathbb{Z}_p$ is made public. The elements r_U must be distinct.

2. The *TA* chooses three random elements $a, b, c \in \mathbb{Z}_p$ (not necessarily distinct), and forms the polynomial

$$f(x, y) = a + b(x + y) + cxy \bmod p.$$

3. For each user U, the *TA* computes the polynomial

$$g_U(x) = f(x, r_U) \bmod p$$

and transmits $g_U(x)$ to U over a secure channel. Note that $g_U(x)$ is a linear polynomial in x, so it can be written as

$$g_U(x) = a_U + b_U x,$$

where

$$a_U = a + b r_U \bmod p \quad \text{and} \quad b_U = b + c r_U \bmod p.$$

4. If U and V want to communicate, then they use the common key

$$K_{U,V} = K_{V,U} = f(r_U, r_V) = a + b(r_U + r_V) + c r_U r_V \bmod p,$$

where U computes

$$K_{U,V} = g_U(r_V)$$

and V computes

$$K_{V,U} = g_V(r_U).$$

---

tribution scheme). Note that the amount of information transmitted by the *TA* is independent of n.

We first present the special case of the *Blom Key Predistribution Scheme* where $k = 1$. Here, the *TA* will transmit two elements of $\mathbb{Z}_p$ to each user over a secure channel. The security achieved is that no individual user, W, say, will be able to determine any information about $K_{U,V}$ if $W \neq U, V$. The *Blom KPS* with $k = 1$ is presented as Protocol 10.2.

An important feature in Protocol 10.2 is that the polynomial f is symmetric: $f(x, y) = f(y, x)$ for all x, y. This property ensures that $g_U(r_V) = g_V(r_U)$, so U and V compute the same key in step 4 of the scheme.

We illustrate the *Blom KPS* with $k = 1$ in the following example.

Example 10.2 Suppose the three users are U, V and W, $p = 17$, and their public elements are $r_U = 12$, $r_V = 7$ and $r_W = 1$. Suppose that the TA chooses $a = 8$, $b = 7$ and $c = 2$, so the polynomial f is

$$f(x, y) = 8 + 7(x + y) + 2xy.$$

The g polynomials are as follows:

$$g_U(x) = 7 + 14x$$

$$g_V(x) = 6 + 4x$$

$$g_W(x) = 15 + 9x.$$

The three keys are thus

$$K_{U,V} = 3$$

$$K_{U,W} = 4$$

$$K_{V,W} = 10.$$

U would compute

$$K_{U,V} = g_U(r_V) = 7 + 14 \times 7 \bmod 17 = 3.$$

while V would compute

$$K_{V,U} = g_V(r_U) = 6 + 4 \times 12 \bmod 17 = 3.$$

We leave the computation of the other keys as an exercise for the reader. ▯

We now prove that no one user can determine the key of two other users.[2]

THEOREM 10.1 *The Blom Key Predistribution Scheme with $k = 1$ is unconditionally secure against any individual user.*

PROOF Let's suppose that user W wants to try to compute the key

$$K_{U,V} = a + b(r_U + r_V) + c\,r_U r_V \bmod p,$$

where $W \neq U$, V. The values r_U and r_V are public, but a, b and c are unknown. W knows the values

$$a_W = a + br_W \bmod p$$

---

[2] Here, we just show that no user W can rule out any possible value of a key $K_{U,V}$. It is in fact possible to prove a stronger result, analogous to a a "perfect secrecy" condition. Such a result would have the form $\mathbf{Pr}[K_{U,V} = K^* | g_W(x)] = \mathbf{Pr}[K_{U,V} = K^*]$ for all $K^* \in \mathbb{Z}_p$.

and

$$bw = b + crw \bmod p,$$

because these are the coefficients of the polynomial $g_W(x)$ that was sent to W by the *TA*.

We will show that the information known by W is consistent with any possible value $K^* \in \mathbb{Z}_p$ of the key $K_{U,V}$ (therefore, W cannot rule out any values for $K_{U,V}$). Consider the following matrix equation (in $\mathbb{Z}_p$):

$$\begin{pmatrix} 1 & r_U + r_V & r_U r_V \\ 1 & r_W & 0 \\ 0 & 1 & r_W \end{pmatrix} \begin{pmatrix} a \\ b \\ c \end{pmatrix} = \begin{pmatrix} K^* \\ a_W \\ b_W \end{pmatrix}.$$

The first equation represents the hypothesis that $K_{U,V} = K^*$; the second and third equations contain the information that W knows about a, b and c from $g_W(x)$. The determinant of the coefficient matrix is

$$r_W{}^2 + r_U r_V - (r_U + r_V) r_W = (r_W - r_U)(r_W - r_V),$$

where all arithmetic is done in $\mathbb{Z}_p$. Because $r_W \neq r_U$, $r_W \neq r_V$, and p is prime, it follows that the coefficient matrix has non-zero determinant (modulo p), and hence the matrix equation has a unique solution (in $\mathbb{Z}_p$) for a, b and c. Therefore, we have shown that any possible value K^* of $K_{U,V}$ is consistent with the information known to W. Hence, W cannot compute $K_{U,V}$. ∎

On the other hand, a coalition of two users, say $\{W, X\}$, will be able to determine any key $K_{U,V}$ where $\{W, X\} \cap \{U, V\} = \emptyset$. W and X together know that

$$a_W = a + b r_W,$$
$$b_W = b + c r_W,$$
$$a_X = a + b r_X, \quad \text{and}$$
$$b_X = b + c r_X.$$

Thus they have four equations in three unknowns, and they can easily compute the unique solution for a, b and c. Once they know a, b and c, they can form the polynomial $f(x, y)$ and compute any key they desire. Hence, we have shown the following:

THEOREM 10.2   *The Blom Key Predistribution Scheme with $k = 1$ can be broken by any coalition of two users.*

It is straightforward to generalize the *Blom Key Predistribution Scheme* to be secure against coalitions of size k. The only thing that changes is that the polynomial $f(x, y)$ has degree equal to k. Therefore, the *TA* uses a polynomial $f(x, y)$

Protocol 10.3: BLOM KPS (ARBITRARY k)

1. A prime number p is made public, and for each user U, an element $r_U \in \mathbb{Z}_p$ is made public. The elements r_U must be distinct.

2. For $0 \leq i, j \leq k$, the *TA* chooses random elements $a_{i,j} \in \mathbb{Z}_p$, such that $a_{i,j} = a_{j,i}$ for all i, j. Then the *TA* forms the polynomial

$$f(x, y) = \sum_{i=0}^{k} \sum_{j=0}^{k} a_{i,j} \, x^i y^j \bmod p.$$

3. For each user U, the *TA* computes the polynomial

$$g_U(x) = f(x, r_U) \bmod p = \sum_{i=0}^{k} a_{U,i} \, x^i$$

and transmits the coefficient vector $(a_{U,0}, \ldots, a_{U,k})$ to U over a secure channel.

4. For any two users U and V, the key $K_{U,V} = f(r_U, r_V)$, where U computes

$$K_{U,V} = g_U(r_V)$$

and V computes

$$K_{V,U} = g_V(r_U).$$

---

having the form

$$f(x, y) = \sum_{i=0}^{k} \sum_{j=0}^{k} a_{i,j} \, x^i y^j \bmod p,$$

where $a_{i,j} \in \mathbb{Z}_p$ ($0 \leq i \leq k, 0 \leq j \leq k$), and $a_{i,j} = a_{j,i}$ for all i, j. Note that the polynomial $f(x, y)$ is symmetric, as before. The remainder of the scheme is unchanged; see Protocol 10.3.

We will show that the *Blom KPS* satisfies the following security properties:

1. No set of k users, say $W_1, \ldots, W_k$, can determine any information about a key for two other users, say $K_{U,V}$.

2. Any set of $k + 1$ users, say $W_1, \ldots, W_{k+1}$, can break the scheme.

First, we consider how $k + 1$ users can break the scheme. A set of users $W_1, \ldots, W_{k+1}$ (collectively) know the polynomials

$$g_{W_i}(x) = f(x, r_{W_i}) \bmod p,$$

$1 \leq i \leq k+1$. Rather than attempt to modify the attack we presented in the case $k = 1$, we will present a more general and elegant approach. This attack makes use of certain formulas for *polynomial interpolation*, which are presented in the next two theorems without proof.

THEOREM 10.3 *(Lagrange interpolation formula) Suppose p is prime, suppose $x_1, x_2, \ldots, x_{m+1}$ are distinct elements in $\mathbb{Z}_p$, and suppose $a_1, a_2, \ldots, a_{m+1}$ are (not necessarily distinct) elements in $\mathbb{Z}_p$. Then there is a unique polynomial $A(x) \in \mathbb{Z}_p[x]$ having degree at most m, such that $A(x_i) = a_i$, $1 \leq i \leq m+1$. The polynomial $A(x)$ is as follows:*

$$A(x) = \sum_{j=1}^{m+1} a_j \prod_{1 \leq h \leq m+1, h \neq j} \frac{x - x_h}{x_j - x_h}.$$

The Lagrange interpolation formula also has a bivariate form, which we state now.

THEOREM 10.4 *(Bivariate Lagrange interpolation formula) Suppose p is prime, suppose that $x_1, x_2, \ldots, x_{m+1}$ are distinct elements in $\mathbb{Z}_p$, and suppose that $a_1(x), a_2(x), \ldots, a_{m+1}(x) \in \mathbb{Z}_p[x]$ are polynomials of degree at most m. Then there is a unique polynomial $A(x, y) \in \mathbb{Z}_p[x, y]$ having degree at most m (in x and y), such that $A(x, y_i) = a_i(x)$, $1 \leq i \leq m+1$. The polynomial $A(x, y)$ is as follows:*

$$A(x, y) = \sum_{j=1}^{m+1} a_j(x) \prod_{1 \leq h \leq m+1, h \neq j} \frac{y - y_h}{y_j - y_h}.$$

We provide an example of bivariate Lagrange interpolation.

Example 10.3   Suppose that $p = 13$, $m = 2$, $y_1 = 1$, $y_2 = 2$, $y_3 = 3$,

$$a_1(x) = 1 + x + x^2,$$
$$a_2(x) = 7 + 4x^2, \quad \text{and}$$
$$a_3(x) = 2 + 9x.$$

Then

$$\frac{(y - 2)(y - 3)}{(1 - 2)(1 - 3)} = 7y^2 + 4y + 3,$$

$$\frac{(y - 1)(y - 3)}{(2 - 1)(2 - 3)} = 12y^2 + 4y + 10, \quad \text{and}$$

$$\frac{(y - 1)(y - 2)}{(3 - 1)(3 - 2)} = 7y^2 + 5y + 1.$$

Therefore,

$$A(x, y) = (1 + x + x^2)(7y^2 + 4y + 3) + (7 + 4x^2)(12y^2 + 4y + 10)$$
$$+ (2 + 9x)(7y^2 + 5y + 1) \bmod 13$$
$$= y^2 + 3y + 10 + 5xy^2 + 10xy + 12x + 3x^2y^2 + 7x^2y + 4x^2.$$

It can easily be verified that $A(x, i) = a_i(x)$, $i = 1, 2, 3$. For example, when $i = 1$, we have

$$A(x, 1) = 1 + 3 + 10 + 5x + 10x + 12x + 3x^2 + 7x^2 + 4x^2 \bmod 13$$
$$= 14 + 27x + 14x^2 \bmod 13$$
$$= 1 + x + x^2.$$

$\square$

It is straightforward to show that the *Blom Key Predistribution Scheme* is insecure if there are $k + 1$ bad guys. A set of $k + 1$ bad users, say $W_1, \ldots, W_{k+1}$, collectively know $k + 1$ polynomials of degree k, namely,

$$g_{W_i}(x) = f(x, r_{W_i}) \bmod p,$$

$1 \leq i \leq k + 1$. Using the bivariate interpolation formula, they can compute $f(x, y)$. This is done exactly as in Example 10.3. After having computed $f(x, y)$, they can compute any key $K_{U, V}$ that they desire.

We can also show that the *Blom Key Predistribution Scheme* is secure against a coalition of k bad guys by a modification of the preceding argument. A set of k bad users, say $W_1, \ldots, W_k$, collectively know k polynomials of degree k, namely,

$$g_{W_i}(x) = f(x, r_{W_i}) \bmod p,$$

$1 \leq i \leq k$. We show that this information is consistent with any possible value of the key. Let K be the real key (whose value is unknown to the coalition), and let K^* be arbitrary. We will show that there is a symmetric polynomial $f^*(x, y)$ that is consistent with the information known to the coalition, and such that the secret key associated with the polynomial $f^*(x, y)$ is K^*. Therefore, the coalition cannot rule out any possible values of the key.

We define the polynomial $f^*(x, y)$ as follows:

$$f^*(x, y) = f(x, y) + (K^* - K) \prod_{1 \leq i \leq k} \frac{(x - r_{W_i})(y - r_{W_i})}{(r_U - r_{W_i})(r_V - r_{W_i})}. \tag{10.1}$$

We list some properties of $f^*(x, y)$:

1. First, it is easy to see that f^* is a symmetric polynomial (i.e., $f(x,y) = f(y,x)$), because $f(x,y)$ is symmetric and the product in (10.1) is also symmetric in x and y.

2. Next, for $1 \leq i \leq k$, it holds that

$$f^*(x, r_{W_i}) = f(x, r_{W_i}) = g_{W_i}(x).$$

This is because every product in (10.1) contains a term equal to 0 when $y = r_{W_i}$, and hence the product is 0.

3. Finally,

$$f^*(r_U, r_V) = f(r_U, r_V) + K^* - K = K^*,$$

because the product in (10.1) is equal to 1.

These three properties establish that, for any possible value K^* of the key, there is a symmetric polynomial $f^*(x,y)$ such that the key $f^*(U, V) = K^*$ and such that the secret information held by the k bad users is unchanged.

Summarizing, we have proven the following theorem.

THEOREM 10.5 *The Blom Key Predistribution Scheme is unconditionally secure against any k users. However, any $k + 1$ users can break the scheme.*

One drawback of the *Blom Key Predistribution Scheme* is that there is a sharp security threshold (namely, the value of k) which must be prespecified. Once more than k users decide to collaborate, the whole scheme can be broken. On the other hand, the *Blom Key Predistribution Scheme* is optimal with respect to its storage requirements: It has been proven that any unconditionally secure key predistribution that is secure against coalitions of size k requires each user's storage to be at least $k + 1$ times the length of a key.

10.4   Key Distribution Patterns

In this section, we discuss a combinatorial method of key predistribution. We have a *TA* and a network of n users, who will be denoted $\mathcal{U} = \{U_1, \ldots, U_n\}$. The *TA* chooses v random keys, say $k_1, \ldots, k_v \in \mathcal{K}$, where $(\mathcal{K}, +)$ is an additive abelian group. The *TA* then gives a (different) subset of keys to each user. (For example, keys could be chosen to be bitstrings of a suitable length, e.g., 128 bits. In this case, $\mathcal{K} = (\mathbb{Z}_2)^{128}$ and the addition operation is just vector addition modulo 2.)

> **Definition 10.1:**   A *key distribution pattern* (or *KDP*) is a public v by n incidence matrix, denoted M, which has entries in $\{0, 1\}$. The matrix M specifies which users are to receive which keys. Namely, user U_j is given the key k_i if and only if $M[i, j] = 1$.

For a key distribution pattern M and a subset of users $P \subseteq \mathcal{U}$, define

$$keys(P) = \{i : M[i, j] = 1 \text{ for all } U_j \in P\}$$

$keys(P)$ records the indices of the keys that are held by all the users in P. Observe that

$$keys(P) = \bigcap_{U_j \in P} keys(U_j).$$

If $keys(P) \neq \emptyset$, then the *group key* for the subset P is defined to be

$$k_P = \sum_{i \in keys(P)} k_i, \tag{10.2}$$

where the sum is computed using the operation "$+$" defined in $(\mathcal{K}, +)$. k_P is called a group key because each member of P can compute k_P, with no interaction required. Most often, a group will consist of two users, but the framework we are describing may permit group keys to be constructed for larger groups, so there is no harm in defining concepts in this extended setting.

We illustrate the above concepts with a small example.

Example 10.4   Suppose $n = 4$, $v = 6$, and the matrix M is as follows:

$$M = \begin{pmatrix} 1 & 1 & 0 & 0 \\ 1 & 0 & 1 & 0 \\ 1 & 0 & 0 & 1 \\ 0 & 1 & 1 & 0 \\ 0 & 1 & 0 & 1 \\ 0 & 0 & 1 & 1 \end{pmatrix}.$$

Then we have:

$$keys(U_1) = \{1, 2, 3\},$$
$$keys(U_2) = \{1, 4, 5\},$$
$$keys(U_1, U_2) = \{1\}.$$

Hence, $k_{\{U_1, U_2\}} = k_1$. $\square$

If M satisfies certain combinatorial properties, then the scheme will be secure against certain coalitions. We investigate this question now. Suppose that P is a subset of participants who possess a group key k_P defined using the formula (10.2), and suppose that F is a coalition that wants to be able to compute k_P by pooling all the information that they hold collectively. If there is a user $U_i \in F \cap P$, then U_i can compute k_P already. Therefore we will assume $F \cap P = \emptyset$.

The coalition F can compute k_P if the following condition holds:

$$keys(P) \subseteq \bigcup_{U_j \in F} keys(U_j). \tag{10.3}$$

This is easy to see, because the coalition F collectively holds all the keys required to be able to compute k_P using the formula (10.2), provided that (10.3) is satisfied.

If (10.3) does not hold, then there is an element

$$i \in keys(P) \setminus \left(\bigcup_{U_j \in F} keys(U_j) \right).$$

The value of k_P is a sum of terms, one of which is k_i. Because the value of k_i is not known to the coalition F, it follows that the coalition has no information (in the sense of perfect secrecy, as discussed in Chapter 2) about the value of the group key k_P. Therefore, (10.3) is a necessary and sufficient condition for a group key k_P to be secure against a coalition F, where $P \cap F = \emptyset$. The security that is provided is unconditional.

The method used in Example 10.4 can be generalized easily. It is always possible to construct a trivial $\binom{n}{2}$ by n matrix M, in which any two users in the resulting KDP have exactly one common key, and each key is given to exactly two users. The group key for any two users is the unique key that they both possess. There are $\binom{n}{2}$ group keys in this KDP, and each user must store $n - 1$ keys. Each group key $k_{\{U_j, U_{j'}\}}$ is secure against the coalition $\mathcal{U} \setminus \{U_j, U_{j'}\}$ of all the remaining users (this coalition has size $n - 2$).

In general, we want to construct KDPs in which the number of keys stored by each user is as small as possible. The previous example provides maximum security but has a large storage requirement. It is sometimes possible to reduce the storage requirement if the security conditions are relaxed. Similar to the *Blom Key Predistribution Scheme*, we might consider a scenario where we only require security against coalitions of a specified size. We consider a particular example to illustrate.

Example 10.5   Suppose $n = 7$, $v = 7$, and the matrix M is as follows:

$$M = \begin{pmatrix} 1 & 1 & 1 & 0 & 1 & 0 & 0 \\ 0 & 1 & 1 & 1 & 0 & 1 & 0 \\ 0 & 0 & 1 & 1 & 1 & 0 & 1 \\ 1 & 0 & 0 & 1 & 1 & 1 & 0 \\ 0 & 1 & 0 & 0 & 1 & 1 & 1 \\ 1 & 0 & 1 & 0 & 0 & 1 & 1 \\ 1 & 1 & 0 & 1 & 0 & 0 & 1 \end{pmatrix}.$$

Now we have:

$$keys(U_1) = \{1, 4, 6, 7\},$$
$$keys(U_2) = \{1, 2, 5, 7\}, \quad \text{and}$$
$$keys(U_1, U_2) = \{1, 7\}.$$

Hence, $k_{\{U_1, U_2\}} = k_1 + k_7$.

No other user has both k_1 and k_7, so $k_{\{U_1, U_2\}}$ is secure against any other individual user. (However, U_3 and U_4 (for example) together can compute $k_{\{U_1, U_2\}}$ by pooling their secret information.) In this example, it can be checked that every pair of users can compute a key that is secure against any other individual user.

This scheme enables the construction of $\binom{7}{2} = 21$ group keys for pairs of users. The total number of distributed keys is 7, which is much less than 21. The number of keys stored by each user is 4, which is less than the $n - 1 = 6$ keys that must be stored by each user in the trivial scheme. ☐

10.4.1 Fiat-Naor Key Distribution Patterns

In this section, we present a construction for a family of key distribution patterns due to Fiat and Naor. These schemes allow the computation of group keys for arbitrary subsets of users in the network. Choose an integer w, where $1 \leq w \leq n$ (w is a security parameter). Then define

$$v = \sum_{i=0}^{w} \binom{n}{i}.$$

A *Fiat-Naor w-KDP* is a v by n matrix M whose rows are the incidence vectors of all subsets of $\mathcal{U}$ having cardinality at least $n - w$.

Given a Fiat-Naor w-KDP, it is easy to see that there is a group key, for any subset $P \subseteq \mathcal{U}$, that is secure against any disjoint coalition F of size at most w. This is easily proven, as follows: $|F| \leq w$, so $|\mathcal{U} \setminus F| \geq n - w$. Hence, there exists a key k_i given to the users in $\mathcal{U} \setminus F$ and to no other user. The set $P \subseteq (\mathcal{U} \setminus F)$, so all users in P have k_i and no user in F has k_i, i.e., (10.3) is not satisfied.

Here is a small example to illustrate.

Example 10.6 Suppose $n = 6$ and $w = 1$. The resulting Fiat-Naor 1-KDP has $v = 7$, and the key distribution pattern M is as follows:

$$M = \begin{pmatrix} 1 & 1 & 1 & 1 & 1 & 1 \\ 1 & 1 & 1 & 1 & 1 & 0 \\ 1 & 1 & 1 & 1 & 0 & 1 \\ 1 & 1 & 1 & 0 & 1 & 1 \\ 1 & 1 & 0 & 1 & 1 & 1 \\ 1 & 0 & 1 & 1 & 1 & 1 \\ 0 & 1 & 1 & 1 & 1 & 1 \end{pmatrix}.$$

Consider the subset of users $P = \{U_1, U_3, U_4\}$. We have

$$keys(U_1) = \{1, 2, 3, 4, 5, 6\},$$
$$keys(U_3) = \{1, 2, 3, 4, 6, 7\},$$
$$keys(U_4) = \{1, 2, 3, 5, 6, 7\}, \quad \text{and}$$
$$keys(U_1, U_3, U_4) = \{1, 2, 3, 6\}.$$

Hence, $k_{\{U_1, U_3, U_4\}} = k_1 + k_2 + k_3 + k_6$, and no other single user can compute this key. □

10.4.2   Mitchell-Piper Key Distribution Patterns

A *Fiat-Naor Key Distribution Pattern* yields a group key, for any subset of users, that is secure against a coalition of size w. In a *Mitchell-Piper Key Distribution Pattern*, there is a group key for every subset of exactly t participants. Again, group keys are required to be secure against coalitions of size w. The case $t = 2$ is of particular interest, because this is the case where keys are associated with pairs of users. In this section, the required properties of the key distribution pattern M are studied and an existence result is proven using a randomized approach called the *probabilistic method*. This powerful technique is due to the famous mathematician Paul Erdös.

First, we require some definitions.

> **Definition 10.2:**   A *set system* is a pair $(X, \mathcal{A})$, where X is a finite set of elements called *points* and $\mathcal{A}$ is a set of subsets of X called *blocks*.
> Denote $X = \{x_1, \ldots, x_v\}$ and denote $\mathcal{A} = \{A_1, \ldots, A_n\}$. $(X, \mathcal{A})$ is a (t, w)-*cover-free family* provided that, for any two disjoint subsets of blocks $P, F \subseteq \mathcal{A}$, where $|P| = t$ and $|F| = w$, it holds that
>
> $$\bigcap_{A_i \in P} A_i \not\subseteq \bigcup_{A_j \in F} A_j.$$
>
> A (t, w)-cover-free family will be denoted as a (t, w)-CFF(v, n) if $|X| = v$ and $|\mathcal{A}| = n$.

Informally, the intersection of t blocks in a (t, w)-cover-free family is never covered by the union of w other blocks.

The *incidence matrix* of a set system $(X, \mathcal{A})$ is the $v \times n$ matrix $M = (m_{i,j})$ in which

$$m_{i,j} = \begin{cases} 1 & \text{if } x_i \in A_j \\ 0 & \text{otherwise} . \end{cases}$$

A *Mitchell-Piper* (t, w)-*KDP* (or more briefly, a (t, w)-*KDP*) is a KDP in which there is a key for every group of t users, and each such key is secure against any disjoint coalition of at most w users. We have the following theorem that links cover-free families and Mitchell-Piper KDPs. The proof is left as an Exercise; it follows in a straightforward fashion from the definitions.

THEOREM 10.6 *Suppose that M is a $v \times n$ key distribution pattern. Then M is a (t, w)-KDP if and only if M is the incidence matrix of a (t, w)-CFF(v, n).*

Observe that the blocks of the cover-free family correspond to the columns of M (i.e., the n blocks in the (t, w)-CFF(v, n) are $keys(U_1), \ldots, keys(U_n)$). For example, the columns of the incidence matrix in Example 10.5 form a $(2, 1)$-CFF$(7, 7)$, $(X, \mathcal{A})$, where

$$X = \{1, \ldots, 7\}, \quad \text{and}$$

$$\mathcal{A} = \{\{1, 4, 6, 7\}, \{1, 2, 5, 7\}, \{1, 2, 3, 6\}, \{2, 3, 4, 7\},$$

$$\{1, 3, 4, 5\}, \{2, 4, 5, 6\}, \{3, 5, 6, 7\}\}.$$

There are many known constructions for cover-free families. In general, given t, w and n, we want to construct a (t, w)-CFF(v, n) for v as small as possible. Here, we present a non-constructive existence result.

Suppose a value v is chosen. Let M be a v by n matrix whose columns are labeled $1, \ldots, n$. We choose the entries in M randomly, as follows: Let $0 < \rho < 1$, and suppose that every entry of M is (independently) defined to be a "1" with probability ρ, and a "0" with probability $1 - \rho$. (We are free to choose the value of ρ however we wish. Later, we will specify an optimized value for ρ, after we analyze the construction more fully.)

The probability that a random matrix M constructed in this fashion is a (t, w)-CFF(v, n) will be shown to be positive, provided that certain numerical conditions on the parameters are satisfied. This will prove (non-constructively) that a (t, w)-CFF(v, n) exists for certain values of t, w, v and n. Note, however, that this approach does not immediately yield an efficient algorithm to construct the matrix M or the associated cover-free family.

Let M be formed as described above. Suppose that $P, F \subseteq \{1, \ldots, n\}$, where $|P| = t$, $|F| = w$ and $P \cap F = \emptyset$. We will say that a given row i of M satisfies the property $\gamma(P, F, i)$ provided that the entries in the columns in P are all equal to "1" and the entries in the columns in F are all equal to "0." This is represented pictorially in Figure 10.1.

Define a random variable

$$X(P, F) = \begin{cases} 0 & \text{if there exists a row } i \text{ such that } \gamma(P, F, i) \text{ is satisfied} \\ 1 & \text{otherwise.} \end{cases}$$

First, we observe that M is not the incidence matrix of a (t, w)-CFF(v, n) if $X(P, F) = 1$. This is because $\gamma(P, F, i)$ does not hold, for any i, for a certain

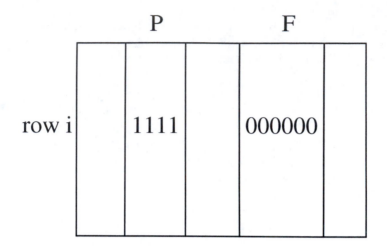

FIGURE 10.1
Property $\gamma(P, F, i)$

choice of P and F. In particular, for every point x_i that is in all of the blocks B_j ($j \in P$), there is at least one block $B_{j'}$ ($j' \in F$) that contains x_i. Hence, the intersection of the t blocks in B_j ($j \in P$) is covered by the union of the w blocks in $B_{j'}$ ($j' \in F$).

On the other hand, it is not hard to see that M is the incidence matrix of a (t, w)-CFF(v, n) if $X(P, F) = 0$ for all choices of P and F. Thus we have the following lemma.

LEMMA 10.7   *A v by n matrix M having entries 0 and 1 is the incidence matrix of a (t, w)-CFF(v, n) if and only if $X(P, F) = 0$ for all choices of P and F such that $P, F \subseteq \{1, \ldots, n\}$, $|P| = t$, $|F| = w$ and $P \cap F = \emptyset$.*

Suppose we fix subsets P and F such that $|P| = t$ and $|F| = w$. The probability that $\gamma(P, F, i)$ is satisfied is easily seen to be equal to $\rho^t (1 - \rho)^w$. Therefore, the probability that $\gamma(P, F, i)$ is satisfied for none of the v choices of i is equal to

$$\left(1 - \rho^t (1 - \rho)^w\right)^v.$$

Hence, the expected (or average) value of the random variable $X(P, F)$ is computed to be

$$\begin{aligned}
\mathbf{Exp}[X(P, F)] &= \mathbf{Pr}[X(P, F) = 0] \times 0 + \mathbf{Pr}[X(P, F) = 1] \times 1 \\
&= \mathbf{Pr}[X(P, F) = 1] \\
&= \left(1 - \rho^t (1 - \rho)^w\right)^v.
\end{aligned}$$

In our analysis, we will want to minimize $\mathbf{Exp}[X(P, F)]$. In order to do this, we should maximize $\rho^t (1 - \rho)^w$. Elementary calculus shows that we should define $\rho = t/(t + w)$; this is the value we will use for ρ from now on.

Next, we define another random variable X by summing all the random variables $X(P, F)$:

$$X = \sum_{\{(P,F):|P|=t,|F|=w,P\cap F=\emptyset\}} X(P, F).$$

Note that Lemma 10.7 asserts that $X = 0$ if and only if M is the incidence matrix of a (t, w)-CFF(v, n).

The random variables $X(P, F)$ are not independent. However, the expected value of a sum of random variables is always the sum of the expected values of the random variables, regardless of whether they are independent. Hence, we can compute the expected value of the random variable X as follows:

$$\mathbf{Exp}[X] = \sum_{\{(P,F):|P|=t,|F|=w,P\cap F=\emptyset\}} \mathbf{Exp}[X(P, F)]$$

$$= \binom{n}{t} \binom{n - t}{w} \left(1 - \rho^t (1 - \rho)^w\right)^v$$

$$< n^{t+w} \left(1 - \frac{t^t w^w}{(t + w)^{t+w}}\right)^v.$$

In the last line, we are substituting $\rho = t/(t + w)$, as well as using the easily verified facts that

$$\binom{n}{t} < n^t$$

and

$$\binom{n - t}{w} < n^w.$$

The random variable X is a sum of non-negative integer values. Therefore, if $X < 1$, then it must be the case that $X = 0$. Suppose that $\mathbf{Exp}[X] < 1$; then $\mathbf{Pr}[X = 0] > 0$. In this situation, it would follow from Lemma 10.7 that a (t, w)-CFF(v, n) exists. In this way, we get an existence result provided that we can choose values for the parameters which causes the inequality $\mathbf{Exp}[X] < 1$ to hold. We proceed to do this now.

Define

$$p_{t,w} = 1 - \frac{t^t w^w}{(t + w)^{t+w}}.$$

Then $\mathbf{Exp}[X] < 1$ if

$$n^{t+w} (p_{t,w})^v < 1$$

$$\Longleftrightarrow \quad (t + w) \log_2 n + v \log_2 (p_{t,w}) < 0$$

$$\Longleftrightarrow \quad v > \frac{(t + w) \log_2 n}{- \log_2 (p_{t,w})}.$$

We record our main result as a theorem, which follows immediately from the discussion above.

THEOREM 10.8   *Suppose that t, w and n are positive integers. Define*

$$p_{t,w} = 1 - \frac{t^t w^w}{(t+w)^{t+w}}.$$

Then a (t, w)-CFF(v, n) exists if

$$v > \frac{(t+w)\log_2 n}{-\log_2(p_{t,w})}.$$

Let's look at a particular example. Suppose we take $t = 2$ and $w = 1$. Thus we are interested in a KDP which yields keys for pairs of users, such that each key is secure against any other individual user. The constant $p_{2,1}$ is computed to be

$$1 - \frac{2^2 1^1}{(2+1)^{2+1}} = \frac{23}{27},$$

and $-1/\log_2(p_{2,1}) \approx 4.323$. Then Theorem 10.8 asserts that a $(2, 1)$-CFF(v, n) exists if $v > 12.97 \log_2 n$.

In general, Theorem 10.8 shows that there exists a (t, w)-CFF(v, n) in which v is $O(\log n)$ (for any fixed t and w). Asymptotically, this is a very good result. Unfortunately, it does not yield an explicit formula or an efficient deterministic algorithm to construct the desired cover-free family. However, some approaches that lead to practical algorithms for the construction of (t, w)-CFF(v, n) are considered in the Exercises.

10.5   Session Key Distribution Schemes

Recall from the introduction of this chapter that the *TA* is assumed to have a shared secret key with every network user in a session key distribution scheme. We will use K_{Alice} to denote Alice's secret key, K_{Bob} is Bob's secret key, etc. In a session key distribution scheme, the *TA* chooses session keys and distributes them on-line in encrypted form, upon request of network users.

Eventually, we will define attack models and adversarial goals for session key distribution. However, it is not easy to formulate precise definitions because SKDS sometimes do not include mutual identification of the users in a session of the scheme. Therefore, we begin by giving a historical tour of some important SKDSs and describing some attacks on them, before we proceed to a more formal treatment of the subject.

Protocol 10.4: NEEDHAM-SCHROEDER SKDS

1. Alice chooses a random number, r_A. Alice sends $ID(Alice)$, $ID(Bob)$ and r_A to the TA.

2. The TA chooses a random session key, K. Then it computes

    $$t_{Bob} = e_{K_{Bob}}(K \parallel ID(Alice))$$

    (which is called a *ticket to Bob*) and

    $$y_1 = e_{K_{Alice}}(r_A \parallel ID(Bob) \parallel K \parallel t_{Bob}),$$

    and it sends y_1 to Alice.

3. Alice decrypts y_1 using her key K_{Alice}, obtaining K and t_{Bob}. Then Alice sends t_{Bob} to Bob.

4. Bob decrypts t_{Bob} using his key K_{Bob}, obtaining K. Then, Bob chooses a random number r_B and computes $y_2 = e_K(r_B)$. Bob sends y_2 to Alice.

5. Alice decrypts y_2 using the session key K, obtaining r_B. Then Alice computes $y_3 = e_K(r_B - 1)$ and she sends y_3 to Bob.

10.5.1 The Needham-Schroeder Scheme

One of the first session key distribution schemes is the *Needham-Schroeder SKDS*, which was proposed in 1978; this scheme is presented in Protocol 10.4. The diagram in Figure 10.2 depicts the five flows in the *Needham-Schroeder SKDS*.

There are also some validity checks required in the *Needham-Schroeder SKDS*, where the term *validity check* refers to verifying that decrypted data has the correct format and contains expected information. (Note there are no message authentication codes being used in the *Needham-Schroeder SKDS*.) These validity checks are as follows:

1. When Alice decrypts y_1, she checks to see that the plaintext $d_{K_{Alice}}(y_1)$ has the form

    $$d_{K_{Alice}}(y_1) = r_A \parallel ID(Bob) \parallel K \parallel t_{Bob}$$

    for some K and t_{Bob}. If this above condition holds, then Alice "accepts," otherwise Alice "rejects" and aborts the session.

2. When Bob decrypts y_3, he checks to see that the plaintext

    $$d_K(y_3) = r_B - 1.$$

    If this condition holds, then Bob "accepts"; otherwise Bob "rejects."

FIGURE 10.2
Information flows in the *Needham-Schroeder SKDS*

TA   A   B

$$\xleftarrow{\quad\quad A, B, r_A \quad\quad}$$

$$t_{Bob} = e_{K_{Bob}}(K \parallel A)$$

$$\xrightarrow{\quad e_{K_{Alice}}(r_A \parallel B \parallel K \parallel t_{Bob}) \quad}$$

$$\xrightarrow{\quad\quad t_{Bob} \quad\quad}$$

$$\xleftarrow{\quad\quad e_K(r_B) \quad\quad}$$

$$\xrightarrow{\quad\quad e_K(r_B - 1) \quad\quad}$$

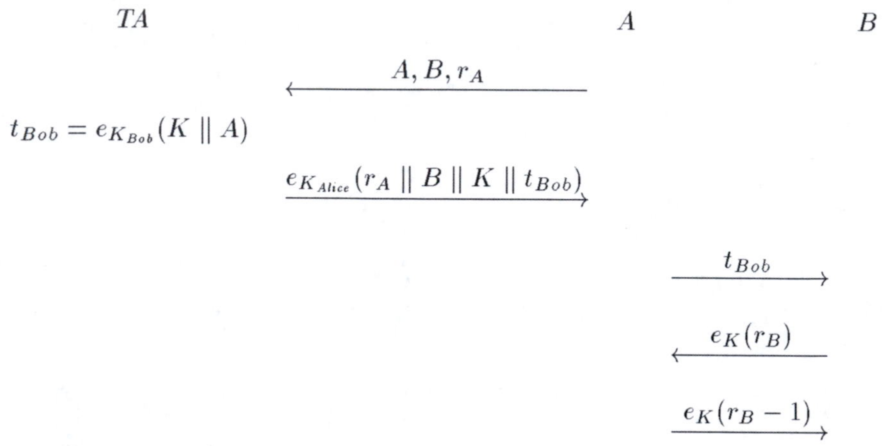

Note that "A" denotes "$ID(Alice)$" and "B" denotes "$ID(Bob)$."

Here is a summary of the main steps in the scheme. In flow 1, Alice asks the *TA* for a session key to communicate with Bob. At this point, Bob might not even be aware of Alice's request. The *TA* transmits the encrypted session key to Alice in flow 2, and Alice sends an encrypted session key to Bob in flow 3. Thus flows 1–3 of *Needham-Schroeder* comprise the session key distribution: the session key K is encrypted using the secret keys of Alice and Bob and it is distributed to both of them. The purpose of flows 4 and 5 is to convince Bob that Alice actually possesses the session key K. This is accomplished by having Alice use the new session key to encrypt the challenge $r_B - 1$; the process is called *key confirmation* (from Alice to Bob).

10.5.2   The Denning-Sacco Attack on the NS Scheme

In 1981, Denning and Sacco discovered an attack on the *Needham-Schroeder SKDS*. We present this attack now. Suppose Oscar records a session, say $\mathcal{S}$, of the *Needham-Schroeder SKDS* scheme between Alice and Bob, and somehow he obtains the session key, K, for the session $\mathcal{S}$. This attack model is called a *known session key attack*. Then Oscar can initiate a new session, say $\mathcal{S}'$, of the *Needham-Schroeder SKDS* with Bob, starting with the third flow of the session

S', by sending the previously used ticket, t_{Bob}, to Bob:

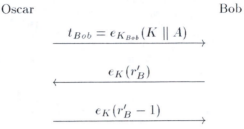

Oscar   Bob

$$t_{Bob} = e_{K_{Bob}}(K \parallel A)$$

$$e_K(r'_B)$$

$$e_K(r'_B - 1)$$

Notice that when Bob replies with $e_K(r'_B)$, Oscar can decrypt this using the known key K, subtract 1, and then encrypt the result. The value $e_K(r'_B - 1)$ is sent to Bob in the last flow of the session S'. Bob will decrypt this and "accept."

Let's consider the consequences of this attack. At the end of the session S' between Oscar and Bob, Bob thinks he has a "new" session key, K, shared with Alice (this is because $ID(Alice)$ occurs in the ticket t_{Bob}). This key K is known to Oscar, but it may not be known to Alice, because Alice might have thrown away the key K after the previous session with Bob, namely S, terminated. Hence, there are two ways in which Bob is deceived by this attack:

1. The key K that is distributed in the session S' is not known to Bob's intended peer, Alice.

2. The key K for the session S' is known to someone other than Bob's intended peer (namely, it is known to Oscar).

10.5.3   Kerberos

Kerberos comprises a popular series of schemes for session key distribution that were developed at MIT in the late 1980s and early 1990s. We provide a simplified treatment of version five of the scheme. This is presented as Protocol 10.5. A diagram depicting the four flows in a session of the scheme is given in Figure 10.3.

As was the case with *Needham-Schroeder*, there are certain validity checks required in *Kerberos*. These are as follows:

1. When Alice decrypts y_1, she checks to see that the plaintext $d_{K_{Alice}}(y_1)$ has the form
$$d_{K_{Alice}}(y_1) = r_A \parallel ID(Bob) \parallel K \parallel L,$$
for some K and L. If this condition does not hold, then Alice "rejects" and aborts the current session.

2. When Bob decrypts y_2 and t_{Bob}, he checks to see that the plaintext $d_K(y_2)$ has the form
$$d_K(y_2) = ID(Alice) \parallel time$$

Protocol 10.5: SIMPLIFIED KERBEROS V5

1. Alice chooses a random number, r_A. Alice sends $ID(Alice)$, $ID(Bob)$ and r_A to the TA.

2. The TA chooses a random session key K and a validity period (or *lifetime*), L. Then it computes a ticket to Bob,

    $$t_{Bob} = e_{K_{Bob}}(K \parallel ID(Alice) \parallel L),$$

    and

    $$y_1 = e_{K_{Alice}}(r_A \parallel ID(Bob) \parallel K \parallel L).$$

    The TA sends t_{Bob} and y_1 to Alice.

3. Alice decrypts y_1 using her key K_{Alice}, obtaining K. Then Alice determines the current time, $time$, and she computes

    $$y_2 = e_K(ID(Alice) \parallel time).$$

    Finally, Alice sends t_{Bob} and y_2 to Bob.

4. Bob decrypts t_{Bob} using his key K_{Bob}, obtaining K. He also decrypts y_2 using the key K, obtaining $time$. Then, Bob computes

    $$y_3 = e_K(time + 1).$$

    Finally, Bob sends y_3 to Alice.

---

and the plaintext $d_{K_{Bob}}(t_{Bob})$ has the form

$$d_{K_{Bob}}(t_{Bob}) = K \parallel ID(Alice) \parallel L,$$

where $ID(Alice)$ is the same in both plaintexts and $time \leq L$. If these conditions hold, then Bob "accepts"; otherwise Bob "rejects."

3. When Alice decrypts y_3, she checks that $d_K(y_3) = time + 1$. If this condition holds, then Alice "accepts"; otherwise Alice "rejects."

Here is a summary of the rationale behind some of the features in *Kerberos*. When a request for a session key is sent by Alice to the TA, the TA will generate a new random session key K. As well, the TA will specify the *lifetime*, L, during which K will be valid. That is, the session key K is to be regarded as a valid key until time L. All this information is encrypted before it is transmitted to Alice.

Alice can use her secret key to decrypt y_1, and thus obtain K and L. She will verify that the current time is within the lifetime of the key and that y_1 contains Alice's random challenge, r_A. She can also verify that y_1 contains $ID(Bob)$,

FIGURE 10.3
The flows in *Kerberos V5*

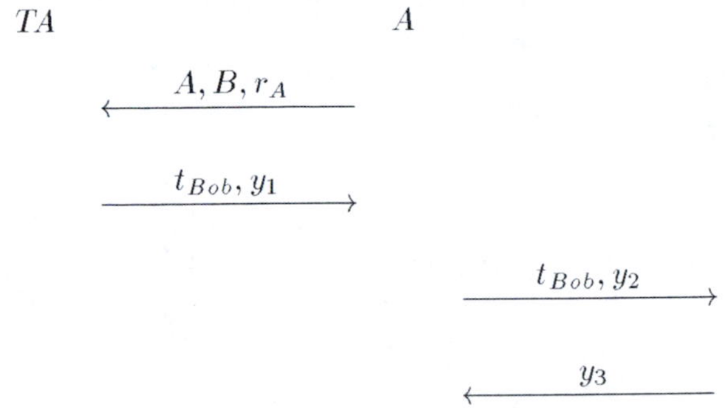

where

$$A = ID(Alice),$$

$$B = ID(Bob),$$

$$t_{Bob} = e_{K_{Bob}}(K \parallel A \parallel L),$$

$$y_1 = e_{K_{Alice}}(r_A \parallel B \parallel K \parallel L)$$

$$y_2 = e_K(A \parallel time), \quad \text{and}$$

$$y_3 = e_K(time + 1).$$

where Bob is Alice's intended peer. These checks prevent Oscar from replaying an "old" y_1 which might have been transmitted by the TA in a previous session.

Next, Alice will relay t_{Bob} to Bob. As well, Alice will use the new session key K to encrypt the current time $time$ and $ID(Alice)$. Then she sends the resulting ciphertext y_2 to Bob.

When Bob receives t_{Bob} and y_2 from Alice, he decrypts t_{Bob} to obtain K, L and $ID(Alice)$. Then he uses the new session key K to decrypt y_2 and he verifies that $ID(Alice)$, as decrypted from t_{Bob} and y_2, are the same. This ensures Bob that the session key encrypted within t_{Bob} is the same key that was used to encrypt y_2. He should also check that $time \leq L$ to verify that the key K has not expired.

Finally, Bob encrypts $time + 1$ using the new session key K and sends the result back to Alice. When Alice receives this message, y_3, she decrypts it using K and verifies that the result is $time + 1$. This ensures Alice that the session key K has been successfully transmitted to Bob, since K was needed in order to produce the message y_3.

The purpose of the lifetime L is to prevent an active adversary from storing "old" messages for retransmission at a later time, as was done in the Denning-Sacco attack on the *Needham-Schroeder SKDS*. One of the drawbacks of *Kerberos* is that all the users in the network should have synchronized clocks, since the current time is used to determine if a given session key K is valid. In practice, it is very difficult to provide perfect synchronization, so some amount of variation in times must be allowed.

We make a few comments comparing *Needham-Schroeder* to *Kerberos*.

1. In *Kerberos*, *mutual key confirmation* is accomplished in flows 3 and 4. By using the new session key K to encrypt $ID(Alice)$, Alice is trying to convince Bob that she knows the value of K. Similarly, when Bob encrypts $time + 1$ using K, he is demonstrating to Alice that he knows the value of K.

2. In *Needham-Schroeder*, information intended for Bob is doubly encrypted: the ticket t_{Bob}, which is already encrypted, is re-encrypted using Alice's secret key. This seems to serve no useful purpose and it adds unnecessary complexity to the scheme. In *Kerberos*, this double encryption was removed.

3. Partial protection against the Denning-Sacco attack is provided in *Kerberos* by verifying that the current time (namely, the value $time$, which is often referred to as a *timestamp*) lies within the lifetime L. Basically, this limits the time period during which a Denning-Sacco type attack can be carried out.

Needham-Schroeder and *Kerberos* have some features that are not generally regarded as useful in present day SKDSs. We discuss these briefly before proceeding to the development of a secure SKDS.

1. Timestamps require reliable, synchronized clocks. Schemes using timestamps are hard to analyze and it is difficult to give convincing security proofs for them. For this reason, it is generally preferred to use random challenges rather than timestamps, if possible.

2. Key confirmation is not necessarily an important attribute of a session key distribution scheme. For example, possession of a key during a session of the SKDS does not imply possession of the key at a later time, when it is actually going to be used. For this reason, it is now often recommended that key confirmation be omitted from SKDSs.

3. In *Needham-Schroeder* and *Kerberos*, encryption is used to provide both secrecy and authenticity. However, it is preferable to use encryption for secrecy and a message authentication code to provide authenticity. For example, in the second flow of *Needham-Schroeder*, we could remove the double encryption and use MACs for authentication, as follows:
The *TA* chooses a random session key K. Then it computes

$$y_1 = (e_{K_{Bob}}(K), MAC_{Bob}(ID(Alice) \,\|\, e_{K_{Bob}}(K))),$$

Protocol 10.6: BELLARE-ROGAWAY SKDS

1. Alice chooses a random number, r_A, and she sends $ID(Alice)$, $ID(Bob)$ and r_A to Bob.

2. Bob chooses a random number, r_B, and he sends $ID(Alice)$, $ID(Bob)$, r_A and r_B to the TA.

3. The TA chooses a random session key K. Then it computes

$$y_B = (e_{K_{Bob}}(K), MAC_{Bob}(ID(Alice) \| ID(Bob) \| r_B \| e_{K_{Bob}}(K)))$$

and

$$y_A = (e_{K_{Alice}}(K), MAC_{Alice}(ID(Bob) \| ID(Alice) \| r_A \| e_{K_{Alice}}(K))).$$

The TA sends y_B to Bob and y_A to Alice.

---

and

$$y_1' = (e_{K_{Alice}}(K), MAC_{Alice}(ID(Bob) \| r_A \| e_{K_{Alice}}(K))).$$

The TA would send y_1 and y_1' to Alice, who would then relay y_1 to Bob.

The revised second flow does not fix the flaw found by Denning and Sacco, however.

4. In order to prevent the Denning-Sacco attack, the flow structure of the scheme must be modified. Any "secure" scheme should involve Bob as an active participant prior to his receiving the session key, in order to prevent Denning-Sacco type replay attacks. The solution requires Alice to contact Bob (or vice versa) before sending a request for a session key to the TA.

10.5.4 The Bellare-Rogaway Scheme

Bellare and Rogaway proposed an SKDS in 1995 and provided a proof of security for their scheme, under certain assumptions. We begin by describing the *Bellare-Rogaway SKDS* in Protocol 10.6. Then we will proceed to a more formal analysis of the scheme, which will require developing rigorous definitions of the attack model and adversarial goals.

Protocol 10.6 has a different flow structure than the schemes we have considered so far. Alice and Bob both choose random challenges, which are sent to the TA. Thus Bob is involved in the scheme before the TA issues the session key. The information that the TA sends to Alice consists of:

1. a session key (encrypted using Alice's secret key), and

2. a MAC for the encrypted session key, the identities of Alice and Bob, and Alice's challenge.

The information sent to Bob is analogous.

Alice and Bob will "accept" if their respective MACs are valid (note that these MACs are computed using secret MAC keys which are known to the *TA*). For example, when Bob receives the encrypted session key, say $y_{B,1}$, and the MAC, say $y_{B,2}$, he verifies that

$$y_{B,2} = MAC_{Bob}(ID(Alice) \parallel ID(Bob) \parallel r_B \parallel y_{B,1}).$$

Observe that no key confirmation is provided in this scheme. When Alice accepts, for example, she does not know if Bob has accepted, or even if Bob has received the message sent by the *TA*. When Alice accepts, it just means that she has received the information she expected, and this information is valid (or, more precisely, the MAC is valid). From Alice's point of view, when she accepts, she believes that she has received a new session key from the *TA*. Moreover, because this session key was encrypted using Alice's secret key, Alice is confident that no one else can compute the session key K from the information that she just received. Of course, Bob should also have received an encryption of the same session key. Alice does not know if this, in fact, transpired, but we will argue that Alice can be confident that no one other than Bob can compute the new session key. The analysis will be similar when the session is examined from Bob's point of view. In other words, we have removed the objective of key confirmation (one-way or two- way) from the SKDS. This is replaced with the somewhat weaker (but still useful) objective that, from the point of view of a participant in the scheme who "accepts," no one other than their intended peer should be able to compute the new session key.

The objective of an adversary will be to cause an honest participant to "accept" in a situation where someone other than the intended peer of that participant knows the value of the session key K. For example, suppose that an honest Alice "accepts" and her intended peer is Bob. The adversary, Oscar, achieves his goal if he (Oscar) can compute the session key, or if some other network user (say Charlie) can compute the session key. On the other hand, Oscar's attack is not considered to be successful if Alice is the only network user who can compute the session key. In this situation, Bob can't compute the session key, but neither can anyone else (except Alice).

Summarizing the above discussion, we will define a *secure session key distribution scheme* to be an SKDS in which the following property holds: if a participant in a session "accepts," then the probability that someone other than that participant's intended peer knows the session key is small.

We now consider how to go about proving that the *Bellare-Rogaway SKDS* is secure. As usual, we make several reasonable assumptions. These include assumptions that Alice, Bob and the *TA* are honest, the encryption scheme and MAC used in the scheme are secure and secret keys are known only to their intended owners, and random challenges are generated using perfect random number generators. (These assumptions are similar to those made in Chapter 9 in the study of identification schemes.) Finally, we assume that the TA generates session keys using a perfect random number generator.

Let us consider various ways in which Oscar can carry out an attack. For each of these possibilities, we argue that Oscar will not be successful, except with a small probability. These possibilities are not all mutually exclusive.

1. Oscar is a passive adversary.

2. Oscar is an active adversary and Alice is a legitimate participant in the scheme. Oscar may impersonate Bob or the TA, and Oscar may intercept and change messages sent during the scheme.

3. Oscar is an active adversary and Bob is a legitimate participant in the scheme. Oscar may impersonate Alice or the TA, and Oscar may intercept and change messages sent during the scheme.

We now go on to analyze the possible attacks enumerated above. In each situation, we discuss what the outcome of the scheme will be, subject to our underlying "reasonable" assumptions.

1. If the adversary is passive, then Alice and Bob will both output "accept" in any session in which they are the two participants. Further, they will both be able to decrypt the same session key, K. No one else (including Oscar) is able to compute K, because the encryption scheme is secure.

2. Suppose Alice is a legitimate participant in the scheme. She wishes to obtain a new session key which will be known only to Bob and to herself. However, Alice does not know if she really is communicating with Bob, because Oscar may be impersonating Bob.

   When Alice receives the message y_A, she checks to see that the MAC is valid. This MAC incorporates Alice's random challenge, r_A, as well as the identities of Alice and Bob, and the encrypted session key $\epsilon_{K_{Alice}}(K)$. This convinces Alice that the MAC was newly computed by the TA, because the TA is the only party other than Alice who knows the key MAC_{Alice}. Furthermore, the random challenge r_A prevents replay of a MAC from a previous session. Finally, including $\epsilon_{K_{Alice}}(K)$ in the MAC prevents an adversary from replacing the session key chosen by the TA with something else. Therefore, Alice can be confident that Bob (her intended peer) is the only other user who is able to decrypt the session key K, even if Oscar has impersonated Bob in the current session of the scheme.

3. Suppose Bob is a legitimate participant in the scheme. He believes he will obtain a new session key which will be known only to Alice and to himself. However, Bob does not know if he really is communicating with Alice, because Oscar may be impersonating Alice.

   When Bob receives the message y_B, he checks to see that the MAC is valid.

This MAC incorporates Bob's random challenge, r_B, as well as the identities of Alice and Bob, and the encrypted session key $e_{K_{Bob}}(K)$. This convinces Bob that the MAC was newly computed by the TA, because the TA is the only party other than Bob who knows the key MAC_{Bob}. Furthermore, the random challenge r_B prevents replay of a MAC from a previous session. Finally, including $e_{K_{Bob}}(K)$ in the MAC prevents an adversary from replacing the session key chosen by the TA with something else. Therefore, Bob can be confident that Alice (his intended peer) is the only other user who is able to decrypt the session key K, even if Oscar has impersonated Alice in the current session of the scheme.

10.6   Notes and References

A recent survey of the key establishment problem has been written by Blundo and D'Arco [53]. For a recent book emphasizing similar topics, see Boyd and Mathuria [67].

The *Blom Key Predistribution Scheme* was presented in [49]. For a generalization of this scheme, see Blundo *et al.* [55].

Mitchell-Piper Key Distribution Patterns are introduced in [242]; see Dyer *et al.* [124] for a description of the randomized construction method we presented in this chapter. Stinson [323] and Stinson and Tran [325] give some deterministic constructions for key distribution patterns.

The *Needham-Schroeder SKDS* is from [251] and the Denning-Sacco attack is from [108]. For information on *Kerberos*, see Kohl and Neuman [201].

The *Bellare-Rogaway SKDS* was described in [25]. Secure session key distribution schemes using public-key cryptography are discussed in Blake-Wilson and Menezes [44].

Exercises

10.1 Suppose that $p = 150001$ and $\alpha = 7$ in the *Diffie-Hellman Key Predistribution Scheme*. (It can be verified that α is a generator of $\mathbb{Z}_p^*$.) Suppose that the private keys of U, V and W are $a_U = 101459$, $a_V = 123967$ and $a_W = 99544$.
   (a) Compute the public keys of U, V and W.
   (b) Show the computations performed by U to obtain $K_{U,V}$ and $K_{U,W}$.
   (c) Verify that V computes the same key $K_{U,V}$ as U does.
   (d) Explain why $\mathbb{Z}_{150001}^*$ is a very poor choice of a setting for the *Diffie-Hellman Key Predistribution Scheme* (notwithstanding the fact that p is too small for the scheme to be secure).

   HINT   Consider the factorizaton of $p - 1$.

10.2 Suppose the *Blom KPS* with $k = 2$ is implemented for a set of five users, U, V, W, X and Y. Suppose that $p = 97$, $r_U = 14$, $r_V = 38$, $r_W = 92$, $r_X = 69$ and $r_Y = 70$. The secret g polynomials are as follows:

$$g_U(x) = 15 + 15x + 2x^2$$
$$g_V(x) = 95 + 77x + 83x^2$$
$$g_W(x) = 88 + 32x + 18x^2$$
$$g_X(x) = 62 + 91x + 59x^2$$
$$g_Y(x) = 10 + 82x + 52x^2.$$

(a) Compute the keys for all $\binom{5}{2} = 10$ pairs of users.

(b) Verify that $K_{U,V} = K_{V,U}$.

10.3 Suppose that the *Blom KPS* is implemented with security parameter k. Suppose that a coalition of k bad users, say $W_1, \ldots, W_k$, pool their secret information. Additionally, assume that a key $K_{U,V}$ is exposed, where U and V are two other users.

(a) Describe how the coalition can determine the polynomial $g_U(x)$ by polynomial interpolation, using known values of $g_U(x)$ at $k + 1$ points.

(b) Having computed $g_U(x)$, describe how the coalition can compute the bivariate polynomial $f(x, y)$ by bivariate polynomial interpolation.

(c) Illustrate the preceding two steps, by determining the polynomial $f(x, y)$ in the sample implementation of the *Blom KPS* where $k = 2$, $p = 34877$, and $r_i = i$ ($1 \le i \le 4$), supposing that

$$g_1(x) = 13952 + 21199x + 19701x^2,$$
$$g_2(x) = 25505 + 24549x + 15346x^2, \quad \text{and}$$
$$K_{3,4} = 9211.$$

10.4 Construct the incidence matrix of a Fiat-Naor 2-KDP for a set of $v = 5$ users.

10.5 (a) Suppose that $(X, \mathcal{A})$ is a (t, w)-CFF(v, n). For every $A \in \mathcal{A}$, define $A^c = X \backslash A$. Prove that $(X, \{A^c : A \in \mathcal{A}\})$ is a (w, t)-CFF(v, n).

(b) By trial and error, construct a $(1, 2)$-CFF$(12, 9)$.

HINT Every block in this CFF will contain exactly four points and every point occurs in exactly three blocks.

(c) Construct a $(2, 1)$-CFF$(12, 9)$.

10.6 In Section 10.4.2, we described a probabilistic approach which provided an existence result for (t, w)-CFF(v, n). It is in fact possible to modify the approach to yield a practical algorithm to construct certain (t, w)-CFF(v, n) with high probability. We explore this approach in this exercise.

Throughout this exercise, let X be the random variable defined in Section 10.4.2. Also, let M be a randomly constructed v by n matrix, having entries in $\{0, 1\}$, where each entry is defined to be 1 with probability ρ. Finally, define $p_{t,w}$ as in Section 10.4.2.

(a) Prove that the probability that M is not the incidence matrix of a (t, w)-CFF(v, n) is at most $\mathbf{Exp}[X]$.

(b) Prove that $\mathbf{Exp}[X] < 2^{-s}$ if

$$v > \frac{(t+w)\log_2 n + s}{-\log_2(p_{t,w})}.$$

(c) Suppose $t = 2$, $w = 1$ and $n = 100$. How big should s and v be in order that there is at least a 99% chance that a randomly constructed M is a $(2,1)$-CFF$(v, 100)$?

10.7 We describe a secret-key based three-party session key distribution scheme in Protocol 10.7. In this scheme, K_{Alice} is a secret key shared by Alice and the TA, and K_{Bob} is a secret key shared by Bob and the TA.

Protocol 10.7: SESSION KEY DISTRIBUTION SCHEME

1. Alice chooses a random number, r_A. Alice sends $ID(Alice)$, $ID(Bob)$ and

$$y_A = e_{K_{Alice}}(ID(Alice) \parallel ID(Bob) \parallel r_A)$$

to Bob.

2. Bob chooses a random number, r_B. Bob sends $ID(Alice)$, $ID(Bob)$, y_A and

$$y_B = e_{K_{Bob}}(ID(Alice) \parallel ID(Bob) \parallel r_B)$$

to the TA.

3. The TA decrypts y_A using the key K_{Alice} and it decrypts y_B using the key K_{Bob}, thus obtaining r_A and r_B. It chooses a random session key, K, and computes

$$z_A = e_{K_{Alice}}(r_A \parallel K)$$

and

$$z_B = e_{K_{Bob}}(r_B \parallel K).$$

z_A is sent to Alice and z_B is sent to Bob.

4. Alice decrypts z_A using the key K_{Alice}, obtaining K; and Bob decrypts z_B using the key K_{Bob}, obtaining K.

(a) State all consistency checks that should be performed by Alice, Bob and the TA during a session of the protocol.

(b) The protocol is vulnerable to an attack if the TA does not perform the necessary consistency checks you described in part (a). Suppose that Oscar replaces $ID(Bob)$ by $ID(Oscar)$, and he also replaces y_B by

$$y_O = e_{K_{Oscar}}(ID(Alice) \parallel ID(Bob) \parallel r_B')$$

in step 2, where r_B' is random. Describe the possible consequences of this attack if the TA does not carry out its consistency checks properly.

(c) In this protocol, encryption is being done to ensure both confidentiality and data integrity. Indicate which pieces of data require encryption for the purposes of confidentiality, and which ones only need to be authenticated. Rewrite the protocol, using MACs for authentication in the appropriate places.

10.8 We describe a public-key protocol, in which Alice chooses a random session key and transmits it to Bob in encrypted form, in Protocol 10.8. (A protocol of this type is called a *key transport scheme*.) In this scheme, K_{Bob} is Bob's public encryption key. Alice and Bob also have private signing keys and public verification keys for a signature scheme.

Protocol 10.8: PUBLIC-KEY KEY TRANSPORT SCHEME

1. Bob chooses a random challenge, r_1. He sends r_1 and **Cert**(Bob) to Alice.
2. Alice verifies Bob's public encryption key, K_{Bob}, on the certificate **Cert**(Bob). Then Alice chooses a random session key, K, and computes

$$z = e_{K_{Bob}}(K).$$

   She also computes

$$y_1 = sig_{Alice}(r_1 \| z \| ID(Bob))$$

   and sends **Cert**($Alice$), z and y_1 to Bob.
3. Bob verifies Alice's public verification key, ver_{Alice}, on the certificate **Cert**($Alice$). Then he verifies that

$$ver_{Alice}(r_1 \| z \| ID(Bob), y_1) = true.$$

   If this is not the case, then Bob "rejects." Otherwise, Bob decrypts z obtaining the session key K, and "accepts." Finally, Bob computes

$$y_2 = sig_{Bob}(z \| ID(Alice))$$

   and sends y_2 to Alice.
4. Alice verifies Bob's public verification key, ver_{Bob}, on the certificate **Cert**(Bob). Then she checks that

$$ver_{Bob}(z \| ID(Alice), y_2) = true.$$

   If so, then Alice "accepts"; otherwise, Alice "rejects."

(a) Determine if the above protocol is a secure mutual identification scheme. If it is, then analyze an active adversary's probability of successfully deceiving Alice or Bob, given suitable assumptions on the security of the signature scheme. If it is not, then demonstrate an attack on the scheme.

(b) What type of key authentication or confirmation is provided by this protocol (from Alice to Bob, and from Bob to Alice)? Justify your answer briefly.

11

Key Agreement Schemes

11.1 Introduction

This chapter is a companion to the previous chapter, where we discussed key pre-distribution schemes and session key distribution schemes. Both of these kinds of key distribution require a trusted authority (TA) to select keys and distribute them to network users. In this chapter, we turn our attention to key agreement schemes (KAS), in which two users can establish a new session key via an interactive protocol which does not require the active participation of a TA. Note that we are mainly discussing key agreement schemes in the public-key setting.

Throughout this chapter, we will use the same terminology and notation as we did in Chapter 10. The reader should review the introductory material pertaining to key agreement that was presented in Section 10.1 before proceeding further.

11.2 Diffie-Hellman Key Agreement

The first and best-known key agreement scheme is the *Diffie-Hellman KAS*. This was actually the very first published realization of public key cryptography, which occurred in 1976. The *Diffie-Hellman KAS* is presented as Protocol 11.1.

Protocol 11.1 is very similar to *Diffie-Hellman Key Predistribution* (Protocol 10.1), which was described in the previous chapter. The difference is that the exponents a_U and a_V of users U and V (respectively) are chosen anew each time the scheme is run, instead of being fixed. Also, there are no long-lived keys in this scheme.

At the end of a session of the *Diffie-Hellman KAS*, U and V have computed the same key,

$$K = \alpha^{a_U a_V} = \mathbf{CDH}(\alpha, b_U, b_V).$$

Here, as usual, **CDH** refers to the **Computational Diffie-Hellman** problem. As-

---

Protocol 11.1: DIFFIE-HELLMAN KAS

The public domain parameters consist of a group $(G, \cdot)$ and an element $\alpha \in G$ having order n.

1. U chooses a_U at random, where $0 \le a_U \le n - 1$. Then she computes

$$b_U = \alpha^{a_U}$$

  and sends b_U to V.

2. V chooses a_V at random, where $0 \le a_V \le n - 1$. Then he computes

$$b_V = \alpha^{a_V}$$

  and sends b_V to U.

3. U computes
$$K = (b_V)^{a_U}$$

  and V computes
$$K = (b_U)^{a_V}.$$

---

suming that the **Decision Diffie-Hellman** problem is intractable, a passive adversary cannot compute any information about K.

It is well-known that the *Diffie-Hellman KAS* has a serious weakness in the presence of an active adversary. The *Diffie-Hellman KAS* is supposed to work like this:

U V

$$\overset{\alpha^{a_U}}{\xrightarrow{\hspace{6cm}}}$$

$$\overset{\alpha^{a_V}}{\xleftarrow{\hspace{6cm}}}$$

Unfortunately, the scheme is vulnerable to an active adversary who uses an *intruder-in-the-middle* attack.[1] The intruder-in-the-middle attack on the *Diffie-Hellman KAS* works in the following way. W will intercept messages between U and V and substitute his own messages, as indicated in Figure 11.1.

At the end of the session, U has actually established the secret key $\alpha^{a_U a_V'}$ with W, and V has established the secret key $\alpha^{a_U' a_V}$ with W. When U tries to encrypt

---

[1]There is an episode of the popular 1950s television comedy *The Lucy Show* in which Vivian Vance is having dinner in a restaurant with a date, and Lucille Ball is hiding under the table. Vivian and her date decide to hold hands under the table. Lucy, trying to avoid detection, holds hands with each of them and they think they are holding hands with each other.

FIGURE 11.1
Intruder-in-the-middle attack

$$U \qquad\qquad W \qquad\qquad V$$

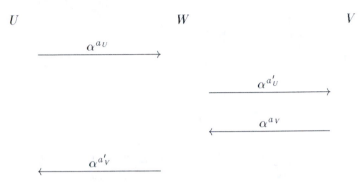

a message to send to V, W will be able to decrypt it but V will not be able to do so. (A similar situation holds if V sends a message to U.)

Clearly, it is essential for U and V to make sure that they are exchanging messages (and keys!) with each other and not with W. Before exchanging keys, U and V might carry out a separate protocol to establish each other's identity, for example by using a secure mutual identification scheme. But this offers no protection against an intruder-in-the-middle attack if W simply remains inactive until after U and V have proved their identities to each other. A more promising approach is to design a key agreement scheme that authenticates the participants' identities at the same time as the key is being established. A KAS of this type will be called an *authenticated* key agreement scheme.

Informally, we define an *authenticated key agreement scheme* to be a key agreement scheme which satisfies the following properties:

mutual identification

> The scheme is a secure mutual identification scheme, as defined in Section 9.3.2: no honest participant in a session of the scheme will "accept" after any flow in which an adversary is active.

key agreement

> If there is no active adversary, then both participants will compute the same new session key K. Moreover, no information about the value of K can be computed by the (passive) adversary.

11.2.1 The Station-to-station Key Agreement Scheme

In this section, we describe an authenticated key agreement scheme which is a modification of the *Diffie-Hellman KAS*. The scheme makes use of certificates which, as usual, are signed by a *TA*. Each user *U* will have a signature scheme with a verification algorithm ver_U and a signing algorithm sig_U. The *TA* also has a signature scheme with a public verification algorithm ver_{TA}. Each user *U* has a certificate

$$\textbf{Cert}(U) = (ID(U), ver_U, sig_{TA}(ID(U), ver_U)),$$

where $ID(U)$ is certain identification information for *U*. (These certificates are the same as the ones described in Section 9.3.1.)

The authenticated key agreement scheme known as the *Station-to-station KAS* (or *STS*, for short) is due to Diffie, Van Oorschot and Wiener. Protocol 11.2 is a slight simplification; it can be used in such a way that it conforms to the ISO 9798-3 schemes.

The basic idea of Protocol 11.2 is to combine the *Diffie-Hellman KAS* with a secure mutual identification scheme, where the exponentiated values b_U and b_V function as the random challenges in the identification scheme. If we follow this recipe, using Protocol 9.14 as the underlying identification scheme, then the result is Protocol 11.2. Roughly speaking, signing the random challenges provides mutual authentication. Furthermore, these challenges, being computed according to the *Diffie-Hellman KAS*, allow both *U* and *V* to compute the same key, $K = \textbf{CDH}(\alpha, b_U, b_V)$.

11.2.2 Security of STS

In this section, we discuss the security properties of the simplified *STS* scheme. For future reference, the information exchanged in a session of the scheme (excluding certificates) is illustrated as follows:

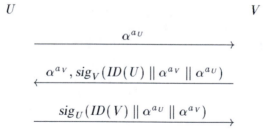

First, let's see how the use of signatures protects against the intruder-in-the-middle attack mentioned earlier. Suppose, as before, that *W* intercepts α^{a_U} and replaces it with $\alpha^{a'_U}$. *W* then receives α^{a_V} and

$$sig_V(ID(U) \,\|\, \alpha^{a_V} \,\|\, \alpha^{a'_U})$$

from *V*. He would like to replace α^{a_V} with $\alpha^{a'_V}$, as before. However, this means that he must also replace the signature by

$$sig_V(ID(U) \,\|\, \alpha^{a'_V} \,\|\, \alpha^{a_U}).$$

Protocol 11.2: SIMPLIFIED STATION-TO-STATION KAS

The public domain parameters consist of a group $(G, \cdot)$ and an element $\alpha \in G$ having order n.

1. U chooses a random number $a_U, 0 \leq a_U \leq n - 1$. Then she computes

$$b_U = \alpha^{a_U}$$

  and she sends **Cert**(U) and b_U to V.

2. V chooses a random number $a_V, 0 \leq a_V \leq n - 1$. Then he computes

$$b_V = \alpha^{a_V}$$

$$K = (b_U)^{a_V}, \quad \text{and}$$

$$y_V = sig_V(ID(U) \parallel b_V \parallel b_U).$$

  Then V sends **Cert**(V), b_V and y_V to U.

3. U verifies y_V using ver_V. If the signature y_V is not valid, then she "rejects" and quits. Otherwise, she "accepts," she computes

$$K = (b_V)^{a_U}, \quad \text{and}$$

$$y_U = sig_U(ID(V) \parallel b_U \parallel b_V),$$

  and she sends y_U to V.

4. V verifies y_U using ver_U. If the signature y_U is not valid, then he "rejects"; otherwise, he "accepts."

Unfortunately for W, he cannot compute V's signature on the string $ID(U) \parallel \alpha^{a_V'} \parallel \alpha^{a_U}$ because he doesn't know V's signing algorithm sig_V. Similarly, W is unable to replace $sig_U(ID(V) \parallel \alpha^{a_U} \parallel \alpha^{a_V'})$ by $sig_U(ID(V) \parallel \alpha^{a_U'} \parallel \alpha^{a_V})$ because he does not know U's signing algorithm.

This situation is illustrated in Figure 11.2, in which the question marks indicate signatures that the adversary is unable to compute. It is the judicious use of signatures that provides for mutual identification of U and V. This in turn thwarts the intruder-in-the-middle attack.

Of course, we want to be convinced that the scheme is secure against all possible attacks, not just one particular attack. However, from the way that the scheme is designed, we can use previous results we have proven to provide a general proof of security of *STS*. In doing so, we need to say more precisely what assurances are provided regarding knowledge of the session key.

FIGURE 11.2
Thwarted intruder-in-the-middle attack on STS

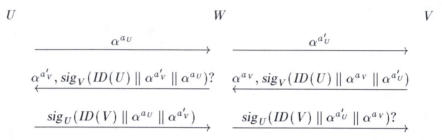

First, we claim that *STS* is a secure mutual identification scheme. This fact can be proven using the methods described in Chapter 9. So, if an adversary is active, he will be detected by the honest participants in the session.

On the other hand, if the adversary is passive, then the session will terminate with both parties "accepting" (provided they behave honestly). That is, U and V successfully identify themselves to each other, and they both compute the key K as in the *Diffie-Hellman KAS*. The adversary cannot compute any information about the key K, assuming the intractability of the **Decision Diffie-Hellman** problem. In summary, an active adversary will be detected, and an inactive adversary is thwarted due to the intractability of the **Decision Diffie-Hellman** problem (exactly as he was in the *Diffie-Hellman KAS*).

Now, using the properties discussed above, let us see what we can infer about the *STS* scheme if U or V "accepts." First, suppose that U "accepts." Because *STS* is a secure mutual identification scheme, U can be confident that she has really been communicating with V (her "intended peer") and that the adversary was inactive before the last flow. Assuming that V is honest and that he has executed the scheme according to its specifications, U can be confident that V can compute the value of K, and that no one other than V can compute the value of K.

Let us consider in a bit more detail why U should believe that V can compute K. The reason for this belief is that U has received V's signature on the values α^{a_U} and α^{a_V}, so it is reasonable for U to infer that V knows these two values. Now, assuming that V executed the scheme according to its specifications, U can infer that V knows the value of a_V. V is able to compute the value of K, provided that he knows the values of α^{a_U} and a_V. Of course, there is no guarantee to U that V has actually computed K at the time when V "accepts."

The analysis from the point of view of V is very similar. If V "accepts," then he is confident that he has really been communicating with U (his intended peer) and that the key K can be computed by U and by no one else. However, there is an asymmetry in the assurances provided to U and to V. When V "accepts," he can be confident that U has already "accepted" (provided that U is honest). However, when U "accepts," she has no way of knowing if V will subsequently "accept," because she does not know if V will actually receive the message being sent to him in the last flow of the session (for example, an adversary might intercept or corrupt this last flow, causing V to "reject"). A similar situation occurred in the setting of mutual identification schemes and was discusseed in Section 9.2.2.

It is useful to define a few variations of properties relating to the users' knowledge of the computed session key, K. Suppose that V is the intended peer of U in a key agreement scheme. Here are three "levels" of assurance regarding key agreement that could be provided to U (or to V):

implicit key authentication

We say that a key agreement scheme provides implicit key authentication to U if U is assured that no one other than V can compute K (in particular, the adversary should not be able to compute K).

implicit key confirmation

We say that a key agreement scheme provides implicit key confirmation if U is assured that V can compute K (assuming that V executed the scheme according to its specifications), and no one other than V can compute K.

explicit key confirmation

We say that a key agreement scheme provides explicit key confirmation if U is assured that V has computed K, and no one other than V can compute K.

We have presented two variations on the idea of key confirmation. The notion of key confirmation discussed in Chapter 10 (in connection with session key distribution schemes) was the "explicit" version. In general, explicit key confirmation is provided by using the newly constructed session key to encrypt a known value (or random challenge). *Kerberos* and *Needham-Schroeder* both attempt to provide explicit key confirmation by exactly this method.

The *STS* scheme does not make immediate use of the new session key, so we don't have explicit key confirmation. However, because both parties sign the exchanged exponentials, we achieve the slightly weaker property of implicit key confirmation. (Furthermore, as we mentioned in Chapter 10, it is always possible to augment any key agreement or key distribution scheme so it achieves explicit key confirmation, if so desired.)

Finally, note that the *Bellare-Rogaway* session key distribution scheme provides implicit key authentication; there is no attempt in that scheme to provide either party with any assurance that their intended peer has received (or can compute) the session key.

Summarizing the discussion in this section, we have established the following theorem.

THEOREM 11.1 *The Station-to-station key agreement scheme is an authenticated key agreement scheme that provides implicit key confirmation to both parties, assuming that the* **Decision Diffie-Hellman** *problem is intractable.*

11.2.3 Known Session Key Attacks

The security result proven in the last section basically considers one session of *STS* in isolation. However, in a realistic scenario involving a network with many users, there could be many sessions of *STS* taking place, involving many different users. In order to make a convincing argument that *STS* is secure, we need to consider the possible influence that different sessions might have on each other.

Therefore, we investigate security under a known session key attack (this attack model was defined in Secton 10.1). In this scenario, the adversary, say $Oscar$, observes several sessions of a key agreement scheme, say $\mathcal{S}_1, \mathcal{S}_2, \ldots, \mathcal{S}_t$. These sessions may be sessions involving other network users, or they may include Oscar himself as one of the participants.

As part of the attack model, Oscar is allowed to request that the session keys for the sessions $\mathcal{S}_1, \mathcal{S}_2, \ldots, \mathcal{S}_t$ be revealed to him. Oscar's goal is to determine a session key (or information about a session key) for some other *target session*, say $\mathcal{S}$, in which Oscar is not a participant. Furthermore, we do not require that the session $\mathcal{S}$ take place after all the sessions $\mathcal{S}_1, \mathcal{S}_2, \ldots, \mathcal{S}_t$ have completed. In particular, we allow parallel session attacks (similar to those considered in the context of identification schemes).

In this section, we study the security of the *STS* key agreement scheme against known session key attacks. First, suppose Oscar observes a session $\mathcal{S}$ between two users U and V. The information transmitted in this session (excluding signatures and certificates) consists of the two values $b_{\mathcal{S},U}$ and $b_{\mathcal{S},V}$. (We are including the name of the session, $\mathcal{S}$, as a subscript to make it clear that these values are associated with a particular session.) Oscar hopes ultimately to be able to determine some information about the value of the key $K_\mathcal{S}$ computed by U and V in the session $\mathcal{S}$. Observe that computing the key $K_\mathcal{S}$ is the same as solving the **Computational Diffie-Hellman** problem for the instance $(b_{\mathcal{S},U}, b_{\mathcal{S},V})$. We denote this relation by the notation $K_\mathcal{S} = \mathbf{CDH}(b_{\mathcal{S},U}, b_{\mathcal{S},V})$.

Once Oscar has the pair $(b_{\mathcal{S},U}, b_{\mathcal{S},V})$, he is free to engage in various other sessions in an attempt to find out some information about $K_\mathcal{S}$. However, we only allow Oscar to request a key for a session $\mathcal{S}'$ from a user in the session $\mathcal{S}'$ who "accepts." Therefore Oscar cannot be active in a session and then request a session key from a user who does not "accept," because *STS* is a secure identification scheme.

However, Oscar can take part in a session $\mathcal{S}'$ as one of the participants, possibly not following the rules of *STS*. In particular, Oscar might transmit a value $b_{\mathcal{S}',Oscar}$ to his peer in the session $\mathcal{S}'$, without knowing the value $a_{\mathcal{S}',Oscar}$ such that $b_{\mathcal{S}',Oscar} = \alpha^{a_{\mathcal{S}',Oscar}}$. In accordance with the known session key attack model, Oscar would be allowed to request that the value of key $K_{\mathcal{S}'}$ be revealed to him. (If Oscar followed the rules of *STS*, then he would be able to compute $K_{S'}$ himself. However, we are considering a situation where Oscar cannot compute $K_{S'}$, but where we allow Oscar to be informed of its value, anyway.)

Suppose that Oscar takes part in such a session S' with a peer W. Then Oscar chooses a value $b_{S',Oscar}$ in any way that he wishes. W chooses a random value $b_{S',W}$ in the subgroup generated by α, by first choosing $a_{S',W}$ uniformly at random and then computing $b_{S',W} = \alpha^{a_{S',W}}$. After the session completes, Oscar requests and is given the value $K_{S'} = \mathbf{CDH}(b_{S',Oscar}, b_{S',W}) = (b_{S',Oscar})^{a_{S',W}}$. Oscar can then record the outcome of the session S' and the value of the session key $K_{S'}$ in the form of a triple of values

$$(b_{S',Oscar}, b_{S',W}, \mathbf{CDH}(b_{S',Oscar}, b_{S',W})).$$

After a number of such sessions, Oscar accumulates a list of triples (i.e., a *trancsript*) $\mathcal{T}$, where each triple $T \in \mathcal{T}$ has the form given above. We assume that Oscar has some polynomial-time algorithm A such that $A(\mathcal{T}, (b_{S,U}, b_{S,V}))$ computes some partial information about the key K_S when $\mathcal{T}$ is constructed by the method described above.

We will argue that the hypothesized algorithm A cannot exist, assuming the intractability of the **DDH** problem. The way that we establish the non-existence of A is to show that it is possible to replace the transcript $\mathcal{T}$ by a simulated transcript $\mathcal{T}_{sim}$ which can be created by Oscar without taking part in any sessions and without requesting that any session keys be revealed to him.

We now show how Oscar can efficiently construct a simluated transcript $\mathcal{T}_{sim}$. Let's consider a typical triple on $\mathcal{T}$, which has the form

$$T = (b_1, b_2, b_3 = \mathbf{CDH}(b_1, b_2)).$$

As mentioned above, b_1 is chosen by Oscar, using whatever method he desires, and b_2 is chosen randomly by Oscar's peer. Then $\mathbf{CDH}(b_1, b_2)$ is revealed to Oscar. Consider the following method of constructing a simulated triple, T_{sim}:

1. Oscar chooses b_1 as before,
2. Oscar chooses a random value a_2 and computes $b_2 = \alpha^{a_2}$,
3. Oscar computes $b_3 = (b_1)^{a_2}$ (observe that $b_3 = \mathbf{CDH}(b_1, b_2)$), and
4. Oscar defines $T_{sim} = (b_1, b_2, b_3)$.

Basically, the simulation replaces a random choice of b_2 made by Oscar's peer with a random choice of b_2 made by Oscar. However, when Oscar chooses b_2 as described above, he can compute the value of b_3 himself.

We claim that a triple T is indistinguishable from a simulated triple T_{sim}. More precisely, it holds that

$$\Pr[T = (b_1, b_2, b_3)] = \Pr[T_{sim} = (b_1, b_2, b_3)]$$

for all triples of the form $(b_1, b_2, b_3 = \mathbf{CDH}(b_1, b_2))$. In fact, this is almost trivial to confirm, because b_1 is chosen exactly the same way in both T and T_{sim}, b_2 is chosen uniformly at random in both T and T_{sim}, and $b_3 = \mathbf{CDH}(b_1, b_2)$ in both T and T_{sim}.

This simulation of triples can be extended to simulate transcripts. The simulated transcript $\mathcal{T}_{sim}$ is built up triple by triple, in the same way as $\mathcal{T}$, except that each triple $T \in \mathcal{T}$ is replaced by a simulated triple T_{sim}. The resulting simulated transcripts are indistinguishable from real transcripts.

Because of this indistinguishability property, it follows immediately that A behaves exactly the same when given a transcript $\mathcal{T}$ as it does when it is given a simulated transcript $\mathcal{T}_{sim}$. That is, the outputs $A(\mathcal{T}_{sim}, (b_{\mathcal{S},U}, b_{\mathcal{S},V}))$ have exactly the same probability distribution as outputs $A(\mathcal{T}, (b_{\mathcal{S},U}, b_{\mathcal{S},V}))$. This means that, whatever Oscar can do using a known session key attack, he can also do using a completely passive attack in which no sessions (other than $\mathcal{S}$) take place. But such an attack is not possible, given that **DDH** is intractable. This contradiction completes our proof, and therfore we have the following theorem.

THEOREM 11.2   *The Station-to-station key agreement scheme is an authenticated key agreement scheme that is secure against known session key attacks and which provides implicit key confirmation to both parties, assuming that the **Decision Diffie-Hellman** problem is intractable.*

11.3   MTI Key Agreement Schemes

Matsumoto, Takashima and Imai have constructed several interesting key agreement schemes by modifying the *Diffie-Hellman KAS*. These schemes, which we call *MTI* schemes, do not require that U and V compute any signatures. They are termed *two-flow key agreement schemes*, because there are only two separate transmissions of information performed in each session of the scheme (one from U to V and one from V to U). In contrast, the *STS KAS* is a three-pass scheme.

We present one of the *MTI* key agreement schemes, namely, the *MTI/A0 KAS*, as Protocol 11.3.

We present a toy example to illustrate the workings of this scheme.

Example 11.1   Suppose $p = 27803$, $n = p - 1$ and $\alpha = 5$. The public domain parameters for the scheme consist of the group $(\mathbb{Z}_p^*, \cdot)$ and α. Here p is prime and α is a generator of $(\mathbb{Z}_p^*, \cdot)$, so the order of α is equal to n.

Assume U chooses secret exponent $a_U = 21131$; then she will compute

$$b_U = 5^{21131} \bmod 27803 = 21420,$$

which is placed on her certificate. As well, assume V chooses secret exponent $a_V = 17555$. Then he will compute

$$b_V = 5^{17555} \bmod 27803 = 17100,$$

Protocol 11.3: MTI/A0 KAS

The public domain parameters consist of a group $(G, \cdot)$ and an element $\alpha \in G$ having order n.

Each user T has a secret exponent a_T, where $0 \leq a_T \leq n - 1$, and a corresponding public value

$$b_T = \alpha^{a_T}.$$

The value b_T is included in T's certificate and is signed by the TA.

1. U chooses r_U at random, $0 \leq r_U \leq n - 1$, and computes

$$s_U = \alpha^{r_U}.$$

   Then U sends **Cert**(U) and s_U to V.

2. V chooses r_V at random, $0 \leq r_V \leq n - 1$, and computes

$$s_V = \alpha^{r_V}.$$

   Then V sends **Cert**(V) and s_V to U.
   Finally, V computes the session key

$$K = s_U{}^{a_V} b_U{}^{r_V},$$

   where he obtains the value b_U from **Cert**(U).

3. U computes the session key

$$K = s_V{}^{a_U} b_V{}^{r_U},$$

   where she obtains the value b_V from **Cert**(V).

At the end of the session, U and V have both computed the same session key

$$K = \alpha^{r_U a_V + r_V a_U}.$$

---

which is placed on his certificate.

Now suppose that U chooses $r_U = 169$; then she will send the value

$$s_U = 5^{169} \bmod 27803 = 6268$$

to V. Suppose that V chooses $r_V = 23456$, then he will send the value

$$s_V = 5^{23456} \bmod 27803 = 26759$$

to U.

Now U can compute the key

$$K_{U,V} = s_V{}^{a_U} b_V{}^{r_U} \bmod p$$

$$= 26759^{21131} 17100^{169} \bmod 27803$$

$$= 21600,$$

and V can compute the (same) key

$$K_{U,V} = s_U{}^{a_V} b_U{}^{r_V} \bmod p$$

$$= 6268^{17555} 21420^{23456} \bmod 27803$$

$$= 21600.$$

☐

For future reference, the information transmitted during a session of the scheme is depicted as follows:

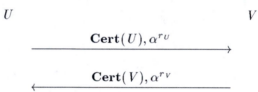

Let's now examine the security of the scheme. It is not too difficult to show that the security of the *MTI/A0* key agreement scheme against a passive adversary is exactly the same as in the *Diffie-Hellman* key agreement scheme — see the Exercises. As with many schemes, proving security in the presence of an active adversary is problematic. We will not attempt to prove anything in this regard, and we limit ourselves to some informal arguments.

Here is one threat we might consider: Without the use of signatures during the scheme, it might appear that there is no protection against an intruder-in-the-middle attack. Indeed, it is possible that W might alter the values that U and V send to each other. In Figure 11.3, we depict one typical scenario that might arise, which is analogous to the original intruder-in-the-middle attack on the *Diffie-Hellman KAS*.

In this situation, U and V will compute different keys: U will compute

$$K = \alpha^{r_U a_V + r'_V a_U},$$

while V will compute

$$K = \alpha^{r'_U a_V + r_V a_U}.$$

However, neither of the computations of keys by U or V can be carried out by W, since they require knowledge of the secret exponents a_U and a_V, respectively.

FIGURE 11.3
Unsuccessful intruder-in-the-middle attack on MTI/A0

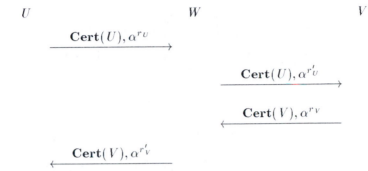

So even though U and V have computed different keys (which will of course be useless to them), neither of these keys can be computed by W, nor can he obtain any information about these keys (assuming the intractability of the **DDH** problem).

If this were the only possible attack on the scheme, then we would be able to say that the scheme provides implicit key authentication. This is because, even in the presence of this attack, both U and V are assured that the other is the only user in the network that could compute the key that they have computed. However, we will show in the next section that there are additional attacks which can be carried out by an adversary in the known session key attack model.

11.3.1 Known Session Key Attacks on MTI/A0

We begin by presenting a parallel session known session key attack on *MTI/A0*. This attack is a known session key attack utilizing a parallel session; hence the awkward terminology. The adversary, W, is an active participant in two sessions: W pretends to be V in a session $\mathcal{S}$ with U; and W pretends to be U in a parallel session $\mathcal{S}'$ with V. The actions taken by W are illustrated in Figure 11.4.

The flows in the two sessions are labeled in the order in which they occur. (1) and (2) represent the initial flows in the sessions $\mathcal{S}'$ and $\mathcal{S}$, respectively. Then the information in flow (1) is copied to flow (3), and the information in flow (2) is copied to flow (4) by W. Since the two sessions are being executed in parallel, we have a parallel session attack.

After the two sessions have completed, W requests the key K for session $\mathcal{S}'$, which he is allowed to do in a known session key attack. Of course, K is also the key for session $\mathcal{S}$, so W achieves his goal of computing the key for a session in which he is an active adversary and for which he has not requested the session

FIGURE 11.4
Known session key attack on MTI/A0

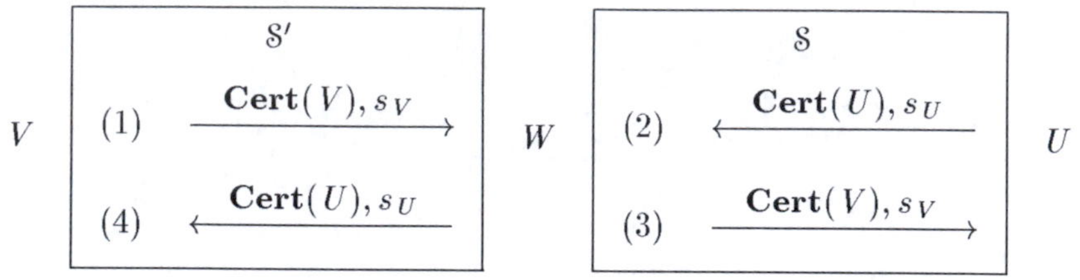

key. This represents a successful attack in the known session key attack model.

The parallel session attack can be carried out because the key is a symmetric function of the inputs provided by the two parties:

$$K((r_U, a_U), (r_V, a_V)) = K((r_V, a_V), (r_U, a_U)).$$

To eliminate the attack, we should destroy this symmetry property. This could be done, for example, by using a hash function h as a *key derivation function*. Suppose that the actual session key K was defined to be

$$K = h(\alpha^{r_U a_V} \| \alpha^{r_V a_U})$$

U (the initiator of the session) would compute

$$K = h(b_V^{r_U} \| s_V^{a_U})$$

while V (the responder of the session) would compute

$$K = h(s_U^{a_V} \| b_U^{r_V}).$$

With this modified method of constructing a session key, the previous attack no longer works. This is because the two sessions $\mathcal{S}$ and $\mathcal{S}'$ now have different keys: the key for session $\mathcal{S}$ is

$$K_{\mathcal{S}} = h(\alpha^{r_U a_V} \| \alpha^{r_V a_U}),$$

while the key for session $\mathcal{S}'$ is

$$K_{\mathcal{S}'} = h(\alpha^{r_V a_U} \| \alpha^{r_U a_V}).$$

If h is a "good" hash function (e.g., if h is a random oracle), then there will be no way for W to compute $K_{\mathcal{S}}$ given $K_{\mathcal{S}'}$, or to compute $K_{\mathcal{S}'}$ given $K_{\mathcal{S}}$.

There is another known session key attack on *MTI/A0* which is called the *Burmester triangle attack*. This attack is depicted in Figure 11.5.

FIGURE 11.5
Burmester triangle attack on MTI/A0

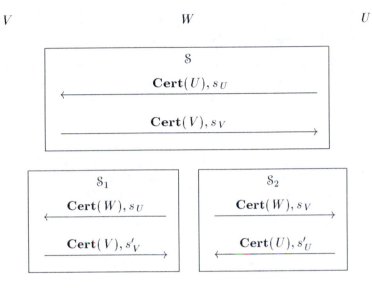

We describe the triangle attack in more detail now. First, W observes a session $\mathcal{S}$ between U and V. Then W participates in two additional sessions $\mathcal{S}_1$ and $\mathcal{S}_2$ with V and U, respectively. In these two sessions, W transmits values s_U and s_V that are copied from $\mathcal{S}$. (Of course, W does not know the exponents r_U and r_V corresponding to s_U and s_V, respectively.) Then, after the sessions $\mathcal{S}_1$ and $\mathcal{S}_2$ have concluded, W requests the keys for these two sessions, which is permitted in a known session key attack.

The session keys K, K_1 and K_2 for the sessions $\mathcal{S}$, $\mathcal{S}_1$ and $\mathcal{S}_2$ (respectively) are as follows:

$$K = \alpha^{r_U a_V + r_V a_U}$$
$$K_1 = \alpha^{r_U a_V + r'_V a_W}$$
$$K_2 = \alpha^{r'_U a_W + r_V a_U}.$$

Given K_1 and K_2, W is able to compute K as follows:

$$K = \frac{K_1 K_2}{(s'_V s'_U)^{a_W}}.$$

Therefore this is a successful known session key attack.

The triangle attack can also be defeated through the use of the key derivation

function described above. It is conjectured that this modified version of *MTI/A0*
is secure against known session key attacks.

11.4   Key Agreement Using Self-certifying Keys

In this section, we describe a method of key agreement, due to Girault, that does
not require certificates. In this scheme, the value of a public key and the identity
of its owner implicitly authenticate each other. We will see a bit later how this is
accomplished.

The *Girault KAS* combines features of both *RSA* and discrete logarithm based
schemes. Suppose $n = pq$, where $p = 2p_1+1, q = 2q_1+1$, and p, q, p_1 and q_1 are
all large primes. Then the multiplicative group $\mathbb{Z}_n^*$ is isomorphic to $\mathbb{Z}_p^* \times \mathbb{Z}_q^*$.
The maximum (multiplicative) order of any element in $\mathbb{Z}_n^*$ is therefore the least
common multiple of $p - 1$ and $q - 1$, or $2p_1q_1$. Let α be an element in $\mathbb{Z}_n^*$ of
order $2p_1q_1$. Then the cyclic subgroup of $\mathbb{Z}_n^*$ generated by α is a suitable setting
for the **Discrete Logarithm** problem provided that p and q are large enough.

In the *Girault KAS*, the factorization of n is known only to the *TA*. The values
n and α are public domain parameters, but p, q, p_1 and q_1 are all secret. The
TA chooses a public RSA encryption exponent, which we will denote by e. The
corresponding decryption exponent, d, is secret (as usual, $d = e^{-1} \bmod \phi(n)$).

Each user U has an identification string $ID(U)$, as in many previous schemes.
A user U obtains a *self-certifying public key*, p_U, from the *TA* as shown in Proto-
col 11.4. Observe that U needs the help of the *TA* to produce the public key p_U.
Note also that

$$b_U = p_U{}^e + ID(U) \bmod n$$

can be computed from p_U and $ID(U)$ using publicly available information.

The *Girault Key Agreement Scheme* is presented as Protocol 11.5. The information transmitted during a session of Protocol 11.5 is depicted as follows:

$$U \quad \frac{ID(U), p_U, \alpha^{r_U} \bmod n}{ID(V), p_V, \alpha^{r_V} \bmod n} \quad V$$

At the end of a session, U and V each have computed the key

$$K = \alpha^{r_U a_V + r_V a_U} \bmod n = b_V{}^{r_U} b_U{}^{r_V} \bmod n.$$

Here is an example of key agreement using the *Girault KAS*.

Example 11.2   Suppose

$$p = 839 \quad \text{and} \quad q = 863;$$

Protocol 11.4: GIRAULT PUBLIC KEY GENERATION

Note: n is public and d is a secret value known only to the TA.

1. U chooses a secret exponent a_U, and computes

$$b_U = \alpha^{a_U} \bmod n.$$

 U gives a_U and b_U to the TA.

2. The TA computes

$$p_U = (b_U - ID(U))^d \bmod n.$$

 The TA gives p_U to U.

---

then
$$n = 724057 \quad \text{and} \quad \phi(n) = 722356.$$

The element $\alpha = 5$ has order $2p_1q_1 = \phi(n)/2$ in $\mathbb{Z}_n^*$. Suppose the TA chooses $d = 125777$ as the RSA decryption exponent; then

$$e = 125777^{-1} \bmod 722356 = 84453.$$

Suppose U has
$$ID(U) = 500021 \quad \text{and} \quad a_U = 111899;$$

then
$$b_U = 488889 \quad \text{and} \quad p_U = 650704.$$

Suppose also that V has
$$ID(V) = 500022 \quad \text{and} \quad a_V = 123456;$$

then
$$b_V = 111692 \quad \text{and} \quad p_V = 683556.$$

Now, U and V want to compute a new session key. Suppose U chooses $r_U = 56381$, which means that $s_U = 171007$. Further, suppose V chooses $r_V = 356935$, which means that $s_V = 320688$.

Then U computes

$$K = 320688^{111899} \left(683556^{84453} + 500022\right)^{56381} \bmod 724057 = 42869$$

and V computes

$$K = 171007^{123456} \left(650704^{84453} + 500021\right)^{356935} \bmod 724057 = 42869.$$

Protocol 11.5: GIRAULT KEY AGREEMENT SCHEME

The public domain parameters consist of n, e and α.

1. U chooses r_U at random and computes

$$s_U = \alpha^{r_U} \bmod n.$$

   U sends $ID(U)$, p_U and s_U to V.

2. V chooses r_V at random and computes

$$s_V = \alpha^{r_V} \bmod n.$$

   V sends $ID(V)$, p_V and s_V to U.

3. U computes

$$K = s_V{}^{a_U} \left(p_V{}^e + ID(V)\right)^{r_U} \bmod n$$

   and V computes

$$K = s_U{}^{a_V} \left(p_U{}^e + ID(U)\right)^{r_V} \bmod n.$$

Therefore, both U and V compute the same key, $K = 42869$.

                                                                $\square$

Let's consider how the self-certifying keys guard against one specific type of attack. Since the values b_U, p_U and $ID(U)$ are not signed by the TA, there is no way for anyone else to verify their authenticity directly. Suppose this information is forged by W (i.e., it is not produced in co-operation with the TA), who wants to masquerade as U. W might star with $ID(U)$ and a fake value p'_U. She can easily compute the quantity $b'_U = (p'_U)^e + ID(U)$, but there is no way for her to compute the exponent a'_U corresponding to b'_U if the **Discrete Logarithm** problem is intractable. W can execute a session of the scheme while pretending to be U; however, without knowing the value of a'_U, the computation of the session key K cannot be performed by W.

The situation is similar if W acts as an intruder-in-the-middle. W will be able to prevent U and V from computing a common key, but W is unable to duplicate the computations of either U or V. Thus the scheme provides implicit key authentication against this attack, as did the *MTI/A0* scheme.

An attentive reader might wonder why U is required to supply the value a_U to the TA. Indeed, the TA can compute p_U directly from b_U, without knowing

FIGURE 11.6
Attack on the Girault KAS

$$U \quad \frac{ID(U), p_U, \alpha^{r_U} \bmod n}{ID(V), p_V, \alpha^{r_V} \bmod n} \quad W \quad \frac{ID(U), p'_U, \alpha^{r'_U} \bmod n}{ID(V), p_V, \alpha^{r_V} \bmod n} \quad V$$

a_U. Actually, the important thing here is that the TA should be convinced that U knows the value of a_U before the TA computes p_U for U.

We illustrate this point by showing how the scheme can be attacked if the TA indiscriminately issues public keys p_U to users without first checking that they possess the value a_U corresponding to their b_U. Suppose W chooses a fake value a'_U and computes the corresponding value

$$b'_U = \alpha^{a'_U} \bmod n.$$

Here is how he can conpute the corresponding public key

$$p'_U = (b'_U - ID(U))^d \bmod n.$$

W will compute

$$b'_W = b'_U - ID(U) + ID(W)$$

and then give b'_W and $ID(W)$ to the TA. Suppose the TA issues the public key

$$p'_W = (b'_W - ID(W))^d \bmod n$$

to W. Using the fact that

$$b'_W - ID(W) \equiv b'_U - ID(U) \pmod{n},$$

it is immediate that

$$p'_W = p'_U.$$

Now, at some later time, suppose U and V execute the scheme, and W substitutes information as shown in Figure 11.6.

Then V will compute the key

$$K' = \alpha^{r'_U a_V + r_V a'_U} \bmod n,$$

whereas U will compute the key

$$K = \alpha^{r_U a_V + r_V a_U} \bmod n.$$

W can compute K' as

$$K' = s_V{}^{a'_U} \left(p_V{}^e + ID(V)\right)^{r'_U} \bmod n.$$

Thus W and V share a key, but V thinks he is sharing a key with U. So W will be able to decrypt messages sent by V to U.

11.5   Encrypted Key Exchange

With the exception of the *Girault KAS*, all the schemes considered so far have made essential use of certificates and/or signature schemes. Key agreement can also be studied in the secret key setting, assuming that the parties involved have a prior shared secret key. (As mentioned in Chapter 10, two parties might want to use a KAS to negotiate a session key, even though they already have a shared long-lived key.)

One particularly interesting scenario is that of a *password-based key agreement scheme*, in which two users, U and V, say, have a prior shared secret called a *password*. We will denote this password by $pwd_{U,V}$. The term "password" suggests that the length of this password is short — too short for it to be used as a cryptographic key. For example, a typical password might be chosen from a set of 2^{20} or 2^{30} possible passwords, whereas a key space in a typical secret-key cryptosystem could have size 2^{128}.

A common situation where passwords might be employed is when a client C is communicating with a server S. The server might maintain a database containing the passwords of all the clients that have registered with it. In this context, we will denote C's password by pwd_C. Note that passwords are usually relatively short so that they can be memorized by human users.

To motivate the schemes we describe in this section, let's first consider a flawed approach which uses a password to encrypt a cryptographic key. Suppose that the server S chooses a random 128-bit session key K and encrypts K using the client's password pwd_C as a key, thereby constructing the encrypted key $y = e_{pwd_C}(K)$. (In this discussion, $e(\cdot)$ and $d(\cdot)$ are respectively encryption and decryption functions for a secret-key cryptosystem.) Then S transmits y to C, who will decrypt y to obtain K. Now that C and S have a session key K, they will use K to encrypt information in a subsequent session.

Unfortunately, an adversary, Oscar, can carry out a *dictionary attack*. This amounts to performing an exhaustive search over the set of all possible passwords. Suppose that it is known that pwd_C is chosen from some set $\mathcal{K}_{pwd}$. Suppose also that Oscar records y and some ciphertext z that was produced using the session key K. Denote $x = d_K(z)$ (note that Oscar is not supplied with the plaintext x in this attack). Then, for every $pwd \in \mathcal{K}_{pwd}$, Oscar can compute $K' = d_{pwd}(y)$ and $x' = d_{K'}(z)$. If $pwd = pwd_C$, then $K' = K$ and $x' = x$. Oscar can normally recognize when x' is "meaningful" plaintext; this will indicate to him that $pwd = pwd_C$.

This attack can be carried out any time Oscar desires, so it is termed an *off-line attack*. It is also efficient, provided that $|\mathcal{K}_{pwd}|$ is not too large.

Once the attack succeeds, Oscar knows C's password. Then, Oscar could impersonate C, or, alternatively, decrypt any future communications of C that use pwd_C to encrypt a session key. The attack is called a "dictionary attack" because Oscar is testing all the possible passwords exhaustively (where, conceptually, we might think of a password as being chosen from a dictionary of possible passwords).

Even though a dictionary attack renders this simple method of encrypting session keys insecure, a password can still be used beneficially (and securely) to encrypt certain data transmitted during a key agreement scheme. We describe a KAS introduced by Bellovin and Merritt that is known as *Encrypted Key Exchange*, or *EKE*, for short. Roughly speaking, *EKE* is a modified version of the *Diffie-Hellman KAS*, in which a password is used to encrypt the exponentials transmitted during a session of the scheme. We present a simplified version of *EKE*, known as *EKE2*, as Protocol 11.6.

As mentioned above, *EKE2* is obtained from the basic *Diffie-Helman KAS* by encrypting exponentials. The resulting session key is $\alpha^{a_U a_V}$, which is the same as in the *Diffie-Hellman KAS*. There is no entity authentication in *EKE2*, but the fact that the exponentials are encrypted prevents an intruder-in-the-middle from being able to carry out a successful attack. More precisely, we claim that *EKE2* provides implicit key authentication in the presence of an active adversary.

We give an informal justification (not a rigorous proof) of this security result, assuming that the **Decision Diffie-Hellman** problem in the subgroup generated by α is intractable. First, let's consider a session from the point of view of U. She receives the identity string $ID(V)$, so V is the intended peer of U. Of course, U has no way of knowing if she is really communicating with V. However, U will use the secret password $pwd_{U,V}$, which is shared with V, to decrypt the value y_V. Then the decrypted value b_V is used by U to construct the session key K.

The session key K computed by U can be computed by a user other than U only if that user knows the value a_V such that

$$\alpha^{a_V} = d_{pwd_{U,V}}(y_V).$$

Assume that an adversary does not have any information about the value of the password $pwd_{U,V}$. Then, even if the value y_V was not constructed by V as specified in the scheme (e.g., it could have been copied from another session and then replayed), it seems reasonable to believe that an adversary would be unable to compute any information about the value of a_V. So U can be confident that no one other than V can compute any information about the session key that she computed. The analysis from the point of view of V is similar, and he is also provided with implicit key authentication.

Protocol 11.6: EKE2

The public domain parameters consist of a group $(G, \cdot)$ and an element $\alpha \in G$ having order n.

Note: U and V have a shared secret password denoted by $pwd_{U,V}$. Also, $e(\cdot)$ and $d(\cdot)$ are (respectively) encryption and decryption functions for a pre-specified secret-key cryptosystem.

1. U chooses a_U at random, where $0 \leq a_U \leq n - 1$. Then she computes

$$b_U = \alpha^{a_U} \quad \text{and} \quad y_U = e_{pwd_{U,V}}(b_U)$$

and she sends $ID(U)$ and y_U to V.

2. V chooses a_V at random, where $0 \leq a_V \leq n - 1$. Then he computes

$$b_V = \alpha^{a_V} \quad \text{and} \quad y_V = e_{pwd_{U,V}}(b_V)$$

and he sends $ID(V)$ and b_V to U.

3. U computes

$$b_V = d_{pwd_{U,V}}(y_V) \quad \text{and} \quad K = (b_V)^{a_U},$$

and V computes

$$b_U = d_{pwd_{U,V}}(y_U) \quad \text{and} \quad K = (b_U)^{a_V}.$$

The discussion above assumed that the adversary is not able to obtain any information about the password $pwd_{U,V}$. Let us consider briefly whether this is a reasonable assumption. Notice that the password $pwd_{U,V}$ is used only to encrypt information (i.e., the two exponentials) that is used to derive the session key. Even if the value of the session key is revealed to the adversary (e.g., in a known session key attack), this will not provide any information about the values of the unencrypted exponentials or the password.

11.6 Conference Key Agreement Schemes

A *conference key agreement scheme* (or, *CKAS*) is a key agreement scheme in which a subset of two or more users in a network can construct a common secret key (i.e., a group key). In this section, we discuss (without proof) two conference key agreement schemes. The first CKAS we present was described in 1994 by

---

Protocol 11.7: BURMESTER-DESMEDT CONFERENCE KAS

The public domain parameters consist of a group $(G, \cdot)$ and an element $\alpha \in G$ having order n.
Note: all subscripts are to be reduced modulo m in this scheme, where m is the number of participants in the scheme.

1. For $0 \le i \le m - 1$, U_i chooses a random number a_i, where $0 \le a_i \le n - 1$. Then he computes

$$b_i = \alpha^{a_i}$$

and he sends b_i to U_{i+1} and U_{i-1}.

2. For $0 \le i \le m - 1$, U_i computes

$$X_i = (b_{i+1}/b_{i-1})^{a_i}.$$

Then U_i broadcasts X_i to the $m - 1$ other users.

3. For $0 \le i \le m - 1$, U_i computes

$$Z = b_{i-1}^{a_i m} X_i^{m-1} X_{i+1}^{m-2} \cdots X_{i-2}^{1}.$$

Then

$$Z = \alpha^{a_0 a_1 + a_1 a_2 + \cdots + a_{m-1} a_0}$$

is the secret conference key which is computed by $U_0, \ldots, U_{m-1}$.

Burmester and Desmedt. We also present the 1996 CKAS due to Steiner, Tsudik and Waidner.

Both of these schemes are modifications of the *Diffie-Hellman KAS* in which m users, say $U_0, \ldots, U_{m-1}$, compute a common secret key. The schemes are set in a subgroup of a finite group in which the **Decision Diffie-Hellman** problem is intractable.

The *Burmester-Desmedt CKAS* is presented as Protocol 11.7. It is not hard to verify that all the participants in a session of this CKAS will compute the same key, Z, provided that the participants behave correctly and there is no active adversary who changes any of the transmitted messages. Suppose we define

$$Y_i = b_i^{a_{i+1}} = \alpha^{a_i a_{i+1}}$$

for all i (where all subscripts are to be reduced modulo m). Then

$$X_i = \left(\frac{b_{i+1}}{b_{i-1}}\right)^{a_i} = \left(\frac{\alpha^{a_{i+1}}}{\alpha^{a_{i-1}}}\right)^{a_i} = \frac{\alpha^{a_{i+1} a_i}}{\alpha^{a_{i-1} a_i}} = \frac{Y_i}{Y_{i-1}}$$

for all i. Then the following equations confirm that the key computation works correctly:

$$b_{i-1}{}^{a_i m} X_i{}^{m-1} X_{i+1}{}^{m-2} \cdots X_{i-2}{}^1$$

$$= Y_{i-1}{}^m \left(\frac{Y_i}{Y_{i-1}}\right)^{m-1} \left(\frac{Y_{i+1}}{Y_i}\right)^{m-2} \cdots \left(\frac{Y_{i-2}}{Y_{i-3}}\right)^1$$

$$= Y_{i-1} Y_i \cdots Y_{i-2}$$

$$= \alpha^{a_{i-1}a_i + a_i a_{i+1} + \cdots + a_{i-2}a_{i-1}}$$

$$= Z.$$

Protocol 11.7 takes place in two "rounds." In the first round, every participant sends a message to his or her two neighbors, where we view the m participants as being arranged in a ring of size m. In the second round, each participant broadcasts one piece of information to everyone else. All the transmissions in each of the two rounds can be done in parallel.

Overall, each participant transmits two pieces of information, and each participant receives $m + 1$ pieces of information during a session of the scheme. This is quite efficient, but it requires the existence of a broadcast channel.

Steiner, Tsudik and Waidner suggested a CKAS which is more "sequential" in nature, but which does not require a broadcast channel. Their scheme is presented as Protocol 11.8.

A session of Protocol 11.8 takes place in two stages. In the first stage, information is transmitted sequentially from U_0 to U_1, from U_1 to U_2, ..., and finally from U_{m-2} to U_{m-1}. For $i \geq 1$, each user U_i receives a list of values from U_{i-1}, computes one new value, and appends it to the list. By the end of the first stage, a list of m values is held by U_{m-1}.

Then stage 2 begins. In stage 2, information is transmitted in the opposite order to stage 1. Each participant in turn computes the session key from the last element in the current list and then modifies the remaining elements in the list. At the end of this stage, every participant has computed the same session key, $Z = \alpha^{a_0 a_1 \cdots a_{m-1}}$. See Figure 11.7 for a diagram illustrating the information transmitted in a session of Protocol 11.8 in which there are four participants.

It is not difficult to count the number of messages transmitted and received by each participant in Protocol 11.8. For example, for $0 \leq i \leq m - 2$, it is easily seen that U_i transmits $2i + 1$ messages, while U_{m-1} transmits $m - 1$ messages. The total number of messages transmitted is $m^2 - m$.

Neither Protocol 11.7 nor Protocol 11.8 provides any kind of authentication. Security against active adversaries would require additional information to be transmitted such as signatures, certificates, etc. It is also not immediately obvious that these schemes are secure even against a passive adversary (as usual, under the assumption that the **Decision Diffie-Hellman** problem is intractable). However, it has in fact been proven that the *Steiner-Tsudik-Waidner Conference KAS* is secure in this setting.

Protocol 11.8: STEINER-TSUDIK-WAIDNER CONFERENCE KAS

The public domain parameters consist of a group $(G, \cdot)$ and an element $\alpha \in G$ having order n.

Stage 1.

U_0 chooses a random number a_0, he computes α^{a_0} and he sends $\mathcal{L}_0 = (\alpha^{a_0})$ to U_1.

For $i = 1, \ldots, m - 2$, U_i receives the list $\mathcal{L}_{i-1}$ from U_{i-1}. Then U_i chooses a random number a_i and he computes $\alpha^{a_0 a_1 \cdots a_i} = (\alpha^{a_0 a_1 \cdots a_{i-1}})^{a_i}$. Then he sends the list $\mathcal{L}_i = \mathcal{L}_{i-1} \parallel \alpha^{a_0 a_1 \cdots a_i}$ to U_{i+1}.

U_{m-1} receives the list $\mathcal{L}_{m-2}$ from U_{m-2}. Then he chooses a random number a_{m-1} and he computes $\alpha^{a_0 a_1 \cdots a_{m-1}} = (\alpha^{a_0 a_1 \cdots a_{m-2}})^{a_{m-1}}$. Then he constructs the list $\mathcal{L}_{m-1} = \mathcal{L}_{m-2} \parallel \alpha^{a_0 a_1 \cdots a_{m-1}}$.

Stage 2.

U_{m-1} extracts the conference key $Z = \alpha^{a_0 a_1 \cdots a_{m-1}}$ from $\mathcal{L}_{m-1}$. For every other element $y \in \mathcal{L}_{m-1}$, U_{m-1} computes the value $y^{a_{m-1}}$. Then U_{m-1} constructs the list of $m - 1$ values

$$\mathcal{M}_{m-1} = (\alpha^{a_{m-1}}, \alpha^{a_0 a_{m-1}}, \ldots, \alpha^{a_0 a_1 a_{m-1}}, \ldots, \alpha^{a_0 a_1 \cdots a_{m-3} a_{m-1}})$$

and he sends $\mathcal{M}_{m-1}$ to U_{m-2}.

For $i = m - 2, \ldots, 1$, U_i receives the list $\mathcal{M}_{i+1}$ from U_{i+1}. He computes the conference key $Z = (\alpha^{a_0 \cdots a_{i-1} a_{i+1} \cdots a_{m-1}})^{a_i}$ from the last element in $\mathcal{M}_{i+1}$. For every other element $y \in \mathcal{L}_{i+1}$, U_i computes the value y^{a_i}. Then U_i constructs the list of i values

$$\mathcal{M}_i = (\alpha^{a_i \cdots a_{m-1}}, \alpha^{a_0 a_i \cdots a_{m-1}}, \alpha^{a_0 a_1 a_i \cdots a_{m-1}}, \ldots, \alpha^{a_0 a_1 \cdots a_{i-2} a_i \cdots a_{m-1}})$$

and he sends $\mathcal{M}_i$ to U_{i-1}.

U_0 receives the list $\mathcal{M}_1$ from U_1. He computes the conference key $Z = (\alpha^{a_1 \cdots a_{m-1}})^{a_0}$ from the (only) element in $\mathcal{M}_1$.

11.7 Notes and References

Diffie and Hellman presented their key agreement scheme in [117]. The idea of key exchange was discovered independently by Merkle [238]. The *Station-to-*

FIGURE 11.7
Information transmitted in the Steiner, Tsudik and Waidner CKAS with four participants

Stage 1	Stage 2	
U_3	U_3	$Z = (\alpha^{a_0 a_1 a_2})^{a_3}$
$\uparrow$	$\downarrow$	
$(\alpha^{a_0}, \alpha^{a_0 a_1}, \alpha^{a_0 a_1 a_2})$	$(\alpha^{a_3}, \alpha^{a_0 a_3}, \alpha^{a_0 a_1 a_3})$	
$\uparrow$	$\downarrow$	
U_2	U_2	$Z = (\alpha^{a_0 a_1 a_3})^{a_2}$
$\uparrow$	$\downarrow$	
$(\alpha^{a_0}, \alpha^{a_0 a_1})$	$(\alpha^{a_2 a_3}, \alpha^{a_0 a_2 a_3})$	
$\uparrow$	$\downarrow$	
U_1	U_1	$Z = (\alpha^{a_0 a_2 a_3})^{a_1}$
$\uparrow$	$\downarrow$	
(α^{a_0})	$(\alpha^{a_1 a_2 a_3})$	
$\uparrow$	$\downarrow$	
U_0	U_0	$Z = (\alpha^{a_1 a_2 a_3})^{a_0}$

station KAS is due to Diffie, van Oorschot and Wiener [118]. For an overview of key agreement schemes based on the **Diffie-Hellman** problems, see Blake-Wilson and Menezes [45].

The schemes of Matsumoto, Takashima and Imai can be found in [228]. The triangle attack is from Burmester [78].

Self-certifying key distribution was introduced by Girault [157]. The scheme he presented was actually a key predistribution scheme; the modification to a key agreement scheme is based on [287].

The *Encrypted Key Exchange KAS* was proposed by Bellovin and Merritt [27]. A formal security proof for a version of this scheme can be found in Bellare, Pointcheval and Rogaway [21].

The *Burmester-Desmedt Conference KAS* is described in [79] and the *Steiner-Tsudik-Waidner Conference KAS* is from [316].

Worthwhile surveys on key agreement schemes have been written by van Tilburg [333], Rueppel and van Oorschot [287] and Blundo and D'Arco [53]. Boyd and Mathuria [67] is a recent monograph that contains a great deal of information on the topics covered in this chapter. Two recent masters theses on key agreement that the reader might wish to consult are Ng [252] and Wang [337].

Exercises

11.1 Suppose that U and V take part in a session of the *Diffie-Hellman KAS* with $p = 27001$ and $\alpha = 101$. Suppose that U chooses $a_U = 21768$ and V chooses $a_V = 9898$. Show the computations performed by both U and V, and determine the key that they will compute.

11.2 Consider the following modification of the *STS KAS*:

Protocol 11.9: MODIFIED STATION-TO-STATION KAS

The public domain parameters consist of a group $(G, \cdot)$ and an element $\alpha \in G$ having order n.

1. U chooses a random number a_U, $0 \leq a_U \leq n - 1$. Then she computes

$$b_U = \alpha^{a_U}$$

and she sends **Cert**(U) and b_U to V.

2. V chooses a random number a_V, $0 \leq a_V \leq n - 1$. Then he computes

$$b_V = \alpha^{a_V}$$

$$K = (b_U)^{a_V}, \quad \text{and}$$

$$y_V = sig_V(b_V \parallel b_U).$$

Then V sends **Cert**(V), b_V and y_V to U.

3. U verifies y_V using ver_V. If the signature y_V is not valid, then she "rejects" and quits. Otherwise, she "accepts," she computes

$$K = (b_V)^{a_U}, \quad \text{and}$$

$$y_U = sig_U(b_U \parallel b_V),$$

and she sends y_U to V.

4. V verifies y_U using ver_U. If the signature y_U is not valid, then he "rejects"; otherwise, he "accepts."

In this modificaton of the protocol, the signatures omit the intended receiver. Show how this renders the protocol insecure, by describing an intruder-in-the-middle attack. Discuss the consequences of this attack, in terms of key authentication properties and how they are violated. (This attack is known as an *unknown key-share attack*.)

11.3 Suppose that U and V carry out the *MTI/A0 KAS* with $p = 30113$ and $\alpha = 52$. Suppose that U has $a_U = 8642$ and he chooses $r_U = 28654$; and V has $a_V = 24673$ and she chooses $r_V = 12385$. Show the computations performed by both U and V, and determine the key that they will compute.

11.4 If a passive adversary tries to compute the key K constructed by U and V by using the *MTI/A0 KAS*, then he is faced with an instance of what we might term the **MTI problem**, which we present as Problem 11.1.

Problem 11.1: **MTI**

Instance: $I = (p, \alpha, \beta, \gamma, \delta, \epsilon)$, where p is prime, $\alpha \in \mathbb{Z}_p^*$ is a primitive
element, and $\beta, \gamma, \delta, \epsilon \in \mathbb{Z}_p^*$.

Question: Compute $\beta^{\log_\alpha \gamma} \delta^{\log_\alpha \epsilon} \mod p$.

Prove that any algorithm that can be used to solve the **MTI** problem can be used
to solve the **Computational Diffie-Hellman** problem, and vice versa. (i.e., give
Turing reductions between these two problems).

11.5 Consider the *Girault KAS*, where $p = 167$, $q = 179$, and hence $n = 29893$.
Suppose $\alpha = 2$ and $e = 11101$.
   (a) Compute d.
   (b) Given that $ID(U) = 10021$ and $a_U = 9843$, compute b_U and p_U. Given
   that $ID(V) = 10022$ and $a_V = 7692$, compute b_V and p_V.
   (c) Show how b_U can be computed from p_U and $ID(U)$ using the public exponent e. Similarly, show how b_V can be computed from p_V and $ID(V)$.
   (d) Suppose that U chooses $r_U = 15556$ and V chooses $r_V = 6420$. Compute
   s_U and s_V, and show how each of U and V compute their common key.

11.6 Discuss whether the property of perfect forward secrecy (which was defined in Section 10.1) is achieved in the following key agreement schemes:
   (a) in the *STS KAS*, assuming that the secret signing keys of one or more users
   are revealed;
   (b) in *MTI/A0*, assuming that secret exponents a_T of one or more users are
   revealed; and
   (c) in the *Girault KAS*, assuming that secret exponents a_T of one or more users
   are revealed.

11.7 The purpose of this question is to perform the required computations in a session
of the *Burmester-Desmedt Conference KAS*. Suppose we take $p = 128047$, $\alpha = 8$
and $n = 21341$. (It can be verified that p is prime and the order of α in $\mathbb{Z}_p^*$ is
equal to n.) Suppose there are $m = 4$ participants, and they choose secret values
$a_0 = 4499$, $a_1 = 9854$, $a_2 = 19887$ and $a_3 = 10002$.
   (a) Compute the values b_0, b_1, b_2 and b_3.
   (b) Compute the values X_0, X_1, X_2 and X_3.
   (c) Show the computations performed by U_0, U_1, U_2 and U_3 to construct the
   conference key Z.

11.8 Show all the computations performed in a session of the *Steiner-Tsudik-Waidner
Conference KAS* involving four participants. Use the same values of p, α, n, a_0,
a_1, a_2 and a_3 as in the previous exercise.

12

Public-key Infrastructure

12.1 Introduction: What is a PKI?

Perhaps the biggest challenge in public-key cryptography is ensuring the authenticity of public keys. If Alice wants to encrypt information to send to Bob, and Bob is someone who Alice does not know personally, how can Alice be sure that Bob's purported public key really is Bob's key (and not Charlie's, for example)? We have already introduced certificates as a tool to help authenticate public keys. A *public-key infrastructure* (or *PKI*) is a secure system that is used to manage and control certificates.

In their excellent book "Understanding PKI, Second Edition," Adams and Lloyd give the following very general definition of a public-key infrastructure:

> *A PKI is the basis of a pervasive security infrastructure whose services are implemented and delivered using public-key concepts and techniques.*

Several aspects of this definition are worth discussing. First is the idea that a PKI is an infrastructure. Ideally, it should function without the active intervention of the user. In various examples, we might describe operations as being performed by a network user such as Alice, but it should be understood that it would often be the software on Alice's computer that is carrying out the relevant tasks. Alice might not even be aware of many of the PKI-related procedures that are taking place.

Another inherent feature of a PKI is that it uses the techniques of public-key cryptography, the most important of which is a signature scheme. If we were to restrict ourselves to symmetric-key cryptography, we would have to assume the existence of prior shared secret keys between two parties that want to communicate. As mentioned before, one of the purposes of public-key cryptography is to eliminate this requirement.

In general, a PKI has many components. The most important components are described briefly, as follows:

certificate issuance

> This refers to the issuing of new certificates to users in a given PKI. Most PKIs have one or more trusted authorities (usually called *certification authorities*) that control the issuing of certificates. Before a certificate is issued, the identity and credentials of the user must be verified by non-cryptographic means, as we already mentioned in Section 9.3.1. Then, and only then, it is appropriate to issue a certificate to a user. At the same time, a secure procedure must be used to generate the public and private keys for the certificate's owner. Once the certificate has been constructed, it must be transmitted to its owner in a secure manner (perhaps by non-cryptographic means).

certificate revocation

> This refers to the revoking of certificates before a normal expiration date which might be specified on the certificate. This could be due to some unforeseen circumstances, such as a private key being lost or stolen, or other fraudulent use of the key.
>
> For example, suppose a certificate is supposed to be valid until December 31, 2020, but the corresponding user's private key is stolen before that date. Then the certificate should be revoked, so that it is no longer considered as valid, and a new certificate issued in its place. (This is analogous to the situation where a credit card is stolen and a new one must be issued to replace it.) The new certificate would replace the stolen key(s) with new, secure key(s). Since there is nothing on the certificate itself to specify if it has been revoked, additional infrastructure is required to recognize revoked certificates.

key backup / recovery / update

> Key backup refers to the secure storage of users' private keys by the administrator of the PKI, in case users lose or forget their private keys.
>
> Key recovery is a protocol that allows a lost or forgotten key to be restored or re-activated. Typically, a user has to prove his or her identity before being allowed to access a stored private key.
>
> Key update could occur when a key has to be changed for some reason, or as a general security precaution. For example, when a certificate is about to expire, a protocol could be used to choose a new replacement key and to generate a new certificate to replace the old one. This key updating protocol could use the old key (before it expires) to encrypt the new key, so the updated certificate could then be transmitted electronically to its owner. This process could be simpler and more efficient than generating new keys and certificates "from scratch" and the distributing them to users via a non-cryptographic secure channel.

timestamping

> For various reasons, the times at which keys are issued, revoked, or updated may be important. For example, certificates usually have fixed length va-

lidity periods. A signature on some data (either in a certificate, or not) that includes a specified time or period of time during which a key is valid is called a timestamp.

Once a PKI is implemented and operational, it allows various applications to be built on top of it. These applications could be termed *PKI-enabled services*. Some examples of PKI-enabled services include the following:

secure communication

Here are some examples currently in common use. Secure e-mail protocols include Secure Multipurpose Internet Mail Extensions (S/MIME) and Pretty Good Privacy (PGP). Secure web service access is provided through Secure Sockets Layer (SSL) or Transport Layer Security (TLS). Secure virtual private networks (VPNs) utilize the Internet Protocol Security (IPsec) protocols. (As an example, we will describe the basic structure of SSL in the next subsection.)

access control

Access control is also known as *privilege management*. It incorporates authentication, authorization and delegation.

An example of access control would be access to a database. Various people might have different levels of access. Whether a particular person has access to specific information on the database might be determined by the status of the person within the organization (this information might be included on a certificate) together with an access policy that determines if the given individual is to be granted authorization to access the information that he or she seeks. Access control may involve some form of user authentication, e.g., via a password or cryptographic identification scheme.

Delegation could be used in a situation where someone might be granted temporary access to the database by a higher-ranking individual.

privacy architecture

A privacy architecture permits the use of anonymous / pseudonymous certificates. These types of certificates could show membership of an individual in a specified class of users, without specifying their particular identitiy. This would permit enhanced versions of access control, for example.

12.1.1 A Practical Protocol: Secure Socket Layer

Practical protocols used in real-life applications may incorporate many cryptographic tools. We use *SSL* (*Secure Socket Layer*) as an illustration. An SSL session is used, for example, to facilitate online purchases from a company's web page using a web browser. Suppose a *client* (Alice) wants to purchase something from a *server* (Bob, Inc.). The main steps in setting up an SSL session are summarized in Figure 12.1.

In more detail, here is what takes place: First, Alice and Bob, Inc. introduce

FIGURE 12.1
Setting up an SSL Session

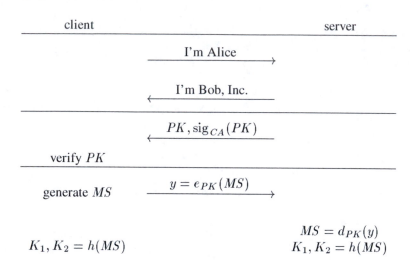

themselves. This is called a "hello," and no cryptographic tools are used in this step. At this time, Alice and Bob, Inc. also agree on the specific cryptographic algorithms they will use in the rest of the protocol.

Next, Bob Inc. authenticates himself to Alice; he sends her a certificate containing a copy of his public key, PK, signed by a trusted certification authority (denoted CA). Alice verifies the CA's signature on PK using the CA's public verification key (which would have been bundled with the web browser software running on Alice's computer).

Now Alice and Bob, Inc. proceed to determine two common secret keys. Alice generates a random master secret, MS, using an appropriate pseudorandom number generator. She encrypts MS using Bob, Inc.'s public key and sends the resulting ciphertext to Bob., Inc. Bob Inc. decrypts the ciphertext, obtaining MS. Now Alice and Bob Inc. independently generate the same two keys K_1 and K_2 from MS (this step probably involves the use of a hash function, denoted h).

Finally, now that Alice and Bob, Inc. have both derived the same two secret keys, they use these keys to authenticate and encrypt the messages they send to each other. The key K_1 would be used to authenticate data using a message authentication code, and the key K_2 would be used to encrypt and decrypt data using a secret key cryptosystem. Therefore the SSL protocol enables secure communication between Alice and Bob, Inc.

It is interesting to note that only the server (Bob, Inc.) is required to supply a certificate during an SSL Session. The client (Alice) may not even have a public key (or a certificate). This is a common state of affairs at present in electronic commerce: companies setting up web pages for business purposes require certificates, but users do not need a certificate in order to make an online purchase. From the company's point of view, the important point is not that Alice is really who she claims to be. It is more important that Alice's credit card number, which is supplied as part of the ensuing financial transaction, is valid. The credit card number and any personal information supplied by Alice will be encrypted (and authenticated, via a MAC) using the keys that are created in the SSL session.

12.2 Certificates

Certificates are the building blocks of PKIs, and they ultimately enable secure and scalable PKIs to be built from them. (As an interesting historical note, the concept of a certificate was introduced in 1978 by Kohnfelder, in his MIT bachelor's thesis.) We have mentioned already that, in its simplest form, a certificate binds an identity to a public key. This is usually done by having a trusted authority (i.e., a certification authority, denoted CA) sign the information on a certificate, as described in Protocol 9.5. It is generally assumed that everyone has access to an authentic copy of the public key of the CA. Hence, a CA's signature on a certificate can be verified, which allows the information on the certificate to be authenticated.[1]

For purposes of illustration, we now describe the format of X.509 v3 certificates, which are a popular type of certificate. X.509 certificates contain the following fields:

1. version number
2. serial number
3. signature algorithm ID
4. issuer name
5. validity period
6. subject name (i.e., the certificate owner)
7. the certificate owner's public key
8. optional fields
9. the CA's signature on all the previous fields

X.509 certificates were originally defined using *X.500 names* for subject names.[2] X.500 names have a hierarchical format, such as

---

[1] Strictly speaking, verifying the CA's signature only allows someone to verify that the certificate was issued by the CA. However, having verified the CA's signature, a user would then believe that the information contained in the certificate is valid provided that he or she trusts the CA to verify that information before signing the certificate.

[2] Other subject naming formats are now permitted in X.509, however.

$$
\begin{array}{rcl}
C & = & US \\
O & = & Microsoft \\
OU & = & Management \\
CN & = & Bill Gates,
\end{array}
$$

where "C" denotes country, "O" denotes organization, "OU" denotes organizational unit and "CN" denotes common name. Subject names are actually encoded numerically, using *object identifiers* (OIDs). So, for example, instead of containing the alphabetic string "Microsoft," the certificate will contain a numerical OID that stands for "Microsoft."

The hierarchical format of X.500 names ensures that everyone has a unique global name. X.500 is intended to be a true directory — analogous to a global "telephone directory" — which would allow X.509 certificates to be looked up and accessed remotely. Unfortunately, no widely deployed X.500 directory exists at the present time.

Another issue is that X.500 names have a different format and structure than *domain name system* (DNS) names and IP (internet protocol) addresses. DNS names are commonly used in e-mail addresses. For example, in the e-mail address

```
dstinson@cacr.math.uwaterloo.ca,
```

the "domain" is

```
cacr.math.uwaterloo.ca.
```

An IP address is a (hierarchical) numerical code that allows network software to contact specific domains. In any local environment, the domain name system maps names for computers to IP addresses. For example, the domain

```
cacr.math.uwaterloo.ca
```

has the IP address

```
129.97.140.130.
```

Most people are accustomed to using names and e-mail addresses in the format described above and are not familiar with X.500 names. So any widespread of use of X.500 names would have to address the question of converting from one format (or naming system) to another.

Various other certificate formats exist, although none is universally accepted at the present time. Some examples include the following:

SPKI

Simple public key infrastructure (SPKI) certificates use local (as opposd to global) names. Thus they emphasize authorizations rather than identities. SPKI certificates are incompatible with X.509 certificates.

PGP

Pretty Good Privacy (PGP) is a user-based e-mail system, based on local names. PGP certificates are incompatible with X.509 certificates.

Algorithm 12.1: CERTIFICATE VALIDATION

1. Verify the integrity and authenticity of the certificate by verifying the signature of the CA on the certificate (we are assuming that the verification key of the CA either is known *a priori*, implicitly trusted, or verified by some "external" information, e.g., by looking it up on an "official" website).

2. Verify that the certificate has not expired (i.e., check that the current date lies within the validity period specified on the certificate).

3. Verify that the certificate has not been revoked.

4. If it is relevant, verify that the usage of the certificate conforms to any policy constraints specified in relevant optional fields on the certificate.

SET

Secure Electronic Transaction (SET) specifications use enhanced (modified) X.509 certificates.

12.2.1 Certificate Life-cycle Management

Certificate life-cycle management has several phases:

1. registration
2. key generation and distribution
3. key backup
4. certificate issuance
5. certificate retrieval
6. certificate validation
7. key update
8. key recovery
9. certificate revocation
10. certificate expiration
11. key history
12. key archive

In this chapter, we will concentrate mainly on methods of certificate validation. (Most of the other operations will not be discussed in any detail here.) In most situations, certificate validation incorporates the operations enumerated in Algorithm 12.1,

Note that it is only the first operation in Algorithm 12.1 that uses cryptography. However, we comment briefly on some of the more popular techniques used to provide assurance that a certificate has not been revoked.

Recall that certificate revocation is required when a certificate is invalid before its normal expiration date. A PKI needs a mechanism to verify that certificates have not been revoked. The most common technique is a *certificate revocation list* (or *CRL*), which is a list of the serial numbers of all certificates that are revoked but not expired. The CRL is prepared and signed by a *CA*. CRLs must be updated periodically, and must be made available in a public directory in order to be useful.

Instead of frequently issuing new CRLs, a more efficient method is to employ *Delta CRLs*. Delta CRLs contain the changes (i.e., the new revocations) that have occurred since the most recent previously issued CRL or delta CRL. For example, a CRL might be issued once per month, with delta CRLs being issued on a daily or weekly basis.

An alternative approach to CRLs is an *online certificate status protocol* (or, *OCSP*). In an OCSP, a certain online server is queried as to the revocation status of a given certificate. This server must maintain, or be able to access, an up-to-date CRL in real time in order to be able to process the queries it receives in a timely manner.

12.3   Trust Models

Often, a certificate will not be signed directly by a trusted *CA*. Instead, it is necessary to follow a *certificate path* from a trusted *CA* to a given certificate. Each certificate in the path should be signed by the owner of the previous certificate in the path. By validating all the certificates in the path, the user can be confident that the last certificate in the path is valid.

A *trust model* specifies rules which determine how a certificate path should be constructed. Here are some examples of trust models that we will discuss:

1. *strict hierarchy*

2. *networked PKIs*

3. *web browser model*

4. *user-centric model* (also known as a *web of trust*).

12.3.1   Strict Hierarchy Model

In a *strict hierarchy*, the *root CA* has a self-signed, self-issued certificate; the root *CA* is called a *trust anchor*. The root *CA* may issue certificates for lower-level *CA*s, and any *CA* can issue certificates for *end users*.

This model is illustrated in Figure 12.2, in which there is a root *CA*, four lower

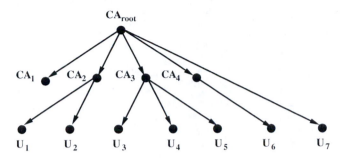

FIGURE 12.2
A Strict Hierarchy

level CAs, and seven end users. (A "real example" would have many more end users, of course.) In this representation, we have a directed graph whose nodes are CAs and end users. A directed edge $x \to y$ means that the entity corresponding to node x has signed a certificate for the entity corresponding to node y. Observe that every end user in Figure 12.2 has a certificate that is signed by one of the five CAs.

An end user, say *Alice*, in order to verify the certificate of another end user, say *Bob*, needs a certificate path from the root CA to *Bob*. (For example, if CA_{root} has signed the certificate of CA_1, who in turn has signed *Bob*'s certificate, then this would comprise a suitable certificate path.) In a strict hierarchy, *Bob* might be expected to store this information on his computer. Then he could provide it to *Alice* or anyone else who requests it. In the example under consideration, *Bob* would send all the certificates in the path

$$CA_{root} \to CA_1 \to Bob$$

to *Alice*.

It is assumed that *Alice* knows the verification key of CA_{root}, say $\text{ver}_{CA_{root}}$. Together with the certificate path provided by Bob, this will allow Alice to carry out *certificate path validation*. The given certificate path is validated as follows:

1. *Alice* validates **Cert**(CA_{root}) using the key $\text{ver}_{CA_{root}}$
2. *Alice* validates **Cert**(CA_1) using the key $\text{ver}_{CA_{root}}$
3. *Alice* extracts the key ver_{CA_1} from **Cert**(CA_1)
4. *Alice* validates **Cert**(Bob) using the key ver_{CA_1}
5. *Alice* extracts *Bob*'s public key(s) from **Cert**(Bob).

Finally, *Alice* must also ensure that the path adheres to the trust model, i.e., that the path originates at CA_{root}, terminates at *Bob*, and (optionally, as in this exam-

FIGURE 12.3
The Mesh Configuration

ple) it contains an interior node which is a lower-level CA.[3]

In this trust mode, an end user is not allowed to sign a certificate belonging to another end user. If a user did create such a certificate, then the certificate should not be regarded as valid because the certificate path would not adhere to the trust model. For example, if Charlie signed Bob's certificate, then the path

$$CA_{root} \to Charlie \to Bob$$

would not be an acceptable path, and Alice could not validate Bob's certificate by using this path.

12.3.2 Networked PKIs

The strict hierarchy may work well within a single organization. Sometimes, however, it may be desirable to "connect" root CAs of two or more different PKI domains, a process sometimes called *PKI networking*. This creates a "super-PKI" made up of users in different domains. The PKI domains within a super-PKI may be heterogeneous. For example, the domains do not all need to be strict hierarchies.

Cross-certification is when one CA signs the certificate of another CA. In the *mesh configuration*, all the root CAs will cross-certify each other. Hence, n root CAs will require $n(n-1)$ cross-certifications in a mesh configuration. An example with three root CAs is depicted in Figure 12.3, where bi-directed edges are used to indicate cross-certification.

An alternative approach is the *hub-and-spoke configuration*, in which the n root CAs each cross-certify independently with a new *hub CA*. The number of cross-certifications that are required is equal to $2n$. Observe that, for $n > 3$, fewer cross-certifications are required using this approach, as compared to the mesh configuration.

We now consider how certificate path discovery could be carried out in these two configurations. First we consider the mesh configuration. In order to validate

---

[3]Note that each certificate should contain information stating the status of the owner as an end user, CA, etc.

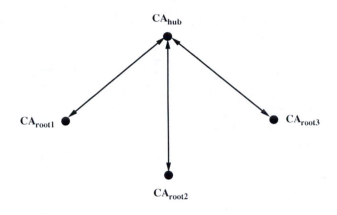

FIGURE 12.4
The Hub-and-spoke Configuration

Bob's certificate, *Alice* must be able to find a path of certificates from her trust anchor to *Bob*. (This process is called *path discovery*.) Suppose that *Alice*'s trust anchor is CA_{root_i} and *Bob*'s trust anchor is CA_{root_j}, where $i \neq j$. Assume that *Bob* sends *Alice* the certificates in the path from CA_{root_j} to *Bob*. Therefore, *Alice* needs to find the certificate belonging to CA_{root_j} (*Bob*'s trust anchor) that has been signed by CA_{root_i} (*Alice*'s trust anchor). Ideally, this certificate could be looked up in a directory which is maintained by CA_{root_i}. Now *Alice* has all the certificates in a path from CA_{root_i} to *Bob*, which allows *Alice* to validate *Bob*'s certificate.

Let's now look at the hub and spoke configuration. Again, we assume that *Bob* sends *Alice* the certificates in the path from CA_{root_j} to *Bob*. Now, *Alice* needs to find the certificate path from CA_{root_i} to CA_{hub} to CA_{root_j}. One "reasonable" way in which this could be done is as follows:

1. First, *Alice* obtains a copy of CA_{hub}'s certificate, that has been signed by CA_{root_i}, from CA_{root_i}.

2. Then, *Alice* looks up the certificate of CA_{root_j} that has been signed by CA_{hub}, in a directory which is maintained by CA_{hub}.

Now *Alice* has all the certificates in a path from CA_{root_i} to *Bob*, which allows *Alice* to validate *Bob*'s certificate.

12.3.3 The Web Browser Model

The *web browser model* is somewhat different from the models considered previously. Most web browsers (e.g., Netscape or Internet Explorer) come pre-configured with a set of "independent" root CAs, all of which are treated by the user of the browser as trust anchors. In Netscape, for example, this list can be found via pull-down menus in

Communicator/Tools/Security Info/Certificates/Signers.

There may be on the order of 100 such root CAs in a typical web browser; however, there is no cross-certification between these root CAs. Conceptually, the trust model can be thought of as a single strict hierarchy with a "virtual" root CA (i.e., the browser itself). In effect, the user trusts the provider of the web browser to only include valid root CAs in the browser.

There are several security issues with the web browser model. First, a user may not have any information about the security of these pre-configured root CAs. The user can edit this list, but most users do not have the technical expertise to do this in a knowledgeable manner. Furthermore, no mechanism exists to revoke a root CA from a web browser. There is no legal relationship or implied contract between a user and the set of CAs configured in a web browser. If a certificate is not signed by a known root CA, the user can choose to accept it anyway via a pop-up window in an SSL session. However, it must be remembered that verifying a certificate of a root CA (using the web browser software) does not in itself imply that the certificate is authentic, because the certificate is self-signed.

Another issue is that of expiration dates. Because there is no automated mechanism to update certificates of root CAs, it is common for a user to receive a message saying that a certificate has expired when he or she wants to institute an SSL session, for example. Then the user is given the opportunity to accept the certificate even though it has expired. The natural response is to accept the certificate, because the only alternative is to not carry out the desired SSL session. This is questionable security practice, and it could be argued that it defeats the purpose of certificates. Also, we might reasonably ask if there is any real point in including expiration dates on certificates, given that these expiration dates frequently are going to be ignored in practice.

12.3.4   Pretty Good Privacy

In the popular e-mail program *Pretty Good Privacy* (or *PGP*), every user is his or her own CA. A PGP certificate contains an e-mail address (ID), a public key (PK), and one or more signatures on this (ID, PK) pair. For example, Alice might create a self-signed certificate

$$\mathbf{Cert}(Alice) = (\mathbf{data}, \mathbf{signatures})$$

that contains the following information:

$$\mathbf{data} = (ID = \texttt{alice@uwaterloo.ca}, PK = 12345)$$

$$\mathbf{signatures} = sig_{Alice}(\mathbf{data}).$$

Algorithm 12.2: COMPUTE KEY LEGITIMACY FIELD

1. The KLF for a user U is assigned the value "valid" if the data of user U is signed by at least one key for which the OTF has the value "trusted" (implicitly or completely); or at least two keys for which the OTF has the value "partially trusted."

2. The KLF is assigned the value "marginally valid" if the user's data is signed by one key for which the OTF has the value "partially trusted."

3. Otherwise, the KLF is assigned the value "invalid."

Later on, other users might also create signatures on the data on Alice's certificate. Thus, for example, if Bob gives Alice such a signature, she will add it to the list of signatures on her certificate:

$$\textbf{signatures} = (sig_{Alice}(\textbf{data}), sig_{Bob}(\textbf{data})).$$

We will see that the signatures on a certificate will help to verify the certificate's authenticity to other users.

Alice keeps a collection of certificates, which she has obtained from various sources, in a data structure called a *keyring*. Associated with each certificate in the keyring is an *owner trust field* (or *OTF*) and a *key legitimacy field* (or *KLF*). Ultimately, the key legitimacy field will specify whether a particular public key will regarded as valid by Alice. The KLF has the value *valid, marginally valid* or *invalid*. It should be noted that a KLF value "invalid" does not mean that the key has been shown to be invalid. It really means just that there is insufficient evidence to show that the key is valid. The KLF is computed by the PGP software using a certain method that we will describe shortly.

The value of the owner trust field is determined by Alice, according to her own judgement; it indicates the extent to which Alice trusts the owner of that key to sign other keys. The OTF has the value *implicitly trusted, completely trusted, partially trusted* or *untrusted*. Alice's own OTF has the value "implicitly trusted." If Alice sets Bob's OTF to the value "completely trusted" in her keyring, she is asserting her belief that:

1. Bob's public key is valid, and
2. Bob is an individual who would be careful not to sign any invalid (ID, PK) pairs.

Once all the OTF values have been set, it is possible to compute the KLF values. These are calculated by the PGP software, using the rules specified in Algorithm 12.2.

Here is an example to show some computations of KLF values.

TABLE 12.1
Alice's keyring

owner	OTF	signatures
Alice	implicit	Alice, Bob, Doris, Eve
Bob	complete	Alice, Bob, Eve, Ginger
Charlie	partial	Bob, Charlie, Eve
Doris	untrusted	Doris, Eve, Janet
Eve	untrusted	Doris, Eve, Fred
Fred	partial	Bob, Eve, Fred
Ginger	partial	Charlie, Eve, Fred, Ginger
Harry	untrusted	Eve, Harry, Irene
Irene	untrusted	Eve, Fred, Harry, Irene
Janet	complete	Alice, Bob, Eve, Janet

Example 12.1   Suppose that the Alice's keyring contains the data given in Table 12.1. The reader can verify that the KLFs are determined as follows:

- Because Alice has signed Bob's and Janet's certificates and Alice trusts her own signatures implicitly, Bob's and Janet's keys are regarded as valid by Alice.

- Because Bob has signed Charlie's and Fred's certificates and Alice trusts Bob's signatures completely, Charlie's and Fred's keys are regarded as valid by Alice.

- Because Janet has signed Doris's certificate and Alice trusts Janet's signatures completely, Doris's key is regarded as valid by Alice.

- Because Charlie and Fred have both signed Ginger's certificate and Alice partially trusts both Charlie and Fred, Ginger's key is regarded as valid by Alice.

- Eve and Irene both have one signature from someone that Alice partially trusts, so these keys are regarded by Alice as marginally valid.

- Harry has no signatures from anyone that Alice trusts at least partially, so Harry's key is regarded as invalid by Alice.

□

There are several possible problems with a PGP-type PKI. First of all, PGP is not a true infrastructure. Users have to set trust levels (OTFs) themselves, and revocation can be done only by the certificate owners. These features require technologically informed users in order to carry out explicit actions. Further, there is no mechanism to revoke bogus certificates. Anyone can create a forged certificate and there is no assurance that forged certificates will be detected.

Another problematic feature of PGP is that it is not easily scalable to large networks or user communities; it works best within a "local" community where most users know each other. Finally, it is difficult to enforce uniform policies or standards, so a PGP-type PKI is not suitable for most large or medium-sized organizations.

12.4 The Future of PKI?

There are many potential difficulties associated with practical large-scale deployments of PKIs. The first (and perhaps most basic) problem is who should be responsible for development, maintenance, and regulation of PKIs. Should PKIs be administered by governments, or is this task better left to industry?

Secondly, what standards should be used in PKIs? Issues relating to standardization include entity naming, certificate formats, cryptographic algorithms, revocation, path discovery, certificate directories, and many others. The present lack of widely-adopted standards makes interoperability of different PKIs difficult to achieve.

A third problem is that of differing PKI needs in different environments. A PKI to serve the purposes of a centralized company would be completely unlike a PKI to enable secure communication among a widely dispersed group of individuals belonging to a common organization. Some PKIs may only require certificates in order to verify public keys of users, while other PKIs may be used primarily for access control or privilege management. It should be clear that no one-size-fits-all solution is possible for PKIs.

Finally, a lack of PKI-compatible applications is slowing the deployment of PKIs. This is a kind of chicken-and-egg problem: developers will not market PKI-compatible applications until widely deployed PKIs exist, but PKIs will not be deployed until there are more applications built using the capabilities of PKIs.

Although it is difficult to predict the future of PKI, it would seem reasonable to focus on modest, achievable expectations, i.e., to recognize that the primary purpose of PKIs is key management, and try to solve that problem first.

12.4.1 Alternatives to PKI

We have already mentioned that the development of large-scale PKIs has proceeded slowly. It is reasonable to ask what alternatives exist. If PKIs are not going to be used to enable verification of public keys, what techniques will be adopted instead? If we eliminate the use of public-key cryptography completely, then we return to a 1970s world where we require an online key server that is implemented using secret-key cryptography. Such a system would not work well in the modern-day world of the Internet,

Basically, the current situation that exists with respect to PKI might be termed an "ad hoc" infrastructure. We see a fairly wide usage of certificates signed by various certification authorities. However, many certificates in use today are expired or are signed by CAs not known to end users (or to the software employed by the end users). Perhaps this state of affairs is "good enough," but is difficult to quantify the real value of an infrastructure which is constantly being used in a manner that is not in accordance with the way in which it is intended (e.g, by allowing the use of public keys even if the certificates containing them are not validated by the user or by the user's software). As an analogy, a seat belt is not very useful if the passengers and driver in the car do not "buckle up."

Another possible alternative is identity-based cryptography, which refers to public-key cryptography where the values of public keys can be computed from the identity of their owner. This is a concept due to Shamir. Identity-based cryptography renders certificates unnecessary, and hence the need for an infrastructure to verify public keys is avoided. We discuss an example of identity-based cryptographic schemes in the next section.

12.5   Identity-based Cryptography

The basic idea of identity-based cryptography is that the public key for a user U is obtained by applying a public hash function h to the user's identity string, $ID(U)$. The corresponding private key would be generated by a central trusted authority (denoted by TA). This private key would then be supplied to the user U after that user proves his or her identity to the TA. The issuing of a private key by the TA replaces the issuing of a certificate. The resulting public key and private key are to be used in an encryption scheme, signature scheme, or other cryptographic scheme. The scheme involves some fixed, public system parameters (including a certain "master key") that will be used by everyone.[4]

Notice that identity-based cryptography removes the need for certificates. However, we still need a convenient and reliable method of associating an identity string with a person. The naming problems that exist in PKI (and which were discussed in Section 12.2) are not alleviated simply by using an identity-based cryptosystem. Other issues that were discussed in relation to PKIs, such as revocation, would also have to be dealt with. (For example, if an e-mail address is to be used as an identity string in an identity-based cryptosystem, what happens when someone changes their e-mail address?)

Designing an identity-based cryptosystem is not an easy exercise. Unfortunately, there does not seem to be an obvious or strightforward way to turn an arbitrary public-key cryptosystem into an identity-based cryptosystem. To illustrate,

---

[4]We have already discussed one type of identity-based protocol, namely, identity-based identification schemes, in Section 9.6.1.

suppose that we tried to transform the *RSA Cryptosystem* into an identity-based cryptosystem in a naive way. We might envisage a situation where the *TA* chooses the RSA modulus $n = pq$; the integer n would then be the public master key. The factors p and q would not be known to anyone except the *TA*; they function as the master private key.

The public RSA key of a user U is an encryption exponent and a private key is a decryption exponent. However, once U has a public key and corresponding private key, then he or she can easily factor n (we showed how to do this in Section 5.7.2). Once U knows the private master key, he can impersonate the *TA* and issue private keys to anyone else, as well as compute anyone else's private key. So this method of creating an identity-based cryptosystem fails utterly.

As can be seen from the above example, identity-based cryptography necessitates devising a system where a user's public and private key cannot be used to determine the private master key of the *TA*.

Here is a detailed description of the required operations in an identity-based (public-key) encryption scheme, and how they should work.

master key generation

The *TA* generates a *master public key* M^{pub} and a corresponding *master private key* M^{priv}. The *master key* is $M = (M^{pub}, M^{priv})$. A hash function h is also public; the master key and the hash function comprise the *system parameters*.

user key generation

When a user U identifies himself to the *TA*, the *TA* uses a function **extract** to compute U's private key K_U^{priv}, as follows:

$$K_U^{priv} = \textbf{extract}(M, K_U^{pub})$$

where U's public key is

$$K_U^{pub} = h(ID(U)).$$

User U's key is $K_U = (K_U^{pub}, K_U^{priv})$.

encryption

U's public key K_U^{pub} defines a public encryption rule e_{K_U} that can be used (by anyone) to encrypt messages sent to U.

decryption

U's private key K_U^{priv} defines a private decryption rule d_{K_U} that U will use to decrypt messages he receives.

12.5.1   The Cocks Identity-based Encryption Scheme

In this section, we discuss the *Cocks Identity-based Cryptosystem*, which is presented as Cryptosystem 12.1.

Cryptosystem 12.1 depends on certain properties of Jacobi symbols. The setting is the same as that of the *BBS Generator*, which was described in Section 8.3. Namely, the scheme is based on arithmetic in $\mathbb{Z}_n$, where $n = pq$, and p and q are distinct primes, each congruent to 3 modulo 4. As before, $\mathbf{QR}(n)$ denotes the set of quadratic residues modulo n:

$$\mathbf{QR}(n) = \left\{ x \in \mathbb{Z}_n : \left(\frac{x}{p}\right) = \left(\frac{x}{q}\right) = 1 \right\}.$$

As well, $\widetilde{\mathbf{QR}}(n)$ denotes the pseudo-squares modulo n:

$$\widetilde{\mathbf{QR}}(n) = \left\{ x \in \mathbb{Z}_n : \left(\frac{x}{p}\right) = \left(\frac{x}{q}\right) = -1 \right\}.$$

The security of the scheme, which will be discussed later, is related to the difficulty of the **Composite Quadratic Residues** problem in $\mathbb{Z}_n$ (Problem 8.1).

Several aspects of Cryptosystem 12.1 require explanation. First, we have stated that the hash function h produces outputs that are always elements in $\mathbf{QR}(n) \cup \widetilde{\mathbf{QR}}(n)$. This is equivalent to saying that $0 < h(x) < n$ and the Jacobi symbol $\left(\frac{h(x)}{n}\right) = 1$ for all $x \in \{0, 1\}^*$. In practice, one might compute $\left(\frac{h(x)}{n}\right)$. If it is equal to -1, then we would multiply $h(x)$ by some fixed value $a \in \mathbb{Z}_n$ having Jacobi symbol equal to -1. This value a could be pre-determined and made public. In any event, we will assume that some method has been specified so that $h(x) \in \mathbf{QR}(n) \cup \widetilde{\mathbf{QR}}(n)$ for all relevant values of x.

The generation of a user's private key is basically a matter of extracting a square root modulo n, as was done in the decryption operation of the *Rabin Cryptosystem*. This computation can be carried out by the *TA* because the *TA* knows the factorization of n.

Note that the *TA* will only compute square roots of numbers having a special, predetermined form, namely, $h(ID(U))$ or $-h(ID(U))$, for a user U, say. This is important, because a square root oracle can be used to factor n, as we showed in Section 5.8.1. This attack cannot be carried out in the context of the *Cocks Identity-based Cryptosystem* because U cannot use the *TA* as an oracle to extract square roots of arbitrary elements of $\mathbb{Z}_n$.

When a user V wishes to encrypt a plaintext $x = \pm 1$ to send to U, V has to generate two random elements of $\mathbb{Z}_n$, both having Jacobi symbols equal to x. This would be done by choosing random elements of $\mathbb{Z}_n$ and computing their Jacobi symbols, until elements with the desired Jacobi symbols are obtained. (Recall that the computation of Jacobi symbols modulo n can be performed efficiently without knowing the factorization of n.) If V wishes to encrypt a long string of plaintext elements, then each element must be encrypted independently, using different random t's.

When U wants to decrypt a ciphertext y, U only requires one of y_1 and y_2. So U chooses the appropriate one and the other one can be discarded. The reason why both y_1 and y_2 are transmitted to U is that V does not know whether U's private key is a square root of K_U^{pub} or a square root of $-K_U^{pub}$.

Cryptosystem 12.1: *Cocks Identity-based Cryptosystem*

Let p, q be two distinct primes such that $p \equiv q \equiv 3 \bmod 4$, and define $n = pq$.

System parameters: The master key $M = (M^{pub}, M^{priv})$, where

$$M^{pub} = n$$

and

$$M^{priv} = (p, q).$$

As well, $h : \{0, 1\}^* \to \mathbb{Z}_n$ is a public hash function with the property that $h(x) \in \mathbf{QR}(n) \cup \widetilde{\mathbf{QR}}(n)$ for all $x \in \{0, 1\}^*$.

User key generation: For a user U, the key $K_U = (K_U^{pub}, K_U^{priv})$, where

$$K_U^{pub} = h(ID(U))$$

and

$$(K_U^{priv})^2 = \begin{cases} K_U^{pub} & \text{if } K_U^{pub} \in \mathbf{QR}(n) \\ -K_U^{pub} & \text{if } K_U^{pub} \in \widetilde{\mathbf{QR}}(n). \end{cases}$$

Encryption: A plaintext is an element in the set $\{1, -1\}$. To encrypt a plaintext element $x \in \{1, -1\}$, the following steps are performed:

1. Choose two random values $t_1, t_2 \in \mathbb{Z}_n$ such that the Jacobi symbols $\left(\frac{t_1}{n}\right) = \left(\frac{t_2}{n}\right) = x$.

2. Compute

$$y_1 = t_1 + K_U^{pub} (t_1)^{-1} \bmod n$$

and

$$y_2 = t_2 - K_U^{pub} (t_2)^{-1} \bmod n.$$

3. The ciphertext $y = (y_1, y_2)$.

Decryption: Given a ciphertext $y = (y_1, y_2)$, y is decrypted as follows:

1. If $(K_U^{priv})^2 = K_U^{pub}$, then define $s = y_1$; otherwise, define $s = y_2$.

2. Compute the Jacobi symbol

$$x = \left(\frac{s + 2K_U^{priv}}{n}\right).$$

3. The decrypted plaintext is x.

Now, let's show that the decryption operation works correctly, i.e., any encryption of x can successfully be decrypted, given the relevant private key. Suppose that U receives a ciphertext (y_1, y_2). Suppose further that $(K_U^{priv})^2 = K_U^{pub}$; then we will show that

$$\left(\frac{y_1 + 2K_U^{priv}}{n}\right) = x.$$

(If $(K_U^{priv})^2 = -K_U^{pub}$, then the decryption process is modified, but the proof of correctness is similar.) The following sequence of equations follows from basic properties of Jacobi symbols:

$$\left(\frac{y_1 + 2K_U^{priv}}{n}\right) = \left(\frac{t_1 + K_U^{pub}(t_1)^{-1} + 2K_U^{priv}}{n}\right)$$

$$= \left(\frac{t_1 + 2K_U^{priv} + (K_U^{priv})^2(t_1)^{-1}}{n}\right)$$

$$= \left(\frac{t_1(1 + 2K_U^{priv}(t_1)^{-1} + (K_U^{priv})^2(t_1)^{-2})}{n}\right)$$

$$= \left(\frac{t_1}{n}\right)\left(\frac{1 + 2K_U^{priv}(t_1)^{-1} + (K_U^{priv})^2(t_1)^{-2}}{n}\right)$$

$$= \left(\frac{t_1}{n}\right)\left(\frac{(1 + K_U^{priv}(t_1)^{-1})^2}{n}\right)$$

$$= \left(\frac{t_1}{n}\right)\left(\frac{1 + K_U^{priv}(t_1)^{-1}}{n}\right)^2$$

$$= \left(\frac{t_1}{n}\right).$$

In the derivation of the last line, we are using the fact that

$$\left(\frac{1 + K_U^{priv}(t_1)^{-1}}{n}\right) = \pm 1,$$

which can be proven easily (see the Exercises).

Next, let's consider the security of the scheme. We will prove that a decryption oracle for Cryptosystem 12.1 can be used to solve the **Composite Quadratic Residues** problem in $\mathbb{Z}_n$. Thus the cryptosystem is provably secure provided that the **Composite Quadratic Residues** problem is intractable.

First, we begin with an important technical lemma.

LEMMA 12.1 *Suppose that $x = \pm 1$ and $\left(\frac{t}{n}\right) = x$, where x and t are unknown. If $(K_U^{priv})^2 \equiv K_U^{pub} \pmod{n}$, then the value*

$$t - K_U^{pub}t^{-1} \bmod n$$

provides no information about x. Similarly, if $(K_U^{priv})^2 \equiv -K_U^{pub} \pmod{n}$, then the value

$$t + K_U^{pub}\, t^{-1} \bmod n$$

provides no information about x.

PROOF Suppose that

$$(K_U^{priv})^2 \equiv K_U^{pub} \pmod{n}$$

and denote

$$y = t - K_U^{pub}\, t^{-1} \bmod n.$$

Then

$$t^2 - ty - K_U^{pub} \equiv 0 \pmod{n},$$

so

$$t^2 - ty - K_U^{pub} \equiv 0 \pmod{p}$$

and

$$t^2 - ty - K_U^{pub} \equiv 0 \pmod{q}.$$

The first congruence has two solutions modulo p, and the product of these two solutions is congruent to $-K_U^{pub}$ modulo p. If the two solutions are r_1 and r_2, then we have

$$\left(\frac{r_2}{p}\right) = \left(\frac{-r_1\, K_U^{pub}}{p}\right) = \left(\frac{-r_1\, (K_U^{priv})^2}{p}\right) = \left(\frac{-r_1}{p}\right) = -\left(\frac{r_1}{p}\right).$$

A similar property holds for the two solutions of the second congruence. If these two solutions are s_1 and s_2, then

$$\left(\frac{s_2}{q}\right) = -\left(\frac{s_1}{q}\right).$$

Now, the congruence modulo n has four solutions for t. It is easy to see that two of the solutions have Jacobi symbol $\left(\frac{t}{n}\right) = 1$ and two of them have $\left(\frac{t}{n}\right) = -1$. Therefore it is not possible to compute any information about the Jacobi symbol $\left(\frac{t}{n}\right)$.

The second part of this lemma is proven in a similar manner. ∎

Now, suppose that COCKS-DECRYPT is a decryption oracle for the *Cocks Identity-based Cryptosystem*. That is, COCKS-DECRYPT(K_U^{pub}, n, y) correctly outputs the value of x whenever y is a valid encryption of x. We will show how to use the algorithm COCKS-DECRYPT to determine if K_U^{pub} is a quadratic residue or a pseudo-square modulo n. Our algorithm is presented as Algorithm 12.3.

We will analyze Algorithm 12.3 informally. First, let's discuss the operations it performs. The input $a \in \mathbf{QR}(n) \cup \widetilde{\mathbf{QR}}(n)$. We treat a as a public key for

---

Algorithm 12.3: COCKS-ORACLE-RESIDUE-TESTING(n, a)

comment: $\left(\frac{a}{n}\right) = 1$

external COCKS-DECRYPT
choose $x \in \{1, -1\}$ randomly
choose a random $t \in \mathbb{Z}_n$ such that $\left(\frac{t}{n}\right) = x$
$y_1 \leftarrow t + a t^{-1} \bmod n$
choose a random $y_2 \in \mathbb{Z}_n^*$
$y \leftarrow (y_1, y_2)$
$x' \leftarrow$ COCKS-DECRYPT(n, a, y)
if $x' = x$
   then return ("$a \in \mathbf{QR}(n)$")
   else return ("$a \in \widetilde{\mathbf{QR}}(n)$")

---

the *Cocks Cryptosystem* and encrypt a random plaintext x. However, we only compute y_1 according to the encryption rule; y_2 is just a random element of $\mathbb{Z}_n^*$. Then we give the pair (y_1, y_2) to the decryption oracle COCKS-DECRYPT. The oracle outputs a decryption x'. Then Algorithm 12.3 reports that a is a quadratic residue modulo n if and only if $x = x'$.

Suppose that $a \in \mathbf{QR}(n)$. Then Lemma 12.1 says that, even if y_2 were computed accoding to the encryption rule, it would provide no information about x. So COCKS-DECRYPT can correctly compute x from y_1 alone. In this case, Algorithm 12.3 correctly states that a is a quadratic residue.

On the other hand, suppose that $a \in \widetilde{\mathbf{QR}}(n)$. Then Lemma 12.1 says that y_1 provides no information about x. Clearly y_2 provides no information about x because y_2 is random. Therefore, the value x' that is returned by COCKS-DECRYPT is going to be equal to x exactly half the time, because x is random and y is independent of x. Therefore the output of Algorithm 12.3 will be correct with probability $1/2$.

The situation is analogous to that of a biased Monte Carlo algorithm. If $x \neq x'$, then we can be sure that $a \in \widetilde{\mathbf{QR}}(n)$. On the other hand, if $x = x'$, we cannot say with certainty that $a \in \mathbf{QR}(n)$; it may be only that COCKS-DECRYPT guessed the value of x correctly. So we should run Algorithm 12.3 several times on the same input. If it always reports that $a \in \mathbf{QR}(n)$, then we have some confidence that this conclusion is correct. The analysis of the probability of correctness of this approach is the same as was done in Section 5.4.

The above discussion assumed that COCKS-DECRYPT always outputs the correct answer if it is given a correctly formed ciphertext. A more complicated analysis shows that we can obtain a (possibly) unbiased Monte Carlo algorithm for **Composite Quadratic Residues** with error probability as small as desired, provided that COCKS-DECRYPT has an error probability less than $1/2$. The process is somewhat similar to the one used in Section 8.3 when we studied the security of the *BBS Generator*.

12.6 Notes and References

"Understanding PKI, Second Edition" [1] , by Adams and Lloyd, is the standard reference for information on public-key infrastructures. Two other recent books that have extensive content relating to PKI and related topics include "Cryptography and Public Key Infrastructure on the Internet" [291] by Schmeh and "Network Security, Private Communication in a Public World, Second Edition" [187], by Kaufman, Perlman and Speciner. For an interesting article critiquing PKI, see Ellison and Schneier [126].

A worthwhile early discussion of PKIs can be found in the 1997 Master's Thesis of Branchaud [68]. Kohnfelder's Bachelor's Thesis [202], which introduced the idea of certificates, was published in 1978.

The concept of identity-based cryptography was introduced by Shamir [298] in 1984. For a recent survey on this topic, see Gagné [151]. Research into identity-based cryptosystems exploded after the publication of the system proposed by Boneh and Franklin [61] in 2002, which was the first really practical system. The *Cocks Identity-based Cryptosystem*, which we presented in Section 12.5.1, was published in 2001 (see [94]). However, this system is not really practical due to the data expansion of the encryption process.

Most recent proposals for identity-based cryptography are based on the idea of pairings. Suppose that $(G, +)$ and $(H, \cdot)$ are abelian groups. A *pairing* is a non-degenerate mapping $e : G \times G \to H$ which satisfies the property of *bilinearity*: $e(a\,g_1, b\,g_2) = e(g_1, g_2)^{ab}$ for all $g_1, g_2 \in G$ and for all integers a, b. Two well-known examples of pairings are the so-called Tate and Weil pairings. In these pairings, G is an elliptic curve and H is a finite field. For a recent paper discussing implementations of pairings in cryptography, see Barreto, Kim, Lynn and Scott [8].

Pairings were used by Menezes, Okamoto and Vanstone [235] to efficiently solve the **Elliptic Curve Discrete Logarithm** problem on supersingular elliptic curves. However, the recent applications of pairings in identity-based cryptography have used pairings to construct cryptosystems. This approach was first suggested by Sakai, Ohgishi and Kasahara [288].

Exercises

12.1 Setting up an SSL session, as described in Figure 12.1, incorporates server-to-client authentication but no client-to-server authentication. Suppose the client (Alice) is intending to purchase something from the server (Bob, Inc.) using her credit card. The protocol in Figure 12.1 is used to derive keys K_1 and K_2 that will be used to encrypt and authenticate Alice's credit card number when it is sent to Bob, Inc. in the ensuing SSL session. Briefly discuss the following points relating to *SSL*.

    (a) Why is it desirable for Alice's web browser to verify the authenticity of Bob's public key?

    (b) In this version of the protocol, there is no way for Bob to authentice Alice during the set-up phase. Does this matter to Bob? Explain.

    (c) The keys K_1 and K_2 are derived from a random number MS supplied by Alice. Why is MS generated by Alice rather than by Bob, Inc.? Are there any potential security risks associated with this method of deriving the keys K_1 and K_2?

12.2 Discuss potential problems in using a networked PKI to connect root CAs of two domains having different trust models.

12.3 Suppose that Alice's PGP keyring contains the following data:

owner	OTF	signatures
Alice	implicit	Alice, Bob
Bob	complete	Alice, Bob, Ginger
Charlie	partial	Bob, Charlie, Janet
Doris	partial	Doris, Eve, Janet
Eve	partial	Doris, Eve, Fred
Fred	partial	Bob, Harry , Fred
Ginger	untrusted	Charlie, Eve, Fred, Ginger
Harry	untrusted	Eve, Harry, Irene
Irene	untrusted	Eve, Fred, Harry, Irene
Janet	untrusted	Alice, Bob, Eve, Janet

Compute the KLFs of all the users in Alice's keyring.

12.4 In the *Cocks Identity-based Cryptosystem*, verify that

$$\left(\frac{1 + K_U^{priv}\,(t_1)^{-1}}{n} \right) = \pm 1.$$

12.5 Suppose the *Cocks Identity-based Cryptosystem* is implemented with master public key $n = 16402692653$, and suppose that a user U has public key $K_U^{pub} = 9305496225$.

    (a) Let $t_1 = 3975333024$ and $t_2 = 4892498575$. Verify that $\left(\frac{t_1}{n}\right) = \left(\frac{t_2}{n}\right) = -1$.

    (b) Encrypt the plaintext $x = -1$ using the "random" values t_1 and t_2, obtaining the ciphertext (y_1, y_2).

    (c) Given that $K_U^{priv} = 96465$, verify that the decryption of (y_1, y_2) is equal to x.

13
Secret Sharing Schemes

13.1 Introduction: The Shamir Threshold Scheme

In a bank, there is a vault which must be opened every day. The bank employs three senior tellers, but they do not trust the combination to any individual teller. Hence, we would like to design a system whereby any two of the three senior tellers can gain access to the vault, but no individual teller can do so. This problem can be solved by means of a secret sharing scheme, the topic of this chapter.

Here is an interesting "real-world" example of this situation: According to *Time Magazine*,[1] control of nuclear weapons in Russia in the early 1990s depended upon a similar "two-out-of-three" access mechanism. The three parties involved were the President, the Defense Minister and the Defense Ministry.

We first study a special type of secret sharing scheme called a threshold scheme. Here is an informal definition.

Definition 13.1: Let t, w be positive integers, $t \leq w$. A (t, w)-*threshold scheme* is a method of sharing a *key K* among a set of w participants (denoted by $\mathcal{P}$), in such a way that any t participants can compute the value of K, but no group of $t - 1$ participants can do so.

We will study the unconditional security of secret sharing schemes. That is, we do not place any limit on the amount of computation that can be performed by any subset of participants.

Note that the examples described above are $(2, 3)$-threshold schemes.

The value of K is chosen by a special participant called the *dealer*. The dealer is denoted by D and we assume $D \notin \mathcal{P}$. When D wants to share the key K among the participants in $\mathcal{P}$, he gives each participant some partial information called a *share*. The shares should be distributed secretly, so no participant knows the share given to another participant.

---

[1] Time Magazine, May 4, 1992, p. 13

Cryptosystem 13.1: *Shamir (t, w)-Threshold Scheme*

Initialization Phase

1. D chooses w distinct, non-zero elements of $\mathbb{Z}_p$, denoted x_i, $1 \leq i \leq w$ (this is where we require $p \geq w + 1$). For $1 \leq i \leq w$, D gives the value x_i to P_i. The values x_i are public.

Share Distribution

2. Suppose D wants to share a key $K \in \mathbb{Z}_p$. D secretly chooses (independently at random) $t - 1$ elements of $\mathbb{Z}_p$, which are denoted $a_1, \ldots, a_{t-1}$.

3. For $1 \leq i \leq w$, D computes $y_i = a(x_i)$, where

$$a(x) = K + \sum_{j=1}^{t-1} a_j x^j \bmod p.$$

4. For $1 \leq i \leq w$, D gives the share y_i to P_i.

At a later time, a subset of participants $B \subseteq \mathcal{P}$ will pool their shares in an attempt to compute the key K. (Alternatively, they could give their shares to a trusted authority which will perform the computation for them.) If $|B| \geq t$, then they should be able to compute the value of K as a function of the shares they collectively hold; if $|B| < t$, then they should not be able to compute K.

We will use the following notation. Let

$$\mathcal{P} = \{P_i : 1 \leq i \leq w\}$$

be the set of w participants. $\mathcal{K}$ is the *key set* (i.e., the set of all possible keys); and $\mathcal{S}$ is the *share set* (i.e., the set of all possible shares).

In this section, we present a method of constructing a (t, w)-threshold scheme, called the *Shamir Threshold Scheme*, which was invented by Shamir in 1979. Let $\mathcal{K} = \mathbb{Z}_p$, where $p \geq w + 1$ is prime. Also, let $\mathcal{S} = \mathbb{Z}_p$. Hence, the key will be an element of $\mathbb{Z}_p$, as will be each share given to a participant. The *Shamir Threshold Scheme* is presented as Cryptosystem 13.1.

In this scheme, the dealer constructs a random polynomial $a(x)$ of degree at most $t - 1$ in which the constant term is the key, K. Every participant P_i obtains a point (x_i, y_i) on this polynomial.

Let's look at how a subset B of t participants can reconstruct the key. This is basically accomplished by means of polynomial interpolation. We will describe a couple of methods of doing this.

Suppose that participants $P_{i_1}, \ldots, P_{i_t}$ want to determine K. They know that

$$y_{i_j} = a(x_{i_j}),$$

$1 \leq j \leq t$, where $a(x) \in \mathbb{Z}_p[x]$ is the (secret) polynomial chosen by D. Since $a(x)$ has degree at most $t - 1$, $a(x)$ can be written as

$$a(x) = a_0 + a_1x + \cdots + a_{t-1}x^{t-1},$$

where the coefficients $a_0, \ldots, a_{t-1}$ are unknown elements of $\mathbb{Z}_p$, and $a_0 = K$ is the key. Since $y_{i_j} = a(x_{i_j})$, $1 \leq j \leq t$, the subset B can obtain t linear equations in the t unknowns $a_0, \ldots, a_{t-1}$, where all arithmetic is done in $\mathbb{Z}_p$. If the equations are linearly independent, there will be a unique solution, and a_0 will be revealed as the key.

Here is a small example to illustrate.

Example 13.1 Suppose that $p = 17$, $t = 3$, and $w = 5$; and the public x-coordinates are $x_i = i$, $1 \leq i \leq 5$. Suppose that $B = \{P_1, P_3, P_5\}$ pool their shares, which are respectively 8, 10, and 11. Writing the polynomial $a(x)$ as

$$a(x) = a_0 + a_1x + a_2x^2,$$

and computing $a(1)$, $a(3)$ and $a(5)$, the following three linear equations in $\mathbb{Z}_{17}$ are obtained:

$$a_0 + a_1 + a_2 = 8$$

$$a_0 + 3a_1 + 9a_2 = 10$$

$$a_0 + 5a_1 + 8a_2 = 11.$$

This system has a unique solution in $\mathbb{Z}_{17}$: $a_0 = 13$, $a_1 = 10$, and $a_2 = 2$. The key is therefore $K = a_0 = 13$. □

Clearly, it is important that the system of t linear equations has a unique solution, as in Example 13.1. There are various ways to show that this is always the case. Perhaps the nicest way to address this question is to appeal to the Lagrange interpolation formula for polynomials, which was presented in Theorem 10.3. This theorem states that the desired polynomial $a(x)$ of degree at most $t - 1$ is unique, and it provides an explicit formula that can be used to compute $a(x)$. The formula for $a(x)$ is as follows:

$$a(x) = \sum_{j=1}^{t} \left(y_{i_j} \prod_{1 \leq k \leq t, k \neq j} \frac{x - x_{i_k}}{x_{i_j} - x_{i_k}} \right) \bmod p.$$

A group B of t participants can compute $a(x)$ by using the interpolation formula. But a simplification is possible, bacause the participants in B do not need to know the whole polynomial $a(x)$. It is sufficient for them to deduce the constant term $K = a(0)$. Hence, they can compute the following expression, which is obtained by substituting $x = 0$ into the Lagrange interpolation formula:

$$K = \sum_{j=1}^{t} \left(y_{i_j} \prod_{1 \le k \le t, k \ne j} \frac{x_{i_k}}{x_{i_k} - x_{i_j}} \right) \bmod p.$$

Suppose we define

$$b_j = \prod_{1 \le k \le t, k \ne j} \frac{x_{i_k}}{x_{i_k} - x_{i_j}} \bmod p,$$

$1 \le j \le t$. (Note that the b_j's can be precomputed, if desired, and their values are not secret.) Then we have

$$K = \sum_{j=1}^{t} b_j y_{i_j} \bmod p.$$

Hence, the key is a linear combination (modulo p) of the t shares.

To illustrate this approach, let's recompute the key from Example 13.1.

Example 13.1 **(Cont.)** The participants $\{P_1, P_3, P_5\}$ can compute b_1, b_2, and b_3 according to the formula given above. For example, they would obtain

$$b_1 = \frac{x_3 x_5}{(x_3 - x_1)(x_5 - x_1)} \bmod 17$$

$$= 3 \times 5 \times (-2)^{-1} \times (-4)^{-1} \bmod 17$$

$$= 4.$$

Similarly, $b_2 = 3$ and $b_3 = 11$. Then, given shares $8, 10$, and 11 (respectively), they would obtain

$$K = 4 \times 8 + 3 \times 10 + 11 \times 11 \bmod 17 = 13,$$

as before.   □

What happens if a subset B of $t - 1$ participants attempt to compute K? Suppose they hypothesize a value $y_0 \in \mathbb{Z}_p$ for the key K. In the *Shamir Threshold Scheme*, the key is $K = a_0 = a(0)$. Recall that the $t - 1$ shares held by B are obtained by evaluating the polynomial $a(x)$ at $t - 1$ elements of $\mathbb{Z}_p$. Now, applying Theorem 10.3 again, there is a unique polynomial $a_{y_0}(x)$ such that

$$y_{i_j} = a_{y_0}(x_{i_j}),$$

$1 \le j \le t - 1$, and such that

$$y_0 = a_{y_0}(0).$$

Cryptosystem 13.2: *Simplified (t, t)-Threshold Scheme*

1. D secretly chooses (independently at random) $t - 1$ elements of $\mathbb{Z}_m$, $y_1, \ldots, y_{t-1}$.

2. D computes

$$y_t = K - \sum_{i=1}^{t-1} y_i \bmod m.$$

3. For $1 \le i \le t$, D gives the share y_i to P_i.

That is, there is a polynomial $a_{y_0}(x)$ that is consistent with the $t-1$ shares known to B and which also has y_0 as the key. Since this is true for any possible value $y_0 \in \mathbb{Z}_p$, it follows that no value of the key can be ruled out, and thus a group of $t - 1$ participants can obtain no information about the key.

For example, suppose that P_1 and P_3 try to compute K, given shares as in Example 13.1. Thus P_1 has the share 8 and P_3 has the share 10. For any possible value y_0 of the key, there is a unique polynomial $a_{y_0}(x)$ that takes on the value 8 at $x = 1$, the value 10 at $x = 3$, and the value y_0 at $x = 0$. Using the interpolation formula, this polynomial is seen to be

$$a_{y_0}(x) = 6y_0(x - 1)(x - 3) + 13x(x - 3) + 13x(x - 1) \bmod 17.$$

The subset $\{P_1, P_3\}$ has no way of knowing which of these polynomials is the correct one, and hence they have no information about the value of K.

13.1.1 A Simplified (t, t)-threshold Scheme

The last topic of this section is a simplified construction for threshold schemes in the special case $w = t$. This construction will work for any key set $\mathcal{K} = \mathbb{Z}_m$, and it has $\mathcal{S} = \mathbb{Z}_m$. (For this scheme, it is not required that m be prime, and it is not necessary that $m \geq w + 1$.) If D wants to share the key $K \in \mathbb{Z}_m$, he carries out the steps in Cryptosystem 13.2.

Observe that the t participants can compute K by the formula

$$K = \sum_{i=1}^{t} y_i \bmod m.$$

Can $t-1$ participants compute K? Clearly, the first $t-1$ participants cannot do so, since they receive $t-1$ independent random numbers as their shares. Consider the $t-1$ participants in the set $\mathcal{P}\backslash\{P_i\}$, where $1 \leq i \leq t-1$. These $t-1$ participants possess the shares

$$y_1, \ldots, y_{i-1}, y_{i+1}, \ldots, y_{t-1}$$

and

$$K - \sum_{i=1}^{t-1} y_i.$$

By summing their shares, they can compute $K - y_i$. However, they do not know the random value y_i, and hence they have no information as to the value of K. Consequently, we have a (t, t)-threshold scheme.

Example 13.2   Suppose that $p = 10$ and $t = 4$ in Cryptosystem 13.2. Suppose also that the shares for the four participants are $y_1 = 7$, $y_2 = 2$, $y_3 = 4$ and $y_4 = 2$. The key is therefore

$$K = 7 + 2 + 4 + 2 \bmod 10 = 5.$$

Suppose that the first three participants try to determine K. They know that $y_1 + y_2 + y_3 \bmod 10 = 3$, but they do not know the value of y_4. There is a one-to-one correspondence between the ten possible values of y_4 and the ten possible values of the key K:

$$y_4 = 0 \Leftrightarrow K = 3, y_4 = 1 \Leftrightarrow K = 4, \ldots, y_4 = 9 \Leftrightarrow K = 2.$$

◻

13.2   Access Structures and General Secret Sharing

In the previous section, we desired that any t of the w participants should be able to determine the key. A more general situation is to specify exactly which subsets of participants should be able to determine the key and which should not. Let Γ be a set of subsets of $\mathcal{P}$; the subsets in Γ are those subsets of participants that should be able to compute the key. Γ is called an *access structuresecret sharing scheme!access structure* and each subset in Γ is called an *authorized subsetsecret sharing scheme!authorized subset*.

Let $\mathcal{K}$ be the key set and let $\mathcal{S}$ be the share set. As before, when a dealer D wants to share a key $K \in \mathcal{K}$, he will give each participant a share from $\mathcal{S}$. At a later time a subset of participants will attempt to determine K from the shares they collectively hold.

> **Definition 13.2:** A *perfect secret sharing scheme* realizing the access structure Γ is a method of sharing a key K among a set of w participants (denoted by $\mathcal{P}$), in such a way that the following two properties are satisfied:
>
> 1. If an authorized subset of participants $B \subseteq \mathcal{P}$ pool their shares, then they can determine the value of K.
>
> 2. If an unauthorized subset of participants $B \subseteq \mathcal{P}$ pool their shares, then they can determine nothing about the value of K.

Suppose that $B \in \Gamma$ and $B \subseteq C \subseteq \mathcal{P}$. Suppose the subset C wants to determine K. Since B is an authorized subset, it can already determine K. Hence, the subset C can determine K by ignoring the shares of the participants in $C \backslash B$. Stated another way, a superset of an authorized set is again an authorized set. What this says is that the access structure should satisfy the *monotone property*:

$$\text{if } B \in \Gamma \text{ and } B \subseteq C \subseteq \mathcal{P}, \text{ then } C \in \Gamma.$$

In the remainder of this chapter, we will assume that all access structures are monotone.

Observe that a (t, w)-threshold scheme realizes the access structure

$$\{B \subseteq \mathcal{P} : |B| \geq t\}.$$

Such an access structure is called a *threshold access structure*. We showed in the previous section that the *Shamir Threshold Scheme* is a perfect secret sharing scheme realizing a threshold access structure.

If Γ is an access structure, then $B \in \Gamma$ is a *minimal authorized subset* if $A \notin \Gamma$ whenever $A \subseteq B$, $A \neq B$. The set of minimal authorized subsets of Γ is denoted Γ_0 and is called the *basis* of Γ. Since Γ consists of all subsets of $\mathcal{P}$ that are supersets of a subset in the basis Γ_0, it follows that Γ is determined uniquely as a function of Γ_0. Expressed mathematically, we have

$$\Gamma = \{C \subseteq \mathcal{P} : B \subseteq C, B \in \Gamma_0\}.$$

Example 13.3 Suppose $\mathcal{P} = \{P_1, P_2, P_3, P_4\}$ and

$$\Gamma_0 = \{\{P_1, P_2, P_4\}, \{P_1, P_3, P_4\}, \{P_2, P_3\}\}.$$

Then

$$\Gamma = \Gamma_0 \cup \{\{P_1, P_2, P_3\}, \{P_2, P_3, P_4\}, \{P_1, P_2, P_3, P_4\}\}.$$

Conversely, given this access structure Γ, it is easy to see that Γ_0 consists of the minimal subsets in Γ. $\square$

In the case of a (t, w)-threshold access structure, the basis consists of all subsets of (exactly) t participants.

13.2.1   The Monotone Circuit Construction

In this section, we will give a conceptually simple and elegant construction due to Benaloh and Leichter that shows that any (monotone) access structure can be realized by a perfect secret sharing scheme. The idea is to first build a monotone circuit that "recognizes" the access structure, and then to build the secret sharing scheme from the description of the circuit. We call this the "Monotone circuit construction."

Suppose we have a *boolean circuit* $\mathbf{C}$, with w boolean inputs, $x_1, \ldots, x_w$ (corresponding to the w participants $P_1, \ldots, P_w$), and one boolean output, y. The circuit consists of "or" gates and "and" gates; we do not allow any "not" gates. Such a circuit is called a *monotone boolean circuit*. The reason for this nomenclature is that changing any input x_i from "0" (false) to "1" (true) can never result in the output y changing from "1" to "0." The circuit is permitted to have arbitrary *fan-in*, but we require *fan-out* equal to 1 (that is, a gate can have arbitrarily many input wires, but only one output wire).

If we specify boolean values for the w inputs of such a monotone circuit, we can define

$$B(x_1, \ldots, x_w) = \{P_i : x_i = 1\},$$

i.e., the subset of $\mathcal{P}$ corresponding to the true inputs. Suppose $\mathbf{C}$ is a monotone circuit, and define

$$\Gamma(\mathbf{C}) = \{B(x_1, \ldots, x_w) : \mathbf{C}(x_1, \ldots, x_w) = 1\},$$

where $\mathbf{C}(x_1, \ldots, x_w)$ denotes the output of $\mathbf{C}$, given inputs $x_1, \ldots, x_w$. Since the circuit $\mathbf{C}$ is monotone, it follows that $\Gamma(\mathbf{C})$ is a monotone set of subsets of $\mathcal{P}$.

It is easy to see that there is a one-to-one correspondence between monotone circuits of this type and *boolean formulae* which contain the operators $\wedge$ ("and") and $\vee$ ("or"), but which do not contain any negations.

If Γ is a monotone set of subsets of $\mathcal{P}$, then it is easy to construct a monotone circuit $\mathbf{C}$ such that $\Gamma(\mathbf{C}) = \Gamma$. One way to do this is as follows. Let Γ_0 be the basis of Γ. Then construct the *disjunctive normal form* boolean formula

$$\bigvee_{B \in \Gamma_0} \left(\bigwedge_{P_i \in B} P_i \right).$$

In Example 13.3, where

$$\Gamma_0 = \{\{P_1, P_2, P_4\}, \{P_1, P_3, P_4\}, \{P_2, P_3\}\},$$

we would obtain the boolean formula

$$(P_1 \wedge P_2 \wedge P_4) \vee (P_1 \wedge P_3 \wedge P_4) \vee (P_2 \wedge P_3). \tag{13.1}$$

Each clause in the boolean formula corresponds to an "and" gate of the associated monotone circuit; the final disjunction corresponds to an "or" gate. The

---

Algorithm 13.1: MONOTONE CIRCUIT CONSTRUCTION(**C**)

$f(W_{out}) \leftarrow K$
while there exists a wire W such that $f(W)$ is not defined
do $\left\{\begin{array}{l}
\text{find a gate } G \text{ of } \mathbf{C} \text{ such that } f(W_G) \text{ is defined, where } W_G \text{ is the} \\
\quad \text{output wire of } G, \text{ but } f(W) \text{ is not defined for any of the input} \\
\quad \text{wires of } G \\
\textbf{if } G \text{ is an "or" gate} \\
\quad \textbf{then } f(W) \leftarrow f(W_G) \text{ for every input wire } W \text{ of } G \\
\quad \textbf{else } \left\{\begin{array}{l}
\text{let the input wires of } G \text{ be } W_1, \ldots, W_t \\
\text{choose (independently at random) } t-1 \text{ elements of } \mathbb{Z}_m, \\
\quad \text{denoted by } y_{G,1}, \ldots, y_{G,t-1} \\
y_{G,t} \leftarrow f(W_G) - \sum_{i=1}^{t-1} y_{G,i} \bmod m \\
\textbf{for } i \leftarrow 1 \textbf{ to } t \\
\quad \textbf{do } f(W_i) \leftarrow y_{G,i}
\end{array}\right.
\end{array}\right.$

---

number of gates in the circuit is $|\Gamma_0| + 1$. This particular circuit has two "levels" (more formally, its depth is two), but this is not a requirement.

Suppose **C** is any monotone circuit that recognizes Γ (note that **C** need not be the circuit described above.) We describe an algorithm which enables D, the dealer, to construct a perfect secret sharing scheme that realizes Γ. This scheme will use as a building block the (t, t)-schemes constructed in Cryptosystem 13.2. Hence, we can take the key set to be $\mathcal{K} = \mathbb{Z}_m$, for any positive integer m.

The algorithm proceeds by assigning a value $f(W) \in \mathcal{K}$ to every wire W in the circuit **C**. Initially, the output wire W_{out} of the circuit is assigned the value K, the key. The algorithm iterates a number of times, working from the bottom of the circuit up to the top, until every wire has a value assigned to it. Finally, each participant P_i is given the list of values $f(W)$ such that W is an input wire of the circuit which receives input x_i. A description of the construction is given as Algorithm 13.1.

Note that, whenever a gate G is an "and" gate having (say) t input wires, we share the "key" $f(W_G)$ for the gate G among its input wires, using a (t, t)-threshold scheme.

Let's carry out this procedure for the access structure of Example 13.3, using the circuit corresponding to the boolean formula (13.1).

Example 13.4 We illustrate the construction in Figure 13.1. Suppose K is the key. The value K is given to each of the three input wires of the final "or" gate. Next, we consider the "and" gate corresponding to the clause $P_1 \wedge P_2 \wedge P_4$. The three input wires are assigned values $a_1, a_2, K - a_1 - a_2$, respectively, where all

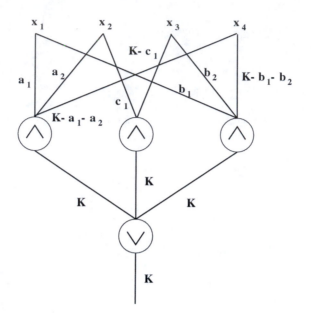

FIGURE 13.1
A monotone circuit

arithmetic is done in $\mathbb{Z}_m$. In a similar way, the three input wires corresponding to $P_1 \wedge P_3 \wedge P_4$ are assigned values $b_1, b_2, K - b_1 - b_2$. Finally, the two input wires corresponding to $P_2 \wedge P_3$ are assigned values $c_1, K - c_1$. Note that a_1, a_2, b_1, b_2 and c_1 are all independent random values in $\mathbb{Z}_m$.

If we look at the shares that the four participants receive, we have the following:

1. P_1 receives $(y_1^1, y_1^2) = (a_1, b_1)$.
2. P_2 receives $(y_2^1, y_2^2) = (a_2, c_1)$.
3. P_3 receives $(y_3^1, y_3^2) = (b_2, K - c_1)$.
4. P_4 receives $(y_4^1, y_4^2) = (K - a_1 - a_2, K - b_1 - b_2)$.

Thus, every participant receives two elements of $\mathbb{Z}_m$ as his or her share.

Let's prove that the scheme is perfect. First, we verify that each basis subset can compute K. The authorized subset $\{P_1, P_2, P_4\}$ can compute

$$K = y_1^1 + y_2^1 + y_4^1 \bmod m = a_1 + a_2 + (K - a_1 - a_2) \bmod m.$$

The subset $\{P_1, P_3, P_4\}$ can compute

$$K = y_1^2 + y_3^1 + y_4^2 \bmod m = b_1 + b_2 + (K - b_1 - b_2) \bmod m.$$

Finally, the subset $\{P_2, P_3\}$ can compute

$$K = y_2^2 + y_3^2 \bmod m = c_1 + (K - c_1) \bmod m.$$

Thus any authorized subset can compute K, so we turn our attention to the unauthorized subsets. Note that we do not need to look at all the unauthorized subsets. For, if B_1 and B_2 are both unauthorized subsets, $B_1 \subseteq B_2$, and B_2 cannot compute K, then neither can B_1 compute K. Define a subset $B \subseteq \mathcal{P}$ to be a *maximal* unauthorized subset if $B_1 \in \Gamma$ for all $B_1 \supseteq B$, $B_1 \neq B$. It suffices to verify that none of the maximal unauthorized subsets can determine any information about K.

Here, the maximal unauthorized subsets are

$$\{P_1, P_2\}, \{P_1, P_3\}, \{P_1, P_4\}, \{P_2, P_4\}, \{P_3, P_4\}.$$

In each case, it is easy to see that K cannot be computed, either because some necessary piece of "random" information is missing, or because all the shares possessed by the subset are random. For example, the subset $\{P_1, P_2\}$ possesses only the random values a_1, b_1, a_2, c_1. As another example, the subset $\{P_3, P_4\}$ possesses the values $b_2, K - c_1, K - a_1 - a_2, K - b_1 - b_2$. Since the values of c_1, a_1, a_2, and b_1 are unknown random values, K cannot be computed.

In each possible case, it can be checked that an unauthorized subset has no information about the value of K. ▯

We can obtain a different scheme realizing the same access structure by using a different circuit. We illustrate by returning again to the access structure of Example 13.3.

Example 13.5 Suppose we convert the formula (13.1) to a *conjunctive normal form* boolean formula:

$$(P_1 \lor P_2) \land (P_1 \lor P_3) \land (P_2 \lor P_3) \land (P_2 \lor P_4) \land (P_3 \lor P_4). \tag{13.2}$$

(The reader can verify that this formula is equivalent to the formula (13.1).) If we implement the scheme using the circuit corresponding to formula (13.2), then we obtain the following:

1. P_1 receives $(y_1^1, y_1^2) = (a_1, a_2)$.
2. P_2 receives $(y_2^1, y_2^2, y_2^3) = (a_1, a_3, a_4)$.
3. P_3 receives $(y_3^1, y_3^2, y_3^3) = (a_2, a_3, K - a_1 - a_2 - a_3 - a_4)$.
4. P_4 receives $(y_4^1, y_4^2) = (a_4, K - a_1 - a_2 - a_3 - a_4)$.

We leave the details for the reader to check. ▯

We now prove that the monotone circuit construction always produces a perfect secret sharing scheme.

THEOREM 13.1   *Let* **C** *be any monotone boolean circuit. Then the monotone circuit construction yields a perfect secret sharing scheme realizing the access structure* $\Gamma(\mathbf{C})$.

PROOF   We proceed by induction on the number of gates in the circuit **C**. If **C** contains only one gate, then the result is fairly trivial: If **C** consists of one "or" gate, then every participant will be given the key. (This scheme realizes the access structure consisting of all non-empty subsets of participants.) Alternatively, if **C** consists of a single "and" gate with t inputs, then the scheme is the (t, t)-threshold scheme presented in Cryptosystem 13.2.

Now, as an induction assumption, suppose that there is an integer $j > 1$ such that, for all circuits **C** with fewer than j gates, the construction produces a scheme that realizes $\Gamma(\mathbf{C})$. Let **C** be a circuit on j gates. Consider the "last" gate, G, in the circuit; again, G could be either an "or" gate or an "and" gate. Let's first consider the case where G is an "or" gate. Denote the input wires to G by W_i, $1 \leq i \leq t$. These t input wires are the outputs of t sub-circuits of **C**, which we denote $\mathbf{C}_i$, $1 \leq i \leq t$. Corresponding to each $\mathbf{C}_i$, we have a (sub-)scheme that realizes the access structure $\Gamma_{\mathbf{C}_i}$, by induction. Now, it is easy to see that

$$\Gamma(\mathbf{C}) = \bigcup_{i=1}^{t} \Gamma_{\mathbf{C}_i}.$$

Since every W_i is assigned the key K, it follows that the scheme realizes $\Gamma(\mathbf{C})$, as desired.

The analysis is similar if G is an "and" gate. In this situation, we have

$$\Gamma(\mathbf{C}) = \bigcap_{i=1}^{t} \Gamma_{\mathbf{C}_i}.$$

Since the key K is shared among the t wires W_i using a (t, t)-threshold scheme, it follows again that the scheme realizes $\Gamma(\mathbf{C})$. This completes the proof.   ∎

Of course, when an authorized subset, B, wants to compute the key, the participants in B need to know the circuit used by D to distribute shares, and which shares correspond to which wires of the circuit. All this information will be public knowledge. Only the actual values of the shares are secret.

The algorithm for reconstructing the key involves combining shares according to the structure of the circuit, with the stipulation that an "and" gate corresponds to summing the values on the input wires modulo m (provided these values are all known), and an "or" gate involves choosing the value on any input wire (with the understanding that all these values will be identical).

Suppose we return to Example 13.4 and we again consider the authorized subset $\{P_1, P_2, P_4\}$. We already demonstrated how this subset can compute K. Our observation above is that the circuit enables the reconstruction to be carried out in a systematic way. Here, we would assign values to six of the eight input wires, namely, the ones emanating from x_1, x_2 and x_4 in Figure 13.1. It can be seen that the leftmost "and" gate has values assigned to all three of its input wires. The sum of the values on these input wires yields K. This computation is, in fact, the same as the one described in Example 13.4.

13.2.2   Formal Definitions

In this section, we will give formal mathematical definitions of a (perfect) secret sharing scheme. We represent a secret sharing scheme by a set of distribution rules. A *distribution rule* is a function

$$f : \mathcal{P} \to \mathcal{S}.$$

A distribution rule represents a possible distribution of shares to the participants, where $f(P_i)$ is the share given to P_i, $1 \leq i \leq w$.

Now, for each $K \in \mathcal{K}$, let $\mathcal{F}_K$ be a set of distribution rules. $\mathcal{F}_K$ will be distribution rules corresponding to the key having the value K. The sets of distribution rules $\mathcal{F}_K$ are public knowledge.

Next, define

$$\mathcal{F} = \bigcup_{K \in \mathcal{K}} \mathcal{F}_K.$$

$\mathcal{F}$ is the complete set of distribution rules of the scheme. If $K \in \mathcal{K}$ is the value of the key that D wishes to share, then D will choose a distribution rule $f \in \mathcal{F}_K$, and use it to distribute shares.

This is a general model in which we can study secret sharing schemes. Any of our existing schemes can be described in this setting by determining the possible distribution rules which the scheme will use. The fact that this model is mathematically precise makes it easier to give definitions and to present rigorous proofs.

It is useful to develop conditions which ensure that a set of distribution rules for a scheme realizes a specified access structure. This will involve looking at certain probability distributions, as we did previously when studying the concept of perfect secrecy in Chapter 2. To begin with, we suppose that there is a probability distribution defined on the set of keys, $\mathcal{K}$. The probability that the key $K \in \mathcal{K}$ is selected by the dealer will be denoted by $\mathbf{Pr}[\mathbf{K} = K]$. Further, given $K \in \mathcal{K}$, D will choose a distribution rule $f \in \mathcal{F}_K$ according to a specified probability distribution; $\mathbf{Pr}[\mathbf{F_K} = f]$ will denote the probability that f is the distribution rule chosen by D (given that K is the key).

Given these probability distributions, it is straightforward to compute the probability distribution on the list of shares given to any subset of participants, B

(authorized or unauthorized). This is done as follows. Suppose $B \subseteq \mathcal{P}$. Define

$$\mathcal{S}(B) = \{f|_B : f \in \mathcal{F}\},$$

where the function $f|_B$ denotes the restriction of the distribution rule f to B. That is, $f|_B : B \to \mathcal{S}$ is defined by

$$f|_B(P_i) = f(P_i)$$

for all $P_i \in B$. Thus, $\mathcal{S}(B)$ is the set of possible distributions of shares to the participants in B.

The probability distribution on $\mathcal{S}(B)$ is computed as follows: Let $g_B \in \mathcal{S}(B)$. Then

$$\mathbf{Pr[S(B)} = g_B] = \sum_{K \in \mathcal{K}} \left(\mathbf{Pr}[K] \times \sum_{\{f \in \mathcal{F}_K : f|_B = g_B\}} \mathbf{Pr[F_K} = f] \right).$$

Also, it is straightforward to see that

$$\mathbf{Pr[S(B)} = g_B | \mathbf{K} = K] = \sum_{\{f \in \mathcal{F}_K : f|_B = g_B\}} \mathbf{Pr[F_K} = f],$$

for all $g_B \in \mathcal{S}(B)$ and all $K \in \mathcal{K}$.

Here is a formal definition of a perfect secret sharing scheme.

Definition 13.3:   Suppose Γ is an access structure and

$$\mathcal{F} = \bigcup_{K \in \mathcal{K}} \mathcal{F}_K$$

is a set of distribution rules. Then $\mathcal{F}$ is a *perfect secret sharing scheme* realizing the access structure Γ provided that the following two properties are satisfied:

1. For any authorized subset of participants $B \subseteq \mathcal{P}$, there do not exist two distribution rules $f \in \mathcal{F}_K$ and $f' \in \mathcal{F}_{K'}$ with $K \neq K'$, such that $f|_B = f'|_B$. (That is, any distribution of shares to the participants in an authorized subset B determines the value of the key.)

2. For any unauthorized subset of participants $B \subseteq \mathcal{P}$ and for any distribution of shares $g_B \in \mathcal{S}_B$, it holds that

$$\mathbf{Pr[K} = K | \mathbf{S(B)} = g_B] = \mathbf{Pr[K} = K]$$

for every $K \in \mathcal{K}$. (That is, the conditional probability distribution on $\mathcal{K}$, given a distribution of shares g_B to an unauthorized subset B, is the same as the *a priori* probability distribution on $\mathcal{K}$. In other words, the distribution of shares to B provides no information as to the value of the key.)

Observe that the second property in Definition 13.3 is very similar to the concept of perfect secrecy presented in Definition 2.3; this similarity is why the resulting secret sharing scheme is termed "perfect."

The probability $\mathbf{Pr}[\mathbf{K} = K | \mathbf{S(B)} = g_B]$ can be computed from probability distributions exhibited above using Bayes' theorem:

$$\mathbf{Pr}[\mathbf{K} = K | \mathbf{S(B)} = g_B] = \frac{\mathbf{Pr}[\mathbf{S(B)} = g_B | \mathbf{K} = K] \times \mathbf{Pr}[\mathbf{K} = K]}{\mathbf{Pr}[\mathbf{S(B)} = g_B]}.$$

Let us now illustrate these definitions by looking at a small example.

Example 13.6 Suppose that the scheme constructed in Example 13.5 is implemented in $\mathbb{Z}_m$. Then we have $\mathcal{S} = (\mathbb{Z}_m)^2 \cup (\mathbb{Z}_m)^3$ and $|\mathcal{F}_K| = m^4$ for every $K \in \mathbb{Z}_m$. For any $K \in \mathbb{Z}_m$, each of the m^4 distribution rules in $\mathcal{F}_K$ is chosen with the same probability, $1/m^4$. The m possible keys need not be chosen equiprobably, however.

For the sake of concreteness, we present the distribution rules for this scheme when $m = 2$. In this case, each of $\mathcal{F}_0$ and $\mathcal{F}_1$ contains 16 equiprobable distribution rules. In order to achieve a concise representation, we replace a binary k-tuple by an integer between 0 and $2^k - 1$. If this is done, then $\mathcal{F}_0$ and $\mathcal{F}_1$ are as depicted in Figure 13.2, where each row represents a distribution rule.

This yields a perfect secret sharing scheme for any probability distribution on the keys. We will not perform all the verifications here, but we will look at a couple of typical cases to illustrate the two properties in Definition 13.3.

The subset $\{P_2, P_3\}$ is an authorized subset. Thus the shares that P_2 and P_3 receive should (together) determine a unique key. It can easily be checked that any distribution of shares to these two participants occurs in a distribution rule in at most one of the sets $\mathcal{F}_0$ and $\mathcal{F}_1$. For example, if P_2 has the share 3 and P_3 has the share 6, then the distribution rule must be the eighth rule in $\mathcal{F}_0$ and thus the key is 0.

On the other hand, $B = \{P_1, P_2\}$ is an unauthorized subset. It is not too hard to see that any distribution of shares to these two participants occurs in exactly one distribution rule in $\mathcal{F}_0$ and in exactly one distribution rule in $\mathcal{F}_1$. That is,

$$\mathbf{Pr}[\mathbf{S(B)} = g_B | \mathbf{K} = K] = \frac{1}{16}$$

for any $g_B \in \mathcal{S}(B)$ and for $K = 0, 1$. Next, we compute

$$\mathbf{Pr}[\mathbf{S(B)} = g_B] = \sum_{K \in \mathcal{K}} \left(\mathbf{Pr}[K] \times \sum_{\{f \in \mathcal{F}_K : f|_B = g_B\}} \mathbf{Pr}[\mathbf{F_K} = f] \right)$$

$$= \sum_{K=0}^{1} \left(\mathbf{Pr}[K] \times \frac{1}{16} \right)$$

$$= \frac{1}{16},$$

$\mathcal{F}_0$			
P_1	P_2	P_3	P_4
0	0	0	0
0	1	1	3
0	2	3	1
0	3	2	2
1	0	4	0
1	1	5	3
1	2	7	1
1	3	6	2
2	4	0	0
2	5	1	3
2	6	3	1
2	7	2	2
3	4	4	0
3	5	5	3
3	6	7	1
3	7	6	2

$\mathcal{F}_1$			
P_1	P_2	P_3	P_4
0	0	1	1
0	1	0	2
0	2	2	0
0	3	3	3
1	0	5	1
1	1	4	2
1	2	6	0
1	3	7	3
2	4	1	1
2	5	0	0
2	6	2	2
2	7	3	3
3	4	5	1
3	5	4	2
3	6	6	0
3	7	7	3

FIGURE 13.2
Distribution rules for a secret sharing scheme

for any $g_B \in \mathcal{S}(B)$.

Finally, we use Bayes' theorem to compute $\mathbf{Pr}[\mathbf{K} = K | \mathbf{S}(\mathbf{B}) = g_B]$:

$$\mathbf{Pr}[\mathbf{K} = K | \mathbf{S}(\mathbf{B}) = g_B] = \frac{\mathbf{Pr}[\mathbf{S}(\mathbf{B}) = g_B | \mathbf{K} = K] \times \mathbf{Pr}[\mathbf{K} = K]}{\mathbf{Pr}[\mathbf{S}(\mathbf{B}) = g_B]}$$

$$= \frac{\frac{1}{16} \times \mathbf{Pr}[\mathbf{K} = K]}{\frac{1}{16}}$$

$$= \mathbf{Pr}[\mathbf{K} = K],$$

so the second property is satisfied for this subset B.

Similar computations can be performed for other authorized and unauthorized sets, and in each case the appropriate property is satisfied. Hence we have a perfect secret sharing scheme. □

13.3 Information Rate and Construction of Efficient Schemes

The results of Section 13.2.1 prove that any monotone access structure can be realized by a perfect secret sharing scheme. We now want to consider the efficiency of the resulting schemes. In the case of a (t, w)-threshold scheme, we can construct a circuit corresponding to the disjunctive normal form boolean formula which will have $1 + \binom{w}{t}$ gates. Each participant will receive $\binom{w-1}{t-1}$ elements of $\mathbb{Z}_m$ as his or her share. This is very inefficient, because a *Shamir (t, w)-Threshold Scheme* enables a key to be shared by giving each participant only one "piece" of information.

In general, we measure the efficiency of a secret sharing scheme by the information rate, which we define now.

Definition 13.4:   Suppose we have a perfect secret sharing scheme realizing an access structure Γ. The *information rate for P_i* is the ratio

$$\rho_i = \frac{\log_2 |\mathcal{K}|}{\log_2 |\mathcal{S}(P_i)|}.$$

(Note that $\mathcal{S}(P_i)$ denotes the set of possible shares that P_i might receive; of course $\mathcal{S}(P_i) \subseteq \mathcal{S}$.) The *information rate* of the scheme is denoted by ρ and is defined as

$$\rho = \min\{\rho_i : 1 \le i \le w\}.$$

The motivation for this definition is as follows. Since the key K comes from a finite set $\mathcal{K}$, we can think of K as being represented by a bit-string of length $\log_2 |\mathcal{K}|$, by using a binary encoding, for example. In a similar way, a share given to P_i can be represented by a bit-string of length $\log_2 |\mathcal{S}(P_i)|$. Intuitively, P_i receives $\log_2 |\mathcal{S}(P_i)|$ bits of information (in his or her share), but the information content of the key is $\log_2 |\mathcal{K}|$ bits. Thus ρ_i is the ratio of the number of bits in a share to the number of bits in the key.

Example 13.7   Let's look at the two schemes from Section 13.2, both of which realize the access structure having basis

$$\Gamma_0 = \{\{P_1, P_2, P_4\}, \{P_1, P_3, P_4\}, \{P_2, P_3\}\}.$$

The scheme produced in Example 13.4 has

$$\rho_i = \frac{\log_2 m}{\log_2 m^2} = \frac{1}{2},$$

$i = 1, \ldots, 4$. Hence, $\rho = 1/2$.

In Example 13.5, we have a scheme with $\rho_1 = \rho_4 = 1/2$ and $\rho_2 = \rho_3 = 1/3$. Hence, $\rho = 1/3$. The first implementation, having higher information rate, is preferable.   ▯

In general, if we construct a scheme from a circuit **C** using the monotone circuit construction, then the information rate can be computed as indicated in the following theorem.

THEOREM 13.2   *Let* **C** *be any monotone boolean circuit. Then there is a perfect secret sharing scheme realizing the access structure* $\Gamma(\mathbf{C})$ *having information rate*

$$\rho = \max\{1/r_i : 1 \le i \le w\},$$

where r_i *denotes the number of input wires to* **C** *carrying the input* x_i.

With respect to threshold access structures, we observe that the *Shamir Threshold Scheme* has information rate 1, which we show below is the optimal value. In contrast, an implementation of a (t, w)-threshold scheme using a disjunctive normal form boolean circuit would have information rate $1/\binom{w-1}{t-1}$, which is much lower (and therefore inferior) if $1 < t < w$.

Obviously, a high information rate is desirable. The first general result we prove is that $\rho \le 1$ in any perfect secret sharing scheme.

THEOREM 13.3   *In any perfect secret sharing scheme realizing an access structure* Γ, $\rho \le 1$.

PROOF   Suppose we have a perfect secret sharing scheme that realizes the access structure Γ. Let $B \in \Gamma_0$ and choose any participant $P_j \in B$. Define $B' = B \backslash \{P_j\}$. Let $g \in \mathcal{S}(B)$. Now, $B' \notin \Gamma$, so the distribution of shares $g|_{B'}$ provides no information about the key. Hence, for each $K \in \mathcal{K}$, there is a distribution rule $g^K \in \mathcal{F}_K$ such that $g^K|_{B'} = g|_{B'}$. Since $B \in \Gamma$, it must be the case that $g^K(P_j) \neq g^{K'}(P_j)$ if $K \neq K'$. Hence, $|\mathcal{S}(P_j)| \ge |\mathcal{K}|$, and thus $\rho_j \le 1$. Therefore, $\rho \le 1$.   ∎

Since $\rho = 1$ is the optimal situation, we refer to such a scheme an *ideal secret sharing scheme*. The *Shamir Threshold Schemes* are ideal schemes. In the next section, we present a construction for ideal schemes that generalizes the *Shamir Threshold Schemes*.

13.3.1   The Vector Space Construction

In this section, we present a construction method for certain ideal schemes known as the *Vector space construction*. This technique is due to Brickell.

Suppose Γ is an access structure, and let $(\mathbb{Z}_p)^d$ denote the vector space of all d-tuples over $\mathbb{Z}_p$, where p is prime and $d \ge 2$. Suppose there exists a function

$$\phi : \mathcal{P} \to (\mathbb{Z}_p)^d$$

which satisfies the property

$$(1, 0, \ldots, 0) \in \langle \phi(P_i) : P_i \in B \rangle \Leftrightarrow B \in \Gamma. \tag{13.3}$$

In other words, the vector $(1, 0, \ldots, 0)$ can be expressed as a linear combination (modulo p) of the vectors in the set $\{\phi(P_i) : P_i \in B\}$ if and only if B is an

---

Cryptosystem 13.3: *Brickell Secret Sharing Scheme*

Input: vectors $\phi(P_1), \ldots, \phi(P_w)$ satisfying Property (13.3).

Initialization Phase

1. For $1 \leq i \leq w$, D gives the vector $\phi(P_i) \in (\mathbb{Z}_p)^d$ to P_i. These vectors are public.

Share Distribution

2. Suppose D wants to share a key $K \in \mathbb{Z}_p$. D defines $a_1 = K$, and he secretly chooses (independently at random) $d-1$ elements of $\mathbb{Z}_p$, $a_2, \ldots, a_d$.

3. For $1 \leq i \leq w$, D computes $y_i = \overline{a} \cdot \phi(P_i)$, where

$$\overline{a} = (a_1, a_2, \ldots, a_d).$$

4. For $1 \leq i \leq w$, D gives the share y_i to P_i.

---

authorized subset. In general, finding such a function is often a matter of trial and error, though we will see some explicit constructions of suitable functions ϕ for certain access structures a bit later.

Assuming that ϕ that satisfies Property (13.3), we are going to construct an ideal secret sharing scheme with $\mathcal{K} = \mathcal{S}(P_i) = \mathbb{Z}_p$, $1 \leq i \leq w$. The distribution rules of the scheme are as follows: for every vector $\overline{a} = (a_1, \ldots, a_d) \in (\mathbb{Z}_p)^d$, define a distribution rule $f_{\overline{a}} \in \mathcal{F}_{a_1}$, where

$$f_{\overline{a}}(x) = \overline{a} \cdot \phi(x)$$

for every $x \in \mathcal{P}$, and the operation "$\cdot$" is the inner product modulo p. Observe that the key is given by

$$K = a_1 = \overline{a} \cdot (1, \ldots, 0).$$

Note that each $\mathcal{F}_K$ contains p^{d-1} distribution rules. We will stipulate that the probability distributions defined on each $\mathcal{F}_K$ are uniform: $\mathbf{Pr}[f] = 1/p^{d-1}$ for every $f \in \mathcal{F}_K$.

Provided that we have found a set of vectors satisfying Property (13.3) for a given access structure, we can proceed to construct the corresponding secret sharing scheme. This scheme is known as a *Brickell Secret Sharing Scheme*, and it has distribution rules as described above. In more detail, the *Brickell Secret Sharing Scheme* is presented as Cryptosystem 13.3.

We have the following result.

THEOREM 13.4   *Suppose ϕ satisfies Property (13.3). Then the sets of distribution rules $\mathcal{F}_K$, $K \in \mathcal{K}$, comprise an ideal secret sharing scheme that realizes Γ.*

PROOF   First, we will show that if B is an authorized subset, then the participants in B can compute K. Since

$$(1, 0, \ldots, 0) \in \langle \phi(P_i) : P_i \in B \rangle,$$

we can write

$$(1, 0, \ldots, 0) = \sum_{\{i : P_i \in B\}} c_i \, \phi(P_i),$$

where each $c_i \in \mathbb{Z}_p$. The share given to P_i is y_i, where

$$y_i = \overline{a} \cdot \phi(P_i),$$

$\overline{a} = (a_1, \ldots, a_d)$ is an unknown vector chosen by D, and $K = a_1$.

By the linearity of the inner product operation, we have

$$K = \overline{a} \cdot (1, 0, \ldots, 0)$$

$$= \overline{a} \cdot \sum_{\{i : P_i \in B\}} c_i \, \phi(P_i)$$

$$= \sum_{\{i : P_i \in B\}} c_i (\overline{a} \cdot \phi(P_i))$$

$$= \sum_{\{i : P_i \in B\}} c_i \, y_i.$$

Thus, it is a simple matter for the participants in B to compute the key as a linear combination of the shares that they hold:

$$K = \sum_{\{i : P_i \in B\}} c_i \, y_i.$$

What happens if B is not an authorized subset? Suppose that B hypothesizes that $K = y_0$, for some $y_0 \in \mathbb{Z}_p$. We will show that any such guess is consistent with the information (i.e., the shares) that they hold.

Denote by e the dimension of the subspace $\langle \phi(P_i) : P_i \in B \rangle$ (note that $e \leq |B|$), and consider the system of equations:

$$\phi(P_i) \cdot \overline{a} = s_i, \forall P_i \in B$$

$$(1, 0, \ldots, 0) \cdot \overline{a} = y_0.$$

This is a system of linear equations (modulo p) in the d unknowns $a_1, \ldots, a_d$. The coefficient matrix has rank $e + 1$, because

$$(1, 0, \ldots, 0) \notin \langle \phi(P_i) : P_i \in B \rangle.$$

Provided that the system of equations is consistent, the solution space will have dimension $d - e - 1$ (for any possible value $y_0 \in \mathbb{Z}_p$). It will then follow that there are precisely p^{d-e-1} distribution rules in each $\mathcal{F}_{y_0}$ that are consistent with a possible distribution g_B of shares to B. By a computation similar to the one performed in Example 13.6, it can be shown that

$$\mathbf{Pr}[\mathbf{K} = y_0 | g_B] = \mathbf{Pr}[\mathbf{K} = y_0]$$

for every $y_0 \in \mathbb{Z}_p$.

Why is the system consistent? The first $|B|$ equations are consistent, because the vector $\overline{a}$ chosen by D is a solution. Further, we have assumed that

$$(1, 0, \ldots, 0) \notin \langle \phi(P_i) : P_i \in B \rangle.$$

Therefore the last equation is consistent with the first $|B|$ equations. This completes the proof. ∎

Here is an example to illustrate the Vector space construction.

Example 13.8 We consider the access structure having basis

$$\{\{P_1, P_2, P_3\}, \{P_1, P_4\}\}.$$

Suppose we take $d = 3$, $p \geq 3$, and define the vectors $\phi(P_i)$ as follows:

$$\phi(P_1) = (0, 1, 0)$$
$$\phi(P_2) = (1, 0, 1)$$
$$\phi(P_3) = (0, 1, -1)$$
$$\phi(P_4) = (1, 1, 0).$$

We verify that Property (13.3) holds. First, we have

$$\phi(P_4) - \phi(P_1) = (1, 1, 0) - (0, 1, 0)$$
$$= (1, 0, 0).$$

Also,

$$\phi(P_2) + \phi(P_3) - \phi(P_1) = (1, 0, 1) + (0, 1, -1) - (0, 1, 0)$$
$$= (1, 0, 0).$$

Hence,

$$(1, 0, 0) \in \langle \phi(P_1), \phi(P_2), \phi(P_3) \rangle$$

and
$$(1, 0, 0) \in \langle \phi(P_1), \phi(P_4) \rangle.$$

Now, it suffices to show that

$$(1, 0, 0) \notin \langle \phi(P_i) : P_i \in B \rangle$$

if B is a maximal unauthorized subset. There are three such subsets B to be considered: $\{P_1, P_2\}$, $\{P_1, P_3\}$, and $\{P_2, P_3, P_4\}$. In each case, we need to establish that a certain system of linear equations has no solution modulo p. For example, suppose that

$$(1, 0, 0) = a_2 \, \phi(P_2) + a_3 \, \phi(P_3) + a_4 \, \phi(P_4),$$

where $a_2, a_3, a_4 \in \mathbb{Z}_p$. This is equivalent to the system

$$a_2 + a_4 = 1$$
$$a_3 + a_4 = 0$$
$$a_2 - a_3 = 0.$$

The system is easily seen to have no solution. We leave the other two subsets B for the reader to consider.

Now we look at the *Brickell Secret Sharing Scheme* that would be implemented using this choice of vectors $\phi(P_i)$, $1 \leq i \leq 4$. Suppose that $p = 127$ and $K = 99$, and the dealer chooses $a_2 = 55$ and $a_3 = 38$. Then the four shares are as follows:

$$y_1 = 55$$
$$y_2 = 10$$
$$y_3 = 17, \quad \text{and}$$
$$y_4 = 27.$$

Suppose that the subset $\{P_1, P_2, P_3\}$ wants to compute K. We showed above that

$$(1, 0, 0) = -\phi(P_1) + \phi(P_2) + \phi(P_3).$$

Hence,

$$K = -y_1 + y_2 + y_3 \bmod p = -55 + 10 + 17 \bmod 127 = 99.$$

□

It is interesting to observe that the *Shamir (t, w)-Threshold Scheme* is a special case of the vector space construction. To see this, define $d = t$ and let

$$\phi(P_i) = (1, x_i, x_i^2, \ldots, x_i^{t-1})$$

for $1 \leq i \leq w$, where x_i is the x-coordinate given to P_i. The resulting scheme is equivalent to the *Shamir Threshold Scheme*; we leave the details to the reader to check.

Here is another general result that is easy to prove. It concerns access structures that have as a basis a collection of pairs of participants that forms a complete multipartite graph. A graph $G = (V, E)$ with vertex set V and edge set E is defined to be a *complete multipartite graph* if the vertex set V can be partitioned into subsets $V_1, \ldots, V_\ell$ such that $\{x, y\} \in E$ if and only if $x \in V_i, y \in V_j$, where $i \neq j$. The sets V_i are called *parts*. The complete multipartite graph is denoted by $K_{n_1, \ldots, n_\ell}$ if $|V_i| = n_i$, $1 \leq i \leq \ell$. A complete multipartite graph $K_{1, \ldots, 1}$ (with ℓ parts) is in fact a *complete graph* and is denoted K_ℓ.

THEOREM 13.5 *Suppose $G = (V, E)$ is a complete multipartite graph. Then there is an ideal scheme realizing the access structure having basis E on participant set V.*

PROOF Let $V_1, \ldots, V_\ell$ be the parts of G. Let $x_1, \ldots, x_\ell$ be distinct elements of $\mathbb{Z}_p$, where $p \geq \ell$. Let $d = 2$. For every participant $v \in V_i$, define $\phi(v) = (x_i, 1)$. It is straightforward to verify that Property (13.3) holds. Then, by Theorem 13.4, we have an ideal scheme. ∎

To further illustrate the application of the Vector space construction, we will consider all the possible access structures on up to four participants. We consider only the *connected access structures*, i.e., those in which the basis cannot be partitioned into two non-empty subsets on disjoint participant sets. For example,

$$\Gamma_0 = \{\{P_1, P_2\}, \{P_3, P_4\}\}$$

can be partitioned as

$$\{\{P_1, P_2\}\} \cup \{\{P_3, P_4\}\}.$$

This partition forms two access structures on disjoint subsets of participants, so we do not consider it.

We list the non-isomorphic connected access structures on two, three, and four participants in Table 13.1. In Table 13.1, the quantities ρ^* denote the maximum possible information rates that can be attained; this is discussed further in Section 13.3.2.

Of the 18 access structures listed in Table 13.1, we can easily obtain ideal schemes for ten of them using general constructions we have described already. These ten access structures are either threshold access structures (for which we have a *Shamir Threshold Scheme*), or they have a basis which is a complete multipartite graph (so Theorem 13.5 can be applied).

One such access structure is #9, whose basis is the complete multipartite graph $K_{1,1,2}$. We illustrate this in the following example.

TABLE 13.1
Access structures for at most four participants

	w	subsets in Γ_0	ρ^*	comments
1.	2	$P_1 P_2$	1	$(2,2)$-threshold
2.	3	$P_1 P_2, P_2 P_3$	1	$\Gamma_0 \cong K_{1,2}$
3.	3	$P_1 P_2, P_2 P_3, P_1 P_3$	1	$(2,3)$-threshold
4.	3	$P_1 P_2 P_3$	1	$(3,3)$-threshold
5.	4	$P_1 P_2, P_2 P_3, P_3 P_4$	$2/3$	Example 13.11
6.	4	$P_1 P_2, P_1 P_3, P_1 P_4$	1	$\Gamma_0 \cong K_{1,3}$
7.	4	$P_1 P_2, P_1 P_4, P_2 P_3, P_3 P_4$	1	$\Gamma_0 \cong K_{2,2}$
8.	4	$P_1 P_2, P_2 P_3, P_2 P_4, P_3 P_4$	$2/3$	Example 13.12
9.	4	$P_1 P_2, P_1 P_3, P_1 P_4, P_2 P_3, P_2 P_4$	1	$\Gamma_0 \cong K_{1,1,2}$
10.	4	$P_1 P_2, P_1 P_3, P_1 P_4, P_2 P_3, P_2 P_4, P_3 P_4$	1	$(2,4)$-threshold
11.	4	$P_1 P_2 P_3, P_1 P_4$	1	Example 13.8
12.	4	$P_1 P_3 P_4, P_1 P_2, P_2 P_3$	$2/3$	
13.	4	$P_1 P_3 P_4, P_1 P_2, P_2 P_3, P_2 P_4$	$2/3$	
14.	4	$P_1 P_2 P_3, P_1 P_2 P_4$	1	Example 13.10
15.	4	$P_1 P_2 P_3, P_1 P_2 P_4, P_3 P_4$	1	
16.	4	$P_1 P_2 P_3, P_1 P_2 P_4, P_1 P_3 P_4$	1	
17.	4	$P_1 P_2 P_3, P_1 P_2 P_4, P_1 P_3 P_4, P_2 P_3 P_4$	1	$(3,4)$-threshold
18.	4	$P_1 P_2 P_3 P_4$	1	$(4,4)$-threshold

Example 13.9 For access structure #9, take $d = 2$, $p \geq 3$, and define ϕ as follows:

$$\phi(P_1) = (0,1)$$
$$\phi(P_2) = (1,1)$$
$$\phi(P_3) = (2,1)$$
$$\phi(P_4) = (2,1).$$

Applying Theorem 13.5, an ideal scheme results.   ▯

Eight access structures remain to be considered. An ideal scheme for access structure #11 was presented in Example 13.8. Access structure #14 is discussed in the next example, Example 13.10, where an ideal scheme is described. It is also possible to use applications of the vector space construction to construct ideal schemes for access structures #15 and #16; see the Exercises.

Example 13.10 For access structure #14, take $d = 3$, $p \geq 2$ and define ϕ as follows:

$$\phi(P_1) = (0, 1, 0)$$
$$\phi(P_2) = (1, 0, 1)$$
$$\phi(P_3) = (0, 1, 1)$$
$$\phi(P_4) = (0, 1, 1).$$

The reader can verify that Property (13.3) is satisfied and hence an ideal scheme results. ⬜

In the next section, we will show that the remaining four access structures cannot be realized by ideal schemes.

13.3.2 An Upper Bound on the Information Rate

Four access structures remain to be considered: #5, #8, #12, and #13. We will show in this section that, for each of these access structures, there does not exist a perfect secret sharing scheme having information rate $\rho > 2/3$.

Denote by $\rho^* = \rho^*(\Gamma)$ the maximum information rate for any perfect secret sharing scheme realizing a specified access structure Γ. The first result we present is an entropy bound that will lead to an upper bound on ρ^* for certain access structures. We have assumed there is a probability distribution on the set of keys $\mathcal{K}$; the entropy of this probability distribution is denoted $H(\mathbf{K})$. We have also discussed the probability distribution on the list of shares $\mathcal{S}(B)$ given to any specified subset of participants $B \subseteq \mathcal{P}$. We will denote the entropy of this probability distribution by $H(\mathbf{B})$.

We begin by giving yet another definition of perfect secret sharing schemes, this time using the language of entropy. This definition is equivalent to Definition 13.3.

Definition 13.5: Suppose Γ is an access structure and $\mathcal{F}$ is a set of distribution rules. Then $\mathcal{F}$ is a *perfect secret sharing scheme* realizing the access structure Γ provided that the following two properties are satisfied:

1. For any authorized subset of participants $B \subseteq \mathcal{P}$, $H(\mathbf{K}|\mathbf{B}) = 0$.
2. For any unauthorized subset of participants $B \subseteq \mathcal{P}$, $H(\mathbf{K}|\mathbf{B}) = H(\mathbf{K})$.

We will require several entropy identities and inequalities. Some of these results were given in Section 2.5 and the rest are proved similarly, so we state them without proof in the following Lemma.

LEMMA 13.6   *Let* $\mathbf{X}$, $\mathbf{Y}$ *and* $\mathbf{Z}$ *be random variables. Then the following hold:*

$$H(\mathbf{X}, \mathbf{Y}) = H(\mathbf{X}|\mathbf{Y}) + H(\mathbf{Y}) \tag{13.4}$$

$$H(\mathbf{X}, \mathbf{Y}|\mathbf{Z}) = H(\mathbf{X}|\mathbf{Y}, \mathbf{Z}) + H(\mathbf{Y}|\mathbf{Z}) \tag{13.5}$$

$$H(\mathbf{X}, \mathbf{Y}|\mathbf{Z}) = H(\mathbf{Y}|\mathbf{X}, \mathbf{Z}) + H(\mathbf{X}|\mathbf{Z}) \tag{13.6}$$

$$H(\mathbf{X}|\mathbf{Y}) \geq 0 \tag{13.7}$$

$$H(\mathbf{X}|\mathbf{Z}) \geq H(\mathbf{X}|\mathbf{Y}, \mathbf{Z}) \tag{13.8}$$

$$H(\mathbf{X}, \mathbf{Y}|\mathbf{Z}) \geq H(\mathbf{Y}|\mathbf{Z}) \tag{13.9}$$

We next prove two preliminary entropy lemmas for secret sharing schemes.

LEMMA 13.7   *Suppose* Γ *is an access structure and* $\mathcal{F}$ *is a set of distribution rules realizing* Γ. *Suppose* $B \notin \Gamma$ *and* $A \cup B \in \Gamma$, *where* $A, B \subseteq \mathcal{P}$. *Then*

$$H(\mathbf{A}|\mathbf{B}) = H(\mathbf{K}) + H(\mathbf{A}|\mathbf{B}, \mathbf{K}).$$

PROOF   From Equations 13.5 and 13.6, we have that

$$H(\mathbf{A}, \mathbf{K}|\mathbf{B}) = H(\mathbf{A}|\mathbf{B}, \mathbf{K}) + H(\mathbf{K}|\mathbf{B})$$

and

$$H(\mathbf{A}, \mathbf{K}|\mathbf{B}) = H(\mathbf{K}|\mathbf{A}, \mathbf{B}) + H(\mathbf{A}|\mathbf{B}),$$

so

$$H(\mathbf{A}|\mathbf{B}, \mathbf{K}) + H(\mathbf{K}|\mathbf{B}) = H(\mathbf{K}|\mathbf{A}, \mathbf{B}) + H(\mathbf{A}|\mathbf{B}).$$

Since, by Property 2 of Definition 13.5, we have

$$H(\mathbf{K}|\mathbf{B}) = H(\mathbf{K}),$$

and, by Property 1 of Definition 13.5, we have

$$H(\mathbf{K}|\mathbf{A}, \mathbf{B}) = 0,$$

the result follows.   ∎

LEMMA 13.8   *Suppose* Γ *is an access structure and* $\mathcal{F}$ *is a set of distribution rules realizing* Γ. *Suppose* $A \cup B \notin \Gamma$, *where* $A, B \subseteq \mathcal{P}$. *Then*

$$H(\mathbf{A}|\mathbf{B}) = H(\mathbf{A}|\mathbf{B}, \mathbf{K}).$$

PROOF   As in Lemma 13.7, we have that

$$H(\mathbf{A}|\mathbf{B}, \mathbf{K}) + H(\mathbf{K}|\mathbf{B}) = H(\mathbf{K}|\mathbf{A}, \mathbf{B}) + H(\mathbf{A}|\mathbf{B}).$$

Since

$$H(\mathbf{K}|\mathbf{B}) = H(\mathbf{K})$$

and

$$H(\mathbf{K}|\mathbf{A}, \mathbf{B}) = H(\mathbf{K}),$$

the result follows. ∎

We now prove the following important theorem, which is due to Capocelli, De Santis, Gargano and Vaccaro.

THEOREM 13.9 *Suppose Γ is an access structure such that*

$$\{W, X\}, \{X, Y\}, \{W, Y, Z\} \in \Gamma$$

and

$$\{W, Y\}, \{X\}, \{W, Z\} \notin \Gamma.$$

Let $\mathcal{F}$ be any perfect secret sharing scheme realizing Γ. Then

$$H(\mathbf{XY}) \geq 3H(\mathbf{K}).$$

PROOF We establish a sequence of inequalities:

$$
\begin{aligned}
H(\mathbf{K}) &= H(\mathbf{Y}|\mathbf{W}, \mathbf{Z}) - H(\mathbf{Y}|\mathbf{W}, \mathbf{Z}, \mathbf{K}) && \text{by Lemma 13.7} \\
&\leq H(\mathbf{Y}|\mathbf{W}, \mathbf{Z}) && \text{by (13.7)} \\
&\leq H(\mathbf{Y}|\mathbf{W}) && \text{by (13.8)} \\
&= H(\mathbf{Y}|\mathbf{W}, \mathbf{K}) && \text{by Lemma 13.8} \\
&\leq H(\mathbf{X}, \mathbf{Y}|\mathbf{W}, \mathbf{K}) && \text{by (13.9)} \\
&= H(\mathbf{X}|\mathbf{W}, \mathbf{K}) + H(\mathbf{Y}|\mathbf{W}, \mathbf{X}, \mathbf{K}) && \text{by (13.5)} \\
&\leq H(\mathbf{X}|\mathbf{W}, \mathbf{K}) + H(\mathbf{Y}|\mathbf{X}, \mathbf{K}) && \text{by (13.8)} \\
&= H(\mathbf{X}|\mathbf{W}) - H(\mathbf{K}) + H(\mathbf{Y}|\mathbf{X}) - H(\mathbf{K}) && \text{by Lemma 13.7} \\
&\leq H(\mathbf{X}) - H(\mathbf{K}) + H(\mathbf{Y}|\mathbf{X}) - H(\mathbf{K}) && \text{by (13.7)} \\
&= H(\mathbf{X}, \mathbf{Y}) - 2H(\mathbf{K}) && \text{by (13.4).}
\end{aligned}
$$

Hence, the result follows. ∎

COROLLARY 13.10 *Suppose that Γ is an access structure that satisfies the hypotheses of Theorem 13.9. Suppose also that the keys are equally probable. Then $\rho \leq 2/3$.*

PROOF   Since the keys are equiprobable, we have

$$H(\mathbf{K}) = \log_2 |\mathcal{K}|.$$

Also, we have that

$$H(\mathbf{X}, \mathbf{Y}) \leq H(\mathbf{X}) + H(\mathbf{Y})$$
$$\leq \log_2 |\mathcal{S}(X)| + \log_2 |\mathcal{S}(Y)|.$$

By Theorem 13.9, we have that

$$H(\mathbf{X}, \mathbf{Y}) \geq 3H(\mathbf{K}).$$

Hence it follows that

$$\log_2 |\mathcal{S}(X)| + \log_2 |\mathcal{S}(Y)| \geq 3 \log_2 |\mathcal{K}|.$$

Now, by the definition of information rate, we have

$$\rho \leq \frac{\log_2 |\mathcal{K}|}{\log_2 |\mathcal{S}(X)|}$$

and

$$\rho \leq \frac{\log_2 |\mathcal{K}|}{\log_2 |\mathcal{S}(Y)|}.$$

It follows that

$$3 \log_2 |\mathcal{K}| \leq \log_2 |\mathcal{S}(X)| + \log_2 |\mathcal{S}(Y)|$$
$$\leq \frac{\log_2 |\mathcal{K}|}{\rho} + \frac{\log_2 |\mathcal{K}|}{\rho}$$
$$= 2\frac{\log_2 |\mathcal{K}|}{\rho}.$$

Hence, $\rho \leq 2/3$.   ■

For the access structures #5, #8, #12, and #13, the hypotheses of Theorem 13.9 are satisfied. Hence, $\rho^* \leq 2/3$ for these four access structures.

We also have the following result concerning ρ^* in the case where the access structure has a basis Γ_0 which is a graph. The proof involves showing that any connected graph which is not a multipartite graph contains an induced subgraph on four vertices that is isomorphic to the basis of access structure #5 or #8. If $G = (V, E)$ is a graph with vertex set V and edge set E, and $V_1 \subseteq V$, then the *induced subgraph* $G[V_1]$ is defined to be the graph (V_1, E_1), where

$$E_1 = \{uv \in E, u, v \in V_1\}.$$

THEOREM 13.11 *Suppose G is a connected graph that is not a complete multipartite graph. Let $\Gamma(G)$ be the access structure having basis E, where E is the edge set of G. Then $\rho^*(\Gamma(G)) \leq 2/3$.*

PROOF We will first prove that any connected graph that is not a complete multipartite graph must contain four vertices w, x, y, z such that the induced subgraph $G[w, x, y, z]$ is isomorphic to either the basis of access structure #5 or #8.

Let G^C denote the complement of G. Since G is not a complete multipartite graph, there must exist three vertices x, y, z such that $xy, yz \in E(G^C)$ and $xz \in E(G)$. Define

$$d = \min\{d_G(y, x), d_G(y, z)\},$$

where d_G denotes the length of a shortest path (in G) between two vertices. Then $d \geq 2$. Without loss of generality, we can assume that $d = d_G(y, x)$ by symmetry. Let

$$y_0, y_1, \ldots, y_{d-1}, x$$

be a path in G, where $y_0 = y$. We have that

$$y_{d-2}z, y_{d-2}x \in E(G^C)$$

and

$$y_{d-2}y_{d-1}, y_{d-1}x, xz \in E(G).$$

It follows that $G[y_{d-2}, y_{d-1}, x, z]$ is isomorphic to the basis of access structure #5 or #8, as desired.

So, we can assume that we have found four vertices w, x, y, z such that the induced subgraph $G[w, x, y, z]$ is isomorphic to either the basis of access structure #5 or #8. Now, let $\mathcal{F}$ be any scheme realizing the access structure $\Gamma(G)$. If we restrict the domain of the distribution rules to $\{w, x, y, z\}$, then we obtain a scheme $\mathcal{F}'$ realizing access structure #5 or #8. It is also obvious that $\rho(\mathcal{F}') \geq \rho(\mathcal{F})$. Since $\rho(\mathcal{F}') \leq 2/3$, it follows that $\rho(\mathcal{F}) \leq 2/3$. This completes the proof. ∎

Since $\rho^* = 1$ for complete multipartite graphs, Theorem 13.11 tells us that it is never the case that $2/3 < \rho^* < 1$ for any access structure whose basis is the edge set of a connected graph.

13.3.3 The Decomposition Construction

We still have four access structures in Table 13.1 to consider. Of course, we can use the monotone circuit construction to produce schemes for these access structures. However, by this method, the best we can do is to obtain information rate $\rho = 1/2$ in each case. (We can get $\rho = 1/2$ in cases #5 and #12 by using a disjunctive normal form boolean circuit. For cases #8 and #13, a disjunctive normal form boolean circuit will yield $\rho = 1/3$, but other monotone circuits exist which allow us to attain $\rho = 1/2$.) However, it is possible to produce schemes with $\rho = 2/3$ for each of these four access structures, by employing constructions that use ideal schemes as building blocks in the construction of larger schemes.

We first present a simple construction of this type called the *Decomposition construction*. First, we need to define an important concept.

Definition 13.6:   Suppose Γ is an access structure having basis Γ_0. Let $\mathcal{K}$ be a specified key set. An *ideal decomposition* of Γ_0 (for the key set $\mathcal{K}$) consists of a set of subsets, $\{\Gamma_1, \ldots, \Gamma_n\}$, such that the following properties are satisfied:

1. $\Gamma_k \subseteq \Gamma_0$ for $1 \leq k \leq n$

2. $\displaystyle\bigcup_{k=1}^{n} \Gamma_k = \Gamma_0$

3. For $1 \leq k \leq n$, there exists an ideal scheme with key set $\mathcal{K}$, on the subset of participants
$$\mathcal{P}_k = \bigcup_{B \in \Gamma_k} B,$$
for the access structure having basis Γ_k.

Given an ideal decomposition of an access structure Γ, we can easily construct a perfect secret sharing scheme, as described in the following theorem.

THEOREM 13.12   *(Decomposition construction) Suppose Γ is an access structure having basis Γ_0. Let $\mathcal{K}$ be a specified key set, and suppose $\{\Gamma_1, \ldots, \Gamma_n\}$ is an ideal decomposition of Γ for the key set $\mathcal{K}$. For every participant P_i, define*

$$R_i = |\{j : P_i \in \mathcal{P}_j\}|.$$

Then there exists a perfect secret sharing scheme realizing Γ, having information rate $\rho = 1/R$, where
$$R = \max\{R_i : 1 \leq i \leq w\}.$$

PROOF   For $1 \leq j \leq n$, we have an ideal scheme realizing the access structure with basis Γ_j, with key set $\mathcal{K}$, having $\mathcal{F}^j$ as its set of distribution rules. We will construct a scheme realizing Γ, with key set $\mathcal{K}$. The set of distribution rules $\mathcal{F}$ is constructed according to the following recipe. Suppose D wants to share a key K. Then, for every j, $1 \leq j \leq n$, he chooses a random distribution rule $f^j \in \mathcal{F}_K^j$ and distributes the resulting shares to the participants in $\mathcal{P}_j$.

We omit the proof that the scheme is perfect. However, it is easy to compute the information rate of the resulting scheme. Since each of the component schemes is ideal, it follows that
$$|\mathcal{S}(P_i)| = |\mathcal{K}|^{R_i},$$

for $1 \leq i \leq w$. So

$$\rho_i = \frac{1}{R_i},$$

and

$$\rho = \frac{1}{\max\{R_i : 1 \leq i \leq w\}},$$

which is what we were required to prove. ∎

Although Theorem 13.12 is useful, it is often preferable to employ a generalization in which we have ℓ ideal decompositions of Γ_0 instead of just one. Each of the ℓ decompositions is used to share a key chosen from $\mathcal{K}$. Thus, we build a scheme with key set $\mathcal{K}^\ell$ (that is, keys are ℓ-tuples). The construction of the scheme and its information rate are as stated in the following theorem.

THEOREM 13.13 (*ℓ-Decomposition Construction*) *Suppose that Γ is an access structure having basis Γ_0 and suppose that $\ell \geq 1$ is an integer. Let $\mathcal{K}$ be a specified key set, and for $1 \leq h \leq \ell$, suppose that $\mathcal{D}_h = \{\Gamma_{h,1}, \ldots \Gamma_{h,n_h}\}$ is an ideal decomposition of Γ_0 for the key set $\mathcal{K}$. Let $\mathcal{P}_{h,j}$ denote the participant set for the access structure $\Gamma_{h,j}$. For every participant P_i, define*

$$R_i = \sum_{h=1}^{\ell} |\{j : P_i \in \mathcal{P}_{h,j}\}|.$$

Then there exists a perfect secret sharing scheme realizing Γ, having information rate $\rho = \ell/R$, where

$$R = \max\{R_i : 1 \leq i \leq w\}.$$

PROOF For $1 \leq h \leq \ell$ and $1 \leq j \leq n_h$, we have an ideal scheme realizing the access structure on basis $\Gamma_{h,j}$, with key set $\mathcal{K}$, having $\mathcal{F}^{h,j}$ as its set of distribution rules.

We construct a scheme realizing Γ, with key set $\mathcal{K}^\ell$. The set of distribution rules $\mathcal{F}$ is constructed according to the following recipe. Suppose D wants to share a key $K = (K_1, \ldots, K_\ell)$. Then for all h and j such that $1 \leq h \leq \ell$ and $1 \leq j \leq n_h$, he chooses a random distribution rule $f^{h,j} \in \mathcal{F}_{K_h}^{h,j}$ and distributes the resulting shares to the participants in $\mathcal{P}_{h,j}$.

The information rate can be computed in a manner similar to that of Theorem 13.12. ∎

Let's look at a couple of examples.

Example 13.11 The basis of access structure #5 is a graph that is not a complete multipartite graph. Therefore we know from Theorem 13.11 that $\rho^* \leq 2/3$.

Let p be any prime, and consider the following two ideal decompositions (the key set for each decomposition will be $\mathbb{Z}_p$):

$$\mathcal{D}_1 = \{\Gamma_{1,1}, \Gamma_{1,2}\},$$

where

$$\Gamma_{1,1} = \{\{P_1, P_2\}\}$$
$$\Gamma_{1,2} = \{\{P_2, P_3\}, \{P_3, P_4\}\},$$

and

$$\mathcal{D}_2 = \{\Gamma_{2,1}, \Gamma_{2,2}\},$$

where

$$\Gamma_{2,1} = \{\{P_1, P_2\}, \{P_2, P_3\}\}$$
$$\Gamma_{2,2} = \{\{P_3, P_4\}\}.$$

Each of the two decompositions consists of a graph K_2 and a graph $K_{1,2}$, so they are indeed ideal decompositions. Either one by itself yields a scheme with $\rho = 1/2$. However, if we "combine" them by applying Theorem 13.13 with $\ell = 2$, then we get a scheme with $\rho = 2/3$, which is optimal.

One implementation of the scheme, using Theorem 13.5, is as follows. D will choose four random elements (independently) from $\mathbb{Z}_p$, say b_{11}, b_{12}, b_{21}, and b_{22}. Given a key $(K_1, K_2) \in (\mathbb{Z}_p)^2$, D distributes shares as follows:

1. P_1 receives b_{11}, b_{21}.
2. P_2 receives $b_{11} + K_1, b_{12}, b_{21} + K_2$.
3. P_3 receives $b_{12} + K_1, b_{21}, b_{22}$.
4. P_4 receives $b_{12}, b_{22} + K_2$.

(All arithmetic is performed in $\mathbb{Z}_p$.)   □

Example 13.12 Consider access structure #8. Again, $\rho^* \leq 2/3$ by Theorem 13.11, and two suitable ideal compositions will yield an (optimal) scheme with $\rho = 2/3$.

Take $\mathcal{K} = \mathbb{Z}_p$ for any prime $p \geq 3$, and define two ideal decompositions to be as follows:

$$\mathcal{D}_1 = \{\Gamma_{1,1}, \Gamma_{1,2}\},$$

where

$$\Gamma_{1,1} = \{\{P_1, P_2\}\}$$
$$\Gamma_{1,2} = \{\{P_2, P_3\}, \{P_2, P_4\}, \{P_3, P_4\}\},$$

and

$$\mathcal{D}_2 = \{\Gamma_{2,1}, \Gamma_{2,2}\},$$

where

$$\Gamma_{2,1} = \{\{P_1, P_2\}, \{P_2, P_3\}, \{P_2, P_4\}\}$$
$$\Gamma_{2,2} = \{\{P_3, P_4\}\}.$$

$\mathcal{D}_1$ consists of a K_2 and a K_3, and $\mathcal{D}_2$ consists of a K_2 and a $K_{1,3}$, so both are ideal decompositions for key set $\mathcal{K}$. Applying Theorem 13.13 with $\ell = 2$, we get a scheme with $\rho = 2/3$.

One implementation, using Theorem 13.5, is as follows. D will choose four random elements (independently) from $\mathbb{Z}_p$, say b_{11}, b_{12}, b_{21}, and b_{22}. Given a key $(K_1, K_2) \in (\mathbb{Z}_p)^2$, D distributes shares as follows:

1. P_1 receives $b_{11} + K_1, b_{21} + K_2$.

2. P_2 receives b_{11}, b_{12}, b_{21}.

3. P_3 receives $b_{12} + K_1, b_{21} + K_2, b_{22}$.

4. P_4 receives $b_{12} + 2K_1, b_{21} + K_2, b_{22} + K_2$.

(All arithmetic is performed in $\mathbb{Z}_p$.) ◻

So far, we have explained all the information in Table 13.1 except for the values of ρ^* for access structures #12 and #13. These values arise from a more general version of the decomposition construction which we do not describe here; see the Notes and References.

13.4 Notes and References

Threshold schemes were invented independently by Blakley [46] and Shamir [297]. Secret sharing for general access structures was first studied in Ito, Saito, and Nishizeki [180]; we based Section 13.2 on the approach of Benaloh and Leichter [28]. The vector space construction is due to Brickell [73]. The entropy bound of Section 13.3.2 is proved in Capocelli *et al.* [83], and some of the other material from this section is found in Blundo *et al.* [54].

Additional discussion on decomposition techniques can be found in Stinson [318] and [320]. Also see van Dijk, Jackson and Martin [331] for a very general version of a decomposition construuction.

In this chapter, we have emphasized a linear-algebraic and combinatorial approach to secret sharing. Some interesting connections with matroid theory can be found in Brickell and Davenport [75]. Secret sharing schemes can also be constructed using geometric techniques. Simmons has done considerable research in this direction; we refer to [305] for an overview of geometric techniques in secret sharing.

Exercises

13.1 Write a computer program to compute the key for the *Shamir* (t, w)-*Threshold Scheme* implemented in $\mathbb{Z}_p$. That is, given t public x-coordinates, $x_1, x_2, \ldots, x_t$, and t y-coordinates $y_1, \ldots, y_t$, compute the resulting key using the Lagrange interpolation formula.

    (a) Test your program if $p = 31847$, $t = 5$ and $w = 10$, with the following shares:

x_1	=	413	y_1	=	25439
x_2	=	432	y_2	=	14847
x_3	=	451	y_3	=	24780
x_4	=	470	y_4	=	5910
x_5	=	489	y_5	=	12734
x_6	=	508	y_1	=	12492
x_7	=	527	y_2	=	12555
x_8	=	546	y_3	=	28578
x_9	=	565	y_4	=	20806
x_{10}	=	584	y_5	=	21462

        Verify that the same key is computed by using several different subsets of five shares.

    (b) Having determined the key, compute the share that would be given to a participant with x-coordinate equal to 10000. (Note that this can be done without computing the whole secret polynomial $a(x)$.)

13.2   (a) Suppose that the following are the nine shares in a $(5, 9)$-*Shamir Threshold Scheme* implemented in $\mathbb{Z}_{94875355691}$:

i	x_i	y_i
1	11	537048626
2	22	89894377870
3	33	65321160237
4	44	18374404957
5	55	24564576435
6	66	87371334299
7	77	60461341922
8	88	10096524973
9	99	81367619987

        Exactly one of these shares is defective (i.e., incorrect). Your task is to determine which share is defective, and then figure out its correct value, as well as the value of the secret.

        The "primitive operations" in your algorithm are polynomial interpolations and polynomial evaluations. Try to minimize the number of polynomial interpolations you perform.

HINT The question can be answered using at most three polynomial interpolations.

(b) Suppose that a (t, w)-*Shamir Threshold Scheme* has exactly one defective share, and suppose that $w - t \geq 2$. Describe how it is possible to determine which share is defective using at most $\lceil \frac{w}{w-t} \rceil$ polynomial interpolations. Why is this problem impossible to solve if $w - t = 1$?

(c) Suppose that a (t, w)-*Shamir Threshold Scheme* has exactly τ defective shares, and suppose that $w \geq (\tau + 1)t$. Describe how it is possible to determine which shares are defective using at most $\tau + 1$ polynomial interpolations.

13.3 For access structures having the following bases, use the monotone circuit construction to construct a secret sharing scheme with information rate $\rho = 1/3$.

(a) $\Gamma_0 = \{\{P_1, P_2\}, \{P_2, P_3\}, \{P_2, P_4\}, \{P_3, P_4\}\}$.

(b) $\Gamma_0 = \{\{P_1, P_3, P_4\}, \{P_1, P_2\}, \{P_2, P_3\}, \{P_2, P_4\}\}$.

(c) $\Gamma_0 = \{\{P_1, P_2\}, \{P_1, P_3\}, \{P_2, P_3, P_4\}, \{P_2, P_4, P_5\}, \{P_3, P_4, P_5\}\}$.

13.4 Use the vector space construction to obtain ideal *Brickell Secret Sharing Schemes* for access structures having the following bases:

(a) $\Gamma_0 = \{\{P_1, P_2, P_3\}, \{P_1, P_2, P_4\}, \{P_3, P_4\}\}$.

(b) $\Gamma_0 = \{\{P_1, P_2, P_3\}, \{P_1, P_2, P_4\}, \{P_1, P_3, P_4\}\}$.

(c) $\Gamma_0 = \{\{P_1, P_2\}, \{P_1, P_3\}, \{P_2, P_3\}, \{P_1, P_4, P_5\}, \{P_2, P_4, P_5\}\}$.

13.5 Use the decomposition construction to obtain secret sharing schemes with specified information rates for the access structures having the following bases:

(a) $\Gamma_0 = \{\{P_1, P_3, P_4\}, \{P_1, P_2\}, \{P_2, P_3\}\}, \rho = 3/5$.

(b) $\Gamma_0 = \{\{P_1, P_3, P_4\}, \{P_1, P_2\}, \{P_2, P_3\}, \{P_2, P_4\}\}, \rho = 4/7$.

14

Multicast Security and Copyright Protection

14.1 Introduction to Multicast Security

The term *multicast* refers to a message that has many designated receivers, i.e., one-to-many communication as opposed to one-to-one communication. This situation arises in the context of a network of users in which it is possible to send messages simultaneously to all of the users. We will not be concerned with the mechanics of doing this; we will just assume this is an inherent property of the network being considered. In this chapter, we are interested in the study of security in the context of a multicast network.

Two "benchmark scenarios" are often considered. These are as follows:

single source broadcast

In this model, we have a single entity broadcasting information to a network of users, which is sometimes called a *group*. Pay-TV provides a possible illustration of this situation. Typically the group is long-lived; however, it may be *dynamic*, i.e., users may join or leave the group over time, and therefore we will require algorithms to add or delete group members. A *key revocation algorithm* will be used to delete or disable keys of a user who leaves the group. This is done through a broadcast, which will allow the other (non-disabled) users to update their keys.

In any broadcast of this type, we will want to ensure confidentiality, and perhaps authenticity, of broadcasted information.

virtual conference

Suppose that a subset of a larger group wishes to form a small virtual conference (e.g., a conference call by telephone, or a meeting of a committee formed from certain participants in the group). Here we have a short-lived group, which is probably static, We generally need to enable multiple sender multicast, because any member of the conference should be able to send information to any of the other members.

We will want to provide confidentiality, and perhaps authenticity, of broad-

FIGURE 14.1
Schematic illustration of encryption in the *Trivial broadcast encryption scheme*

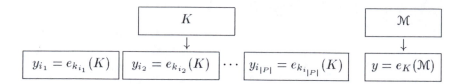

casted information; this is usually accomplished by first establishing a temporary session key that is known only to the members of the group. Depending on the requirements of the conference, it may or may not be desirable to provide source authentication during the conference. In fact, anonymity may be required for some purposes, such as secret voting. However, we need to ensure that a sender is always a member of the specified group.

In the next sections, we will be describing cryptographic schemes that are useful in the scenarios described above.

14.2   Broadcast Encryption

In a *broadcast encryption scheme* (or *BES*), a trusted authority (TA) wants to broadcast an encrypted message to a subset P of a network $\mathcal{U}$ of n users. The set P, which comprises the intended recipients of the broadcast, is usually called a *privileged subset*.

For example, a pay-TV movie, $\mathcal{M}$, might be encrypted (using a block cipher) with a secret key K, i.e., $y = e_K(\mathcal{M})$. A broadcast encryption scheme is used to encrypt K in such a way that only the members of P can determine K. Note that the BES is used to encrypt a short key, rather than a long movie, because of the message expansion that is typically required in a BES (we will say more about this later). A user U_i not in the privileged subset may have access to the broadcast, but this should not allow U_i to compute the encryption key K.

Note that the privileged set P is, in general, not known before the scheme is initialized. In fact, once a scheme is set up, it can potentially be used to broadcast messages to various privileged sets over a period of time.

We begin by presenting the *Trivial Broadcast Encryption Scheme*; see Cryptosystem 14.1.

In the setup phase for the *Trivial BES*, the TA gives each user U_i in the network a different key. Let P be a privileged subset of users. Then, for every user $U_i \in P$, the TA encrypts K with the user key k_i, by computing $y_i = e_{k_i}(K)$. It is clear

---

Cryptosystem 14.1: *Trivial Broadcast Encryption Scheme*

Key predistribution (Setup phase): The TA gives a secret key k_i to user U_i, for all users $U_i \in \mathcal{U}$.

Key encryption: Let P be the privileged set. The TA encrypts K with the key k_i, for all i such that $U_i \in P$.

Message encryption: The message $\mathcal{M}$ is encrypted with the secret key K, i.e., $y = e_K(\mathcal{M})$. The broadcast consists of P, y, and the list of encrypted keys:

$$b_P = (e_{k_i}(K) : U_i \in P).$$

The general stucture of this process is illustrated in Figure 14.1. In this diagram, the privileged set is $P = \{U_{i_1}, \ldots, U_{i_{|P|}}\}$.

---

that a user $U_i \in P$ can decrypt y_i to obtain K, and then use K to decrypt y and obtain $\mathcal{M}$. A user not in P cannot decrypt any of the y_i's, and hence y cannot be decrypted by that user.

The broadcast b_P is a $|P|$-tuple of encrypted keys, so the *broadcast message expansion* is said to be equal to $|P|$. The *Trivial BES* has low storage requirements (there is only one key per user) and high security (because no coalition of non-privileged users can compute K). However, the *Trivial BES* has a very high message expansion.

In general, we want to find good tradeoffs between the parameters of a BES. For example, we may be willing to tolerate a reduced security level and increased storage, provided that the broadcast message expansion is lower than what is achieved in the *Trivial BES*.

We now present a high-level view of the design technique we will use in constructing efficient broadcast encryption schemes. Suppose $P \subseteq \mathcal{U}$ is the privileged subset, and let w denote the maximum size of a coalition (w is the *security parameter* for the BES). The integers r and v are additional parameters of the BES whose values will be specified later. The *General Broadcast Encryption Scheme* (*General BES*) involves a set-up phase consisting of key predistribution, followed by three steps that are performed every time a message is broadcasted. A general overview of the scheme is given in Cryptosystem 14.2.

The *General BES* uses a secret sharing operation that is not present in the *Trivial BES*. Basically, the key K is split into shares and each share is encrypted. The encrypted shares are then broadcasted.

Efficient realizations of the *General BES* can be constructed by using the *Fiat-Naor Key Distribution Patterns* that were described in Section 10.4.1. We will

---

Cryptosystem 14.2: *General Broadcast Encryption Scheme*

Key predistribution (Setup phase): The *TA* distributes keying material for v key predistribution schemes to the users in the network $\mathcal{U}$.

Secret sharing: The *TA* chooses a secret key K and it splits K into shares $s_1, \ldots, s_v$ using an (r, v)-threshold scheme.

Share encryption/decryption: Let P be the privileged set. For $1 \leq i \leq v$, the *TA* encrypts the share s_i with a key k_i from the ith KPS, in such a way that the following conditions are satisfied:

1. Every user $U_j \in P$ can compute at least r of the keys $k_1, \ldots, k_v$ (hence U_j can decrypt r shares of K, and then reconstruct K).

2. Any coalition F, such that $F \cap P = \emptyset$ and $|F| \leq w$, can compute at most $r - 1$ of the keys $k_1, \ldots, k_v$ (hence F can decrypt at most $r - 1$ shares of K, and therefore F cannot obtain any information about K).

Message encryption: The message $\mathcal{M}$ is encrypted with the secret key K, i.e., $y = e_K(\mathcal{M})$. The broadcast consists of P (the privileged set), y (the encrypted message), and the list of encrypted shares:

$$b_P = (e_{k_i}(s_i) : 1 \leq i \leq v).$$

The encryption process used in the *General BES* is illustrated in Figure 14.2.

---

construct a collection of v different Fiat-Naor 1-KDPs, each of which is defined on a certain subset of participants. (Recall that a 1-KDP defines keys that are secure against individual adversaries.)

A v by n incidence matrix, M, having entries from $\{0, 1\}$, is used to indicate which users are associated with which KDPs. The v KDPs are denoted $\mathcal{F}_1, \ldots, \mathcal{F}_v$, and user U_j is given keys from the KDP $\mathcal{F}_i$ if and only if $M[i, j] = 1$. The following notation will be useful: For $1 \leq i \leq v$, let

$$users(i) = \{U_j : M[i, j] = 1\}$$

and for $1 \leq j \leq n$, let

$$schemes(j) = \{i : M[i, j] = 1\}.$$

That is, $users(i)$ records the users associated with the ith KDP, and $schemes(j)$ specifies the KDPs that user U_j belongs to. We will see a bit later that the matrix M should satisfy certain properties if the resulting *General BES* is to be secure.

FIGURE 14.2
Schematic illustration of encryption in the *General Broadcast Encryption Scheme*

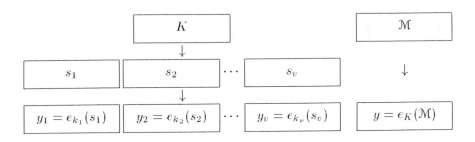

Algorithm 14.1: GENERAL BES MESSAGE ENCRYPTION

Input: The key K, the privileged set P, and the message $\mathcal{M}$.

1. The TA uses the share generation algorithm of an (r, v)-threshold scheme to construct v shares of the secret K, denoted $s_1, \ldots, s_v$.

2. For $1 \leq i \leq v$, the TA computes the key k_i to be the group key for the subset of users $P \cap users(i)$ in the scheme $\mathcal{F}_i$.

3. For $1 \leq i \leq v$, the TA computes $b_i = e_{k_i}(s_i)$.

4. The TA computes $y = e_K(\mathcal{M})$.

Output: The encrypted message y and the broadcast b_P.

We now discuss the process of share encryption in the *General BES* in more detail. First, we will require that $|schemes(j)| = r$ for every j, $1 \leq j \leq n$ (i.e., every user is associated with exactly r of the v schemes). The TA encrypts K (the encryption key) as in the high-level description of the *General BES*; this is described in more detail in Algorithm 14.1.

Decryption of the broadcasted message by a user in the privileged set P is straightforward. On input P, y and $b_P = (b_1, \ldots, b_v)$, a user $U_j \in P$ performs the operations described in Algorithm 14.2.

We still need to consider the security of the *General BES*. First, however, we present a small example to illustrate the encryption and decryption steps.

Example 14.1 Suppose that we have seven users in our network. We will construct a *General BES* that uses seven KDPs. Thus we have $n = 7$ and $v = 7$. The

---

Algorithm 14.2: GENERAL BES MESSAGE DECRYPTION

Input: The privileged set P, the encrypted message y and the broadcast b_P.

1. For all $i \in schemes(j)$, U_j constructs the group key k_i for the subset of users $P \cap users(i)$ in the scheme $\mathcal{F}_i$.

2. For all $i \in schemes(j)$, U_j computes $s_i = d_{k_i}(b_i)$.

3. U_j uses the share reconstruction algorithm of the (r, v)-threshold scheme to compute K from the r shares in the set $\{s_i : i \in schemes(j)\}$.

4. U_j computes $\mathcal{M} = d_K(y)$.

Output: The decrypted message $\mathcal{M}$.

---

incidence matrix M is defined as follows:

$$M = \begin{pmatrix} 1 & 1 & 0 & 1 & 0 & 0 & 0 \\ 0 & 1 & 1 & 0 & 1 & 0 & 0 \\ 0 & 0 & 1 & 1 & 0 & 1 & 0 \\ 0 & 0 & 0 & 1 & 1 & 0 & 1 \\ 1 & 0 & 0 & 0 & 1 & 1 & 0 \\ 0 & 1 & 0 & 0 & 0 & 1 & 1 \\ 1 & 0 & 1 & 0 & 0 & 0 & 1 \end{pmatrix}.$$

It is easy to verify that every user is associated with $r = 3$ schemes, and every one of the seven KDPs is defined on a subset of three of the seven users.

In the set-up phase, a total of seven 1-KDPs are constructed. Each $\mathcal{F}_i$ ($1 \leq i \leq 7$) has four keys, and each user in $users(i)$ receives three out of these four keys. The keys are distributed according to the following rule: Suppose that $users(i) = \{\alpha, \beta, \gamma\}$. Then a key ℓ_i is given to α, β and γ; a key $\ell_{i,\alpha}$ is given to β and γ; a key $\ell_{i,\beta}$ is given to α and γ; and a key $\ell_{i,\gamma}$ is given to α and β. This is done for each of the seven schemes.

It follows that a total of nine keys will be given to each user, as indicated in the following table:

U_1	U_2	U_3	U_4	U_5	U_6	U_7
ℓ_1	ℓ_1	ℓ_2	ℓ_1	ℓ_2	ℓ_3	ℓ_4
$\ell_{1,2}$	$\ell_{1,1}$	$\ell_{2,2}$	$\ell_{1,1}$	$\ell_{2,2}$	$\ell_{3,3}$	$\ell_{4,4}$
$\ell_{1,4}$	$\ell_{1,4}$	$\ell_{2,5}$	$\ell_{1,2}$	$\ell_{2,3}$	$\ell_{3,4}$	$\ell_{4,5}$
ℓ_5	ℓ_2	ℓ_3	ℓ_3	ℓ_4	ℓ_5	ℓ_6
$\ell_{5,5}$	$\ell_{2,3}$	$\ell_{3,4}$	$\ell_{3,3}$	$\ell_{4,4}$	$\ell_{5,1}$	$\ell_{6,2}$
$\ell_{5,6}$	$\ell_{2,5}$	$\ell_{3,6}$	$\ell_{3,6}$	$\ell_{4,7}$	$\ell_{5,5}$	$\ell_{6,6}$
ℓ_7	ℓ_6	ℓ_7	ℓ_4	ℓ_5	ℓ_6	ℓ_7
$\ell_{7,3}$	$\ell_{6,6}$	$\ell_{7,1}$	$\ell_{4,5}$	$\ell_{5,1}$	$\ell_{6,2}$	$\ell_{7,1}$
$\ell_{7,7}$	$\ell_{6,7}$	$\ell_{7,7}$	$\ell_{4,7}$	$\ell_{5,6}$	$\ell_{6,7}$	$\ell_{7,3}$

Now suppose that the TA wants to broadcast a message to the set $P = \{U_1, U_2, U_3\}$. The following will be the keys used in the seven Fiat-Naor KDPs:

i	$users(i) \cap P$	k_i
1	$\{U_1, U_2\}$	$\ell_1 + \ell_{1,4}$
2	$\{U_2, U_3\}$	$\ell_2 + \ell_{2,5}$
3	$\{U_3\}$	$\ell_3 + \ell_{3,4} + \ell_{3,6}$
4	$\emptyset$	$\ell_4 + \ell_{4,4} + \ell_{4,5} + \ell_{4,7}$
5	$\{U_1\}$	$\ell_5 + \ell_{5,5} + \ell_{5,6}$
6	$\{U_2\}$	$\ell_6 + \ell_{6,6} + \ell_{6,7}$
7	$\{U_1, U_3\}$	$\ell_7 + \ell_{7,7}$

In this example, we have $r = 3$ and $v = 7$, so the TA could use a *Shamir* $(3, 7)$-*Threshold Scheme* implemented in $\mathbb{Z}_p$, where p is any prime (see Section 13.1 for a description of the *Shamir Threshold Scheme*). Let $s_1, \ldots, s_7$ be the seven shares of the secret K, where $K \in \mathbb{Z}_p$, as determined by share generation algorithm of the threshold scheme. The broadcast $b_P = (b_1, \ldots, b_7)$ consists of the following values:

$$b_1 = e_{\ell_1 + \ell_{1,4}}(s_1),$$

$$b_2 = e_{\ell_2 + \ell_{2,5}}(s_2),$$

$$b_3 = e_{\ell_3 + \ell_{3,4}}(s_3),$$

$$b_4 = e_{\ell_4 + \ell_{4,4} + \ell_{4,5} + \ell_{4,7}}(s_4),$$

$$b_5 = e_{\ell_5 + \ell_{5,5} + \ell_{5,6}}(s_5),$$

$$b_6 = e_{\ell_6 + \ell_{6,6} + \ell_{6,7}}(s_6), \quad \text{and}$$

$$b_7 = e_{\ell_7 + \ell_{7,7}}(s_7).$$

It is clear that U_1, U_2 and U_3 each can compute three out of the seven k_i's: U_1 can compute k_1, k_5 and k_7; U_2 can compute k_1, k_2 and k_6; and U_3 can compute k_2, k_3 and k_7. □

Suppose that F is a coalition of users that is disjoint from the privileged set P. In order for the *General BES* to be secure against the coalition F, it must be the case that the coalition F cannot compute more than $r - 1$ of the keys $k_1, \ldots, k_v$. (This implies that F can decrypt at most $r - 1$ of the shares. Because the threshold scheme has a threshold value of r, it follows that F cannot compute any information about K.)

The keys in the list $k_1, \ldots, k_v$ that can be computed by F can be determined by examining the incidence matrix. The security of the scheme will be guaranteed provided that M satisfies certain combinatorial properties. Basically, everything depends on the construction of the group keys. Recall that, for $1 \leq i \leq v$, the scheme $\mathcal{F}_i$ is a *Fiat-Naor 1-KDP* defined on the subset $users(i)$. In general, to set up the scheme $\mathcal{F}_i$, the *TA* distributes keys as follows:

1. a key ℓ_i is given to every user in $users(i)$, and
2. for every $U_j \in users(i)$, a key $\ell_{i,j}$ is given to every user in $users(i)\backslash\{U_j\}$.

Then, the group key for the subset $users(i) \cap P$ in the KDP $\mathcal{F}_i$ is

$$k_i = \ell_i + \sum_{\{j:U_j \in users(i)\backslash P\}} \ell_{i,j}.$$

The security property of the Fiat-Naor 1-KDP ensures that no individual user not in the set $users(i) \cap P$ can compute k_i. However, a subset of two or more users in $users(i)\backslash P$ can compute k_i. Therefore, the coalition F can compute a group key k_i if and only if $|F \cap users(i)| \geq 2$.

We continue the previous example:

Example 14.1 **(Cont.)** We recall the incidence matrix M, together with the sets $users(i)$, $i = 1, \ldots, 7$:

$$M = \begin{pmatrix} 1 & 1 & 0 & 1 & 0 & 0 & 0 \\ 0 & 1 & 1 & 0 & 1 & 0 & 0 \\ 0 & 0 & 1 & 1 & 0 & 1 & 0 \\ 0 & 0 & 0 & 1 & 1 & 0 & 1 \\ 1 & 0 & 0 & 0 & 1 & 1 & 0 \\ 0 & 1 & 0 & 0 & 0 & 1 & 1 \\ 1 & 0 & 1 & 0 & 0 & 0 & 1 \end{pmatrix} \qquad \begin{aligned} users(1) &= \{U_1, U_2, U_4\} \\ users(2) &= \{U_2, U_3, U_5\} \\ users(3) &= \{U_3, U_4, U_6\} \\ users(4) &= \{U_4, U_5, U_7\} \\ users(5) &= \{U_1, U_5, U_6\} \\ users(6) &= \{U_2, U_6, U_7\} \\ users(7) &= \{U_1, U_3, U_7\} \end{aligned}$$

We can prove that this *General BES* is secure against coalitions of size $w = 2$. The security depends on the following property of the incidence matrix M, which can easily be verified: Given any subset of two users, say $\{U_j, U_{j'}\}$, there is exactly one KDP $\mathcal{F}_i$ such that $\{U_j, U_{j'}\} \subseteq users(i)$. Now, let $F = \{U_j, U_{j'}\}$ be an arbitrary coalition of size two that is disjoint from the privileged set P. Because of the property mentioned above, it follows immediately that the coalition F can compute only one of the seven group keys, $k_1, \ldots, k_7$. Hence, F can decrypt only one of the seven encrypted shares. The threshold used in the secret sharing scheme is $r = 3$, so the coalition does not have enough shares to determine any information about the secret.

Note that this argument is valid for any privileged set P and any coalition of size two. Therefore, the *General BES* is secure against coalitions of size two. □

In general, M is required to be a v by n incidence matrix in which there are exactly r "1"s in each column. Let $(\mathcal{U}, \mathcal{A})$ be the set system in which the blocks are formed by the rows of M, which are just the subsets $users(i)$. (Note that in Section 10.4.2 we considered certain set systems — namely, cover-free families

— associated with the columns of KDPs.) There are n points and v blocks in the set system $(\mathcal{U}, \mathcal{A})$. Every point (i.e., user) is in exactly r blocks (i.e., schemes). Suppose the set system has the property that every pair of points occurs in at most λ blocks, for some positive integer λ (we have $\lambda = 1$ in the above example). Then a coalition F of size w can compute at most $\lambda\binom{w}{2}$ group keys, because there are $\binom{w}{2}$ 2-subsets of F, each of which can compute at most λ of the group keys. Therefore, the *General BES* is secure against coalitions of size w if $r > \lambda\binom{w}{2}$.

In general, we want to construct an incidence matrix such that the associated set system $(\mathcal{U}, \mathcal{A})$ satisfies the following properties:

1. $|\mathcal{U}| = n$ (there are n points),

2. $|\mathcal{A}| = v$ (there are v blocks),

3. every point occurs in exactly r blocks,

4. every pair of points occurs in at most λ blocks, and

5. $r > \lambda\binom{w}{2}$.

We will refer to a set system that satisfies properties 1–4 as an (n, v, r, λ)-*broadcast key distribution pattern*, or an (n, v, r, λ)-*BKDP*.

As an example, consider the sets $users(i)$ $(i = 1, \ldots, 7)$, which were presented in the continuation of Example 14.1. These seven sets comprise the blocks in a $(7, 7, 3, 1)$-BKDP.

The following theorem states that an (n, v, r, λ)-BKDP yields a *General BES* with certain properties.

THEOREM 14.1   *Suppose there is an (n, v, r, λ)-BKDP, and suppose that w is a positive integer such that $\binom{w}{2} < r/\lambda$. Then there is a broadcast encryption scheme for a network of n users which satisfies the following properties:*

1. *an encrypted message is secure against coalitions of size w,*

2. *the message expansion is equal to v, and*

3. *each user is required to store at most $r + \lambda(n - 1)$ keys.*

PROOF   Let $(\mathcal{U}, \mathcal{A})$ be an (n, v, r, λ)-BKDP. The only part of this theorem that has not already been addressed in the preceding discussion is the statement regarding storage requirements. Consider a user U_j. For all $i \in schemes(j)$, user U_j receives keys from $\mathcal{F}_i$. In fact, U_j is given precisely $|users(i)|$ keys from the scheme $\mathcal{F}_i$ whenever $i \in schemes(j)$.

The total number of keys that U_j receives, expressed in terms of the set system $(\mathcal{U}, \mathcal{A})$, is

$$\sum_{\{i : i \in schemes(j)\}} |users(i)|.$$

Property 3 of the set system implies that

$$\sum_{\{i : i \in schemes(j)\}} 1 = r.$$

It is also not hard to see that properties 1 and 4 imply that

$$\sum_{\{i:i\in schemes(j)\}} (|users(i)| - 1) \le \lambda(n - 1).$$

Combining the two previous relations, we have that

$$\sum_{\{i:i\in schemes(j)\}} |users(i)| \le r + \lambda(n - 1),$$

as claimed.   ∎

Clearly, for the sake of efficiency, we want r, λ and v to be as small as possible (given n and w). In particular, we want to minimize the value of v. (As well r/λ has to exceed $\binom{w}{2}$.) We now present a polynomial-based construction for BKDP whose parameters yield useful and efficient schemes.

THEOREM 14.2   *Suppose that q is prime and let d be an integer such that $2 \le d \le q$. Then there exists a $(q^d, q^2, q, d - 1)$-BKDP.*

PROOF   We construct the incidence matrix M of a $(q^d, q^2, q, d - 1)$-BKDP. The columns of M are labeled by d-tuples $(a_0, \ldots, a_{d-1}) \in (\mathbb{Z}_q)^d$ (note that a d-tuple corresponds in an obvious way to a polynomial in $\mathbb{Z}_q[x]$ having degree at most $d - 1$). The rows of M are labeled by ordered pairs $(x, y) \in (\mathbb{Z}_q)^2$. The entries of M are defined as follows:

$$M\big((x, y), (a_0, \ldots, a_{d-1})\big) = 1 \Leftrightarrow \sum_{i=0}^{d-1} a_i x^i \equiv y \pmod{q}.$$

As a first observation, it is clear that $r = q$, because every polynomial takes on a unique y-value, given any x-value.

It remains for us to compute the parameter λ. Suppose we have two columns of M, denoted $a = (a_0, \ldots, a_{d-1})$ and $a' = (a_0', \ldots, a_{d-1}')$. These define two polynomials, say $a(x)$ and $a'(x)$, respectively. We want to determine an upper bound on

$$\lambda_{a,a'} = |\{(x, y) : a(x) = a'(x) = y\}| = |\{x : a(x) = a'(x)\}|.$$

Clearly, $\lambda_{a,a'} \le d - 1$, because the values at d points determine a unique polynomial of degree at most $d - 1$. Hence, we can take $\lambda = d - 1$.   ∎

Summarizing the above discussion, we obtain the following parameters for the polynomial-based construction. Let q be prime and let $d < q$. Then there exists an incidence matrix M of a BKDP with $n = q^d$, $v = q^2$, $r = q$, and $\lambda = d - 1$.

The resulting BES is secure against coalitions of size at most w provided that $q > (d-1)\binom{w}{2}$, i.e., if

$$d < 1 + \frac{q}{\binom{w}{2}}.$$

Note that this condition implies the weaker condition $d < q$.

Example 14.2   We present a $(9, 9, 3, 1)$-BKDP. We take $q = 3$ and $d = 2$ in Theorem 14.2. Then $n = v = 9$, and the incidence matrix M is as follows:

(x, y)	0	1	2	x	$1+x$	$2+x$	$2x$	$1+2x$	$2+2x$
$(0,0)$	1	0	0	1	0	0	1	0	0
$(0,1)$	0	1	0	0	1	0	0	1	0
$(0,2)$	0	0	1	0	0	1	0	0	1
$(1,0)$	1	0	0	0	0	1	0	1	0
$(1,1)$	0	1	0	1	0	0	0	0	1
$(1,2)$	0	0	1	0	1	0	1	0	0
$(2,0)$	1	0	0	0	1	0	0	0	1
$(2,1)$	0	1	0	0	0	1	1	0	0
$(2,2)$	0	0	1	1	0	0	0	1	0

We now look at a numerical example to illustrate how appropriate parameters can be chosen.

Example 14.3   Suppose we want to have security against coalitions of size 3, so we take $w = 3$. For any prime q, we can let $d = \lfloor 1 + \frac{q}{3} \rfloor$. Then we get a *General BES* for $n = q^d$ users, based on $v = q^2$ *Fiat-Naor* 1-*KDPs*, that is secure against coalitions of size 3.

Here are some sample parameters:

q	d	v	n
7	3	49	343
13	5	169	371,293
19	7	361	893,871,739

The main observation here is that n (the size of the network supported by the scheme) increases extremely quickly as q increases. On the other hand, the size of the broadcast depends on v, which grows much more slowly.

14.2.1   An Improvement using Ramp Schemes

In the *General BES*, recall that each user in P can decrypt r shares, and the coalition F can (collectively) decrypt at most $\lambda\binom{w}{2}$ shares, where $\lambda\binom{w}{2} < r$. The use of an (r, v)-threshold scheme ensures that the secret K cannot be computed, given $r - 1$ shares. If $\lambda\binom{w}{2} < r - 1$, then the security provided by the threshold scheme is stronger than what is actually required. In this situation, it is possible to replace the threshold scheme by a related scheme known as a *ramp scheme*, which is defined below. We can maintain the same security level, while decreasing the message expansion.

Definition 14.1:   Let t', t'' and v be non-negative integers, where $t' < t'' \leq v$. There is a dealer, denoted D, and v users, denoted $P_1, \ldots, P_v$. The dealer has a secret value $K \in \mathcal{K}$ and he uses a *share generation algorithm* to split K into v *shares*, denoted $y_1, \ldots, y_v$. Each share $y_i \in \mathcal{S}$, where $\mathcal{S}$ is a specified finite set. A $(t', t,'' v)$-*ramp scheme* satisfies the following two properties:

1. a *reconstruction algorithm* can be used to reconstruct the secret, given any t'' of the v shares, and

2. no set of at most t' shares reveals any information as to the value of the secret.

It is worth noting that a $(t - 1, t, v)$-ramp scheme is exactly the same thing as a (t, v)-threshold scheme. It is perhaps not surprising that the *Shamir Threshold Scheme* can be generalized to produce a ramp scheme. This scheme, which is presented as Cryptosystem 14.3, is called the *Shamir Ramp Scheme*.

The main difference between the *Shamir Threshold Scheme* and the *Shamir Ramp Scheme* is that the ramp scheme allows for larger keys, whenever $t' + 1 < t''$. More precisely, because the key space in the ramp scheme consists of all t_0-tuples of elements from $\mathbb{Z}_p$, where $t_0 = t'' - t'$, the key length in the ramp scheme is longer by a factor of t_0 (as compared to the threshold scheme).

Share reconstruction in the *Shamir Ramp Scheme* is almost identical to the process used in the *Shamir Threshold Scheme*. It is carried out in the following manner: Suppose users $P_{i_1}, \ldots, P_{i_{t''}}$ want to determine K. They know that $y_{i_j} = a(x_{i_j})$, $1 \leq j \leq t''$. Since $a(x)$ is a polynomial of degree at most $t'' - 1$, they can compute

$$a(x) = \sum_{j=0}^{t''-1} a_j\, x^j$$

using Lagrange interpolation, which was described in Section 13.1. Then the key is revealed to be $K = (a_0, \ldots, a_{t_0-1})$.

Now we consider the security condition. Suppose t' users, say $P_{i_1}, \ldots, P_{i_{t'}}$, pool their shares in an attempt to determine some information about K. Let $(a_0', \ldots, a_{t_0-1}')$ be a guess for the secret. It can be shown that there exists a unique polynomial $a'(x)$ of degree at most $t'' - 1$ such that

Cryptosystem 14.3: *Shamir $(t', t,'' v)$-Ramp Scheme*

Initialization Phase

1. Let $p \geq v + 1$ be a prime. Define $t_0 = t'' - t'$, let $\mathcal{K} = (\mathbb{Z}_p)^{t_0}$, and let $\mathcal{S} = \mathbb{Z}_p$. The values $x_1, x_2, \ldots, x_v$ are defined to be v distinct non-zero elements of $\mathbb{Z}_p$. D gives x_i to P_i, for all i, $1 \leq i \leq v$. The x_i's are public information.

Share Distribution

2. Suppose D wants to share a key $K = (a_0, \ldots, a_{t_0-1}) \in (\mathbb{Z}_p)^{t_0}$. First, D chooses $a_{t_0}, \ldots, a_{t''-1}$ independently and uniformly at random from $\mathbb{Z}_p$. Then, D defines

$$a(x) = \sum_{j=0}^{t''-1} a_j x^j.$$

   Note that $a(x) \in \mathbb{Z}_p[x]$ is a random polynomial of degree at most $t'' - 1$, such that the first t_0 coefficients comprise the secret, K.

3. For $1 \leq i \leq v$, D constructs the share $y_i = a(x_i)$.

4. For $1 \leq i \leq v$, D gives the share y_i to P_i.

1. the first t_0 coefficients of $a'(x)$ are $a'_0, \ldots, a'_{t_0-1}$, and

2. $y_{i_j} = a'(x_{i_j})$, $1 \leq j \leq t'$.

(Basically, this is because there are t' remaining coefficients of $a'(x)$, and the value of $a'(x)$ is known at t' points.) Therefore, no information about K can be computed by a set of at most t' users.

Here is a small example to illustrate.

Example 14.4 Suppose that $p = 17$, $t' = 1$, $t'' = 3$, and $v = 5$. Then $t_0 = 2$, so the secret will be an ordered pair, i.e., $K = (a_0, a_1) \in \mathbb{Z}_{17} \times \mathbb{Z}_{17}$.

Suppose that the public x-co-ordinates are $x_i = i$, $1 \leq i \leq 5$, and suppose that D wants to share the secret $K = (6, 9)$. Because $t'' = 3 = t_0 + 1$, D chooses one random value $a_2 \in \mathbb{Z}_{17}$, say $a_2 = 12$. Then the polynomial $a(x)$ is defined to be

$$a(x) = 6 + 9x + 12x^2 \bmod 17.$$

The shares $y_i = a(x_i)$, $i = 1, \ldots, 5$, are as follows:

$$y_1 = 10, y_2 = 4, y_3 = 9, y_4 = 13, \text{ and } y_5 = 11.$$

Any subset of three participants can compute $a(x)$ by Lagrange interpolation. Hence they can compute the secret $K = (6, 9)$.

On the other hand, any one participant, say P_i, has no information about the ordered pair $K = (a_0, a_1)$. The reader can verify that any "guess" for K is consistent with the share held by P_i.

The definition of a ramp scheme does not specify what should happen if two participants try to compute K. In this example, it can be shown that there are exactly 17 ordered pairs (a_0, a_1) that are consistent with the shares held by any two given participants, say P_i and P_j. In other words, the number of "possible" secrets can be reduced from 17^2 to 17, given the values of two shares.   ⬜

Now we show how ramp schemes can be used in the *General BES* in place of threshold schemes. The relevant observation is that the *General BES* remains secure if a $\left(\lambda\binom{w}{2}, r, v\right)$-ramp scheme is used to construct shares of the secret, in place of an (r, v)-threshold scheme. Everything else remains the same, but the the size of the key K can be increased. With this modification, we have that $K \in (\mathbb{Z}_p)^{t_0}$, where

$$t_0 = r - \lambda \binom{w}{2}.$$

Shares are still elements of $\mathbb{Z}_p$, however, so the message expansion has been decreased by a factor of t_0, from v to v/t_0. The properties of the modified scheme are summarized in Theorem 14.3, which is an improvement of Theorem 14.1.

THEOREM 14.3   *Suppose there is an (n, v, r, λ)-BKDP, and suppose that w is a positive integer such that $\binom{w}{2} < r/\lambda$. Then there is a broadcast encryption scheme for a network of n users which satisfies the following properties:*

1.   *an encrypted message is secure against coalitions of size w,*

2.   *the message expansion is equal to*

$$\frac{v}{r - \lambda\binom{w}{2}},$$

   and

3.   *each user is required to store at most $r + \lambda(n - 1)$ keys.*

Example 14.5   We illustrate this approach by referring once again to the incidence matrix presented in Example 14.1. We already observed that this incidence matrix provides a $(7, 7, 3, 1)$-BKDP. Applying Theorem 14.3, we see that the message expansion is reduced by a factor of $r - \lambda\binom{w}{2} = 3 - 1 = 2$, as compared to the original version of the *General BES* that was presented in Throrem 14.1.

In a bit more detail, here is how the message size is reduced. Recall that the broadcast is an encrypted version of the key K. K is used to encrypt a large amount of data, but K could be a 128-bit key, for example. In the original formulation of the *General BES*, we would take p to be a 129-bit prime in order to share the 128-bit secret K using a $(3, 7)$-threshold scheme. Then the broadcast b_P consists of a 7-tuple of elements chosen from $\mathbb{Z}_p$.

In the modified scheme, the secret K is an ordered pair. In order to accommodate a 128-bit secret, it suffices to choose p to be a 65-bit prime when the $(1, 3, 7)$-ramp scheme is implemented. The broadcast b_P consists of a 7-tuple of elements chosen from $\mathbb{Z}_p$, but now p is a prime of length 65 bits rather than 129 bits. So the broadcast b_p is about half as long as before. As well, all of the users' keys are only half as long as before, since each key will now be 65 bits in length. (The encrypted content, $e_K(\mathcal{M})$, does not change, however.) ⬜

14.3 Multicast Re-keying

In the next sections, we consider the setting of a long-lived dynamic group, say $\mathcal{U}$, with single source broadcast. An on-line subscription service is one example of this. Here, a TA wants to broadcast to every user in the group, but members may join or leave the group over time.

We might make use of a *multicast re-keying scheme* in this scenario. Communications to the group are encrypted with a single *group key*, and every user has a copy of the group key. Users may also have additional *long-lived keys* (or LL-keys), that are used to update the system as the group evolves over time. The system is initialized in a *key predistribution phase*, during which the TA gives LL-keys and an initial group key to the users in the network.

When a new user joins the group, that user is given a copy of the current group key, as well as appropriate long-lived keys; this is called a *user join operation*. When a user U leaves the group, a *user revocation operation* is necessary in order to remove the user from the group. The user revocation operation will establish a new group key for the remaining users, namely, all the users in $\mathcal{U} \setminus \{U\}$ (this is called *re-keying*). In addition, updating of LL-keys may be required as part of the user revocation operation.

Criteria used to evaluate multicast re-keying schemes include the following:

communication and storage complexity
　　This includes the size of broadcasts required for key updating, and the size (and number) of secret LL-keys that have to be stored by users.

security
　　Here, we mainly consider security against revoked users and coalitions of revoked users. Note that a revoked user has more information than someone who never belonged to the group in the first place. Hence, if we achieve security against revoked users, then this automatically implies security against "outsiders."

flexibility of user revocation

Flexibility and efficiency of user revocation operations is an important consideration. For example, it might be the case that users must be revoked one at a time. In some schemes, however, *multiple user revocation* may be possible (up to some specified number of revoked users). This would be more convenient, because users would need to update their keys less frequently.

flexibility of user join

There are several possible variations. In some systems, it may be that any number of new users may be added easily to the system. In other systems, it might be the case that the entire system has to be re-initialized in order to add new users (this would be thought of as a one-time system). Obviously, a flexible and efficient user join operation is desirable in the situation where it is expected that new users will want to join the group.

efficiency of updating LL-keys

Here there are also many possibilities. Perhaps no updating is required (i.e., the LL-keys are static). On the other hand, LL-keys might require updating by an efficient update operation (e.g., via a broadcast). In the worst case, the entire system would have to be reinitialized after a user revocation (basically, this would mean that the system does not accommodate revocation).

One possible approach to multicast re-keying would be to use a broadcast encryption scheme, such as the schemes discussed in the previous section. However, these schemes were designed for a different purpose, namely, broadcasting to an arbitrary subset of users in a large group. Here, we have a situation where we are usually only revoking a small number of users from the group, so the "privileged set" would typically include almost all of the users in the group. The particular requirements of multicast re-keying mean that schemes specifically tailored to this problem are more efficient than general tools such as broadcast encryption.

We will discuss three approaches which can be used to solve this problem, which are summarized as follows:

1. *Blacklisting Schemes* are re-keying schemes that use $(1, w)$-cover-free families. They were proposed by Kumar, Rajagopalan and Sahai.

2. *Naor-Pinkas Schemes* are re-keying schemes that are based on threshold schemes.

3. The *Logical Key Hierarchy* is a tree-based re-keying scheme. It was suggested (independently) by Wallner, Harder and Agee, and by Wong and Lam.

We will discuss each of these three approaches in turn, in the next subsections.

14.3.1   The Blacklisting Scheme

It is possible to use a $(1, w)$-CFF (cover-free family) to revoke a set of w users. Recall from Section 10.4.2 that a $(1, w)$-CFF is a set system having the property that no block is a subset of the union of w other blocks. Here, each block is a set of keys given to a user in the network (as in Section 10.4.2). More precisely, LL-keys are distributed according to an incidence matrix derived from a $(1, w)$-CFF(v, n). So there are n users and a total of v LL-keys; each user receives a subset of the LL-keys.

Suppose that F is a subset of users who are to be revoked. Assume that $|F| = w$. the TA chooses a random value K', which denotes the new group key for the non-revoked users, namely, the users in $\mathcal{U} \backslash F$. Then, for every

$$i \notin \bigcup_{U_j \in F} keys(U_j),$$

the TA computes $y_i = e_{k_i}(K')$ and broadcasts y_i. This is called the *Blacklisting Scheme*.

It can be verified that no user $U_j \in F$ can compute K' (even if the users in F combine all their information). This is because K' is not encrypted with any key held by any member of the coalition F. So the security of the scheme is evident.

Further, because $|F| \leq w$, it follows that every user $U_h \notin F$ can compute K'. To see this, observe that

$$keys(U_h) \not\subseteq \bigcup_{U_j \in F} keys(U_j)$$

(this is ensured by the $(1, w)$ cover-free property). Hence, for every user $U_h \notin F$, there is a key held by U_h that is used to encrypt K'. Therefore U_h can compute K' by decrypting the appropriate ciphertext.

We illustrate with a small example.

Example 14.6   In Example 10.5, we presented the incidence matrix of a $(2, 1)$-CFF$(7, 7)$. Suppose we take the complement of this incidence matrix (i.e., we interchange "1"s with "0"s). Then, it follows from the Exercises of Chapter 10 that we obtain the incidence matrix of a $(1, 2)$-CFF$(7, 7)$, $(X, \mathcal{B})$, where

$$X = \{1, \ldots, 7\}, \quad \text{and}$$
$$\mathcal{B} = \{\{2, 3, 5\}, \{3, 4, 6\}, \{4, 5, 7\}, \{1, 5, 6\},$$
$$\{2, 6, 7\}, \{1, 3, 7\}, \{1, 2, 4\}\}.$$

This means that U_1 gets keys k_2, k_3, k_5; U_2 gets keys k_3, k_4, k_6; etc.

This is a *Blacklisting Scheme* with $w = 2$. For example, suppose that U_2 and U_5 are to be revoked. Then the new group key K' would be encrypted using the

---

Algorithm 14.3: REVOKE($F_1, \ldots, F_T; K_1, \ldots, K_T$)

$F \leftarrow \emptyset$
for $i \leftarrow 1$ **to** T

$\quad$ **do** $\begin{cases} F \leftarrow F \cup \{F_i\} \\ \text{at stage } i, \text{ broadcast the group key } K_i \text{ to } \mathcal{U} \backslash F \end{cases}$

---

keys not in $\{k_3, k_4, k_6\} \cup \{k_2, k_6, k_7\}$. That is, the broadcast consists of the two values $e_{k_1}(K')$ and $e_{k_5}(K')$.

Observe that all the remaining users can decrypt one of these two values, because U_1, U_3 and U_4 all have key k_5 while U_6 and U_7 have the key k_1. On the other hand, neither of U_2 and U_5 hold k_1 or k_5. Therefore these two users cannot compute the new group key, K'.   □

The *Blacklisting Scheme* has the property that the LL-keys are static. Using techniques described in Section 10.4.2, it follows that there exists a $(1, w)$-CFF(v, n) in which v is $O(\log n)$ (i.e., the total number of LL-keys in the system is $O(\log n)$). Therefore, the resulting broadcast also has size $O(\log n)$.

If it is desired, users can be revoked in stages (up to a total of w revoked users) over some period of time. Suppose we want to revoke the users in F_i at stage i, $1 \leq i \leq T$. We assume that $|F_1| + \cdots + |F_T| \leq w$ and $F_1, \ldots, F_T$ are disjoint. Let K_i denote the group key at stage i, $1 \leq i \leq T$. Encryptions of the T group keys are broadcasted as shown in Algorithm 14.3.

With the *Blacklisting Scheme*, there doesn't appear to be a convenient way to update LL-keys, so the scheme must be re-initialized after a total of w users have been revoked.

14.3.2   The Naor-Pinkas Re-keying Scheme

The *Naor-Pinkas Rekeying Scheme* is based on the *Shamir Threshold Scheme* (see Section 13.1). The basic version of the *Naor-Pinkas Scheme* is presented as Cryptosystem 14.4.

The scheme distributes shares of the new group key K' during the initialization phase of the system. K' is "pre-positioned," and the broadcast serves to activate it. It is easy to see that the scheme works as it should: after the broadcast, every user not in F has $w + 1$ shares, which permits the new key K' to be computed. On the other hand, the users in F are not able to compute K, because they (collectively) hold only w shares of K'.

In this scheme, each user stores $O(1)$ information (namely, a share of the new key). The broadcast has size $O(w)$. It is possible to revoke $w' < w$ users by

---

Cryptosystem 14.4: *Naor-Pinkas Rekeying Scheme*

Initialization Phase

1. Denote $n = |\mathcal{U}|$. The TA uses a *Shamir $(w + 1, n)$-Threshold Scheme* to construct n shares, say $y_1, \ldots, y_n$, of a new group key, K'. Every user U_i is given the share y_i in the initialization phase.

Revocation

2. Let $|F| = w$ be the set of users to be revoked. The TA broadcasts the w shares y_i for all $U_i \in F$.

---

broadcasting the shares of the w' revoked users, along with $w - w'$ newly created shares for the same secret, where the new shares do not correspond to any existing user.

The basic *Naor-Pinkas Re-keying Scheme* does not allow users to be revoked in stages. However, there is a "re-usable" version of the *Naor-Pinkas Re-keying Scheme*, which allows multiple user revocation, up to a total of w revoked users. We describe this modified scheme now.

Suppose that keys and the shares in the threshold scheme are defined in a subgroup G of $\mathbb{Z}_p^*$, having prime order q, in which the **Decision Diffie-Hellman** problem is intractable. Let α be a generator of G. The TA constructs shares of a value $K \in \mathbb{Z}_q$ using a *Shamir $(w + 1, n)$-Threshold Scheme* implemented in $\mathbb{Z}_q$. Recall from Section 13.1 that K can be reconstructed using the formula

$$K = \sum_{j=1}^{w+1} b_j y_{i_j} \bmod q,$$

where the b_j's are interpolation coefficients defined as follows:

$$b_j = \prod_{1 \le k \le w+1, k \ne j} \frac{x_{i_k}}{x_{i_k} - x_{i_j}}.$$

Then, it follows that

$$\alpha^K = \prod_{j=1}^{w+1} \alpha^{b_j y_{i_j}} \bmod p.$$

Hence, for any r, we have that

$$\alpha^{rK} = \prod_{j=1}^{w+1} \alpha^{r b_j y_{i_j}} \bmod p.$$

Suppose that the TA broadcasts α^r along with w "exponentiated shares," namely, $\gamma_j = \alpha^{r y_{i_j}}$ $(1 \le j \le w)$. A non-revoked user, say $U_{i_{w+1}}$, can compute his own exponentiated share, as follows:

$$\gamma_{w+1} = \alpha^{r y_{i_{w+1}}} = (\alpha^r)^{y_{i_{w+1}}}.$$

Then $U_{i_{w+1}}$ can compute α^{rK} using the formula

$$\alpha^{rK} = \prod_{j=1}^{w+1} (\alpha^{r y_{i_j}})^{b_j} \bmod p.$$

Finally, α^{rK} is the new group key.

Example 14.7   Suppose that $q = 503$ and $p = 6q + 1 = 3019$. Let $\alpha = 64$. It can be verified that $\alpha^{503} \equiv 1 \pmod{3019}$, so α has order q.

Assume that $w = 2$ and the TA chooses the secret polynomial

$$a(x) = 109 + 215x + 307x^2 \bmod 503;$$

thus $K = 109$.

Suppose that there are five users in the network, and the TA assigns the public x-coordinates to be $x_1 = 15$, $x_2 = 30$, $x_3 = 45$, $x_4 = 60$ and $x_5 = 75$. The five shares are then $y_1 = 480$, $y_2 = 173$, $y_3 = 194$, $y_4 = 40$ and $y_5 = 214$.

Now the TA chooses a random number, say $r = 423$; then the new group key is $\alpha^{rK} \bmod p = 2452$. Assume that U_2 and U_4 are the users to be revoked. Then the TA broadcasts the following information:

$$\alpha^r \bmod p = 1341$$

$$\alpha^{r y_2} \bmod p = 2457, \quad \text{and}$$

$$\alpha^{r y_4} \bmod p = 24.$$

Let's verify that U_1 can compute the new group key. First, U_1 computes his exponentiated share to be

$$1341^{480} \bmod 3019 = 701.$$

Then he computes the interpolation coefficients

$$\frac{30}{30 - 15} \times \frac{60}{60 - 15} \bmod 503 = 338$$

$$\frac{15}{15 - 30} \times \frac{60}{60 - 30} \bmod 503 = 501, \quad \text{and}$$

$$\frac{15}{15 - 60} \times \frac{30}{30 - 60} \bmod 503 = 168.$$

Finally, U_1 computes the new group key:

$$701^{338} \, 2457^{501} \, 24^{168} \bmod 3019 = 2452.$$

⬜

FIGURE 14.3
A binary tree with 16 leaf nodes

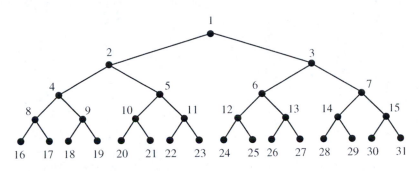

By choosing random values of r for each broadcast, a series of new group keys can be set up. As in previous schemes, coalitions of more than w revoked users can compute group keys. Therefore the total number of revoked users must not exceed w.

14.3.3 Logical Key Hierarchy

In this section, we describe a tree-based re-keying scheme known as the *Logical Key Hierarchy*.

Suppose the number of users, n, satisfies $2^{d-1} < n \le 2^d$. Construct a binary tree, say $\mathcal{T}$, of depth d, having exactly n leaf nodes. All the levels of the tree will be filled, except (possibly) for the last level. The n leaf nodes of $\mathcal{T}$ correspond to the n users. For every user U, let U also denote the (leaf) node corresponding to the user U.

There is a key associated with every node in $\mathcal{T}$ (i.e. there is a different key for every leaf node and every internal node). For every node X, let $k(X)$ denote the key for node X. Then $k(R)$ is the group key, where R is the root node of $\mathcal{T}$. Every user U is given the $d+1$ keys corresponding to the nodes of $\mathcal{T}$ that lie on the unique path from U to R in $\mathcal{T}$. Therefore every user has $O(\log n)$ keys.

Example 14.8 A binary tree with $d = 4$ and $n = 16$, having nodes labeled $1, 2 \ldots, 2^{d+1} - 1 = 31$, is depicted in Figure 14.3. The 16 users are named $16, \ldots, 31$. The group key is $k(1)$ and the keys given to user 25 are $k(1)$, $k(3)$, $k(6)$, $k(12)$ and $k(25)$. □

We will assume that the nodes of $\mathcal{T}$ are labeled so they satisfy the following properties:

1. For $0 \leq \ell \leq d - 1$, the 2^ℓ nodes at depth ℓ are labeled (in order) $2^\ell, 2^\ell + 1, \ldots, 2^{\ell+1} - 1$.

2. The n leaf nodes of $\mathcal{T}$ receive distinct labels chosen from the set $\{2^d, 2^d + 1, \ldots, 2^{d+1} - 1\}$.

3. The parent of node j ($j \neq 1$) is node $\lfloor \frac{j}{2} \rfloor$.

4. The left child of node j is node $2j$ and the right child of node j is node $2j + 1$ (assuming that one or both of these children exist).

5. The sibling of node j ($j \neq 1$) is node $j + 1$, if j is even; or node $j - 1$, if j is odd (assuming that the sibling exists).

It is easy to check that nodes of the tree in Figure 14.3 are labeled in the above-described manner.

Now, assuming that $\mathcal{T}$ is labeled according to the specified rules, it is convenient to store the keys using a simple array-based data structure for binary trees. The keys, corresponding to the nodes of $\mathcal{T}$, can be stored in the form of an array, whose elements are $K[1], \ldots, K[2^{d+1} - 1]$. It is a simple matter to find parents, children and siblings of nodes in this data structure.

Now we can describe the basic user revocation operation in the *Logical Key Hierarchy*. Suppose that we wish to remove user U. Let $\mathcal{P}(U)$ denote the set of nodes in the unique path from a leaf node U to the root node R (recall that R has the label 1). It is necessary to change the keys corresponding to the d nodes in $\mathcal{P}(U) \backslash \{U\}$. For each node $X \in \mathcal{P}(U) \backslash \{U\}$, let $k'(X)$ denote the new key for node X. Let $\mathbf{sib}(\cdot)$ denote the sibling of a given node, and let $\mathbf{par}(\cdot)$ denote the parent of a given node. Then, the following $2d - 1$ items are broadcasted by the TA:

1. $e_{k(\mathbf{sib}(U))}(k'(\mathbf{par}(U)))$

2. $e_{k(\mathbf{sib}(X))}(k'(\mathbf{par}(X)))$ and $e_{k'(X)}(k'(\mathbf{par}(X)))$, for all nodes $X \in \mathcal{P}(U)$, $X \neq U, R$.

We claim that this broadcast allows any non-revoked user V to update all the keys in the intersection $\mathcal{P}(U) \cap \mathcal{P}(V)$. Perhaps the most convincing way to demonstrate this is to consider an example.

Example 14.9   Suppose the TA wants to revoke user $U = 22$. The path $\mathcal{P}(U) = \{22, 11, 5, 2, 1\}$. The TA creates new keys k'_{11}, k'_5, k'_2 and k'_1. The siblings of the nodes in $\mathcal{P}(U)$ are $\{23, 10, 4, 3\}$. The broadcast consists of

$$e_{k(23)}(k'(11)) \quad e_{k(10)}(k'(5)) \quad e_{k(4)}(k'(2)) \quad e_{k(3)}(k'(1))$$
$$e_{k'(11)}(k'(5)) \quad e_{k'(5)}(k'(2)) \quad e_{k'(2)}(k'(1)).$$

This example is depicted in Figure 14.4, where labels of the nodes receiving new keys are boxed and encryptions of new keys are indicated by arrows.

Let's consider how user 23 would update her keys. First she can use her key $k(23)$ to decrypt $e_{k(23)}(k'(11))$; in this way, she computes $k'(11)$. Next, she uses

FIGURE 14.4
A broadcast updating a binary tree

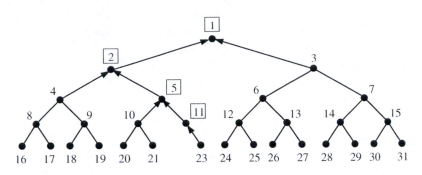

$k'(11)$ to compute $k'(5)$. Then, she uses $k'(5)$ to compute $k'(2)$; and, finally, she uses $k'(2)$ to compute $k'(1)$. ▯

The depth of the tree used in the *Logical Key Hierarchy* is d. Since d is $\Theta(\log n)$, it follows that every user stores $O(\log n)$ keys and the broadcast has size $O(\log n)$. These quantities are larger than the comparable values for the schemes considered previously. However, because the LL-keys are updated each time a user is revoked, there is no limit on the number of users that can be revoked over time. That is, any number of users can be revoked, without affecting the security of the system.

Simultaneous revocation of more than one user can be done, but it is somewhat complicated (see the Exercises). New users can be added to the *Logical Key Hierarchy*, whenever the current number of users is less than 2^d, by assigning the new user to the leftmost "unoccupied" leaf node of the tree. When the number of users exceeds 2^d, one more level of nodes in the tree must be created. This increases the depth of the tree by one, and it allows the number of users to be doubled.

14.4 Copyright Protection

Protection against copyright violation is an important, but very difficult, challenge in the Internet age. Digital content can easily be copied and transmitted over computer networks. Content may be encrypted before it is transmitted — for example, broadcast encryption protects encrypted content (i.e., unauthorized users cannot decrypt it). However, all content must eventually be decrypted before it will be intelligible to an end user. After content is decrypted, it can potentially be copied.

Hardware-based solutions, such as tamper-resistant hardware, for example, can provide a limited amount of protection. Other approaches include algorithms (and coding methods) which enable *tracing* to be carried out. This allows content to be traced to its rightful owner, which discourages people from unauthorized copying of digital data. In this section, we describe some types of "codes" that can be used for tracing.

Before continuing further, it is useful to distinguish some different types of copyright violation. There are many potential threats. Here are two threats that we introduce as typical examples.

illegal content redistribution
As mentioned above, encrypted content is invariably decrypted once it gets to its authorized destination. Decrypted content can then be copied and transmitted to others, for example in an illegal *pirate broadcast*.

illegal key redistribution
Assuming that content is encrypted, there must be a mechanism for the content to be decrypted by an end user. The keys used to decrypt the content may be copied and distributed to other users. Alternatively, these keys may be combined to create a new *pirate decoder*, which can subsequently be used to decrypt encrypted content illegally.

14.4.1   Fingerprinting

We first address the problem of illegal content redistribution. Suppose that every copy of some digital data, D, contains a unique *fingerprint*, F. For example, there might be 1 megabyte of binary data, and a fingerprint might consist of 100 "special" bits "hidden" in the data in a manner which is hard to detect. (Sometimes the process of embedding hidden identifying data is called *watermarking*.)

In this scenario, the vendor can maintain a database that keeps track of all the different fingerprints, as well as the rightful owners of the corresponding copies of the data D. Then any exact copy of the data can be traced back to its owner. Unfortunately, there are some serious flaws with this approach. For example, if the fingerprint is easily recognized, then it can be modified or destroyed, thus making the data impossible to trace. A second threat is that coalitions may be able to recognize fingerprints or parts of fingerprints — even if individual users cannot do so — and then create a new copy of the data with the fingerprint destroyed.

Here is a more precise mathematical model which will facilitate studying this problem. For concreteness, suppose that each copy of the data consists of L bits of *content*, say C, and an ℓ-bit *fingerprint*, F; hence, the data has the form $D = (C, F)$. All the data is represented over some fixed alphabet. For example, binary data uses the alphabet $\{0, 1\}$. We will assume that all copies of the data have the same content but different fingerprints, so we have $D_1 = (C, F_1), D_2 =$

FIGURE 14.5
The marking assumption

(C, F_2), etc. Furthermore, we will assume that the fingerprint bits[1] always occur in the same (secret) positions in all copies of the data; e.g., bits $b_{i_1}, \ldots, b_{i_\ell}$ are fingerprint bits.

Fingerprinting problems are usually studied assuming that a certain *marking assumption* holds. This assumption is stated as follows:

> *Given some number of copies of the data, say $D_1, D_2, \ldots, D_w$, the only bits that can be identified as fingerprint bits by a coalition are those bits b such that $D_i[b] \neq D_j[b]$ for some i, j.*

In other words, we are assuming that the fingerprints are hidden well enough that no particular bit can be identified as a fingerprint bit by a coalition of bad guys unless the coalition possesses two copies of the data in which the bit in question takes on different values.

The diagram in Figure 14.5 illustrates the idea behind the marking assumption. This diagram contains two grids made up of black and white "pixels." It can be verified that there are exactly three "pixels" in which the two grids differ. According to the marking assumption, only these three pixels can be recognized as fingerprint bits.

Let's consider the kinds of attacks that a coalition can carry out, assuming that the marking assumption holds. A bit of thought shows that the marking assumption implies that the actual content is irrelevant, and the problem reduces to studying combinatorial properties of the set of fingerprints. As described above, given w copies of the data, some bits can be identified as fingerprint bits. Then a new "pirate" copy of the data can be constructed, by setting values of these identified fingerprint bits from one of the copies of the data in an arbitrary fashion. The resulting data is $D' = (C, F')$, where F' is a newly created *hybrid fingerprint*.

---

[1] We will use the term "fingerprint bits" to denote the positions in which the fingerprints occur. The term "bits" suggests that the data has a binary form, but we will use this term even if the data is defined over a non-binary alphabet.

The fundamental question is whether a hybrid fingerprint can be "traced" if the fingerprints are constructed in a suitable way.

We now define the notion of hybrid fingerprints precisely.

Definition 14.2:   An (ℓ, n, q)-*code* is a subset $\mathcal{C} \subseteq Q^\ell$ such that $|Q| = q$ and $|\mathcal{C}| = n$. That is, we have n *codewords*, each of which is an ℓ-tuple of elements from the alphabet Q. A codeword is the same thing as a fingerprint. Let $\mathcal{C}_0 \subseteq \mathcal{C}$ (i.e., $\mathcal{C}_0$ is a subset of codewords). Define $\mathbf{desc}(\mathcal{C}_0)$ to consist of all ℓ-tuples $\mathbf{f} = (f_1, \ldots, f_\ell)$ such that, for all $1 \leq i \leq \ell$, there exists a codeword $\mathbf{c} = (c_1, \ldots, c_\ell) \in \mathcal{C}_0$ such that $f_i = c_i$. $\mathbf{desc}(\mathcal{C}_0)$ consists of all the hybrid fingerprints that can be constructed from the fingerprints in $\mathcal{C}_0$; it is called the *descendant code* of $\mathcal{C}_0$.

Finally, for any $\mathbf{c} \in \mathcal{C}_0$ and for any $\mathbf{f} \in \mathbf{desc}(\mathcal{C}_0)$, then we say that $\mathbf{c}$ is a *parent* of $\mathbf{f}$ in the code $\mathbf{desc}(\mathcal{C}_0)$.

For example, suppose that

$$\mathcal{C}_0 = \{(1, 1, 2), (2, 3, 2)\}.$$

In the descendant code, the first co-ordinate can be 1 or 2, the second co-ordinate can be 1 or 3, and the last co-ordinate must be 2. Therefore, it is easy to see that

$$\mathbf{desc}(\{(1, 1, 2), (2, 3, 2)\}) = \{(1, 1, 2), (2, 3, 2), (1, 3, 2), (2, 1, 2)\}.$$

In this example, the descendant code consists of the two original codewords and two new hybrid fingerprints.

For an integer $w \geq 2$, the w-*descendant code* of $\mathcal{C}$, denoted $\mathbf{desc}_w(\mathcal{C})$, consists of the following set of ℓ-tuples:

$$\mathbf{desc}_w(\mathcal{C}) = \bigcup_{\mathcal{C}_0 \subseteq \mathcal{C}, |\mathcal{C}_0| \leq w} \mathbf{desc}(\mathcal{C}_0).$$

The w-descendant code consists of all hybrid fingerprints that could be produced by a coalition of size at most w.

14.4.2   Identifiable Parent Property

We now turn to the "inverse" process, namely trying to determine the coalition that constructed a hybrid fingerprint. Suppose that $\mathbf{f} \in \mathbf{desc}_w(\mathcal{C})$. We define the set of *suspect coalitions* for $\mathbf{f}$ as follows:

$$\mathbf{susp}_w(\mathbf{f}) = \{\mathcal{C}_0 \subseteq \mathcal{C} : |\mathcal{C}_0| \leq w, \mathbf{f} \in \mathbf{desc}(\mathcal{C}_0)\}.$$

$\mathbf{susp}_w(\mathbf{f})$ consists of all the coalitions of size at most w that could have produced the hybrid fingerprint $\mathbf{f}$ by following the process described above. Ideally,

$\mathbf{susp}_w(\mathbf{f})$ would consist of one and only one set. In this case, we would have some evidence that this subset in fact created the hybrid fingerprint (of course, we can never rule out the possibility that some other coalition, necessarily of size exceeding w, is in fact the guilty subset).

Even if $\mathbf{susp}_w(\mathbf{f})$ consists of more than one set, we still may be able to extract some useful information by looking at the sets in $\mathbf{susp}_w(\mathbf{f})$. For example, suppose that there exists a codeword $\mathbf{c} \in \mathcal{C}$ such that $\mathbf{c} \in \mathcal{C}_0$ for all $\mathcal{C}_0 \in \mathbf{susp}_w(\mathbf{f})$. Any such codeword can be identified as guilty (under the assumption that the coalition has size at most w), even if we are not able to identify the complete guilyy subset.

The above-mentioned property can be stated in an equivalent form as follows:

$$\bigcap_{\mathcal{C}_0 \in \mathbf{susp}_w(\mathbf{f})} \mathcal{C}_0 \neq \emptyset. \tag{14.1}$$

We say that $\mathcal{C}$ is a *w-identifiable parent property code* (or *w-IPP code*) provided that (14.1) is satisfied for all $\mathbf{f} \in \mathbf{desc}_w(\mathcal{C})$. Further, in a w-IPP code, if

$$\mathbf{c} \in \bigcap_{\mathcal{C}_0 \in \mathbf{susp}_w(\mathbf{f})} \mathcal{C}_0,$$

then $\mathbf{c}$ is called an *identifiable parent* of $\mathbf{f}$.

Example 14.10 We present a $(3, 6, 3)$ code, and consider coalitions of size at most two:

$$\mathbf{c}_1 = (0, 1, 1), \quad \mathbf{c}_2 = (1, 0, 1), \quad \mathbf{c}_3 = (1, 1, 0),$$
$$\mathbf{c}_4 = (2, 0, 2), \quad \mathbf{c}_5 = (1, 0, 2), \quad \mathbf{c}_6 = (2, 1, 0).$$

Consider the hybrid fingerprint $\mathbf{f}_1 = (1, 1, 1)$. It is not difficult to verify that

$$\mathbf{susp}_2(\mathbf{f}_1) = \{\{1, 2\}, \{1, 3\}, \{2, 3\}, \{1, 5\}, \{2, 6\}\}.$$

This hybrid fingerprint $\mathbf{f}_1$ violates property (14.1), so the code is not a 2-IPP code. On the other hand, consider $\mathbf{f}_2 = (0, 1, 2)$. Here it can be seen that

$$\mathbf{susp}_2(\mathbf{f}_2) = \{\{1, 4\}, \{1, 5\}\}.$$

Observe that property (14.1) is satisfied for the hybrid fingerprint $\mathbf{f}_2$. Hence, $\mathbf{c}_1$ is an identifiable parent of $\mathbf{f}_2$ (under the assumption that a coalition of size at most two created $\mathbf{f}_2$), because

$$\{1, 4\} \cap \{1, 5\} = \{1\}.$$

$\square$

Example 14.11   We present a $(3, 7, 5)$ 2-IPP code:

$$c_1 = (0, 0, 0), \quad c_2 = (0, 1, 1), \quad c_3 = (0, 2, 2), \quad c_4 = (1, 0, 3),$$
$$c_5 = (2, 0, 4), \quad c_6 = (3, 3, 0), \quad c_7 = (4, 4, 0).$$

We show that the property (14.1) holds for all relevant hybrid fingerprints $\mathbf{f}$. Suppose that $\mathbf{f} = (f_1, f_2, f_3)$ is a hybrid fingerprint created by a coalition of size two. If any co-ordinate of $\mathbf{f}$ is non-zero, then at least one parent of $\mathbf{f}$ can be identified, as indicated in the following exhaustive list of possibilities:

$$f_1 = 1 \Rightarrow c_4; \quad f_1 = 2 \Rightarrow c_5; \quad f_1 = 3 \Rightarrow c_6; \quad f_1 = 4 \Rightarrow c_7$$
$$f_2 = 1 \Rightarrow c_2; \quad f_2 = 2 \Rightarrow c_3; \quad f_2 = 3 \Rightarrow c_6; \quad f_2 = 4 \Rightarrow c_7$$
$$f_3 = 1 \Rightarrow c_2; \quad f_3 = 2 \Rightarrow c_3; \quad f_3 = 3 \Rightarrow c_4; \quad f_3 = 4 \Rightarrow c_5.$$

Finally, if $\mathbf{f} = (0, 0, 0)$, then c_1 is an identifiable parent.   ▯

14.4.3   2-IPP Codes

In general, it is not an easy task to do any of the following:

1. construct a w-IPP code;

2. verify whether a given code is a w-IPP code; or

3. find an efficient algorithm to identify a parent, given an ℓ-tuple in the w-descendant code of a w-IPP code.

In reference to the third task, it is of particular interest to design w-IPP codes for which efficient parent-identifying algorithms can be constructed.

   In this section, we will pursue these questions in the easiest case, $w = 2$. We will provide a nice characterization of 2-IPP codes which involves certain kinds of hash families, which we introduce next.

Definition 14.3:   An (n, m, w)-*perfect hash family* is a set of functions, say $\mathcal{F}$, such that $|X| = n$, $|Y| = m$, $f : X \rightarrow Y$ for each $f \in \mathcal{F}$, and for any $X_1 \subseteq X$ such that $|X_1| = w$, there exists at least one $f \in \mathcal{F}$ such that $f|_{X_1}$ is one-to-one.[2] When $|\mathcal{F}| = N$, an (n, m, w)-perfect hash family will be denoted by $\mathrm{PHF}(N; n, m, w)$.

A $\mathrm{PHF}(N; n, m, w)$ can be depicted as an $n \times N$ array with entries from Y, having the property that in any w rows there exists at least one column such that the w entries in the given w rows are distinct. Here the columns of the array are labeled by the functions in $\mathcal{F}$, the rows are labeled by the elements in X, and the entry in row x and column f of the array is $f(x)$.

---

[2] The notation $f|_{X_1}$ denotes the restriction of the function f to the subset X_1 of the domain. The requirement that $f|_{X_1}$ is one-to-one means that $f(x) \neq f(x')$ for all $x, x' \in X_1$ such that $x \neq x'$.

Perfect hash families have been widely studied in the context of information retrieval algorithms. However, as we shall see, perfect hash families have close connections to w-IPP codes.

Here is a related concept.

Definition 14.4: An $(n, m, \{w_1, w_2\})$-*separating hash family* is a set of functions, say $\mathcal{F}$, such that $|X| = n$, $|Y| = m$, $f : X \to Y$ for each $f \in \mathcal{F}$, and for any $X_1, X_2 \subseteq X$ such that $|X_1| = w_1$, $|X_2| = w_2$ and $X_1 \cap X_2 = \emptyset$, there exists at least one $f \in \mathcal{F}$ such that

$$\{f(x) : x \in X_1\} \cap \{f(x) : x \in X_2\} = \emptyset.$$

The notation $\text{SHF}(N; n, m, \{w_1, w_2\})$ will be used to denote an $(n, m, \{w_1, w_2\})$-separating hash family with $|\mathcal{F}| = N$.

An $\text{SHF}(N; n, m, \{w_1, w_2\})$ can be depicted as an $n \times N$ array with entries from the set Y, having the property that in any w_1 rows and any w_2 disjoint rows there exists at least one column such that the entries in the given w_1 rows are distinct from the entries in the given w_2 rows.

The following example serves to illustrate the two previous definitions.

Example 14.12 Consider the following seven by three array:

0	0	0
0	1	1
0	2	2
1	0	3
2	0	4
3	3	0
4	4	0

It can be verified that the above array is simultaneously a PHF$(3; 7, 5, 3)$ and an SHF$(3; 7, 5, \{2, 2\})$.

We note, however, that the array is not a PHF$(3; 7, 5, 4)$. Consider rows $1, 2, 4$ and 6. None of the three columns contain distinct entries in all four of the given rows. ▯

We will now derive an efficient algorithm to determine if a given (ℓ, n, q) code, say $\mathcal{C}$, is a 2-IPP code. Suppose the codewords are written in the form of an n by ℓ array, say $A(\mathcal{C})$, and suppose that $A(\mathcal{C})$ is not a PHF$(\ell; n, q, 3)$. Then there exist three rows, r_1, r_2, r_3 of A that violate the perfect hash family property. For every column c, let f_c be an element that is repeated (i.e., it occurs in at least two of the three given rows r_1, r_2, r_3 in column c). Now, define $\mathbf{f} = (f_1, \ldots, f_\ell)$. Clearly

$$\{r_1, r_2\}, \{r_1, r_3\}, \{r_2, r_3\} \in \mathbf{susp}_2(\mathbf{f}).$$

Therefore, $\mathcal{C}$ is not a 2-IPP code, because the intersection of these three 2-subsets is the empty set.

Next, suppose that $A(\mathcal{C})$ is not an SHF$(\ell; n, q, \{2, 2\})$. Then there exist two sets of two rows of $A(\mathcal{C})$, say $\{r_1, r_2\}$ and $\{r_3, r_4\}$, that violate the separating hash family property. For every column c, let f_c be an element that occurs in column c in one of rows r_1 and r_2, and again in column c in one of rows r_3 and r_4. Define $\mathbf{f} = (f_1, \ldots, f_\ell)$. Clearly,

$$\{r_1, r_2\}, \{r_3, r_4\} \in \mathbf{susp}_2(\mathbf{f}).$$

Therefore, $\mathcal{C}$ is not a 2-IPP code, because the intersection of these two 2-subsets is the empty set.

From the above discussion, we see that a necessary condition for $\mathcal{C}$ to be a 2-IPP code is that $A(\mathcal{C})$ is simultaneously a PHF$(\ell; n, q, 3)$ and an SHF$(\ell; n, q, \{2, 2\})$. The converse is also true (see the Exercises), and therefore we have the following theorem.

THEOREM 14.4   *An (ℓ, n, q) code $\mathcal{C}$ is a 2-IPP code if and only if $A(\mathcal{C})$ is simultaneously a PHF$(\ell; n, q, 3)$ and an SHF$(\ell; n, q, \{2, 2\})$.*

As a corollary, an $(\ell, n, 2)$ code cannot be a 2-IPP code. For $n \geq 3$, it follows from Theorem 14.4 that an (ℓ, n, q) code, $\mathcal{C}$, can be tested to determine if it is a 2-IPP code in polynomial time as a function of n.

Now we consider identification of parents in a 2-IPP code. Suppose that $\mathcal{C}$ is a 2-IPP code and $\mathbf{f} \in \mathbf{desc}_2(\mathcal{C}) \backslash \mathcal{C}$. Thus $\mathbf{f}$ is not a codeword, and there is at least one subset of two codewords for which $\mathbf{f}$ is in the descendant subcode. The fact that $\mathcal{C}$ is a 2-IPP code severely constrains the possible structure of $\mathbf{susp}_2(\mathbf{f})$. It can be shown that one of two possible scenarios must hold:

1. either $\mathbf{susp}_2(\mathbf{f})$ consists of a single set of two codewords, or
2. $\mathbf{susp}_2(\mathbf{f})$ consists of a two or more sets of two codewords, all of which contain a fixed codeword. For example,

$$\mathbf{susp}_2(\mathbf{f}) = \{\{\mathbf{c}_1, \mathbf{c}_2\}, \{\mathbf{c}_1, \mathbf{c}_3\}, \{\mathbf{c}_1, \mathbf{c}_4\}\}$$

would fall into this case.

In the first case, we can identify both parents of $\mathbf{f}$. In the second case, we can identify one parent (namely, $\mathbf{c}_1$, in the example provided).

In a 2-IPP code, we only consider suspect coalitions of size two. Given $\mathbf{f}$, we can examine all the $\binom{n}{2}$ subsets of two codewords. For each 2-subset $\{\mathbf{c}, \mathbf{d}\}$, we can check to see if $\mathbf{f} \in \mathbf{desc}(\{\mathbf{c}, \mathbf{d}\})$. This will yield an algorithm having complexity $\Theta(n^2)$ which will identify a parent in an arbitrary 2-IPP code.

There are many constructions for 2-IPP codes. We present a simple and efficient construction for certain 2-IPP codes with $\ell = 3$, which is due to Hollmann, van Lint, Linnartz and Tolhuizen. Suppose that $r \geq 2$ is an integer, let $q = r^2 + 2r$, and define

$$S = \{1, \ldots, r\} \quad (|S| = r)$$
$$M = \{r+1, \ldots, 2r\} \quad (|M| = r)$$
$$L = \{2r+1, \ldots, q\} \quad (|L| = r^2)$$
$$\mathcal{C}_1 = \{(s_1, s_2, rs_1 + s_2 + r) : s_1, s_2 \in S\} \subseteq S \times S \times L$$
$$\mathcal{C}_2 = \{(m, sr + m, s) : m \in M, s \in S\} \subseteq M \times L \times S$$
$$\mathcal{C}_3 = \{(rm_1 + m_2 - r^2, m_1, m_2) : m_1, m_2 \in M\} \subseteq L \times M \times M.$$

Example 14.13 We construct a $(3, 27, 15)$ 2-IPP code by following the recipe given above. We have $r = 3$, $S = \{1, 2, 3\}$, $M = \{4, 5, 6\}$, and $L = \{7, \ldots, 15\}$. $\mathcal{C}_1$, $\mathcal{C}_2$ and $\mathcal{C}_3$ each consist of nine codewords, as indicated here:

$$\begin{array}{lll}
\mathbf{c}_1 = (1,1,7), & \mathbf{c}_2 = (1,2,8), & \mathbf{c}_3 = (1,3,9), \\
\mathbf{c}_4 = (2,1,10), & \mathbf{c}_5 = (2,2,11), & \mathbf{c}_6 = (2,3,12), \\
\mathbf{c}_7 = (3,1,13), & \mathbf{c}_8 = (3,2,14), & \mathbf{c}_9 = (3,3,15), \\
\hline
\mathbf{c}_{10} = (4,7,1), & \mathbf{c}_{11} = (5,8,1), & \mathbf{c}_{12} = (6,9,1), \\
\mathbf{c}_{13} = (4,10,2), & \mathbf{c}_{14} = (5,11,2), & \mathbf{c}_{15} = (6,12,2), \\
\mathbf{c}_{16} = (4,13,3), & \mathbf{c}_{17} = (5,14,3), & \mathbf{c}_{18} = (6,15,3), \\
\hline
\mathbf{c}_{19} = (7,4,4), & \mathbf{c}_{20} = (8,4,5), & \mathbf{c}_{21} = (9,4,6), \\
\mathbf{c}_{22} = (10,5,4), & \mathbf{c}_{23} = (11,5,5), & \mathbf{c}_{24} = (12,5,6), \\
\mathbf{c}_{25} = (13,6,4), & \mathbf{c}_{26} = (14,6,5), & \mathbf{c}_{27} = (15,6,6).
\end{array}$$

We claim that $\mathcal{C}_1 \cup \mathcal{C}_2 \cup \mathcal{C}_3$ is a 2-IPP code with $n = 3r^2$. Furthermore, this code has an $O(1)$ time algorithm to find an identifiable parent. Actually, we show how to find an identifiable parent (which will prove implicitly that the code is a 2-IPP code). The main steps in a parent-identifying algorithm are as follows:

1. If $\mathbf{f} = (f_1, f_2, f_3)$ has a co-ordinate in L, then a parent is easily identified. For example, suppose that $f_2 = 13$. Then $3s + m = 13$, where $s \in \{1, 2, 3\}$ and $m \in \{4, 5, 6\}$. Hence $s = 3$ and $m = 4$, and therefore $(4, 13, 3)$ is an identifiable parent.

2. If $\mathbf{f}$ has no co-ordinate in L, then it is possible to compute $i \neq j$ such that the two parents of $\mathbf{f}$ are in $\mathcal{C}_i$ and $\mathcal{C}_j$. The parent that contributed two co-ordinates to $\mathbf{f}$ can then be identified.

    For example, suppose that $\mathbf{f} = (1, 3, 2)$. The parents of $\mathbf{f}$ must be from $\mathcal{C}_1$ and $\mathcal{C}_2$. The parent from $\mathcal{C}_1$ contributes f_1 and f_2, and hence $(1, 3, 9)$ is an identifiable parent.

□

The reasoning used in the above example to identify a parent will work for any code in this family. The complexity of the resulting parent-identification algorithm is independent of n (i.e., it has complexity $O(1)$).

Summarizing the results of this section, we have the following theorem.

THEOREM 14.5 *For all integers $r \geq 2$ there exists a $(3, 3r^2, r^2 + 2r)$-code that is a 2-IPP code. Furthermore, this code has a parent-identifying algorithm having complexity $O(1)$.*

14.5 Tracing Illegally Redistributed Keys

Suppose that every user in a network is given a "decoder box" that allows encrypted broadcasts to be decrypted. That is, we have a broadcast encryption scheme in which every user can decrypt the broadcast. In general, every decoder box contains a different collection of keys. In such a scenario, a coalition of w bad guys might create a *pirate decoder* by combining keys from their decoder boxes. Any such pirate decoder will be able to decrypt broadcasts.

The set of keys in each decoder box can be thought of as a codeword in a certain code, and the keys in a pirate decoder can be thought of as a codeword in the w-descendant code. If the code is traceable (e.g., if it satisfies the w-IPP property), then a pirate decoder can be traced back to at least one member of the coalition that created it. Thus, if a pirate decoder is confiscated, then at least one of the guilty parties can be determined.

Let us briefly discuss the broadcast encryption scheme that will be used. This is a simpler type of BES than those considered earlier, because any subset of keys of sufficient size (and having the correct structure) will be able to decrypt broadcasts.

First, the TA chooses ℓ sets of keys, denoted $\mathcal{K}_1, \ldots, \mathcal{K}_\ell$, where each $\mathcal{K}_i$ consists of q keys chosen from $\mathbb{Z}_m$, for some fixed m. For $1 \leq i \leq \ell$, let $\mathcal{K}_i = \{k_{i,j} : 1 \leq j \leq q\}$. Now, a decoder box contains ℓ keys, one from each set $\mathcal{K}_i$.

The secret key $K \in \mathbb{Z}_m$ (which is used to encrypt the broadcast content, $\mathcal{M}$) is split into ℓ shares using an (ℓ, ℓ) threshold scheme (we use the threshold scheme described in Section 13.1.1). The shares are denoted $s_1, \ldots, s_\ell$, where

$$s_1 + \cdots + s_\ell \equiv K \pmod{m}.$$

Then K is used to encrypt $\mathcal{M}$, and for $1 \leq i \leq \ell$, every $k_{i,j}$ is used to encrypt s_i. The entire broadcast consists of the following information:

$$y = e_K(\mathcal{M}) \quad \text{and} \quad (e_{k_{i,j}}(s_i) : 1 \leq i \leq \ell, 1 \leq j \leq q).$$

Observe that this broadcast encryption scheme follows the general model depicted in Figure 14.2.

Every user who possesses a decoder box can perform the following operations:

1. decrypt all ℓ shares of K,
2. reconstruct K from the ℓ shares, and
3. use the key K to decrypt y, thus obtaining $\mathcal{M}$.

Each decoder box corresponds to a codeword $\mathbf{c} \in Q^\ell$, where $Q = \{1, \ldots, q\}$, in an obvious way:

keys in decoder box	codeword
$\{k_{1,j_1}, k_{2,j_2}, \ldots, k_{\ell,j_\ell}\}$	$(j_1, j_2, \ldots, j_\ell)$.

Denote by $\mathcal{C}$ the set of codewords corresponding to all the decoder boxes in the scheme. The keys in a pirate decoder form a codeword in the w-descendant code $\mathbf{desc}_w(\mathcal{C})$.

There is a special class of w-IPP codes that have very efficient tracing algorithms. These tracing algorithms are based on the idea of "nearest neighbor decoding" that is used in error-correcting codes. First, we require a couple of definitions.

Let $\mathbf{dist}(\mathbf{c}, \mathbf{d})$ denote the *hamming distance* between two vectors $\mathbf{c}, \mathbf{d} \in Q^\ell$. That is,

$$\mathbf{dist}(\mathbf{c}, \mathbf{d}) = |\{i : \mathbf{c}_i \neq \mathbf{d}_i\}|.$$

For $\mathbf{f} \in \mathbf{desc}_w(\mathcal{C})$, a *nearest neighbor* to $\mathbf{f}$ is any codeword $\mathbf{c} \in \mathcal{C}$ such that $\mathbf{dist}(\mathbf{f}, \mathbf{c})$ is as small as possible. We will denote any nearest neighbor to $\mathbf{f}$ by $\mathbf{nn}(\mathbf{f})$ (there can be more than one nearest neighbor). Computing $\mathbf{nn}(\mathbf{f})$ is called *nearest neighbor decoding*. Observe that nearest neighbor decoding can be carried out in time $O(n)$ by exhaustive search over all the codewords in the code.[3]

The code $\mathcal{C}$ is said to be a w-*TA code* if the following property holds for all $\mathbf{f} \in \mathbf{desc}_w(\mathcal{C})$:

$$\mathbf{nn}(\mathbf{f}) \in \bigcap_{\mathcal{C}_0 \in \mathbf{susp}_w(\mathbf{f})} \mathcal{C}_0. \tag{14.2}$$

In other words, a w-TA code is a w-IPP code in which nearest neighbor decoding always yields an identifiable parent.

Here is a small example to illustrate.

---

[3]More efficient algorithms for nearest neighbor decoding might exist, depending on the structure of the code.

Example 14.14   We present a certain $(5, 16, 4)$ code:

$$
\begin{aligned}
&\mathbf{c}_1 = (1,1,1,1,1) &&\mathbf{c}_2 = (1,2,2,2,2) \\
&\mathbf{c}_3 = (1,3,3,3,3) &&\mathbf{c}_4 = (1,4,4,4,4) \\
&\mathbf{c}_5 = (2,1,2,3,4) &&\mathbf{c}_6 = (2,2,1,4,3) \\
&\mathbf{c}_7 = (2,3,4,1,2) &&\mathbf{c}_8 = (2,4,3,2,1) \\
&\mathbf{c}_9 = (3,1,4,2,3) &&\mathbf{c}_{10} = (3,2,3,1,4) \\
&\mathbf{c}_{11} = (3,3,2,4,1) &&\mathbf{c}_{12} = (3,4,1,3,2) \\
&\mathbf{c}_{13} = (4,1,3,4,2) &&\mathbf{c}_{14} = (4,2,4,3,1) \\
&\mathbf{c}_{15} = (4,3,1,2,4) &&\mathbf{c}_{16} = (4,4,2,1,3).
\end{aligned}
$$

It can be proven that this code is a 2-TA code. Therefore nearest neighbor decoding can be used to identify parents.

Consider the vector $\mathbf{f} = (2, 3, 2, 4, 4)$. This is a vector in the 2-descendant code. If we compute the distance from $\mathbf{f}$ to all the codewords, then we see that

$$\mathbf{dist}(\mathbf{f}, \mathbf{c}_5) = \mathbf{dist}(\mathbf{f}, \mathbf{c}_{11}) = 2$$

and

$$\mathbf{dist}(\mathbf{f}, \mathbf{c}_i) \geq 3$$

for all $i \neq 5, 11$. Hence $\mathbf{c}_5$ and $\mathbf{c}_{11}$ are both identifiable parents of $\mathbf{f}$.   ⬜

One sufficient condition for a code to be a w-TA code is for it to have a large minimum distance between distinct codewords. Therefore, we define

$$\mathbf{dist}(\mathcal{C}) = \min\{\mathbf{dist}(\mathbf{c}, \mathbf{d}) : \mathbf{c}, \mathbf{d} \in \mathcal{C}, \mathbf{c} \neq \mathbf{d}\}.$$

The next theorem provides a useful, easily tested, condition relating to TA codes.

THEOREM 14.6   *Suppose that $\mathcal{C}$ is an (ℓ, n, q)-code in which*

$$\mathbf{dist}(\mathcal{C}) > \ell \left(1 - \frac{1}{w^2}\right).$$

Then $\mathcal{C}$ is a w-TA code.

PROOF   We will use the following notation. Denote $d = \mathbf{dist}(\mathcal{C})$. For any vectors, $\mathbf{c}, \mathbf{d}$, define

$$\mathbf{match}(\mathbf{c}, \mathbf{d}) = \ell - \mathbf{dist}(\mathbf{c}, \mathbf{d}).$$

Now, suppose that $\mathbf{c} = \mathbf{nn}(\mathbf{f})$ and suppose that $\mathcal{C}_0 \in \mathbf{susp}_w(\mathbf{f})$. We need to prove that $\mathbf{c} \in \mathcal{C}_0$.

First, because $\mathbf{f} \in \mathbf{desc}(\mathcal{C}_0)$, it follows that

$$\sum_{\mathbf{c}' \in \mathcal{C}_0} \mathbf{match}(\mathbf{f}, \mathbf{c}') \geq \ell.$$

Then, because $|\mathcal{C}_0| \le w$, it follows that there exists a codeword $\mathbf{c}' \in \mathcal{C}_0$ such that

$$\mathbf{match}(\mathbf{f}, \mathbf{c}') \ge \frac{\ell}{w}.$$

Hence,

$$\mathbf{match}(\mathbf{f}, \mathbf{c}) \ge \frac{\ell}{w},$$

because $\mathbf{c}$ is the nearest neighbor to $\mathbf{f}$.

Next, let $\mathbf{b} \in \mathcal{C} \backslash \mathcal{C}_0$. Because $\mathbf{f} \in \mathbf{desc}(\mathcal{C}_0)$, we have that

$$\mathbf{match}(\mathbf{f}, \mathbf{b}) \le \sum_{\mathbf{c}' \in \mathcal{C}_0} \mathbf{match}(\mathbf{c}', \mathbf{b})$$

$$\le w(\ell - d).$$

Now, notice that $d > \ell(1 - 1/w^2)$ is equivalent to

$$w(\ell - d) < \frac{\ell}{w}.$$

Therefore, it follows that $\mathbf{match}(\mathbf{f}, \mathbf{b}) < \mathbf{match}(\mathbf{f}, \mathbf{c})$ for all codewords $\mathbf{b} \notin \mathcal{C}_0$. Hence, $\mathbf{c} \in \mathcal{C}_0$, and we have proven that the code is a w-TA code. ∎

We close this section by describing an easy construction for certain w-TA codes.[4] Suppose q is prime and $t < q$. Define the set $\mathcal{P}(q, t)$ to consist of all polynomials $a(x) \in \mathbb{Z}_q[x]$ having degree at most $t - 1$. For a positive integer $\ell < q$, define

$$\mathcal{C}(q, \ell, t) = \{(a(0), a(1), \dots, a(\ell - 1)) : a(x) \in \mathcal{P}(q, t)\}.$$

We claim that $\mathcal{C} = \mathcal{C}(q, \ell, t)$ is an (ℓ, q^t, q) code such that $\mathbf{dist}(\mathcal{C}) = \ell - t + 1$. This is easy to see, because any two polynomials of degree not exceeding $t - 1$ can agree on at most $t - 1$ points.

Suppose we define

$$t = \left\lceil \frac{\ell}{w^2} \right\rceil;$$

then

$$t < \frac{\ell}{w^2} + 1.$$

Therefore,

$$\mathbf{dist}(\mathcal{C}) > \ell\left(1 - \frac{1}{w^2}\right).$$

Hence, by Theorem 14.6, we have a w-TA code with $n = q^{\lceil \frac{\ell}{w^2} \rceil}$.

Summarizing, we obtain the following.

---

[4]The codes we describe are known as *Reed-Solomon codes*.

THEOREM 14.7 *Suppose q is prime, $\ell \leq q$ and $w \geq 2$ is an integer. Then there is an $\left(\ell, q^{\lceil \frac{\ell}{w^2} \rceil}, q\right)$-code that is a w-TA code.*

14.6 Notes and References

For an overall view of multicast security as discussed in this chapter, see Canetti, Garay, Itkis, Micciancio, Naor and Pinkas [82].

The idea of broadcast encryption was introduced in 1991 by Berkovitz [29]. Since that time, there have been many published works on this topic. In Section 14.2, we are following the approach described in Fiat and Naor [134], Stinson [323], Stinson and Tran [325], and Stinson and Wei [327]. The *General Broadcast Encryption Scheme* is from [323]; it is a generalization of the scheme described in [134]. For more information on ramp schemes, see Jackson and Martin [182].

Using cover-free families for blacklisting was suggested by Kumar, Rajagopalan and Sahai [206]. Broadcast anti-jamming systems can be constructed using similar techniques; see Desmedt *et al.* [113]. The *Naor-Pinkas Scheme* is from [248]. The *Logical Key Hierarchy* is (independently) due to Wallner, Harder and Agee [336] and Wong and Lam [348].

We mention a few relevant papers on fingerprinting and related topics. Boneh and Shaw [62] introduced the model used for fingerprinting in a cryptographic context; IPP codes were defined by Hollmann, van Lint, Linnartz and Tolhuizen [177]. Much of Section 14.4.3 is based on [177]. Chor, Fiat, Naor and Pinkas introduced traitor tracing for broadcast encryption schemes; see [92]. For additional information on these types of codes, see Stinson and Wei [326], Staddon, Stinson and Wei [314] and Barg, Blakley and Kabatiansky [6], as well as the thorough and well-written survey paper by Blackburn [41]. Theorem 14.6 is from [92] and Theorem 14.7 was proven in [314]. MacWilliams and Sloane [224] is a standard reference for coding theory; for a recent textbook, see Huffman and Pless [178].

Exercises

14.1 Construct the incidence matrix of a $(25, 25, 5, 1)$-BKDP. How big can w be if this BKDP is used to implement a broadcast encryption scheme as described in Theorem 14.1?

14.2 Suppose we want to construct a broadcast encryption scheme using Theorems 14.1 and 14.2. Assume that values for the parameters n and w are given. Suppose we choose values of q and d such that $q \approx w^2 \log_2 n$ and $d \approx \log_2 n$. Prove that these choices of parameters will lead to a broadcast encryption scheme that is secure against coalitions of size w. Also, give an estimate for v in terms of n and w.

14.3 Suppose we implement a *Shamir* $(2, 4, 6)$-*Ramp Scheme* in $\mathbb{Z}_p$, where $p = 128047$. Suppose that $x_1 = 100$, $x_2 = 200$, $x_3 = 300$, $x_4 = 400$, $x_5 = 500$, $x_6 = 600$, $y_1 = 102016$, $y_2 = 119297$, $y_3 = 58975$, $y_4 = 87929$, $y_5 = 116944$, and $y_6 = 56805$.

    (a) Show how the subset $\{P_1, P_4, P_5, P_6\}$ will compute the secret K.

    (b) Determine all the possible secrets consistent with the shares held by P_2, P_3 and P_4. Express your solution in the form $\{(a_0, a_1) : a_1 = c + da_0 \bmod p\}$ for suitable constants c and d.

14.4 We defined perfect hash families (PHF) in Section 14.4.3. PHF can be used for broadcast encryption; Cryptosystem 14.5 summarizes how this can be done.

Cryptosystem 14.5: *PHF-based Broadcast Encryption Scheme*

1. Given a PHF$(N; n, m, w)$, construct a certain Nm by n incidence matrix, M.
2. Use M to set up Nm Fiat-Naor 1-KDPs, as described in the *General BES* Denote these schemes as $\mathcal{F}_{i,j}$, $1 \le i \le N$, $1 \le j \le m$.
3. Split the key $K \in \mathbb{Z}_p$ into N shares, using the (N, N)threshold scheme (described in Section 13.1.1) in which the secret is the modulo p sum of the shares. Denote these shares as $s_1, \ldots, s_N$.
4. Let P be the subset to which K is being broadcasted. For all i, j, let $k_{i,j}$ be the group key for $P \cap users(\mathcal{F}_{i,j})$. Use $k_{i,j}$ to encrypt s_i, for all i, j.
5. Broadcast the Nm encryptions of the shares.

    (a) Describe how to construct the Nm by n incidence matrix from the PHF.

    (b) Describe how each user in P can decrypt the broadcast, finding K.

    (c) Explain why the scheme is secure against coalitions of size w.

    (d) The following array is a PHF$(4; 9, 3, 3)$:

0	0	0	0
0	1	1	1
0	2	2	2
1	0	1	2
1	1	2	0
1	2	0	1
2	0	2	1
2	1	0	2
2	2	1	0

Describe in detail how the scheme is set up (and list the keys held by each user); how the group keys would be constructed for the set $P = \{1, 2, 3, 4\}$; and how the broadcast is formed.

14.5 Suppose that the re-usable *Naor Pinkas Scheme* is used to broadcast a new group key to a set $\mathcal{U} \backslash F$, where F is a subset of w revoked users and $\mathcal{U} = \{U_1, \ldots, U_n\}$ is the set of all n users.

    Recall that the scheme is implemented in a subgroup of $\mathbb{Z}_p^*$ having order q (say $\langle \alpha \rangle$). The integers p and q are both prime. In this scheme, a secret value $K \in \mathbb{Z}_q$ is used to broadcast series of group keys, $\alpha^{Kr_1}, \alpha^{Kr_2}, \ldots$ These group keys can be computed by non-revoked participants without knowing the value of K.

(a) Suppose the scheme is implemented with the following domain parameters:

$$p = 46919749253781397857 9427,$$

$$q = 260665255385551 \quad \text{and}$$

$$\alpha = 21600950668468895 1924147.$$

Suppose that a user U_i is given the pair $(x_i, y_i) = (122, 202688224274771)$ as her share. Suppose also that the TA broadcasts the following ordered pairs (x_j, γ_j), where each $\gamma_j = \alpha^{ry_j}$:

$$(22, 44911778774799764175082),$$
$$(33, 43669764237759759952 9623),$$
$$(55, 42313937256594578121 7729),$$
$$(66, 13045304476619415320 0365) \quad \text{and}$$
$$(88, 92280508826597132975 88).$$

The TA also broadcasts $\gamma = \alpha^r = 23968850375048072851 9031$. Describe how user U will compute the new group key $K^* = \alpha^{Kr} \bmod p$ efficiently, and determine the value of K^*.

(b) If more than w users are revoked over a period of time (possibly in stages), then these users can collaborate to determine the values of any new group keys. Describe in detail how a set of $w + 1$ revoked users can do this.

14.6 Suppose we want to simultaneously revoke r users, say $U_{i_1}, \ldots, U_{i_r}$, in the *Logical Key Hierarchy*. Assuming that the tree depth is equal to d and the tree nodes are labelled as described Section 14.3.3, we can assume that $2^d \le U_{i_1} < \cdots < U_{i_r} \le 2^{d+1} - 1$.

(a) Informally describe an algorithm that can be used to determine which keys in the tree need to be updated.

(b) Describe the broadcast that is used to update the keys. Which keys are used to encrypt the new, updated keys?

(c) Illustrate your algorithm by describing the updated keys and the broadcast if users 18, 23 and 29 are to be revoked in a tree with depth $d = 4$ (this tree is depicted in Figure 14.3). How much smaller is the broadcast in this case, as compared to the three broadcasts that would be required to revoke these three users one at a time in the basic *Logical Key Hierarchy*?

14.7 Prove the "if" part of Theorem 14.4; i.e., that an (ℓ, n, q) code $\mathcal{C}$ is a 2-IPP code if $A(\mathcal{C})$ is simultaneously a PHF($\ell; n, q, 3$) and an SHF($\ell; n, q, \{2, 2\}$).

14.8 (a) Consider the $(3, 3r^2, r^2 + 2r)$ 2-IPP code $\mathcal{C}$ that was described in Section 14.4.3. Give a complete description of a $O(1)$ time algorithm TRACE, which takes as input a triple $\mathbf{f} = (f_1, f_2, f_3)$ and attempts to determine an identifiable parent of $\mathbf{f}$. If $\mathbf{f} \in \mathcal{C}$, then TRACE($\mathbf{f}$) $= \mathbf{f}$; if $\mathbf{f} \in \mathbf{desc}_2(\mathcal{C}) \backslash \mathcal{C}$, then TRACE($\mathbf{f}$) should find an identifiable parent of $\mathbf{f}$; and TRACE($\mathbf{f}$) should return the output "fail," if $\mathbf{f} \notin \mathbf{desc}_2(\mathcal{C})$.

(b) Illustrate the execution of your algorithm in the case $r = 10$ ($q = 120$) for the following triples: $(13, 11, 17)$; $(44, 9, 14)$; $(18, 108, 9)$.

14.9 We describe a $(4, r^3, r^2)$ 2-IPP code due to Hollman, van Lint, Linnartz and Tolhuizen. The alphabet is $Q = \mathbb{Z}_r \times \mathbb{Z}_r$. The code $\mathcal{C} \subseteq Q^4$ consists of the following set of r^3 4-tuples:

$$\{((a, b), (a, c), (b, c), (a + b \bmod r, c)) : a, b, c \in \mathbb{Z}_r\}.$$

(a) Give a complete description of a $O(1)$ time algorithm TRACE, which takes as input a 4-tuple $\mathbf{f} = ((\alpha_1, \alpha_2), (\beta_1, \beta_2), (\gamma_1, \gamma_2), (\delta_1, \delta_2))$ and attempts to determine an identifiable parent of $\mathbf{f}$. The output of TRACE should be as follows:

- if $\mathbf{f} \in \mathcal{C}$, then TRACE($\mathbf{f}$) $= \mathbf{f}$;
- if $\mathbf{f} \in \mathbf{desc}_2(\mathcal{C}) \backslash \mathcal{C}$, then TRACE($\mathbf{f}$) should find one identifiable parent of $\mathbf{f}$; and
- TRACE($\mathbf{f}$) should return the output "fail," if $\mathbf{f} \notin \mathbf{desc}_2(\mathcal{C})$.

In order for the algorithm to be an $O(1)$ time algorithm, there should be no linear searches, for example. You can assume that an arithmetic operation can be done in $O(1)$ time, however.

HINT In designing the algorithm, you will need to consider several cases. Many of the cases (and resulting subcases) are quite similar, however. You could initially divide the problem into the following four cases:

- $\alpha_1 \neq \beta_1$
- $\alpha_2 \neq \gamma_1$
- $\beta_2 \neq \gamma_2$
- $\alpha_1 = \beta_1, \alpha_2 = \gamma_1$ and $\beta_2 = \gamma_2$.

(b) Illustrate the execution of your algorithm in detail in the case $r = 100$ for each of the following 4-tuples $\mathbf{f}$:

$$((37, 71), (37, 96), (71, 96), (12, 96))$$
$$((25, 16), (83, 54), (16, 54), (41, 54))$$
$$((19, 11), (19, 12), (11, 15), (30, 12))$$
$$((32, 40), (32, 50), (50, 40), (82, 30))$$

14.10 Consider the 3-TA code constructed by applying Theorem 14.7 with $\ell = 19$ and $q = 101$. This code is a $(19, 101^3, 101)$-code.

(a) Write a computer program to construct the 101^3 codewords in this code.

(b) Given the vector

$$\mathbf{f} = (14, 66, 46, 56, 13, 31, 50, 30, 77, 32, 0, 93, 48, 37, 16, 66, 24, 42, 9)$$

in the 3-descendant code, compute a parent of $\mathbf{f}$ using nearest neighbor decoding.

Further Reading

There are now many books and monographs on various aspects of cryptography. Here are several textbooks and monographs which provide fairly general coverage of cryptography:

- *Invitation to Cryptology*, by T. H. Barr [7];
- *Decrypted Secrets, Methods and Maxims of Cryptology, Second Edition*, by F. L. Bauer [9];
- *Cipher Systems, The Protection of Communications*, by H. Beker and F. Piper [13];
- *Cryptology*, by A. Beutelspacher [32];
- *Introduction to Cryptography with Java^{TM} Applets*, by D. Bishop [38];
- *Introduction to Cryptography*, by J. A. Buchmann [77];
- *Codes and Ciphers: Julius Caesar, the Enigma, and the Internet*, by R. Churchhouse [93];
- *Introduction to Cryptography: Principles and Applications*, by H. Delfs and H. Knebl [107];
- *Cryptography and Data Security*, by D. E. R. Denning [109];
- *User's Guide to Cryptography and Standards*, by A. W. Dent and C. J. Mitchell [110];
- *Practical Cryptography*, by N. Ferguson and B. Schneier [132];
- *Making, Breaking Codes: An Introduction to Cryptology*, by P. Garrett [153];
- *Foundations of Cryptography: Basic Tools*, by O. Goldreich [160];
- *Foundations of Cryptography: Volume II, Basic Aplications*, by O. Goldreich [161];
- *The Codebreakers*, by D. Kahn [185];
- *Network Security. Private Communication in a Public World, Second Edition*, by C. Kaufman, R. Perlman and M. Speciner [187];
- *Code Breaking, A History and Exploration*, by R. Kippenhahn [192];

- *A Course in Number Theory and Cryptography, Second Edition*, by N. Koblitz [197];
- *Cryptography, A Primer*, by A. G. Konheim [203];
- *Basic Methods of Cryptography*, by J. C. A. van der Lubbe [222];
- *Modern Cryptography: Theory and Practice*, by W. Mao [225];
- *Cryptography Decrypted*, by H. X. Mel and D. Baker [232];
- *Handbook of Applied Cryptography*, by A. J. Menezes, P. C. Van Oorschot and S. A. Vanstone [237];
- *An Introduction to Cryptography*, by R. A. Mollin [244];
- *Fundamentals of Computer Security*, by J. Pieprzyk, T. Hardjono and J. Seberry [268];
- *Cryptography: A Very Short Introduction*, by F. Piper and S. Murphy [269];
- *Internet Security: Cryptographic Principles, Algorithms and Protocols*, by M. Y. Rhee [280];
- *White-hat Security Arsenal: Tackling the Threats*, by A. D. Rubin [285];
- *Data Privacy and Security*, by D. Salomon [290];
- *Cryptography and Public Key Infrastructure on the Internet*, by K. Schmeh [291];
- *Applied Cryptography, Protocols, Algorithms and Source Code in C, Second Edition*, by B. Schneier [292];
- *Contemporary Cryptology, The Science of Information Integrity*, G. J. Simmons, ed. [306];
- *The Code Book: The Evolution Of Secrecy From Mary, To Queen Of Scots To Quantum Cryptography*, by S. Singh [307];
- *Cryptography: An Introduction*, by N. Smart [309];
- *Cryptography and Network Security: Principles and Practice, Third Edition* by W. Stallings [315];
- *Introduction to Cryptography with Coding Theory*, by W. Trappe and L. C. Washington [330].

Monographs on specialized areas of cryptography are mentioned in Notes at the end of relevant chapters in this book.

Although cryptography is a very broad subject, it is just a small part of the study of "security," which includes computer security, network security and software security. Here are four recommended references that can provide an overview of threats and security:

- *Security Engineering: A Guide to Building Dependable Distributed Systems*, by R. Anderson [3];
- *Introduction to Computer Security*, by M. Bishop [39];

- *Secrets and Lies: Digital Security in a Networked World*, by B. Schneier [293];
- *Malicious Cryptography: Exposing Cryptovirology*, by A. L. Young and M. Yung [351].

The International Association for Cryptologic Research (or IACR) sponsors three annual cryptology conferences: CRYPTO, EUROCRYPT, and ASIACRYPT.

CRYPTO has been held since 1981 in Santa Barbara. The proceedings of CRYPTO have been published annually since 1982:

CRYPTO '82 [90]	CRYPTO '83 [88]	CRYPTO '84 [47]
CRYPTO '85 [347]	CRYPTO '86 [257]	CRYPTO '87 [273]
CRYPTO '88 [162]	CRYPTO '89 [69]	CRYPTO '90 [236]
CRYPTO '91 [129]	CRYPTO '92 [74]	CRYPTO '93 [319]
CRYPTO '94 [112]	CRYPTO '95 [96]	CRYPTO '96 [198]
CRYPTO '97 [186]	CRYPTO '98 [205]	CRYPTO '99 [345]
CRYPTO '00 [15]	CRYPTO '01 [190]	CRYPTO '02 [352]
CRYPTO '03 [59]	CRYPTO '04 [148]	

EUROCRYPT has been held annually since 1982, and except for 1983 and 1986, its proceedings have been published, as follows:

EUROCRYPT '82 [30]	EUROCRYPT '84 [31]	EUROCRYPT '85 [266]
EUROCRYPT '87 [89]	EUROCRYPT '88 [170]	EUROCRYPT '89 [277]
EUROCRYPT '90 [103]	EUROCRYPT '91 [105]	EUROCRYPT '92 [286]
EUROCRYPT '93 [174]	EUROCRYPT '94 [111]	EUROCRYPT '95 [168]
EUROCRYPT '96 [229]	EUROCRYPT '97 [150]	EUROCRYPT '98 [256]
EUROCRYPT '99 [317]	EUROCRYPT '00 [275]	EUROCRYPT '01 [264]
EUROCRYPT '02 [194]	EUROCRYPT '03 [33]	EUROCRYPT '04 [81]

ASIACRYPT (originally, AUSCRYPT) has been held since 1991. Its conference proceedings have also been published:

AUSCRYPT '90 [295]	ASIACRYPT '91 [179]	AUSCRYPT '92 [296]
ASIACRYPT '94 [267]	ASIACRYPT '96 [191]	ASIACRYPT '98 [259]
ASIACRYPT '99 [211]	ASIACRYPT '00 [261]	ASIACRYPT '01 [66]
ASIACRYPT '02 [353]	ASIACRYPT '03 [210]	ASIACRYPT '04 [215]

Bibliography

[1] C. ADAMS AND S. LLOYD. *Understanding PKI: Concepts, Standards, and Deployment Considerations, Second Edition*. Addison Wesley, 2003.

[2] W. ALEXI, B. CHOR, O. GOLDREICH AND C. P. SCHNORR. RSA and Rabin functions: certain parts are as hard as the whole. *SIAM Journal on Computing*, **17** (1988), 194–209.

[3] R. ANDERSON. *Security Engineering: A Guide to Building Dependable Distributed Systems*. John Wiley and Sons, 2000.

[4] H. ANTON. *Elementary Linear Algebra, Eighth Edition*. John Wiley and Sons, 2000.

[5] E. BACH AND J. SHALLIT. *Algorithmic Number Theory, Volume 1: Efficient Algorithms*. The MIT Press, 1996.

[6] A. BARG, G. R. BLAKLEY AND G. A. KABATIANSKY. Digital fingerprinting codes: problem statements, constructions, identification of traitors. *IEEE Transactions on Information Theory*, **49** (2003), 852–865.

[7] T. H. BARR. *Invitation to Cryptology*. Prentice Hall, 2002.

[8] P. S. L. M. BARRETO, H. Y. KIM, B. LYNN AND M. SCOTT. Efficient algorithms for pairing-based cryptosystems. *Lecture Notes in Computer Science*, **2442** (2002), 354–368. (CRYPTO 2002.)

[9] F. L. BAUER. *Decrypted Secrets, Methods and Maxims of Cryptology, Second Edition*. Springer, 2000.

[10] P. BEAUCHEMIN AND G. BRASSARD. A generalization of Hellman's extension to Shannon's approach to cryptography. *Journal of Cryptology*, **1** (1988), 129–131.

[11] P. BEAUCHEMIN, G. BRASSARD, C. CRÉPEAU, C. GOUTIER AND C. POMERANCE. The generation of random numbers that are probably prime. *Journal of Cryptology*, **1** (1988), 53–64.

[12] A. BEIMEL AND B. CHOR. Interaction in key distribution schemes. *Lecture Notes in Computer Science*, **773** (1994), 444–455. (CRYPTO '93.)

[13] H. BEKER AND F. PIPER. *Cipher Systems, The Protection of Communica-

tions. John Wiley and Sons, 1982.

[14] M. BELLARE. Practice-oriented provable-security. In *Lectures on Data Security*, pages 1–15. Springer, 1999.

[15] M. BELLARE (ED.) *Advances in Cryptology – CRYPTO 2000 Proceedings. Lecture Notes in Computer Science*, vol. 1880, Springer, 2000.

[16] M. BELLARE, R. CANETTI AND H. KRAWCZYK. Keying hash functions for message authentication. *Lecture Notes in Computer Science*, **1109** (1996), 1–15. (CRYPTO '96.)

[17] M. BELLARE, S. GOLDWASSER AND D. MICCIANCIO. "Pseudorandom" number generation within cryptographic algorithms: the DSS case. *Lecture Notes in Computer Science*, **1294** (1997), 277–292. (CRYPTO '97.)

[18] M. BELLARE, R. GUERIN AND P. ROGAWAY. XOR MACs: new methods for message authentication using finite pseudorandom functions. *Lecture Notes in Computer Science*, **963** (1995), 15–28. (CRYPTO '95.)

[19] M. BELLARE, J. KILIAN AND P. ROGAWAY. The security of the cipher block chaining message authentication code. *Journal of Computer and System Sciences*, **61** (2000), 362–399.

[20] M. BELLARE AND A. PALACIO. GQ and Schnorr identification schemes: proofs of security against impersonation under active and concurrent attacks. *Lecture Notes in Computer Science*, **2442** (2002), 162–177. (CRYPTO 2002.)

[21] M. BELLARE, D. POINTCHEVAL AND P. ROGAWAY. Authenticated key exchange secure against dictionary attacks. *Lecture Notes in Computer Science*, **1807** (2000), 139–155. (EUROCRYPT 2000.)

[22] M. BELLARE AND P. ROGAWAY. Random oracles are practical: a paradigm for designing efficient protocols. In *First ACM Conference on Computer and Communications Security*, pages 62–73. ACM Press, 1993.

[23] M. BELLARE AND P. ROGAWAY. Entity authentication and key distribution. *Lecture Notes in Computer Science*, **773** (1994), 232–249. (CRYPTO '93.)

[24] M. BELLARE AND P. ROGAWAY. Optimal asymmetric encryption. *Lecture Notes in Computer Science*, **950** (1995), 92–111. (EUROCRYPT '94.)

[25] M. BELLARE AND P. ROGAWAY. Provably secure session key distribution: the three party case. In *27th Annual ACM Symposium on Theory of Computing*, pages 57–66. ACM Press, 1995.

[26] M. BELLARE AND P. ROGAWAY. The exact security of digital signatures: how to sign with RSA and Rabin. *Lecture Notes in Computer Science*, **1070** (1996), 399–416. (EUROCRYPT '96.)

[27] S. BELLOVIN AND M. MERRITT. Encrypted key exchange: password-

based protocols secure against dictionary attacks. In *Proceedings of the IEEE Symposium on Research in Security and Privacy*, pages 72–84. IEEE Press, 1992.

[28] J. BENALOH AND J. LEICHTER. Generalized secret sharing and monotone functions. *Lecture Notes in Computer Science*, **403** (1990), 27–35. (CRYPTO '88.)

[29] S. BERKOVITZ. How to broadcast a secret. *Lecture Notes in Computer Science*, **547** (1991), 535–541. (EUROCRYPT '91.)

[30] T. BETH (ED.) *Cryptography Proceedings, 1982. Lecture Notes in Computer Science*, vol. 149, Springer, 1983.

[31] T. BETH, N. COT AND I. INGEMARSSON (EDS.) *Advances in Cryptology: Proceedings of EUROCRYPT '84. Lecture Notes in Computer Science*, vol. 209, Springer, 1985.

[32] A. BEUTELSPACHER. *Cryptology*. Mathematical Association of America, 1994.

[33] E. BIHAM (ED.) *Advances in Cryptology – EUROCRYPT 2003 Proceedings. Lecture Notes in Computer Science*, vol. 2656, Springer, 2003.

[34] E. BIHAM AND A. SHAMIR. Differential cryptanalysis of DES-like cryptosystems. *Journal of Cryptology*, **4** (1991), 3–72.

[35] E. BIHAM AND A. SHAMIR. *Differential Cryptanalysis of the Data Encryption Standard*. Springer, 1993.

[36] E. BIHAM AND A. SHAMIR. Differential cryptanalysis of the full 16-round DES. *Lecture Notes in Computer Science*, **740** (1993), 494–502. (CRYPTO '92.)

[37] E. BIHAM AND R. CHEN. Near-Collisions of SHA-0. *Lecture Notes in Computer Science*, **3152** (2004), 290–305. (CRYPTO 2004.)

[38] D. BISHOP. *Introduction to Cryptography with Java^{TM} Applets*. Jones and Bartlett, 2003.

[39] M. BISHOP. *Introduction to Computer Security*. Addison-Wesley, 2004.

[40] J. BLACK, S. HALEVI, H. KRAWCZYK, T. KROVETZ AND P. ROGAWAY. UMAC: fast message authentication via optimized universal hash functions. *Lecture Notes in Computer Science*, **1666** (1999), 234–251. (CRYPTO '99.)

[41] S. R. BLACKBURN. Combinatorial schemes for protecting digital content. In *Surveys in Combinatorics 2003*, Cambridge University Press, 2003, pp. 43–78.

[42] I. BLAKE, G. SEROUSSI AND N. SMART. *Elliptic Curves in Cryptography*. Cambridge University Press, 1999.

[43] I. BLAKE, G. SEROUSSI AND N. SMART, EDS. *Advances in Elliptic Curve Cryptography*. Cambridge University Press, 2005.

[44] S. BLAKE-WILSON AND A. J. MENEZES., Entity authentication and authenticated key transport protocols employing asymmetric techniques. *Lecture Notes in Computer Science*, **1361** (1998), 137–158. (Fifth International Workshop on Security Protocols.)

[45] S. BLAKE-WILSON AND A. J. MENEZES., Authenticated Diffie-Hellman key agreement protocols. *Lecture Notes in Computer Science*, **1556** (1999), 339–361. (Selected Areas in Cryptography '98.)

[46] G. R. BLAKLEY. Safeguarding cryptographic keys. *Federal Information Processing Standard Conference Proceedings*, **48** (1979), 313–317.

[47] G. R. BLAKLEY AND D. CHAUM (EDS.) *Advances in Cryptology: Proceedings of CRYPTO '84. Lecture Notes in Computer Science*, vol. 196, Springer, 1985.

[48] D. BLEICHENBACHER AND U. M. MAURER. Directed acyclic graphs, one-way functions and digital signatures. *Lecture Notes in Computer Science*, **839** (1994), 75–82. (CRYPTO '94.)

[49] R. BLOM. An optimal class of symmetric key generation schemes. *Lecture Notes in Computer Science*, **209** (1985), 335–338. (EUROCRYPT '84.)

[50] L. BLUM, M. BLUM AND M. SHUB. A simple unpredictable random number generator. *SIAM Journal on Computing*, **15** (1986), 364–383.

[51] M. BLUM AND S. GOLDWASSER. An efficient probabilistic public-key cryptosystem that hides all partial information. *Lecture Notes in Computer Science*, **196** (1985), 289–302. (CRYPTO '84.)

[52] M. BLUM AND S. MICALI. How to generate cryptographically strong sequences of pseudo-random bits. *SIAM Jounal on Computing*, **13** (1984), 850–864.

[53] C. BLUNDO AND P. D'ARCO. The key establishment problem. *Lecture Notes in Computer Science*, **2946** (2004), 44–90. (Foundations of Security Analysis and Design II.)

[54] C. BLUNDO, A. DE SANTIS, D. R. STINSON AND U. VACCARO. Graph decompositions and secret sharing schemes. *Lecture Notes in Computer Science*, **658** (1993), 1–24. (EUROCRYPT '92.)

[55] C. BLUNDO, A. DE SANTIS, A. HERZBERG, S. KUTTEN, U. VACCARO AND M. YUNG. Perfectly-secure key distribution for dynamic conferences. *Lecture Notes in Computer Science*, **740** (1993), 471–486. (CRYPTO '92.)

[56] D. BONEH. The decision Diffie-Hellman problem. *Lecture Notes in Computer Science*, **1423** (1998), 48–63. (Proceedings of the Third Algorithmic Number Theory Symposium.)

[57] D. BONEH. Twenty years of attacks on the RSA cryptosystem. *Notices of the American Mathematical Society*, **46** (1999), 203–213.

[58] D. BONEH. Simplified OAEP for the RSA and Rabin functions. *Lecture Notes in Computer Science*, **2139** (2001), 275–291. (CRYPTO 2001.)

[59] D. BONEH (ED.) *Advances in Cryptology – CRYPTO 2003 Proceedings. Lecture Notes in Computer Science*, vol. 2729, Springer, 2003.

[60] D. BONEH AND G. DURFEE. Cryptanalysis of RSA with private key d less than $N^{0.292}$. *IEEE Transactions on Information Theory*, **46** (2000), 1339–1349.

[61] D. BONEH AND M. FRANKLIN. Identity-based encryption from the Weil pairing. *Lecture Notes in Computer Science*, **2139** (2001), 213–229. (CRYPTO 2001.)

[62] D. BONEH AND J. SHAW. Collusion-secure fingerprinting for digital data, *IEEE Transactions on Information Theory*, **44** (1998), 1897–1905.

[63] F. BORNEMANN. PRIMES is in P: a breakthrough for "everyman". *Notices of the American Mathematical Society*, **50** (2003), 545–552.

[64] J. N. E. BOS AND D. CHAUM. Provably unforgeable signatures. *Lecture Notes in Computer Science*, **740** (1993), 1–14. (CRYPTO '92.)

[65] J. BOYAR. Inferring sequences produced by pseudo-random number generators. *Journal of the Association for Computing Machinery*, **36** (1989), 129–141.

[66] C. BOYD, (ED.) *Advances in Cryptology – ASIACRYPT 2001 Proceedings. Lecture Notes in Computer Science*, vol. 2248, Springer, 2001.

[67] C. BOYD AND A. MATHURIA *Protocols for Authentication and Key Establishment.* Springer, 2003.

[68] M. BRANCHAUD. *A Survey of Public-Key Infrastructures.* Masters Thesis, McGill University, 1997.

[69] G. BRASSARD (ED.) *Advances in Cryptology – CRYPTO '89 Proceedings. Lecture Notes in Computer Science*, vol. 435, Springer, 1990.

[70] G. BRASSARD AND P. BRATLEY. *Fundamentals of Algorithmics.* Prentice Hall, 1995.

[71] R. P. BRENT. An improved Monte Carlo factorization method. *BIT*, **20** (1980), 176–184.

[72] D. M. BRESSOUD AND S. WAGON. *A Course in Computational Number Theory.* Springer, 2000.

[73] E. F. BRICKELL. Some ideal secret sharing schemes. *Journal of Combinatorial Mathematics and Combinatorial Computing*, **9** (1989), 105–113.

[74] E. F. BRICKELL (ED.) *Advances in Cryptology – CRYPTO '92 Proceedings. Lecture Notes in Computer Science*, vol. 740, Springer, 1993.

[75] E. F. BRICKELL AND D. M. DAVENPORT. On the classification of ideal secret sharing schemes. *Journal of Cryptology*, **4** (1991), 123–134.

[76] E. F. BRICKELL AND K. S. MCCURLEY. An interactive identification scheme based on discrete logarithms and factoring. *Journal of Cryptology*, **5** (1992), 29–39.

[77] J. A. BUCHMANN. *Introduction to Cryptography*. Springer, 2001.

[78] M. BURMESTER. On the risk of opening distributed keys. *Lecture Notes in Computer Science*, **839** (1994), 308–317 (CRYPTO '94.)

[79] M. BURMESTER AND Y. DESMEDT. A secure and efficient conference key distribution system. *Lecture Notes in Computer Science*, **950** (1994), 275–286 (EUROCRYPT '94.)

[80] M. BURMESTER, Y. DESMEDT AND T. BETH. Efficient zero-knowledge identification schemes for smart cards. *The Computer Journal*, **35** (1992), 21–29.

[81] C. CACHIN AND J. CAMENISCH (EDS.) *Advances in Cryptology – EURO-CRYPT 2004 Proceedings*. Lecture Notes in Computer Science, vol. 3027, Springer, 2004.

[82] R. CANETTI, J. GARAY, G. ITKIS, D. MICCIANCIO, M. NAOR AND B. PINKAS. Multicast security: A taxonomy and some efficient cnstructions. In *Proceedings of INFOCOM '99*, pages 708–716. IEEE Press, 1999.

[83] R. M. CAPOCELLI, A. DE SANTIS, L. GARGANO AND U. VACCARO. On the size of shares for secret sharing schemes. *Journal of Cryptology*, **6** (1993), 157–167.

[84] J. L. CARTER AND M. N. WEGMAN. Universal classes of hash functions. *Journal of Computer and System Sciences*, **18** (1979), 143–154.

[85] F. CHABAUD AND A. JOUX. Differential collisions in SHA-0. *Lecture Notes in Computer Science*, **1462** (1998), 56–71. (CRYPTO '98.)

[86] F. CHABAUD AND S. VAUDENAY. Links between differential and linear cryptanalysis. *Lecture Notes in Computer Science*, **950** (1995), 356–365. (EUROCRYPT '94.)

[87] M. CHATEAUNEUF, A. C. H. LING AND D. R. STINSON. Slope packings and coverings, and generic algorithms for the discrete logarithm problem. *Journal of Combinatorial Designs*, **11** (2003), 36–50.

[88] D. CHAUM (ED.) *Advances in Cryptology: Proceedings of CRYPTO '83*. Plenum Press, 1984.

[89] D. CHAUM AND W. L. PRICE (EDS.) *Advances in Cryptology – EURO-CRYPT '87 Proceedings*. Lecture Notes in Computer Science, vol. 304, Springer, 1988.

[90] D. CHAUM, R. L. RIVEST AND A. T. SHERMAN (EDS.) *Advances in Cryptology: Proceedings of CRYPTO '82*. Plenum Press, 1983.

[91] D. CHAUM AND H. VAN ANTWERPEN. Undeniable signatures. *Lecture Notes in Computer Science*, **435** (1990), 212–216. (CRYPTO '89.)

[92] B. CHOR, A. FIAT, M. NAOR AND B. PINKAS. Tracing traitors. *IEEE Transactions on Information Theory*, **46** (2000), 893–910.

[93] R. CHURCHHOUSE. *Codes and Ciphers: Julius Caesar, the Enigma, and the Internet*. Cambridge, 2002.

[94] C. COCKS. An identity based encryption scheme based on quadratic residues. *Lecture Notes in Computer Science*, **2260** (2001), 360–363. (Eighth IMA International Conference on Cryptography and Coding.)

[95] D. COPPERSMITH. The data encryption standard (DES) and its strength against attacks. *IBM Journal of Research and Development*, **38** (1994), 243–250.

[96] D. COPPERSMITH (ED.) *Advances in Cryptology – CRYPTO '95 Proceedings. Lecture Notes in Computer Science*, vol. 963, Springer, 1995.

[97] D. COPPERSMITH, H. KRAWCZYK AND Y. MANSOUR. The shrinking generator. *Lecture Notes in Computer Science*, **773** (1994), 22–39. (CRYPTO '93.)

[98] N. T. COURTOIS AND J. PIEPRZYK. Cryptanalysis of block ciphers with overdefined systems of equations. *Lecture Notes in Computer Science*, **2501** (2002), 267–287. (ASIACRYPT 2002.)

[99] J. DAEMEN, L. KNUDSEN AND V. RIJMEN. The block cipher Square. *Lecture Notes in Computer Science*, **1267** (1997), 149–165. (Fast Software Encryption '97.)

[100] J. DAEMEN AND V. RIJMEN. The block cipher Rijndael. *Lecture Notes in Computer Science*, **1820** (2000), 288-296. (Smart Card Research and Applications.)

[101] J. DAEMEN AND V. RIJMEN. *The Design of Rijndael. AES - The Advanced Encryption Standard*. Springer, 2002.

[102] I. B. DAMGÅRD. A design principle for hash functions. *Lecture Notes in Computer Science*, **435** (1990), 416–427. (CRYPTO '89.)

[103] I. B. DAMGÅRD (ED.) *Advances in Cryptology – EUROCRYPT '90 Proceedings. Lecture Notes in Computer Science*, vol. 473, Springer, 1991.

[104] I. DAMGÅRD, P. LANDROCK AND C. POMERANCE. Average case error estimates for the strong probable prime test. *Mathematics of Computation*, **61** (1993), 177–194.

[105] D. W. DAVIES (ED.) *Advances in Cryptology – EUROCRYPT '91 Proceedings. Lecture Notes in Computer Science*, vol. 547, Springer, 1991.

[106] J. M. DELAURENTIS. A further weakness in the common modulus protocol for the RSA cryptosystem. *Cryptologia*, **8** (1984), 253–259.

[107] H. DELFS AND H. KNEBL. *Introduction to Cryptography: Principles and Applications*. Springer, 2002.

[108] D. E. DENNING AND G. M. SACCO. Timestamps in key distribution pro-

tocols. *Communications of the ACM*, **24** (1981), 533–536.

[109] D. E. R. DENNING. *Cryptography and Data Security*. Addison-Wesley, 1982.

[110] A. W. DENT AND C. J. MITCHELL. *User's Guide to Cryptography and Standards*. Artech House, 2005.

[111] A. DE SANTIS (ED.) *Advances in Cryptology – EUROCRYPT '94 Proceedings. Lecture Notes in Computer Science*, vol. 950, Springer, 1995.

[112] Y. G. DESMEDT (ED.) *Advances in Cryptology – CRYPTO '94 Proceedings. Lecture Notes in Computer Science*, vol. 839, Springer, 1994.

[113] Y. DESMEDT, R. SAFAVI-NAINI, H. WANG, L. BATTEN, C. CHARNES AND J. PIEPRZYK. Broadcast anti-jamming systems. *Computer Networks* **35** (2001), 223–236.

[114] D. DE WALEFFE AND J.-J. QUISQUATER. Better login protocols for computer networks. *Lecture Notes in Computer Science*, **741** (1993), 50–70. (Computer Security and Industrial Cryptography, State of the Art and Evolution, 1991.)

[115] W. DIFFIE. The first ten years of public-key cryptography. In *Contemporary Cryptology, The Science of Information Integrity*, pages 135–175. IEEE Press, 1992.

[116] W. DIFFIE AND M. E. HELLMAN. Multiuser cryptographic techniques. *Federal Information Processing Standard Conference Proceedings*, **45** (1976), 109–112.

[117] W. DIFFIE AND M. E. HELLMAN. New directions in cryptography. *IEEE Transactions on Information Theory*, **22** (1976), 644–654.

[118] W. DIFFIE, P. C. VAN OORSCHOT AND M. J. WIENER. Authentication and authenticated key exchanges. *Designs, Codes and Cryptography*, **2** (1992), 107–125.

[119] H. DOBBERTIN. The status of MD5 after a recent attack. *CryptoBytes*, **2** No. 2 (1996), 1–6.

[120] H. DOBBERTIN. Cryptanalysis of MD4. *Journal of Cryptology*, **11** (1998), 253–271.

[121] M. DWORKIN. *Recommendation for Block Cipher Modes of Operation*. National Institute of Standards and Technology (NIST) Special Publication 800-38A, 2001.

[122] M. DWORKIN. *Recommendation for Block Cipher Modes of Operation: The CMAC Mode for Authentication*. National Institute of Standards and Technology (NIST) Special Publication 800-38B, 2005 (draft).

[123] M. DWORKIN. *Recommendation for Block Cipher Modes of Operation: The CCM Mode for Authentication and Confidentiality*. National Institute of Standards and Technology (NIST) Special Publication 800-38C, 2004.

[124] M. DYER, T. FENNER, A. FRIEZE AND A. THOMASON. On key storage in secure networks. *Journal of Cryptology*, **8** (1995), 189–200.

[125] T. ELGAMAL. A public key cryptosystem and a signature scheme based on discrete logarithms. *IEEE Transactions on Information Theory*, **31** (1985), 469–472.

[126] C. ELLISON AND B. SCHNEIER. Ten risks of PKI: what you're not being told about public key infrastructure. *Computer Security Journal* **16**(1) (2000), 1–7.

[127] A. ENGE. *Elliptic Curves and their Applications to Cryptography: an Introduction*. Kluwer Academic Publishers, 1999.

[128] U. FEIGE, A. FIAT AND A. SHAMIR. Zero-knowledge proofs of identity. *Journal of Cryptology*, **1** (1988), 77–94.

[129] J. FEIGENBAUM (ED.) *Advances in Cryptology – CRYPTO '91 Proceedings. Lecture Notes in Computer Science*, vol. 576, Springer, 1992.

[130] H. FEISTEL. Cryptography and computer privacy. *Scientific American*, **228**(5) (1973), 15–23.

[131] N. FERGUSON, J. KELSEY, S. LUCKS, B. SCHNEIER, M. STAY, D. WAGNER AND D. WHITING. Improved cryptanalysis of Rijndael. *Lecture Notes in Computer Science*, **1978** (2001), 1213–230. (Fast Software Encryption 2000.)

[132] N. FERGUSON AND B. SCHNEIER. *Practical Cryptography*. John Wiley and Sons, 2003.

[133] N. FERGUSON, R. SCHROEPPEL AND D. WHITING. A simple algebraic representation of Rijndael. *Lecture Notes in Computer Science*, **2259** (2001), 103–111. (Selected Areas in Cryptography 2001.)

[134] A. FIAT AND M. NAOR. Broadcast encryption. *Lecture Notes in Computer Science*, **773** (1994), 480–491. (CRYPTO '93.)

[135] A. FIAT AND A. SHAMIR. How to prove yourself: practical solutions to identification and signature problems. *Lecture Notes in Computer Science*, **263** (1987), 186–194. (CRYPTO '86.)

[136] *Data Encryption Standard (DES)*. Federal Information Processing Standard (FIPS) Publication 46, 1977.

[137] *DES Modes of Operation*. Federal Information Processing Standard (FIPS) Publication 81, 1980.

[138] *Guidelines for Implementing and Using the NBS Data Encryption Standard*. Federal Information Processing Standard (FIPS) Publication 74, 1981.

[139] *Computer Data Authentication*. Federal Information Processing Standard (FIPS) Publication 113, 1985.

[140] *Secure Hash Standard*. Federal Information Processing Standard (FIPS)

Publication 180, 1993.

[141] *Secure Hash Standard.* Federal Information Processing Standard (FIPS) Publication 180-1, 1995.

[142] *Secure Hash Standard.* Federal Information Processing Standard (FIPS) Publication 180-2, 2002.

[143] *Digital Signature Standard.* Federal Information Processing Standard (FIPS) Publication 186, 1994.

[144] *Digital Signature Standard.* Federal Information Processing Standard (FIPS) Publication 186-2, 2000.

[145] *Entity Authentication Using Public Key Cryptography.* Federal Information Processing Standard (FIPS) Publication 196, 1997.

[146] *Advanced Encryption Standard.* Federal Information Processing Standard (FIPS) Publication 197, 2001.

[147] *The Keyed-Hash Message Authentication Code.* Federal Information Processing Standard (FIPS) Publication 198, 2002.

[148] M. FRANKLIN (ED.) *Advances in Cryptology – CRYPTO 2004 Proceedings. Lecture Notes in Computer Science,* vol. 3152, Springer, 2004.

[149] E. FUJISAKI, T. OKAMOTO, D. POINTCHEVAL AND J. STERN. RSA-OAEP is secure under the RSA assumption. *Lecture Notes in Computer Science,* **2139** (2001), 260–274. (CRYPTO 2001.)

[150] W. FUMY (ED.) *Advances in Cryptology – EUROCRYPT '97 Proceedings. Lecture Notes in Computer Science,* vol. 1233, Springer, 1997.

[151] M. GAGNÉ. Identity-based encryption: a survey. *CryptoBytes* **6**(1), (2003), 10–19.

[152] S. GALBRAITH AND A. MENEZES. Algebraic curves and cryptography. *Finite Fields and their Applications,* to appear.

[153] P. GARRETT. *Making, Breaking Codes: An Introduction to Cryptology.* Prentice-Hall, 2001.

[154] J. VON ZUR GATHEN AND J. GERHARD. *Modern Computer Algebra.* Cambridge University Press, 1999.

[155] R. GENNARO. An improved pseudo-random generator based on discrete log. *Lecture Notes in Computer Science,* **1880** (2000), 469–481. (CRYPTO 2000.)

[156] E. N. GILBERT, F. J. MACWILLIAMS AND N. J. A. SLOANE. Codes which detect deception. *Bell Systems Technical Journal,* **53** (1974), 405–424.

[157] M. GIRAULT. Self-certified public keys. *Lecture Notes in Computer Science,* **547** (1991), 490–497. (EUROCRYPT '91.)

[158] C. M. GOLDIE AND R. G. E. PINCH. *Communication Theory.* Cambridge

University Press, 1991.

[159] O. GOLDREICH. *Modern Cryptography, Probabilistic Proofs and Pseudo-randomness*. Springer, 1999.

[160] O. GOLDREICH. *Foundations of Cryptography: Basic Tools*. Cambridge University Press, 2001.

[161] O. GOLDREICH. *Foundations of Cryptography: Volume II, Basic Aplications*. Cambridge University Press, 2004.

[162] S. GOLDWASSER (ED.) *Advances in Cryptology – CRYPTO '88 Proceedings. Lecture Notes in Computer Science*, vol. 403, Springer, 1990.

[163] S. GOLDWASSER AND S. MICALI. Probabilistic encryption. *Journal of Computer and Systems Science*, **28** (1984), 270–299.

[164] S. GOLDWASSER, S. MICALI AND P. TONG. Why and how to establish a common code on a public network. In *23rd Annual Symposium on the Foundations of Computer Science*, pages 134–144. IEEE Press, 1982.

[165] J. D. GOLIĆ. Correlation analysis of the shrinking generator. *Lecture Notes in Computer Science*, **2139** (2001), 440–457. (CRYPTO 2001.)

[166] D. M. GORDON AND K. S. MCCURLEY. Massively parallel computation of discrete logarithms. *Lecture Notes in Computer Science*, **740** (1993), 312–323. (CRYPTO '92.)

[167] L. C. GUILLOU AND J.-J. QUISQUATER. A practical zero-knowledge protocol fitted to security microprocessor minimizing both transmission and memory. *Lecture Notes in Computer Science*, **330** (1988), 123–128. (EUROCRYPT '88.)

[168] L. C. GUILLOU AND J.-J. QUISQUATER (EDS.) *Advances in Cryptology – EUROCRYPT '95 Proceedings. Lecture Notes in Computer Science*, vol. 921, Springer, 1995.

[169] C. G. GUNTHER Alternating step generators controlled by de Bruijn sequences. *Lecture Notes in Computer Science*, **304** (1988), 88–92. (EUROCRYPT '87.)

[170] C. G. GUNTHER (ED.) *Advances in Cryptology – EUROCRYPT '88 Proceedings. Lecture Notes in Computer Science*, vol. 330, Springer, 1988.

[171] D. R. HANKERSON, A. J. MENEZES AND S. A. VANSTONE. *Guide to Elliptic Curve Cryptography*. Springer, 2004.

[172] J. HÅSTAD, A. W. SCHRIFT AND A. SHAMIR. The discrete logarithm modulo a composite hides $O(n)$ bits. *Journal of Computer and Systems Science*, **47** (1993), 376–404.

[173] M. E. HELLMAN. A cryptanalytic time-memory trade-off. *IEEE Transactions on Information Theory*, **26** (1980), 401–406.

[174] T. HELLESETH (ED.) *Advances in Cryptology – EUROCRYPT '93 Proceedings. Lecture Notes in Computer Science*, vol. 765, Springer, 1994.

[175] H. M. HEYS. A tutorial on linear and differential cryptanalysis. *Cryptologia*, **26** (2002), 189–221.

[176] H. M. HEYS AND S. E. TAVARES. Substitution-permutation networks resistant to differential and linear cryptanalysis. *Journal of Cryptology*, **9** (1996), 1–19.

[177] H. D. L. HOLLMANN, J. H. VAN LINT, J-P. LINNARTZ AND L. M. G. M. TOLHUIZEN. On codes with the identifiable parent property, *Journal of Combinatorial Theory A*, **82** (1998), 121–133.

[178] W. C. HUFFMAN AND V. PLESS. *Fundamentals of Error-Correcting Codes*. Cambridge University Press, 2003.

[179] H. IMAI, R. L. RIVEST AND T. MATSUMOTO (EDS.) *Advances in Cryptology – ASIACRYPT '91 Proceedings. Lecture Notes in Computer Science*, vol. 739, Springer, 1993.

[180] M. ITO, A. SAITO AND T. NISHIZEKI. Secret sharing scheme realizing general access structure. *Proceedings IEEE Globecom '87*, pages 99–102, 1987.

[181] T. IWATA AND K. KUROSAWA. OMAC: one-key CBC MAC. *Lecture Notes in Computer Science*, **2887** (2003), 129–153. (Fast Software Encryption 2003.)

[182] W.-A. JACKSON AND K. M. MARTIN. A combinatorial interpretation of ramp schemes. *Australasian Journal of Combinatorics*, **14** (1996), 51–60.

[183] D. JOHNSON, A. MENEZES AND S. VANSTONE. The elliptic curve digital signature algorithm (ECDSA). *International Journal on Information Security*, **1** (2001), 36–63.

[184] P. JUNOD. On the complexity of Matsui's attack. *Lecture Notes in Computer Science*, **2259** (2001), 199–211. (Selected Areas in Cryptography 2001.)

[185] D. KAHN. *The Codebreakers*. Scribner, 1996.

[186] B. KALISKI, JR. (ED.) *Advances in Cryptology – CRYPTO '97 Proceedings. Lecture Notes in Computer Science*, vol. 1294, Springer, 1997.

[187] C. KAUFMAN, R. PERLMAN AND M. SPECINER. *Network Security. Private Communication in a Public World, Second Edition*. Prentice Hall, 2002.

[188] L. KELIHER, H. MEIJER AND S. TAVARES. New method for upper bounding the maximum average linear hull probability for SPNs. *Lecture Notes in Computer Science*, **2045** (2001), 420–436. (EUROCRYPT 2001.)

[189] L. KELIHER, H. MEIJER AND S. TAVARES. Improving the upper bound on the maximum average linear hull probability for Rijndael. *Lecture Notes in Computer Science*, **2259** (2001), 112–128. (Selected Areas in Cryptography 2001.)

[190] J. KILIAN (ED.) *Advances in Cryptology – CRYPTO 2001 Proceedings. Lecture Notes in Computer Science*, vol. 2139, Springer, 2001.

[191] K. KIM AND T. MATSUMOTO (EDS.) *Advances in Cryptology – ASIACRYPT '96 Proceedings. Lecture Notes in Computer Science*, vol. 1163, Springer, 1996.

[192] R. KIPPENHAHN. *Code Breaking, A History and Exploration.* Overlook Press, 1999.

[193] L. R. KNUDSEN. Contemporary block ciphers. *Lecture Notes in Computer Science*, **1561** (1999), 105–126. (Lectures on Data Security.)

[194] L. KNUDSEN (ED.) *Advances in Cryptology – EUROCRYPT 2002 Proceedings. Lecture Notes in Computer Science*, vol. 2332, Springer, 2002.

[195] D. E. KNUTH. *The Art of Computer Programming, Volume 2, Seminumerical Algorithms, Second Edition.* Addison-Wesley, 1998.

[196] N. KOBLITZ. Elliptic curve cryptosystems. *Mathematics of Computation*, **48** (1987), 203–209.

[197] N. KOBLITZ. *A Course in Number Theory and Cryptography, Second Edition.* Springer, 1994.

[198] N. KOBLITZ (ED.) *Advances in Cryptology – CRYPTO '96 Proceedings. Lecture Notes in Computer Science*, vol. 1109, Springer, 1996.

[199] N. KOBLITZ. *Algebraic Aspects of Cryptography.* Springer, 1998.

[200] N. KOBLITZ, A. MENEZES AND S. VANSTONE. The state of elliptic curve cryptography. *Designs, Codes and Cryptography*, **19** (2000), 173–193.

[201] J. KOHL AND C. NEUMAN. *The Kerberos Network Authentication Service (V5).* Network Working Group Request for Comments 1510, 1993.

[202] L. M. KOHNFELDER. *Towards a practical public-key cryptosystem.* Bachelor's Thesis, MIT, 1978.

[203] A. G. KONHEIM. *Cryptography, A Primer.* John Wiley and Sons, 1981.

[204] E. KRANAKIS. *Primality and Cryptography.* John Wiley and Sons, 1986.

[205] H. KRAWCZYK, (ED.) *Advances in Cryptology – CRYPTO '98 Proceedings. Lecture Notes in Computer Science*, vol. 1462, Springer, 1998.

[206] R. KUMAR, S. RAJAGOPALAN AND A. SAHAI. Coding constructions for blacklisting problems without computational assumptions. *Lecture Notes in Computer Science*, **1666** (1999), 609–623. (CRYPTO '99.)

[207] K. KUROSAWA, T. ITO AND M. TAKEUCHI. Public key cryptosystem using a reciprocal number with the same intractability as factoring a large number. *Cryptologia*, **12** (1988), 225–233.

[208] J. C. LAGARIAS. Pseudo-random number generators in cryptography and number theory. In *Cryptology and Computational Number Theory*, pages 115–143. American Mathematical Society, 1990.

[209] X. LAI, J. L. MASSEY AND S. MURPHY. Markov ciphers and differential cryptanalysis. *Lecture Notes in Computer Science*, **547** (1992), 17–38. (EUROCRYPT '91.)

[210] C. S. LAIH (ED.) *Advances in Cryptology – ASIACRYPT 2003 Proceedings. Lecture Notes in Computer Science*, vol. 2894, Springer, 2003.

[211] K. Y. LAM, E. OKAMOTO AND C. XING (EDS.) *Advances in Cryptology – ASIACRYPT '99 Proceedings. Lecture Notes in Computer Science*, vol. 1716, Springer, 1999.

[212] S. LANDAU. Standing the test of time: the data encryption standard. *Notices of the American Mathematical Society*, **47** (2000), 341–349.

[213] S. LANDAU. Communications security for the twenty-first century: the advanced encryption standard. *Notices of the American Mathematical Society*, **47** (2000), 450–459.

[214] S. LANDAU. Polynomials in the nation's service: using algebra to design the Advanced Encryption Standard. *American Mathematical Monthly*, **111** (2004), 89–117.

[215] P. J. LEE (ED.) *Advances in Cryptology – ASIACRYPT 2004 Proceedings. Lecture Notes in Computer Science*, vol. 3329, Springer, 2004.

[216] A. K. LENSTRA. Integer factoring. *Designs, Codes and Cryptography*, **19** (2000), 101–128.

[217] A. K. LENSTRA AND H. W. LENSTRA, JR. (EDS.) *The Development of the Number Field Sieve. Lecture Notes in Mathematics*, vol. 1554. Springer, 1993.

[218] A. K. LENSTRA AND H. W. LENSTRA, JR. Algorithms in number theory. In *Handbook of Theoretical Computer Science, Volume A: Algorithms and Complexity*, pages 673–715. Elsevier Science Publishers, 1990.

[219] A. K. LENSTRA AND E. R. VERHEUL. Selecting cryptographic key sizes. *Journal of Cryptology*, **14** (2001), 255–293.

[220] R. LIDL AND H. NIEDERREITER. *Finite Fields, Second Edition*. Cambridge University Press, 1997.

[221] D. L. LONG AND A. WIGDERSON. The discrete log hides $O(\log n)$ bits. *SIAM Jounal on Computing*, **17** (1988), 363–372.

[222] J. C. A. VAN DER LUBBE. *Basic Methods of Cryptography*. Cambridge, 1998.

[223] M. LUBY. *Pseudorandomness and Cryptographic Applications*. Princeton University Press, 1996.

[224] F. J. MACWILLIAMS AND N. J. A. SLOANE. *The Theory of Error-correcting Codes*, North-Holland, 1977.

[225] W. MAO. *Modern Cryptography: Theory and Practice*. Prentice-Hall, 2004.

[226] M. MATSUI. Linear cryptanalysis method for DES cipher. *Lecture Notes in Computer Science*, **765** (1994), 386–397. (EUROCRYPT '93.)

[227] M. MATSUI. The first experimental cryptanalysis of the data encryption standard. *Lecture Notes in Computer Science*, **839** (1994), 1–11. (CRYPTO '94.)

[228] T. MATSUMOTO, Y. TAKASHIMA AND H. IMAI. On seeking smart public-key distribution systems. *Transactions of the IECE (Japan)*, **69** (1986), 99–106.

[229] U. MAURER (ED.) *Advances in Cryptology – EUROCRYPT '96 Proceedings. Lecture Notes in Computer Science*, vol. 1070, Springer, 1996.

[230] U. MAURER AND S. WOLF. The Diffie-Hellman protocol. *Designs, Codes and Cryptography*, **19** (2000), 147–171.

[231] R. MCELIECE. *Finite Fields for Computer Scientists and Engineers*. Kluwer Academic Publishers, 1987.

[232] H. X. MEL AND D. BAKER. *Cryptography Decrypted*. Addison Wesley, 2001.

[233] A. J. MENEZES. *Elliptic Curve Public Key Cryptosystems*. Kluwer Academic Publishers, 1993.

[234] A. J. MENEZES AND N. KOBLITZ. A survey of public-key cryptosystems. *SIAM Review*, **46** (2004), 599–634.

[235] A. J. MENEZES, T. OKAMOTO AND S. A. VANSTONE. Reducing elliptic curve logarithms to logarithms in a finite field. *IEEE Transactions on Information Theory*, **39** (1993), 1639–1646.

[236] A. J. MENEZES AND S. A. VANSTONE (EDS.) *Advances in Cryptology – CRYPTO '90 Proceedings. Lecture Notes in Computer Science*, vol. 537, Springer, 1991.

[237] A. J. MENEZES, P. C. VAN OORSCHOT AND S. A. VANSTONE. *Handbook of Applied Cryptography*. CRC Press, 1996.

[238] R. C. MERKLE. Secure communications over insecure channels. *Communications of the ACM*, **21** (1978), 294–299.

[239] R. C. MERKLE. One way hash functions and DES. *Lecture Notes in Computer Science*, **435** (1990), 428–446. (CRYPTO '89.)

[240] G. L. MILLER. Riemann's hypothesis and tests for primality. *Journal of Computer and Systems Science*, **13** (1976), 300–317.

[241] V. MILLER. Uses of elliptic curves in cryptography. *Lecture Notes in Computer Science*, **218** (1986), 417–426. (CRYPTO '85.)

[242] C. J. MITCHELL AND F. C. PIPER. Key storage in secure networks. *Discrete Applied Mathematics*, **21** (1988), 215–228.

[243] C. J. MITCHELL, F. PIPER AND P. WILD. Digital signatures. In *Contem-*

porary Cryptology, The Science of Information Integrity, pages 325–378. IEEE Press, 1992.

[244] R. A. MOLLIN. *An Introduction to Cryptography*. Chapman & Hall/CRC, 2001.

[245] R. A. MOLLIN. *RSA and Public-key Cryptography*. Chapman & Hall/CRC, 2003.

[246] J. H. MOORE. Protocol failures in cryptosystems. In *Contemporary Cryptology, The Science of Information Integrity*, pages 541–558. IEEE Press, 1992.

[247] S. MURPHY AND M. J. B. ROBSHAW. Essential algebraic structure within the AES. *Lecture Notes in Computer Science*, **2442** (2002), 1–16. (CRYPTO 2002.)

[248] M. NAOR AND B. PINKAS. Efficient trace and revoke schemes. *Lecture Notes in Computer Science* **1962** (2000), 1–20. (Financial Cryptography 2000.)

[249] V. I. NECHAEV. On the complexity of a deterministic algorithm for a discrete logarithm. *Math. Zametki*, **55** (1994), 91–101.

[250] J. NECHVATAL, E. BARKER, L. BASSHAM, W. BURR, M. DWORKIN, J. FOTI AND E. ROBACK. *Report on the Development of the Advanced Encryption Standard (AES)*. October 2, 2000. Available from http://csrc.nist.gov/encryption/aes/.

[251] R. M. NEEDHAM AND M. D. SCHROEDER. Using encryption for authentication in large networks of computers. *Communications of the ACM*, **21** (1978), 993–999.

[252] E. M. NG. *Security Models and Proofs for Key Establishment Protocols*. Masters Thesis, University of Waterloo, 2005.

[253] P. Q. NGUYEN AND I. E. SHPARLINSKI. The insecurity of the digital signature algorithm with partially known nonces. *Journal of Cryptology*, **15** (2002), 151–176.

[254] K. NYBERG. Differentially uniform mappings for cryptography. *Lecture Notes in Computer Science*, **765** (1994), 55–64. (EUROCRYPT '93.)

[255] K. NYBERG. Linear approximation of block ciphers. *Lecture Notes in Computer Science*, **950** (1995), 439–444. (EUROCRYPT '94.)

[256] K. NYBERG (ED.) *Advances in Cryptology – EUROCRYPT '98 Proceedings. Lecture Notes in Computer Science*, vol. 1403, Springer, 1998.

[257] A. M. ODLYZKO (ED.) *Advances in Cryptology – CRYPTO '86 Proceedings. Lecture Notes in Computer Science*, vol. 263, Springer, 1987.

[258] A. M. ODLYZKO. Discrete logarithms: the past and the future. *Designs, Codes, and Cryptography*, **19** (2000), 129–145.

[259] K. OHTA AND D. PEI (EDS.) *Advances in Cryptology – ASIACRYPT*

'98 Proceedings. Lecture Notes in Computer Science, vol. 1514, Springer, 1998.

[260] T. OKAMOTO. Provably secure and practical identification schemes and corresponding signature schemes. *Lecture Notes in Computer Science*, **740** (1993), 31–53. (CRYPTO '92.)

[261] T. OKAMOTO (ED.) *Advances in Cryptology – ASIACRYPT 2000 Proceedings. Lecture Notes in Computer Science*, vol. 1976, Springer, 2000.

[262] T. P. PEDERSEN. Signing contracts and paying electronically. *Lecture Notes in Computer Science*, **1561** (1999), 134–157. (Lectures on Data Security.)

[263] R. PERALTA. Simultaneous security of bits in the discrete log. *Lecture Notes in Computer Science*, **219** (1986), 62–72. (EUROCRYPT '85.)

[264] B. PFITZMANN (ED.) *Advances in Cryptology – EUROCRYPT 2001 Proceedings. Lecture Notes in Computer Science*, vol. 2045, Springer, 2001.

[265] B. PFITZMANN. *Digital Signature Schemes – General Framework and Fail-Stop Signatures. Lecture Notes in Computer Science*, vol. 1100, Springer, 1996.

[266] F. PICHLER (ED.) *Advances in Cryptology – EUROCRYPT '85 Proceedings. Lecture Notes in Computer Science*, vol. 219, Springer, 1986.

[267] J. PIEPRYZK AND R. SAFAVI-NAINI (EDS.) *Advances in Cryptology – ASIACRYPT '94 Proceedings. Lecture Notes in Computer Science*, vol. 917, Springer, 1995.

[268] J. PIEPRYZK, T. HARDJONO AND J. SEBERRY *Fundamentals of Computer Security.* Springer, 2003.

[269] F. PIPER AND S. MURPHY *Cryptography: A Very Short Introduction.* Oxford, 2002.

[270] S. C. POHLIG AND M. E. HELLMAN. An improved algorithm for computing logarithms over $GF(p)$ and its cryptographic significance. *IEEE Transactions on Information Theory*, **24** (1978), 106–110.

[271] D. POINTCHEVAL AND J. STERN. Security arguments for signature schemes and blind signatures. *Journal of Cryptology*, **13** (2000), 361–396.

[272] J. M. POLLARD. Monte Carlo methods for index computation (mod p). *Mathematics of Computation*, **32** (1978), 918–924.

[273] C. POMERANCE (ED.) *Advances in Cryptology – CRYPTO '87 Proceedings. Lecture Notes in Computer Science*, vol. 293, Springer, 1988.

[274] B. PRENEEL. The state of cryptographic hash functions. *Lecture Notes in Computer Science*, **1561** (1999), 158–182. (Lectures on Data Security.)

[275] B. PRENEEL (ED.) *Advances in Cryptology – EUROCRYPT 2000 Proceedings. Lecture Notes in Computer Science*, vol. 1807, Springer, 2000.

[276] B. PRENEEL AND P. C. VAN OORSCHOT. On the security of iterated message authentication codes. *IEEE Transactions on Information Theory*, **45** (1999), 188–199.

[277] J.-J. QUISQUATER AND J. VANDEWALLE (EDS.) *Advances in Cryptology – EUROCRYPT '89 Proceedings. Lecture Notes in Computer Science*, vol. 434, Springer, 1990.

[278] M. O. RABIN. Digitized signatures and public-key functions as intractable as factorization. *MIT Laboratory for Computer Science Technical Report*, LCS/TR-212, 1979.

[279] M. O. RABIN. Probabilistic algorithms for testing primality. *Journal of Number Theory*, **12** (1980), 128–138.

[280] M. Y. RHEE. *Internet Security: Cryptographic Principles, Algorithms and Protocols*. John Wiley & Sons, 2003.

[281] R. L. RIVEST. The MD4 message digest algorithm. *Lecture Notes in Computer Science*, **537** (1991), 303–311. (CRYPTO '90.)

[282] R. L. RIVEST. The MD5 message digest algorithm. Internet Network Working Group RFC 1321, April 1992.

[283] R. L. RIVEST, A. SHAMIR, AND L. ADLEMAN. A method for obtaining digital signatures and public key cryptosystems. *Communications of the ACM*, **21** (1978), 120–126.

[284] K. H. ROSEN. *Elementary Number Theory and its Applications, Fourth Edition*. Addison-Wesley, 1999.

[285] A. D. RUBIN. *White-hat Security Arsenal: Tackling the Threats*. Addison Wesley, 2001.

[286] R. A. RUEPPEL (ED.) *Advances in Cryptology – EUROCRYPT '92 Proceedings. Lecture Notes in Computer Science*, vol. 658, Springer, 1993.

[287] R. A. RUEPPEL AND P. C. VAN OORSCHOT. Modern key agreement techniques. *Computer Communications*, 1994.

[288] R. SAKAI, K. OHGISHI AND M. KASAHARA. Cryptosystems based on pairing. Presented at the *Symposium on Cryptography and Information Security*, Okinawa, Japan, 2000.

[289] A. SALOMAA. *Public-Key Cryptography*. Springer, 1990.

[290] D. SALOMON. *Data Privacy and Security*. Springer, 2003.

[291] K. SCHMEH. *Cryptography and Public Key Infrastructure on the Internet*. John Wiley and Sons, 2001.

[292] B. SCHNEIER. *Applied Cryptography, Protocols, Algorithms and Source Code in C, Second Edition*. John Wiley and Sons, 1995.

[293] B. SCHNEIER. *Secrets and Lies: Digital Security in a Networked World*. John Wiley and Sons, 2000.

[294] C. P. SCHNORR. Efficient signature generation by smart cards. *Journal of Cryptology*, **4** (1991), 161–174.

[295] J. SEBERRY AND J. PIEPRZYK (EDS.) *Advances in Cryptology – AUSCRYPT '90 Proceedings. Lecture Notes in Computer Science*, vol. 453, Springer, 1990.

[296] J. SEBERRY AND Y. ZHENG (EDS.) *Advances in Cryptology – AUSCRYPT '92 Proceedings. Lecture Notes in Computer Science*, vol. 718, Springer, 1993.

[297] A. SHAMIR. How to share a secret. *Communications of the ACM*, **22** (1979), 612–613.

[298] A. SHAMIR. Identity-based cryptosystems and signature schemes. *Lecture Notes in Computer Science*, **196** (1985), 47–53. (Advances in Cryptology – CRYPTO '84.)

[299] C. E. SHANNON. A mathematical theory of communication. *Bell Systems Technical Journal*, **27** (1948), 379–423, 623–656.

[300] C. E. SHANNON. Communication theory of secrecy systems. *Bell Systems Technical Journal*, **28** (1949), 656–715.

[301] V. SHOUP. Lower bounds for discrete logarithms and related problems. *Lecture Notes in Computer Science*, **1233** (1997), 256–266. (EURO-CRYPT '97.)

[302] V. SHOUP. OAEP reconsidered. *Lecture Notes in Computer Science*, **2139** (2001), 239–259. (CRYPTO 2001.)

[303] J. H. SILVERMAN AND J. TATE. *Rational Points on Elliptic Curves*. Springer, 1992.

[304] G. J. SIMMONS. A survey of information authentication. In *Contemporary Cryptology, The Science of Information Integrity*, pages 379–419. IEEE Press, 1992.

[305] G. J. SIMMONS. An introduction to shared secret and/or shared control schemes and their application. In *Contemporary Cryptology, The Science of Information Integrity*, pages 441–497. IEEE Press, 1992.

[306] G. J. SIMMONS (ED.) *Contemporary Cryptology, The Science of Information Integrity*. IEEE Press, 1992.

[307] S. SINGH. *The Code Book: The Evolution Of Secrecy From Mary, To Queen Of Scots To Quantum Cryptography*. Doubleday, 1999.

[308] N. P. SMART. The discrete logarithm problem on elliptic curves of trace one. *Journal of Cryptology*, **12** (1999), 193–196.

[309] N. SMART. *Cryptography: An Introduction*. McGraw-Hill, 2002.

[310] M. E. SMID AND D. K. BRANSTAD. The data encryption standard; past and future. In *Contemporary Cryptology, The Science of Information Integrity*, pages 43–64. IEEE Press, 1992.

[311] M. E. SMID AND D. K. BRANSTAD. Response to comments on the NIST proposed digital signature standard. *Lecture Notes in Computer Science*, **740** (1993), 76–88. (CRYPTO '92.)

[312] J. SOLINAS. Efficient arithmetic on Koblitz curves. *Designs, Codes and Cryptography*, **19** (2000), 195–249.

[313] R. SOLOVAY AND V. STRASSEN. A fast Monte Carlo test for primality. *SIAM Journal on Computing*, **6** (1977), 84–85.

[314] J. N. STADDON, D. R. STINSON AND R. WEI. Combinatorial properties of frameproof and traceability codes. *IEEE Transactions on Information Theory*, **47** (2001), 1042–1049.

[315] W. STALLINGS. *Cryptography and Network Security: Principles and Practice, Third Edition* Prentice Hall, 2002.

[316] M. STEINER, G. TSUDIK AND M. WAIDNER. Diffie-Hellman key distribution extended to group communication. In *Proceedings of the 3rd ACM Conference on Computer and Communications Security*, pages 31–37. ACM Press, 1996.

[317] J. STERN (ED.) *Advances in Cryptology – EUROCRYPT '99 Proceedings. Lecture Notes in Computer Science*, vol. 1592, Springer, 1999.

[318] D. R. STINSON. An explication of secret sharing schemes. *Designs, Codes and Cryptography*, **2** (1992), 357–390.

[319] D. R. STINSON (ED.) *Advances in Cryptology – CRYPTO '93 Proceedings. Lecture Notes in Computer Science*, vol. 773, Springer, 1994.

[320] D. R. STINSON. Decomposition constructions for secret sharing schemes. *IEEE Transactions on Information Theory*, **40** (1994), 118–125.

[321] D. R. STINSON. Universal hashing and authentication codes. *Designs, Codes and Cryptography*, **4** (1994), 369–380.

[322] D. R. STINSON. On the connections between universal hashing, combinatorial designs and error-correcting codes. *Congressus Numerantium*, **114** (1996), 7–27.

[323] D. R. STINSON. On some methods for unconditionally secure key distribution and broadcast encryption. *Designs, Codes and Cryptography*, **12** (1997), 215–243.

[324] D. R. STINSON. Some observations on the theory of cryptographic hash functions. *Designs, Codes and Cryptography*, to appear.

[325] D. R. STINSON AND TRAN VAN TRUNG. Some new results on key distribution patterns and broadcast encryption. *Designs, Codes and Cryptography*, **14** (1998), 261–279.

[326] D. R. STINSON AND R. WEI. Combinatorial properties and constructions of traceability schemes and frameproof codes. *SIAM Journal on Discrete Mathematics* **11** (1998), 41–53.

[327] D. R. STINSON AND R. WEI. An application of ramp schemes to broadcast encryption. *Information Processing Letters*, **69** (1999), 131–135.

[328] E. TESKE. On random walks for Pollard's rho method. *Mathematics of Computation*, **70** (2001), 809–825.

[329] E. THOMÉ. Computation of discrete logarithms in $\mathbb{F}_{2^{607}}$. *Lecture Notes in Computer Science*, **2248** (2001), 107–124. (ASIACRYPT 2001.)

[330] W. TRAPPE AND L. C. WASHINGTON. *Introduction to Cryptography with Coding Theory.* Prentice Hall, 2002.

[331] M. VAN DIJK, W.-A. JACKSON AND K. M. MARTIN. A general decomposition construction for incomplete secret sharing schemes. *Designs, Codes and Cryptography*, **15** (1998), 301–321.

[332] E. VAN HEYST AND T. P. PEDERSEN. How to make efficient fail-stop signatures. *Lecture Notes in Computer Science*, **658** (1993), 366–377. (EUROCRYPT '92.)

[333] J. VAN TILBURG. Secret-key exchange with authentication. *Lecture Notes in Computer Science*, **741** (1993), 71–86. (Computer Security and Industrial Cryptography, State of the Art and Evolution, ESAT Course, May 1991.)

[334] U. VAZIRANI AND V. VAZIRANI. Efficient and secure pseudorandom number generation. In *Proceedings of the 25th Annual Symposium on the Foundations of Computer Science*, pages 458–463. IEEE Press, 1984.

[335] S. S. WAGSTAFF, JR. *Cryptanalysis of Number Theoretic Ciphers.* Chapman & Hall/CRC, 2003.

[336] D. M. WALLNER, E. J. HARDER AND R. C. AGEE. Key management for multicast: issues and architectures. *Internet Request for Comments* 2627, June, 1999.

[337] H. B. WANG. *Desired Features and Design Methodologies of Secure Authenticated Key Exchange Protocols in the Public-key Infrastructure Setting.* Masters Thesis, University of Waterloo, 2004.

[338] X. WANG, D. FENG, X. LAI AND H. YU. Collisions for hash functions MD4, MD5, HAVAL-128 and RIPEMD. *Cryptology ePrint Archive*, Report 2004/199, http://eprint.iacr.org/.

[339] X. WANG, Y. L. YIN AND H. YU. Collision search attacks on SHA-1. Preprint, February 13, 2005.

[340] L. C. WASHINGTON. *Elliptic Curves: Number Theory and Cryptography.* Chapman & Hall/CRC, 2003.

[341] M. N. WEGMAN AND J. L. CARTER. New hash functions and their use in authentication and set equality. *Journal of Computer and System Sciences*, **22** (1981), 265–279.

[342] D. WELSH. *Codes and Cryptography.* Oxford Science Publications, 1988.

[343] M. J. WIENER. Cryptanalysis of short RSA secret exponents. *IEEE Transactions on Information Theory*, **36** (1990), 553–558.

[344] M. J. WIENER. Efficient DES key search. Technical report TR-244, School of Computer Science, Carleton University, Ottawa, Canada, May 1994 (also presented at CRYPTO '93 Rump Session).

[345] M. J. WIENER, (ED.) *Advances in Cryptology – CRYPTO '99 Proceedings. Lecture Notes in Computer Science*, vol. 1666, Springer, 1999.

[346] H. C. WILLIAMS. A modification of the RSA public-key encryption procedure. *IEEE Transactions on Information Theory*, **26** (1980), 726–729.

[347] H. C. WILLIAMS (ED.) *Advances in Cryptology – CRYPTO '85 Proceedings. Lecture Notes in Computer Science*, vol. 218, Springer, 1986.

[348] C. K. WONG AND S. S. LAM. Digital signatures for flows and multicasts. *IEEE/ACM Transactions on Networking*, **7** (1999), 502–513.

[349] S. Y. YAN. *Number Theory for Computing*. Springer, 2000.

[350] A. YAO. Theory and applications of trapdoor functions. In *Proceedings of the 23rd Annual Symposium on the Foundations of Computer Science*, pages 80–91. IEEE Press, 1982.

[351] A. L. YOUNG AND M. YUNG. *Malicious Cryptography: Exposing Cryptovirology*. Joohn Wiley & Sons, 2004.

[352] M. YUNG (ED.) *Advances in Cryptology – CRYPTO 2002 Proceedings. Lecture Notes in Computer Science*, vol. 2442, Springer, 2002.

[353] Y. ZHENG (ED.) *Advances in Cryptology – ASIACRYPT 2002 Proceedings. Lecture Notes in Computer Science*, vol. 2501, Springer, 2002.

Index